Lecture Notes in Computer Science 7134

Commenced Publication in 1973
Founding and Former Series Editors:
Gerhard Goos, Juris Hartmanis, and Jan

T0092340

Editorial Board

David Hutchison
Lancaster University, UK

Takeo Kanade
Carnegie Mellon University, Pittsburgh, PA, USA

Josef Kittler
University of Surrey, Guildford, UK

Jon M. Kleinberg
Cornell University, Ithaca, NY, USA

Alfred Kobsa
University of California, Irvine, CA, USA

Friedemann Mattern
ETH Zurich, Switzerland

John C. Mitchell
Stanford University, CA, USA

Moni Naor
Weizmann Institute of Science, Rehovot, Israel

Oscar Nierstrasz
University of Bern, Switzerland

C. Pandu Rangan
Indian Institute of Technology, Madras, India

Bernhard Steffen
TU Dortmund University, Germany

Madhu Sudan
Microsoft Research, Cambridge, MA, USA

Demetri Terzopoulos
University of California, Los Angeles, CA, USA

Doug Tygar
University of California, Berkeley, CA, USA

Gerhard Weikum
Max Planck Institute for Informatics, Saarbruecken, Germany

Kristján Jónasson (Ed.)

Applied Parallel and Scientific Computing

10th International Conference, PARA 2010
Reykjavík, Iceland, June 6-9, 2010
Revised Selected Papers, Part II

 Springer

Volume Editor

Kristján Jónasson
University of Iceland
School of Engineering and Natural Sciences
Department of Computer Science
Hjardarhagi 4, 107 Reykjavík, Iceland
E-mail: jonasson@hi.is

ISSN 0302-9743   e-ISSN 1611-3349
ISBN 978-3-642-28144-0   e-ISBN 978-3-642-28145-7
DOI 10.1007/978-3-642-28145-7
Springer Heidelberg Dordrecht London New York

Library of Congress Control Number: 2012930011

CR Subject Classification (1998): G.1-4, F.1-2, D.1-3, J.1, C.2

LNCS Sublibrary: SL 1 – Theoretical Computer Science and General Issues

Typesetting: Camera-ready by author, data conversion by Scientific Publishing Services, Chennai, India

Printed on acid-free paper

Springer is part of Springer Science+Business Media (www.springer.com)

Preface

The tenth Nordic conference on applied parallel computing, Para 2010: State of the Art in Scientific and Parallel Computing, was held in Reykjavík, Iceland during June 6–9, 2010. The topics of the conference were announced to include software, hardware, algorithms, tools, environments, as well as applications of scientific and high-performance computing. The conference was hosted by the School of Engineering and Natural Sciences of the University of Iceland, and the conference venue was in the School of Education of the University of Iceland. Three companies in Reykjavík supported the conference financially: the video game developer CCP, Microsoft Íslandi, and Opin kerfi (Hewlett Packard distributor for Iceland).

The series of Para meetings began in 1994. The Danish Computing Centre for Research and Education (UNI-C) and the Department of Informatics and Mathematical Modelling of the Technical University of Denmark (IMM/DTU) in Lyngby, Denmark, organized a series of workshops on Applied Parallel Computing, named Para94, Para95 and Para96. Jerzy Waśniewski, senior researcher at DTU, initiated these workshops and Jack Dongarra, professor at the University of Tennessee, became involved during an extended visit to Lyngby. He played a key part in promoting the meetings internationally. Since 1998, the workshops have become a Nordic effort, but both Jerzy and Jack have continued to be an integral part of the meetings. In fact Jerzy has been a keen advocate of holding a Para conference in Iceland. The themes and locations of the Para meetings have been:

PARA94 Parallel Scientific Computing, Lyngby, Denmark
PARA95 Physics, Chemistry and Engineering Science, Lyngby, Denmark
PARA96 Industrial Problems and Optimization, Lyngby, Denmark
PARA 1998 Large Scale Scientific and Industrial Problems, Umeå, Sweden
PARA 2000 New Paradigms for HPC in Industry and Academia, Bergen, Norway
PARA 2002 Advanced Scientific Computing, Helsinki, Finland
PARA 2004 State of the Art in Scientific Computing, Copenhagen, Denmark
PARA 2006 State of the Art in Scientific and Parallel Computing, Umeå, Sweden
PARA 2008 State of the Art in Scientific and Parallel Computing, Trondheim, Norway
PARA 2010 State of the Art in Scientific and Parallel Computing, Reykjavík, Iceland

The Para 2010 conference included five keynote lectures, one tutorial, 11 minisymposia consisting of a total of 90 presentations, 39 other contributed presentations organized under 10 separate topics, four poster presentations, and eight presentations from industry. Except for the keynote lectures, that were 45 minutes long each, the presentations were organized in five tracks or parallel streams, with 25-minute slots for each presentation, including discussion. The total number of presentations was thus 147. There were altogether 187 participants from 20 countries:

Denmark 9	Canada 1	Poland 16
Finland 4	Czech Republic 3	Russia 2
Iceland 38	France 12	Spain 7
Norway 13	Germany 32	Switzerland 1
Sweden 17	Italy 1	Turkey 1
Australia 2	Japan 4	USA 20
Austria 2	Netherlands 2	

There were volcanic eruptions in Eyjafjallajökull in southern Iceland from March until June 2010 disrupting international flights, and these may have had an adverse effect on participation.

Extended abstracts (in most cases four pages long) of all the minisymposium and contributed presentations were made available on the conference website, http://vefir.hi.is/para10, and in addition a book of short abstracts (also available on the website) was handed out at the conference.

After the conference the presentation authors were invited to submit manuscripts for publication in these peer-reviewed conference proceedings. The reviewing process for the articles appearing here was therefore performed in two stages. In the first stage the extended abstracts were reviewed to select contributions to be presented at the conference, and in the second stage the full papers submitted after the conference were reviewed. As a general rule three referee reports per paper were aimed for, and in most cases these were successfully obtained. However, in cases where it proved difficult to find three willing referees, acquiring only two reports was deemed acceptable.

Fred G. Gustavson, emeritus scientist at IBM Research, New York, and professor at Umeå University, and Jerzy Waśniewski gave a tutorial on matrix algorithms in the new many core era. Fred celebrated his 75th birthday on May 29, 2010, and the Linear Algebra Minisymposium was held in his honor. The material of the tutorial is covered in Fred Gustavson's article in these proceedings.

A conference of this size requires considerable organization and many helping hands. The role of the minisymposium organizers was very important. They reviewed and/or organized reviewing of contributions to their respective minisymposia, both the original extended abstracts and the articles for these proceedings, and in addition they managed the minisymposium sessions at the conference. Several members of the local Organizing Committee helped with the reviewing of other contributed extended abstracts: Elínborg I. Ólafsdóttir, Hjálmtýr Hafsteinsson, Klaus Marius Hansen, Ólafur Rögnvaldsson, Snorri Agnarsson and Sven Þ. Sigurðsson. Other colleagues who helped with this task were Halldór Björnsson, Kristín Vogfjörð and Viðar Guðmundsson.

The editor of these proceedings organized the reviewing of manuscripts falling outside minisymposia, as well as manuscripts authored by the minisymposium organizers themselves. There were 56 such submissions. The following people played a key role in helping him with this task: Sven Þ. Sigurðsson, Julien

Langou, Bo Kågström, Sverker Holmgren, Michael Bader, Jerzy Waśniewski, Klaus Marius Hansen, Kimmo Koski and Halldór Björnsson. Many thanks are also due to all the anonymous referees, whose extremely valueable work must not be forgotten.

The conference bureau Your Host in Iceland managed by Inga Sólnes did an excellent job of organizing and helping with many tasks, including conference registration, hotel bookings, social program, financial management, and maintaining the conference website. Apart from Inga, Kristjana Magnúsdóttir of Your Host was a key person and Einar Samúelsson oversaw the website design. Ólafía Lárusdóttir took photographs for the conference website. The baroque group Custos and the Tibia Trio, both led by recorder player Helga A. Jónsdóttir, and Helgi Kristjánsson (piano) provided music for the social program. Ólafur Rögnvaldsson helped to secure financial support from industry. Jón Blöndal and Stefán Ingi Valdimarsson provided valuable TeX help for the editing of the proceedings.

Finally, I wish to devote a separate paragraph to acknowledge the help of my colleague Sven Þ. Sigurðsson, who played a key role in helping with the conference organization and editing of the proceedings through all stages.

October 2011   Kristján Jónasson

Organization

PARA 2010 was organized by the School of Engineering and Natural Sciences of the University of Iceland.

Steering Committee

Jerzy Waśniewski, Chair, Denmark
Kaj Madsen, Denmark
Anne C. Elster, Norway
Petter Bjørstad, Norway
Hjálmtýr Hafsteinsson, Iceland
Kristján Jónasson, Iceland

Juha Haatja, Finland
Kimmo Koski, Finland
Björn Engquist, Sweden
Bo Kågström, Sweden
Jack Dongarra, Honorary Chair, USA

Local Organizing Committee

Kristján Jónasson, Chair
Sven Þ. Sigurðsson, Vice Chair
Ólafur Rögnvaldsson, Treasurer
Ari Kr. Jónsson
Ebba Þóra Hvannberg
Elínborg Ingunn Ólafsdóttir
Hannes Jónsson

Helmut Neukirchen
Hjálmtýr Hafsteinsson
Jan Valdman
Klaus Marius Hansen
Sigurjón Sindrason
Snorri Agnarsson
Tómas Philip Rúnarsson

Sponsoring Companies

CCP, Reykjavík – video game developer
Microsoft Íslandi, Reykjavík
Opin kerfi, Reykjavík – Hewlett Packard in Iceland

PARA 2010 Scientific Program

Keynote Presentations

Impact of Architecture and Technology for Extreme Scale on Software and Algorithm Design
> *Jack Dongarra*, University of Tennessee and Oak Ridge National Laboratory

Towards Petascale for Atmospheric Simulation
> *John Michalakes*, National Center for Atmospheric Research (NCAR), Boulder, Colorado

Algorithmic Challenges for Electronic-Structure Calculations
> *Risto M. Nieminen*, Aalto University School of Science and Technology, Helsinki

Computational Limits to Nonlinear Inversion
> *Klaus Mosegaard*, Technical University of Denmark

Efficient and Reliable Algorithms for Challenging Matrix Computations Targeting Multicore Architectures and Massive Parallelism
> *Bo Kågström*, Umeå University

Tutorial

New Algorithms and Data Structures for Matrices in the Multi/Many Core Era
> *Fred G. Gustavson*, Umeå University and Emeritus Scientist at IBM Research, New York, and *Jerzy Waśniewski*, Danish Technical University

General Topics

Cloud Computing (1 presentation)
HPC Algorithms (7 presentations and 1 poster)
HPC Programming Tools (4 presentations)
HPC in Meteorology (3 presentations)
Parallel Numerical Algorithms (8 presentations and 1 poster)
Parallel Computing in Physics (2 presentations and 1 poster)
Scientific Computing Tools (10 presentations)
HPC Software Engineering (2 presentations and 1 poster)
Hardware (1 presentation)
Presentations from Industry (8 presentations)

Minisymposia

Simulations of Atomic Scale Systems (15 presentations)
  Organized by *Hannes Jónsson*, University of Iceland

Tools and Environments for Accelerator-Based Computational Biomedicine (6 presentations)
  Organized by *Scott B. Baden*, University of California, San Diego

GPU Computing (9 presentations)
  Organized by *Anne C. Elster*, NTNU, Trondheim

High-Performance Computing Interval Methods (6 presentations)
  Organized by *Bartlomiej Kubica*, Warsaw University of Technology

Real-Time Access and Processing of Large Data Sets (6 presentations)
  Organized by *Helmut Neukirchen*, University of Iceland and *Michael Schmelling*, Max Planck Institute for Nuclear Physics, Heidelberg

Linear Algebra Algorithms and Software for Multicore and Hybrid Architectures, in honor of Fred Gustavson on his 75th birthday (10 presentations)
  Organized by *Jack Dongarra*, University of Tennessee and *Bo Kågström*, Umeå University

Memory and Multicore Issues in Scientific Computing – Theory and Practice (6 presentations)
  Organized by *Michael Bader*, Universität Stuttgart and *Riko Jacob*, Technische Universität München

Multicore Algorithms and Implementations for Application Problems (9 presentations)
  Organized by *Sverker Holmgren*, Uppsala University

Fast PDE Solvers and A Posteriori Error Estimates (8 presentations)
  Organized by *Jan Valdman*, University of Iceland and *Talal Rahman*, University College Bergen

Scalable Tools for High-Performance Computing (12 presentations)
  Organized by *Luiz DeRose*, Cray Inc. and *Felix Wolf*, German Research School for Simulation Sciences

Distributed Computing Infrastructure Interoperability (4 presentations)
  Organized by *Morris Riedel*, Forschungszentrum Jülich

Speakers and Presentations

For a full list of authors and extended abstracts, see http://vefir.hi.is/para10.

Abrahamowicz, Michal: Alternating conditional estimation of complex constrained models for survival analysis

Abramson, David: Scalable parallel debugging: Challenges and solutions

Agnarsson, Snorri: Parallel programming in Morpho

Agullo, Emmanuel: Towards a complexity analysis of sparse hybrid linear solvers

Aliaga, José I.: Parallelization of multilevel ILU preconditioners on distributed-memory multiprocessors

Anzt, Hartwig: Mixed precision error correction methods for linear systems – Convergence analysis based on Krylov subspace methods

Aqrawi, Ahmed Adnan: Accelerating disk access using compression for large seismic datasets on modern GPU and CPU

Arbenz, Peter: A fast parallel poisson solver on irregular domains

Bader, Michael: Memory-efficient Sierpinski-order traversals on dynamically adaptive, recursively structured triangular grids

Bartels, Soeren: A posteriori error estimation for phase field models

Belsø, Rene: Structural changes within the high-performance computing (HPC) landscape

Bientinesi, Paolo: The algorithm of multiple relatively robust representations for multicore processors

Bjarnason, Jón: Fighting real time – The challenge of simulating large-scale space battles within the Eve architecture

Blaszczyk, Jacek Piotr: Aggregated pumping station operation planning problem (APSOP) for large-scale water transmission system

Bohlender, Gerd: Fast and exact accumulation of products

Borkowski, Janusz: Global asynchronous parallel program control for multicore processors

Bozejko, Wojciech: Parallelization of the tabu search algorithm for the hybrid flow shop problem

Breitbart, Jens: Semiautomatic cache optimizations using OpenMP

Brian J. N. Wylie: Performance engineering of GemsFDTD computational electromagnetics solver

Britsch, Markward: The computing framework for physics analysis at LHCb

Brodtkorb, André R.: State of the art in heterogeneous computing

Buttari, Alfredo: Fine granularity sparse QR factorization for multicore-based systems

Cai, Xiao-Chuan: A parallel domain decomposition algorithm for an inverse problem in elastic materials

Cai, Xing: Detailed numerical analyses of the Aliev-Panfilov model on GPGPU

Cambruzzi, Sandro: The new features of Windows HPC Server 2008 V3 and Microsoft's HPC strategy

Cankur, Reydan: Parallel experiments on PostgreSQL (poster)

Casas, Marc: Multiplexing hardware counters by spectral analysis

Cheverda, Vladimir A.: Simulation of seismic waves propagation in multiscale media: Impact of cavernous/fractured reservoirs

Cheverda, Vladimir A.: Parallel algorithm for finite difference simulation of acoustic logging

Contassot-Vivier, Sylvain: Impact of asynchronism on GPU accelerated parallel iterative computations

Cytowski, Maciej: Analysis of gravitational wave signals on heterogeneous architecture

Danek, Tomasz: GPU accelerated wave form inversion through Monte Carlo sampling

Davidson, Andrew: Toward techniques for auto-tuning GPU algorithms

DeRose, Luiz: Automatic detection of load imbalance

Doll, Jimmie D.: Recent developments in rare-event Monte Carlo methods

Dongarra, Jack: Impact of architecture and technology for extreme scale on software and algorithm design (keynote lecture)

Dongarra, Jack: LINPACK on future manycore and GPU-based systems

Dubcova, Lenka: Automatic hp-adaptivity for inductively heated incompressible flow of liquid metal

Einarsdóttir, Dóróthea M.: Calculation of tunneling paths and rates in systems with many degrees of freedom

Ekström, Ulf Egil: Automatic differentiation in quantum chemistry

Elster, Anne C.: Current and future trends in GPU computing

Elster, Anne C.: Visualization and large data processing – State of the art and challenges

Fjukstad, Bård: Interactive weather simulation and visualization on a display wall with manycore compute nodes

Fujino, Seiji: Performance evaluation of IDR(s)-based Jacobi method

Gagunashvili, Nikolai: Intellectual data processing for rare event selection using a RAVEN network

Gepner, Pawel: Performance evaluation of Intel® Xeon® 7500 family processors for HPC

Gerndt, Michael: Performance analysis tool complexity

Gjermundsen, Aleksander: LBM vs. SOR solvers on GPU for real-time fluid simulations

Goerling, Andreas: Novel density-functional methods for ground and excited states of molecules and first steps towards their efficient implementation

Greiner, Gero: Evaluating non-square sparse bilinear forms on multiple vector pairs in the I/O-model

Gross, Lutz: Algebraic upwinding with flux correction in 3D numerical simulations in geosciences

Guðjónsson, Halldór Fannar: HPC and the Eve cluster game architecture

Gustafsson, Magnus: Communication-efficient Krylov methods for exponential integration in quantum dynamics

Gustavson, Fred G.: New Algorithms and data structures for matrices in the multi/manycore era, parts 1, 2, 4 (tutorial)

Gustavson, Fred G.: Enduring linear algebra

Henkelman, Graeme: Accelerating molecular dynamics with parallel computing resources

Hess, Berk: Molecular simulation on multicore clusters and GPUs

Holm, Marcus: Implementing Monte Carlo electrostatics simulations on heterogeneous multicore architectures

Jacobson, Emily R.: A lightweight library for building scalable tools

Jakl, Ondrej: Solution of identification problems in computational mechanics – Parallel processing aspects

Jenz, Domenic: The computational steering framework steereo

Jiang, Steve: GPU-based computational tools for online adaptive cancer radiotherapy

Jónsson, Kristján Valur: Using stackless python for high-performance MMO architecture

Kågström, Bo: Efficient and reliable algorithms for challenging matrix computations targeting multicore architectures and massive parallelism (keynote lecture)

Kamola, Mariusz: Software environment for market balancing mechanisms development, and its application to solving more general problems in parallel way

Karlsson, Lars: Fast reduction to Hessenberg form on multicore architectures

Khan, Malek Olof: Molecular simulations on distributed heterogeneous computing nodes

Kimpe, Dries: Grids and HPC: Not as different as you might think?

Kirschenmann, Wilfried: Multi-target vectorization with MTPS C++ generic library

Kjelgaard Mikkelsen, Carl Christian: Parallel solution of banded and block bidiagonal linear systems

Klüpfel, Peter: Minimization of orbital-density-dependent energy functionals

Knüpfer, Andreas: Rank-specific event tracing for MPI – Extending event tracing to cope with extreme scalability

Kraemer, Walter: High-performance verified computing using C-XSC

Kreutz, Jochen: Black-Scholes and Monte Carlo simulation on accelerator architectures

Krog, Øystein E.: Fast GPU-based fluid simulations using SPH

Kubica, Bartlomiej: Using the second-order information in Pareto-set computations of a multi-criteria problem

Kubica, Bartlomiej: Cache-oblivious matrix formats for computations on interval matrices

Köstler, Harald: Optimized fast wavelet transform utilizing a multicore-aware framework for stencil computations

Lacoursiere, Claude: Direct sparse factorization of blocked saddle point matrices

Langlois, Philippe: Performance evaluation of core numerical algorithms: A tool to measure instruction level parallelism

Niewiadomska-Szynkiewicz, Ewa: Software environment for parallel optimization of complex systems

Niewiadomska-Szynkiewicz, Ewa: A software tool for federated simulation of wireless sensor networks and mobile ad hoc networks

Pedersen, Andreas: Atomistic dynamics using distributed and grid computing

Pizzagalli, Laurent: Computation of transition states for extended defects in materials science: Issues and challenges from selected exemples

Rahman, Talal: A fast algorithm for a constrained total variation minimization with application to image processing

Remón, Alfredo: Accelerating model reduction of large linear systems with graphics processors

Riedel, Morris: The European middleware initiative: Delivering key technologies to distributed computing infrastructures

Riedel, Morris: Grid infrastructure interoperability in EU FP7th Euforia project

Roman Wyrzykowski: Towards efficient execution of erasure codes on multicore architectures

Ruud, Kenneth: Parallelization and grid adaptation of the Dalton quantum chemistry program

Rögnvaldsson, Ólafur: On-demand high-resolution weather forecast for search-and-rescue (SAR)

Saga, Kazushige: Grid interoperation in the RENKEI grid middleware

Schifano, Sebastiano F.: Monte Carlo simulations of spin systems on multicore processors

Schmelling, Michael: Boosting data analysis for the LHC experiments

Schnupp, Michael: Experimental performance of I/O-optimal sparse matrix dense vector multiplication algorithms within main memory

Schwanke, Christoph: Parallel particle-in-cell Monte Carlo algorithm for simulation of gas discharges under PVM and MPI (poster)

Shende, Sameer Suresh: Improving the scalability of performance evaluation tools

Signell, Artur: An efficient approximative method for generating spatially correlated multivariate random normals in parallel

Skovhede, Kenneth: CSP channels for CELL-BE

Skúlason, Egill: Simulations of atomic scale transitions at charged interfaces

Strey, Alfred: Implementation of clustering algorithms on manycore architectures

Strzelczyk, Jacek: Parallel kriging algorithm for unevenly spaced data

Stussak, Christian: Parallel computation of bivariate polynomial resultants on graphics processing units

Tillenius, Martin: An efficient task-based approach for solving the n-body problem on multicore architectures

Tudruj, Marek: Distributed Java programs initial mapping based on extremal optimization

Tudruj, Marek: Scheduling parallel programs with architecturally supported regions

Tudruj, Marek: Streaming model computation of the FDTD problem (poster)

Table of Contents – Part II

Part II – Minisymposium Papers

Simulations of Atomic Scale Systems

Tools and Environments for Accelerator Based Computational Biomedicine

GPU Computing

High Performance Computing Interval Methods

Real-Time Access and Processing of Large Data Sets

Linear Algebra Algorithms and Software for Multicore and Hybrid Architectures in Honor of Fred Gustavson on His 75th Birthday

Memory and Multicore Issues in Scientific Computing - Theory and Practice

Multicore Algorithms and Implementations for Application Problems

Fast PDE Solvers and a Posteriori Error Estimates

Scalable Tools for High Performance Computing

Table of Contents – Part I

Part I – Keynote Papers and General Topics

Keynote Papers

General Topics

Cloud Computing

HPC Algorithms

HPC Programming Tools

HPC in Meteorology

Parallel Numerical Algorithms

Parallel Computing in Physics

HPC Software Engineering

Free Energy Monte Carlo Simulations on a Distributed Network

Luke Czapla[*], Alexey Siretskiy, John Grime, and Malek O. Khan

Uppsala Universitet - Department of Physical and Analytical Chemistry
czapla@biomaps.rutgers.edu

Abstract. While the use of enhanced sampling techniques and parallel computing to determine potentials of mean force is in widespread use in modern Molecular Dynamics and Monte Carlo simulation studies, there have been few methods that efficiently combine heterogeneous computer resources of varying quality and speeds in realizing a single simulation result on a distributed network. Here, we apply an algorithm based on the Monte Carlo method of Wang and Landau within a client-server framework, in which individual computing nodes report a histogram of regions of phase space visited and corresponding updates to a centralized server at regular intervals entirely asynchronously. The server combines the data and reports the sum to all nodes so that the overall free energy determination scales linearly with the total amount of resources allocated. We discuss our development of this technique and present results for molecular simulations of DNA.

1 Introduction

Observing and interpreting the phenomena that occur in complex model systems of interacting particles is a major computational challenge. Monte Carlo sampling [13] is often an efficient technique in statistical mechanics for observing the behavior of these model systems, sampling configurational states according to their relative contribution to the ensemble partition function. However, difficult problems in effective sampling arise when transitions between very probable states of the system can only be obtained by sampling through improbable states of the system. Similarly, Molecular Dynamics simulations are another widespread technique used in computer simulation, but they are similarly limited in addressing these difficulties, due to the limited sampling times that can be achieved in a realistic computation time.

A number of enhanced sampling techniques have been proposed over the years [6,7,18] to attempt to circumvent the problems encountered with these simulation methods. Often, a major goal is estimating the potential of mean force (PMF), or relative free energy of states with respect to the value of a specific reaction coordinate or coordinates. Free energy simulation is an invaluable technique to characterize a wide range of phenomena in model systems, such as molecular clustering, ligand-target binding, and the folded landscape that may occur under

---

[*] Corresponding author.

K. Jónasson (Ed.): PARA 2010, Part II, LNCS 7134, pp. 1–12, 2012.

certain conditions for semi-flexible polyelectrolytes such as DNA [1] as well as in RNA molecules and protein-nucleic acid assemblies of biological relevance. For dense systems, Molecular Dynamics simulation is sometimes more efficient, but Monte Carlo is more general in allowing sampling of a reaction coordinate in the absence of a definition of a force acting along this reaction coordinate. Here we adapt a common Monte Carlo methodology to the problem of free energy simulations for the case of intermediate particle densities, for a model in which ions in solution are represented explicitly but the effect of the solvent is modeled as an effective electrostatic dielectric constant which models the screening due to solvent reorientation and polarizability. Importantly, the method presented here is not model dependent and it is possible to use this approach to optimize any type of system.

While the free energy is directly proportional to the probability density of a system to be in a given state with the parameter of interest, in many instances, large free energy maxima representing energetically unfavorable states may exist as a barrier to efficient and complete sampling of a parameter space using the standard Metropolis Monte Carlo procedure [13] in a reasonable time for computation. In order to circumvent such difficulties, an efficient technique for generating a uniform density of states, as first introduced by Wang and Landau [19], has previously been applied to determining the free-energy profile of molecular systems as a function of a system parameter chosen as the reaction coordinate [2,9]. Unlike some other methods that generate a flat histogram, this method is both scalable to large systems and easily parallelized [9], by use of both multiple simulations for different ranges of the parameter space and multiple random walks for the same range of the parameter space of interest.

Here, to efficiently perform random walks on individual nodes that share information conjointly to generate the free energy with respect to a given system parameter(s), a distributed architecture is proposed. A centralized server to receive all incoming communication requests is implemented, based on observation that many network resources are secured from incoming network traffic, in order to routinely update the node-summed potential of mean force (PMF) and to update the individual nodes. Incoming requests to the server are handled asynchronously, adding information about sites visited in the individual random walks, and providing information about the node-summed PMF to these nodes during the routine reporting intervals.

A few examples of model calculations from DNA and polyelectrolyte physics are presented, illustrating the timing of the simplest implementations as well as the implementation and use of the algorithm in a realistic network environment, combining separate resources to perform a single calculation. With the Wang-Landau implementation and communication framework described here, difficult problems that are impossible to feasibly simulate using standard sampling techniques are efficiently performed in parallel with no required synchronicity. We demonstrate, with a few targetted calculation examples, how the technique may effectively collect statistics about states that are extremely rare to encounter in standard Monte Carlo simulations.

2   Method

The Wang-Landau procedure for generating the density of states has been widely applied to many types of problems in molecular simulation. In our current implementation, a system parameter such as the absolute end-to-end distance or radius of gyration of a polymer chain is used to characterize the free-energy landscape of the polymer with respect to variation of this parameter. In the simplest implementation, a range of interest for a single parameter is studied to determine the potential of mean force, using a single random walk that determines the frequency of visiting each value in the range of the system parameter of interest by dividing the space of values into discrete bins and determining the total probability of configurations occurring in each bin. Typically, bins are of uniform size for simplicity, but this behavior can be defined non-uniformly depending on phenomena under consideration - in some cases, only one region is of interest and coarse-graining other bins to determine the overall probability of these states may be more efficient.

In this algorithm, every time the current configurational state of the random walk falls within a given bin x, a biasing function $U^*(x)$ is updated according to the current value of the simulation parameter ΔU^*, corresponding to a penalty value for biasing the simulation away from this bin. Initially, the value of $U^*(x)$ is zero for all bins, and the parameter ΔU^* is chosen to be fairly large, such as $0.001 k_B T$. Every time a configurational state occurs in simulation, the value of $U^*(x)$ for the corresponding bin is updated according to:

$$U^*_{new}(x) = U^*(x) + \Delta U^* . \tag{1}$$

The Metropolis criteria used to determine the acceptance of a proposed system move to a configuration specified by coordinates $\mathbf{r}'$ is determined based on the total potential plus bias energy of the proposed configuration, $U(\mathbf{r}') + U^*(x(\mathbf{r}'))$, relative to this value for the previous configuration with coordinates $\mathbf{r}$. This acceptance probability is written as follows (β is the Boltzmann factor $\frac{1}{k_B T}$):

$$p = \min(1, \exp\left[-\beta \left\{ (U(\mathbf{r}') + U^*(x(\mathbf{r}'))) - (U(\mathbf{r}) + U^*(x(\mathbf{r}))) \right\} \right]) . \tag{2}$$

Rejection of a proposed change and returning to the previous configuration with coordinates $\mathbf{r}$ includes also adding the penalty update value ΔU^* again to the corresponding bin $x(\mathbf{r})$, updating $U^*(x(\mathbf{r}))$. In addition, configurations that fall outside of any system parameter bin (for simulations where a limited range of possible system parameter space is examined) are discarded as having infinite potential energy, treating the move also as a rejection and returning to the previous system configuration $\mathbf{r}$ [16].

This procedure continues until all bins x are visited by a number of configurations $p(x)$ that is within a given Δp^* of the average number of configurations per bin, for example, $\Delta p^* = 20\%$. At this point, the probability of visiting each bin is uniform to within the error given by this parameter, with respect to the current penalty update value ΔU^*. Then, the number of visits to each bin, $p(x)$, is reset to zero and the biasing potential is retained, but proceeds to be updated according to adding a smaller value of ΔU^* for each visit. A new histogram $p(x)$ is then collected for this new penalty update value.

The value of ΔU^* is updated according to some function that monotonically decreases to zero, for example, $\Delta U^*_{new} = \Delta U^*/4$. For this particular update choice, every four visits to a given system parameter bin with this new penalty update value will increase the biasing function the same amount as one visit in the previous "sweep", i.e., with the previous ΔU^* value. Each iteration concludes when all the values of $p(x)$ are within the same Δp^* of the mean $p(x)$ value of bin visits (the tolerance Δp^* is the same for every sweep in the present implementation).

The whole procedure proceeds until the value of ΔU^* decreases below some tolerance, such as $\Delta U^* < 10^{-6} k_B T$. At this point, ΔU^* is set to zero (no more updates are made to the biasing function $U^*(x)$), and a final new histogram $p(x)$ is collected for the random walk with the resultant biasing function. This concluding random walk obeys the detailed balance criteria, thus the sampling that is performed is equivalent to umbrella sampling with the ideal biasing potential that generates a flat histogram with respect to the system parameter of interest x. The fluctuations in the logarithm of the number of visits to each bin with this final sampling (times the factor $k_B T$) may be used to estimate the statistical error in the resulting $U^*(x)$.

In order to allow multiple random walks on separate computing nodes to efficiently determine the value of $U^*(x)$ in parallel for each sweep, a centralized server is implemented (the present implementation communicates using TCP/IP sockets). Alternatively, one of the nodes can assume the role as this centralized server, with redundancy achieved automatically through sharing of the data, allowing another node to pick up as the central server if this node is disconnected. As a new node joins the network, it requests from this server the current value of the simulation parameter ΔU^*, the system parameter(s) x to measure, as well as the range of values of x to consider and the manner in which this range is divided into discrete bins. In addition, upon joining the computational network, the node receives the current value of $U^*(x)$ for each bin, which is the sum of all previous updates made by the other nodes in the current sweep and any other sweeps that have already been performed.

The node then performs a random walk with the current bias function, making new updates to $U^*(x)$ according to the current penalty value ΔU^*, and reports to the server only these new visits $p(x)$ and the value of ΔU^* used. In order to take advantage of this architecture, nodes must communicate fairly frequently, for example, reporting to the server every ten to one hundred moves. Speed-up due to parallelization is often only achieved when updates to $U^*(x)$ are communicated relatively frequently, thus enabling nodes to rapidly explore the energy landscape by combining information about visited values of the system parameter space and the corresponding increases to the biasing function. However, in some cases this may not be a requirement if simulations are often trapped in separate regions of high probability and need to make many updates to $U^*(x)$ before sampling outside regions. Also, in cases of large simulation systems, the time needed to perform ten to one hundred moves may be much greater, significantly reducing the overhead involved in adding more nodes.

The server examines the value of ΔU^* for all incoming data, in order to ensure that there was no update to the next sweep of the calculation since the last incoming report from this node; this also verifies that the packet is correctly delivered in an unreliable network. If the data corresponds to the current sweep, the server adds the new visits of this node to the node-summed $U^*(x)$ and $p(x)$, and reports this summed $U^*(x)$ to the client, which the client then uses as the current value of the bias function. In addition, if $p(x)$ is then flat to within the chosen accuracy Δp^*, the server updates the penalty value ΔU^* and reports this new value to the client to use when updating $U^*(x)$. Otherwise, if the server has proceeded to the next sweep since the previous reporting from this node, it ignores the incoming data and informs the client of the new value of ΔU^* to use when updating bins, and reports the current node-summed value of $U^*(x)$ without including this invalid data.

When the server updates the penalty value ΔU^* below the chosen tolerance, it finally reports to all nodes for incoming requests and begins collecting a final histogram that is flat to the accuracy of the last sweep. At this point, sampling continues but no more updates are made to $U^*(x)$. Statistics for the value of some other system parameter y can then be obtained at this point with the proper detailed balance condition, by measuring the value of this parameter as a function of the system parameter x for these random walks and computing a Boltzmann-weighted sum over all of the system parameter bins, as expressed:

$$\langle y \rangle = \frac{\sum\limits_{x} \langle y \rangle_x \exp(\beta U^*(x))}{\sum\limits_{x} \exp(\beta U^*(x))} . \tag{3}$$

Finally, the potential of mean force or free energy of the system as a function of the system parameter x (to an arbitrary constant C) for the discretized region of interest is given as:

$$F(x) = -U^*(x) + C . \tag{4}$$

A multi-dimensional extension of this algorithm is trivially obtained by performing updates over a multi-dimensional grid that includes bins for multiple system parameters, e.g., two-dimensional sampling over two unique system parameters x_1 and x_2 can be performed by dividing the range into discrete bins on a 2D grid, in order to determine $F(x_1, x_2) = -U^*(x_1, x_2) + C$. Essentially, it is analogous to a one-dimensional histogram with total number of bins X equal to the product of number of bins for each parameter, or $X = x_1 x_2$ elements.

3   Models and Results

3.1   Semi-flexible Charged Bead Models of Polyelectrolytes for Polymer Folding

In this section we explore the performance of our technique using a homogeneous network of computing nodes. The system under consideration is a charged polymer (freely-jointed model) with counterions to maintain the system's electroneutrality within a rigid cell of radius, R (see Figure 1, *left*). No particle or segment of the chain can go outside the cell. Beads of the chain are connected with rigid bars. Moves of the chain and counterions are self-avoiding, so neither ions nor chain beads can overlap.

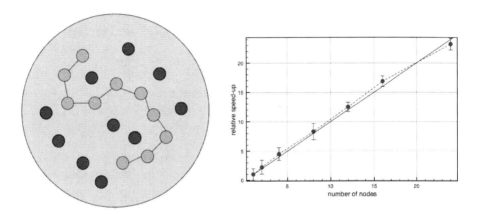

Fig. 1. (*Left*) The polymer model used in calculations evaluating the algorithm performace on a homogeneous network of computing nodes. (*Right*) The relative speed-up (calculated proportionally to the inverse of computation time) for different numbers of nodes in the synchronous MPI implementation.

All the electrostatics are calculated using Coulomb's Law, $U(r_{ij}) = Ke^2 \frac{Z_i Z_j}{\epsilon r_{ij}}$. K is a constant, e is the electron charge, r_{ij} is the distance between particles i and j, Z_i and Z_j are the valencies of particles i and j, and ϵ is the dielectric constant of a media. The original version of our algorithm was implemented on a homogenous network using MPI [9]. Every node contributed the same amount of updates to the flat histogram. Details about the model and implementation have been published earlier [8].

As the algorithm has stochastic behavior, care needs to be taken when discussing the performance. Specifically, this means that the same result (within statistical errors) can be obtained with very different calculation times. The number of steps needed to obtain a result will be different from run to run. The exact reasons of this behavior is under intensive investigation[11,14]. In order to calculate the efficiency together with error estimates, we ran every simulation 10 times.

As illustrated by the graph in Figure 1 (right), parallelization reduces the execution time significantly, in this case from an execution time of 550 minutes on a single computing node to 22 minutes on 24 nodes. The relative speed-up is $\nu_p = t_1/t_p$, where t_1 is the execution time for the program on one core and t_p is the execution time for p cores. The figure shows that the calculation scales almost linearly at least up to 24 nodes.

3.2   Double-Helical Models of DNA and Charged Solute Interactions

While the semi-flexible ion bead models of polyelectrolytes are very simplified models that are feasible for large-scale polymer folding simulations, a more realistic account of the geometry of a polymer such as double-stranded DNA may give clearer insight into the effect of electrostatic interactions in inducing localized bending in short polymers, such as observed in protein-DNA interactions induced by the *E. coli* bacterial nucleoid protein HU [5], which contains positive charge due its high content of the cationic amino acid Lysine and binds to double-helical DNA with little to no sequence specificity [17]. To attempt to address basic questions about the role of electrostatic mechanisms in localized DNA bending, a more detailed but coarse-grained base-pair step parameter model that incorporates the flexibility of DNA as modeled in previous work [4], and extended to consider the shape and charge of double-stranded DNA and approximate positions of anionic -1 phosphate charges along the grooves of the double helix, is incorporated into ion simulations of the electrostatic environment around the polymer.

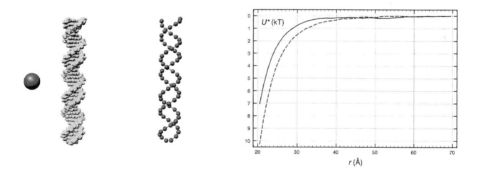

Fig. 2. *(Left)* Model of DNA-solute interaction calculation. The central axis of DNA is aligned along the z axis with the central base pair at the origin. A solute is free to move within the plane defined by $z = 0$ where the central base pair plane is positioned, and the DNA-solute potential of mean force is calculated relative to the solute's distance from the origin, r. The electrostatic interactions between DNA and ions and solute are represented by approximate positions of the DNA phosphates, as illustrated in red. *(Right)* Result for two simulations using the asynchronous distributed method, one in the presence of monovalent cation salt *(dashed line)* and one in a mixed monovalent/divalent cation salts *(solid line)*. The value of the resulting $U^*(r)$ is plotted with the axis reversed to represent the resulting PMF, shown in $k_B T$ energy units.

Here, the cell model is used, placing a DNA molecule containing 35 base pairs within a cell of radius 100 Å, with the middle base pair centered at the origin of the cell. The DNA radius is modeled as a flexible hard cylinder of radius 11 Å for purposes of excluding ions from the space occupied by the polymer. Sixty-eight neutralizing monovalent ions of +1 charge with standard 2 Å radius are placed in the cell to neutralize the 68 phosphate charges (34 per strand), along with additional salt with a total positive charge of 350. Simulations are performed at a temperature $T = 300K$, where 1 kcal/mol of energy is approximately equal to $1.6k_BT$.

In one system, the positive charge is entirely composed of monovalent cations, while in the other system there are 250 monovalent and 50 divalent cations, and both systems include 350 monovalent anions. In addition, the charged solute of charge +10 and radius 5 Å is placed within the plane defined by $z = 0$, and another 10 monovalent anions are added into the cell in order to maintain complete electroneutrality. These two simulation conditions roughly represent a concentration of 100 and 140 mM monovalent salt, similar to conditions often encountered *in vivo*. The ionic strength of the systems is equal in both cases, but the second case incorporates some of this charge as divalent cations (at a low concentration representing at most 20 mM), which bind tightly to double-stranded DNA in both analytical counterion condensation theory and in our simulation observations. This model also uses the Coulomb interaction potentials for all electrostatic interactions (the DNA-solute interaction is purely electrostatic), and the cell model is identical to that in the previous section and illustrated there in Figure 1.

In order to model the flexibility of double-helical DNA in a canonical BDNA-like structure, an elastic potential in terms of base-pair step parameters from previous work [3,4] is applied, where θ_1, θ_2, and θ_3 are the tilt, roll, and twist, respectively, of each of 34 base-pair steps of the 35 base-pair DNA:

$$U = \frac{kT}{2}(\theta_1^2 + 0.02\theta_2^2 + 0.06\theta_3^3) \,. \tag{5}$$

The phosphates move with the DNA steps at a position relative to the center $\mathbf{M}$ of the mid-basis of each base-pair step, in terms of displacements along the short, long and normal unit axis vectors ($\mathbf{x}$, $\mathbf{y}$, and $\mathbf{z}$) described in these same works [3]. These approximate positions of phosphates p_1 and p_2 for each base-pair step are deduced from analysis of high-resolution X-ray and NMR structures for Protein Databank (PDB) entries containing BDNA-like DNA (Dr. Andrew V. Colasanti, personal communication) and are given as:

$$p_1 = -3\mathbf{x} + 8.9\mathbf{y} - 0.4\mathbf{z} \,, \quad p_2 = -3\mathbf{x} - 8.9\mathbf{y} + 0.4\mathbf{z} \,. \tag{6}$$

Each calculation is performed by 16 independent nodes, reporting data asynchronously using the client-server architecture. Calculations begin with $\Delta U^* = (6.4 * 10^{-3})k_BT$ and procede for 5 iterations, with each successive iteration at one-fourth of the previous penalty value, down to a minimum of $\Delta U^* = (2.5 * 10^{-6})k_B T$. Each move series consists of a random displacement of each ion up to a distance of 5 Å, followed by a global pivot of each base pair by a bend and twist of up to 1 degree each, followed by random displacement of the solute up to 1 Å. The solute is constrained in the $z = 0$ plane of the central base pair so that motions only in the x and y coordinates are allowed. The potential of mean force representing the distance between this solute and the center of this base pair is computed in 50 bins ranging from $r = 20$ Å and $r = 50$ Å, and the final obtained PMF which produces equal probability of visiting each bin is then normalized by the area A of each circular disc-like region and reported in terms of the computed probability density $\rho = p/A$ as $-k_B T \ln \rho$. In addition, the average bending angle of DNA is calculated as a function of the DNA-solute distance r, and the average x and y coordinates of the solute are also computed for each r bin.

Figure 2 illustrates the model of DNA and solute, and the electrostatic representation of DNA (in red) on the left, and the graph on the right reports the resulting PMF for the cases with monovalent and with monovalent and divalent cations, to illustrate the effect of salt on modulating solute binding. The calculation also indicates (results not shown) that the solute binds stronger to DNA when divalent are absent, also inducing greater DNA bending of nearly 45 degrees instead of nearly 38 degrees in the mixed salt solution at the closest distances of approach. In both cases, the solute-DNA attraction is capable of inducing some additional DNA bending over the average 33 degree bend found for when the solute is far from DNA. Statistics on the solute position also indicate that the solute approaches the center of the DNA model more often in the major groove of DNA, the side opposite of phosphates neighboring the central base pair (at positive x values).

The two asynchronous calculations presented here each took roughly 48 total hours (each node of 16 total ran for 48 CPU-hours per calculation). Detailed timing calculations for the asynchronous method used here on distributed nodes, for comparison to the presented timings obtained using the previous synchronous MPI method on single clusters, is currently a work in progress.

4   Discussion

For modeling the energetics of charged systems of particles, mean-field approximations of the electrostatic interaction based on Poisson-Boltzmann theory are in many cases inaccurate and fail to predict attractive interactions that may manifest under appropriate conditions. Polyelectrolytes such as DNA exhibit complex interactions with counterions, which strongly associate with the surface of a charged polymer like DNA, a phenomenon originally described by Manning and Oosawa [12], and termed counterion condensation. The explicit treatment of ions considered here and found extensively in past literature [15] allows for a more accurate depiction of the association between a polyelectrolyte and the ions in a solution environment.

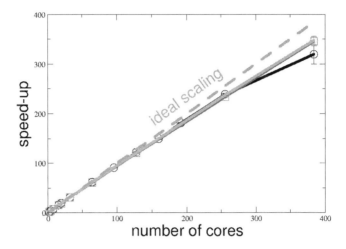

Fig. 3. Scaling up to as many as 384 nodes within the original MPI implementation. The ideal speed-up is shown in the dashed line, while the three solid lines represent the resulting speed-up for three different sets of polyelectrolyte simulations.

Our results show that for polyelectrolyte models, the build up of counterions in shielding electrostatic interactions in the vicinity of the large charge density of the polymer is poorly described by a Boltzmann distribution of ions, due to strong electrostatic interactions and the correlation of ions built up around the polymer. While the counterion condensation theory gives a much more accurate account of the observed correlations in our ion simulations for polyelectrolytes, it also does not, in its simple form, quantitatively provide an accurate fit to the observed interactions for flexible polymers, such as complex attractive forces between segments of the polymer chain at close distances, or the induced local bending in a short double-stranded DNA molecule due to the presence of a charged solute. These kinds of longstanding questions can be addressed by large-scale simulations that are facilitated by the methods presented here.

The scaling of the algorithm with number of nodes from work with the previous synchronous method of performing the described Wang-Landau Monte Carlo method with MPI is shown in Figure 3. The asynchronous distributed method has advantages in that individual computations do not need to be on the same cluster, and computations do not have to wait to synchronize. In practice, the performance of the two is identical, but the distributed method allows for much more robust scaling due to being able to efficiently combine many more resources and is thus able to achieve a much greater number of nodes in practice. Some of the small losses of ideal scaling with hundred of CPUs observed with the earlier method in Figure 3 may be due to synchronization delays.

Another advantage of the asynchronous distributed method presented here is in the maximization of the use of computer resources. The distributed Wang-Landau approach allows for dynamic resizing of an ongoing simulation, such that unallocated computer resources may be allocated to contribute to the calculation for just a fraction of the total simulation time. This allows for maximizing the total amount of computer time dedicated to the computation under the constraints of scheduling issues often encountered with large supercomputing resources.

The applications of distributed and cloud computing to problems in molecular simulation using the free energy methods presented here are endless. In addition, the ideas presented here may be adapted to Molecular Dynamics simulations using similar techniques to Wang-Landau Monte Carlo sampling, such as Metadynamics [10], to similarly obtain free energies from biasing functions and evaluate histograms in parallel over the distributed client-server framework described in this work. With the approach presented here, the goal of truly massive-scale simulations for answering difficult questions in complex systems may be realized.

References

1. Bloomfield, V.A.: DNA condensation by multivalent cations. Biopolymers 44(3), 269–282 (1997)
2. Calvo, F.: Sampling along reaction coordinates with the Wang-Landau method. Molecular Physics 100, 3421–3427 (2002)
3. Coleman, B.D., Olson, W.K., Swigon, D.: Theory of sequence-dependent DNA elasticity. Journal of Chemical Physics 118, 7127–7140 (2003)
4. Czapla, L., Swigon, D., Olson, W.K.: Sequence-dependent effects in the cyclization of short DNA. Journal of Chemical Theory and Computation 2(3), 685–695 (2006)
5. Czapla, L., Swigon, D., Olson, W.K.: Effects of the nucleoid protein HU on the structure, flexibility, and ring-closure properties of DNA deduced from Monte Carlo simulations. J. Mol. Biol. 382(2), 353–370 (2008)
6. Ferrenberg, A.M., Swendsen, R.H.: Optimized Monte Carlo data analysis. Phys. Rev. Lett. 63(12), 1195–1198 (1989)
7. Henin, J., Fiorin, G., Chipot, C., Klein, M.L.: Exploring multidimensional free energy landscapes using time-dependent biases on collective variables. Journal of Chemical Theory and Computation 6, 35–47 (2010)
8. Khan, M.O., Chan, D.Y.C.: Effect of chain stiffness on polyelectrolyte condensation. Macromolecules 38(7), 3017–3025 (2005)
9. Khan, M.O., Kennedy, G., Chan, D.Y.C.: A scalable parallel Monte Carlo method for free energy simulations of molecular systems. J. Comput. Chem. 26(1), 72–77 (2005)
10. Laio, A., Gervasio, F.L.: Metadynamics: a method to simulate rare events and reconstruct the free energy in biophysics, chemistry and material science. Reports on Progress in Physics 71(12), 126601 (2008)
11. Lee, H.K., Okabe, Y., Landau, D.P.: Convergence and refinement of the Wang-Landau algorithm. Technical Report cond-mat/0506555 (2005)
12. Manning, G.S.: Counterion binding in polyelectrolyte theory. Accounts of Chemical Research 12(12), 443–449 (1979)
13. Metropolis, N., Rosenbluth, A.W., Rosenbluth, M.N., Teller, A.H., Teller, E.: Equation of state calculations by fast computing machines. The Journal of Chemical Physics 21(6), 1087–1092 (1953)
14. Morozov, A.N., Lin, S.H.: Accuracy and convergence of the Wang-Landau sampling algorithm. Phys. Rev. E Stat. Nonlin. Soft. Matter Phys. 76(2 pt. 2), 026701 (2007)

15. Olmsted, M.C., Bond, J.P., Anderson, C.F., Record Jr., M.T.: Grand canonical Monte Carlo molecular and thermodynamic predictions of ion effects on binding of an oligocation (L8+) to the center of DNA oligomers. Biophys J. 68(2), 634–647 (1995)
16. Schulz, B.J., Binder, K., Muller, M., Landau, D.P.: Avoiding boundary effects in Wang-Landau sampling. Phys. Rev. E Stat. Nonlin. Soft. Matter Phys. 67(6 pt. 2), 067102 (2003)
17. van Noort, J., Verbrugge, S., Goosen, N., Dekker, C., Dame, R.T.: Dual architectural roles of HU: formation of flexible hinges and rigid filaments. Proc. Natl. Acad. Sci. U S A 101(18), 6969–6974 (2004)
18. Voter, A.F.: Parallel replica method for dynamics of infrequent events. Physical Review B 57, 13985–13988 (1998)
19. Wang, F., Landau, D.P.: Efficient, multiple-range random walk algorithm to calculate the density of states. Phys. Rev. Lett. 86(10), 2050–2053 (2001)

Numerical Investigation of the Cumulant Expansion for Fourier Path Integrals

Nuria Plattner[1], Sharif Kunikeev[2], David L. Freeman[2], and Jimmie D. Doll[1]

[1] Department of Chemistry
Brown University
Providence, RI 02912
[2] Department of Chemistry
University of Rhode Island
Kingston, RI 02881

Abstract. Recent developments associated with the cumulant expansion of the Fourier path integral Monte Carlo method are illustrated numerically using a simple one-dimensional model of a quantum fluid. By calculating the Helmholtz free energy of the model we demonstrate that 1) recently derived approximate asymptotic expressions for the cumulants requiring only one-dimensional quadrature are both accurate and viable, 2) expressions through third-cumulant order are significantly more rapidly convergent than either the primitive Fourier method or the partial average method, and 3) the derived cumulant convergence orders can be verified numerically.

Keywords: path integral, Monte Carlo, cumulant expansion.

1 Introduction

Computational algorithms[5,24] using path integral approaches[13,16] for quantum statistical mechanics have been used for in excess of 30 years, and the techniques have become part of the standard set of methods available to practitioners of computational chemistry and physics.[1,10,4] One approach to numerical path integration for quantum statistical mechanics is often called discrete path integration (DPI), and begins with the quantum density matrix in coordinate representation (we assume a one-dimensional system for simplicity), $\rho(x, x') = \langle x'|e^{-\beta \hat{H}}|x\rangle$, where x and x' are particle coordinates, $\hat{H}$ is the system Hamiltonian operator, and β is the inverse temperature. By writing $\exp\left(-\beta\hat{H}\right) = \left[\exp\left(-\beta\hat{H}/P\right)\right]^P$ and introducing the resolution of the identity P times into the density matrix element, and by using a Trotter product,[27] approximations to the full density matrix are generated that become exact in the limit of infinite P. We refer to the intermediate coordinate states as path variables, and a key concern is the size of P required so that the approximations do not introduce systematic errors in computed averages that are larger than the statistical noise generated during the simulation. It has been proven[4] that

K. Jónasson (Ed.): PARA 2010, Part II, LNCS 7134, pp. 13–22, 2012.

DPI converges to the exact density matrix as P^{-2}, and Suzuki[25] has shown that there is no simple approach to improve these convergence characteristics. For the special case of intermolecular forces represented by pair potentials, there is empirical evidence[4] that alternative formulations using accurate pair propagators enhance the convergence rate to P^{-3}. Other work[19,15,7,6,3,2,23] has shown even better convergence characteristics by including higher-order commutators and generating improved short-time propagators, and recent work on many-body systems has been promising.[23]

An alternative to the DPI method is often called the Fourier path integral (FPI) approach.[20,10] The FPI method starts with the Feynman-Kac expression for the density matrix[13,16]

$$\rho(x, x') = \int \mathcal{D}x(u) \exp\{-S[x(u)]\} \tag{1}$$

where $\int \mathcal{D}x(u)$ represents an integral over all paths connecting x to x' in imaginary time u and S is the action

$$S[x(u)] = \beta \int_0^1 du \left[\frac{m}{2} \dot{x}^2 + V[x(u)] \right] \tag{2}$$

with V the system potential energy. In the Fourier method each path $x(u)$ is expanded in a Fourier series about a straight-line path connecting x to x',

$$x(u) = x + (x' - x)u + \sigma \sum_{k=1}^{\infty} a_k \Lambda_k(u) \tag{3}$$

with $\sigma = \sqrt{\beta \hbar^2 / m}$, $\Lambda_k(u) = \sqrt{2} \sin(k\pi u)/(k\pi)$ and m the mass of the particle. The integration over all paths is replaced by an integral with respect to the infinite set of Fourier coefficients, $\{a_k\}$. In actual simulations the Fourier series is truncated at some upper index K with the K Fourier coefficients being the path variables. As with the DPI method where the convergence with respect to P is an important issue, the convergence of computed properties with respect to the Fourier index K is central.

In the current work we focus on the FPI method and use a one-dimensional model to illustrate recent developments[18,17] using a cumulant expansion that significantly enhance the convergence properties of simulations with respect to K in a manner that is amenable to practical applications. The contents of the remainder of this paper are as follows. In Section 2 we present the key theoretical developments associated with the cumulant expansion for Fourier path integrals. In Section 3 we describe the one-dimensional model system used to illustrate our key results, and we present numerical data illustrating the principal findings of Ref. [17]. We summarize and discuss our results in Section 4.

2 Theory

As shown elsewhere[12] the primitive FPI method converges as K^{-1}, and the partial averaging method[11,8] has been introduced to enhance this convergence rate. It is useful to think of the partial averaging method as the first term in a cumulant series for the density matrix,[9,17] and we now describe that cumulant series and its convergence properties.

The cumulant expansion exploits a series of terms generated by the tail series $(k > K)$ in the Fourier expansion. Defining

$$x_K(u) = x + (x' - x)u + \sum_{k=1}^{K} a_k \Lambda_k(u) \tag{4}$$

and

$$x_{K+1}(u) = \sum_{k=K+1}^{\infty} a_k \Lambda_k(u) \tag{5}$$

the density matrix can be expressed in terms of a tail integral (TI)

$$\rho(x, x') = \prod_{k=1}^{K} \left(\int_{-\infty}^{\infty} \frac{da_k}{\sqrt{2\pi}} e^{-a_k^2/2} \right) \langle \exp(-\beta\overline{V}) \rangle_{TI} \tag{6}$$

where the path-averaged potential energy is defined by

$$\overline{V} = \int_0^1 du V[x(u)] \tag{7}$$

and the average with respect to the TI is defined by

$$\langle \exp(-\beta\overline{V}) \rangle_{TI} = \prod_{k=K+1}^{\infty} \left(\int_{-\infty}^{\infty} \frac{da_k}{\sqrt{2\pi}} e^{-a_k^2/2} \right) \exp(-\beta\overline{V}). \tag{8}$$

By expanding the exponential on the right hand side of Eq. (8) in a power series, the TI can be expressed in terms of a series of moments

$$\langle \exp(-\beta\overline{V}) \rangle_{TI} = 1 + \sum_{p=1}^{\infty} \frac{(-\beta)^p}{p!} \mu_p \tag{9}$$

where the moments are defined by

$$\mu_p = \langle \overline{V}^p \rangle_{TI}. \tag{10}$$

The cumulant expansion is obtained by replacing the sum of moments by an exponential

$$\langle \exp(-\beta\overline{V}) \rangle_{TI} = \exp(\nu_c) \tag{11}$$

where

$$\nu_c = \ln \left(1 + \sum_{p=1}^{\infty} \frac{(-\beta)^p}{p!} \mu_p \right) \tag{12}$$

$$= \sum_{k=1}^{\infty} \frac{(-\beta)^k}{k!} \mu_{ck}, \tag{13}$$

and μ_{ck} is called the kth-order cumulant.

Using the cumulant expansion the density matrix can be written

$$\rho(x, x') = \prod_{k=1}^{K} \left(\int_{-\infty}^{\infty} \frac{da_k}{\sqrt{2\pi}} e^{-a_k^2/2} \right) \exp(-\beta\mu_{c1} + (\beta^2/2)\mu_{c2} - (\beta^3/6)\mu_{c3} + \ldots). \tag{14}$$

If we truncate the exponential on the right hand side of Eq. (14) at the first-order cumulant, we obtain the partial average approximation to the full density matrix. As has been understood for some time,[21,17] the partial average method converges asymptotically as K^{-3}. Implementation of the partial averaged method requires that the potential involved have a readily available Gaussian transform. Because many potential functions used in simulations have either no finite Gaussian transform or have a Gaussian transform not expressible in closed form, the partial average method has typically been used only in approximate forms with decreased convergence rates. The Lennard-Jones potential is an example of an interaction that is not amenable to full partial averaging. While the Lennard-Jones interaction appears to be problematic, we have recently shown[18] that we can fit the Lennard-Jones potential to functions having closed-form Gaussian transforms providing results having negligible systematic errors. The current work uses a one-dimensional model that exploits the successful use of Gaussian-fit potentials.

In a recent publication[17] we have demonstrated analytically that the cumulant expansion truncated at order p converges to the exact density matrix as $K^{-(2p+1)}$. Consequently, the first-order cumulant (partial averaging) converges asymptotically as K^{-3}, the second-order cumulant converges as K^{-5} and so on. Previously,[9] we have examined the possibility of including the second-order cumulant. Our past efforts have been hampered by the need to evaluate numerically nested multiple integrals with respect to the imaginary time variable u. In our most recent work,[17] by examining asymptotic representations of the higher-order cumulants, we have developed expressions in the asymptotic limit through third-cumulant order requiring u integrations in only a single dimension. Through the third-order cumulant these asymptotic expressions retain an asymptotic convergence rate of, at least, K^{-5}. A goal of the current work is to provide a numerical demonstration that we reach the asymptotic convergence rate rapidly so that our methods are viable.

Space limitations make it impossible to produce the proof of the assertions in the previous paragraph, but formal proofs can be found in Ref. [17]. We do find it useful to give the asymptotic forms for the second and third-order cumulants.

Using Eqs. (77) and (98) from Ref. [17], we have

$$\mu_{c2} = \frac{\sigma^2}{3\pi^4 K^3} \left\{ 4V'(x)^2 + \frac{\sigma^2}{2} \int_0^1 du (V''_{PA}[x_K(u)])^2 \right\} + \frac{2(-1)^K \sigma^2}{\pi^4 K^4} V'(x)^2 + O(K^{-5}) \quad (15)$$

and

$$\mu_{c3} = \frac{\sigma^4}{5\pi^6 K^5} \left\{ \sigma^2 \int_0^1 du (V''_{PA}[x_K(u)])^3 + 6V''(x)(V'(x))^2 \right\} + O(K^{-6}) \quad (16)$$

where to simplify the current discussion we have limited the expressions to one-dimensional systems, V_{PA} represents the partial-average potential,[11,8] and the expressions apply only to diagonal density matrix elements. We have also corrected a typographical error from Ref. [17] in Eq. (16) where the partial averaged rather than the bare potential is included inside the u integration on the right hand side. Only one-dimensional integrals appear in Eqs. (15) and (16) making the expressions numerically tractable.

We have included terms through third cumulant order in Eq. (14), because truncation at second order can lead to partition functions that are not finite. While partial averaging is well behaved, the second-order cumulant appears in the exponent with a positive sign. In regions of space where the potential is highly repulsive, the second-order term can dominate the first-order cumulant resulting in a divergent partition function. The inclusion of the third-order cumulant (having a negative sign) eliminates the divergence. Consequently, we have well-behaved results when both Eqs. (15) and (16) are introduced into Eq. (14). Numerical evidence supporting our assertion of the utility of these methods appears in the next section.

3 Results

To explore numerically the utility of the asymptotic representations of the density matrix through third-cumulant order, we have also generated data for a one-dimensional system where we have used the asymptotic expressions given in Eqs. (15) and (16). The one-dimensional system chosen is a Gaussian fit to the Lennard-Jones cage potential studied previously[14] as a model for particles in a liquid. The explicit form used is

$$V(x) = V_g(a - x) + V_g(a + x) \quad (17)$$

where

$$V_g(x) = 4\epsilon \left(a_1 \exp\left\{ -c_1 \left(\frac{x}{\sigma}\right)^2 \right\} + a_2 \exp\left\{ -c_2 \left(\frac{x}{\sigma}\right)^2 \right\} \right) \quad (18)$$

with associated parameters $a_1 = 3177.6792398702582$, $a_2 = -1.4800039095374943$, $c_1 = 8.993289559255235$, $c_2 = 1.2988205358018494$, and Lennard-Jones parameters chosen for neon, $\epsilon = 35.6K$ and $\sigma = 2.749$ Å. The cage parameter, a, is taken to be $2^{1/6}\sigma$, the minimum of the one-dimensional Lennard-Jones potential. The Helmholtz free energy, A expressed in units of the Boltzmann constant k, at

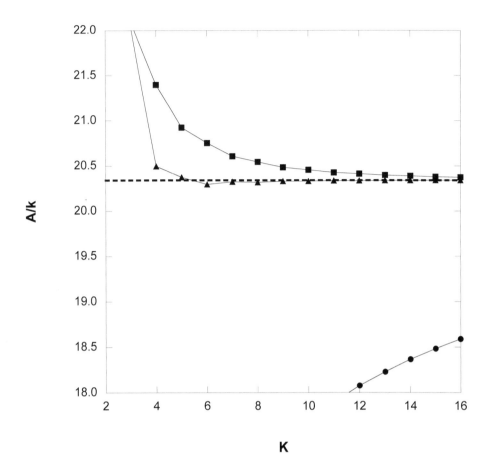

Fig. 1. The free energy of the cage potential in units of the Boltzmann constant k as a function of Fourier index K. The solid circles represent the primitive Fourier path integral data, the squares represent data computed with the first-order cumulant and the triangles have been obtained using the asymptotic expressions through cumulant order 3. As discussed in the text, the dashed line represents the numerically exact result determined using numerical matrix multiplication. The solid lines connecting the data points are included as a visual aid and have no independent meaning.

3.56 K (0.1 ϵ) as a function of the Fourier index K is represented in Fig. 1. To calculate A, we have first determined the diagonal density matrix as a function of coordinate using Eq.(14) followed by a quadrature evaluation of

$$A = -kT \ln \int_{-\infty}^{\infty} dx \rho(x, x). \tag{19}$$

The horizontal dashed line in Fig. 1 represents the exact result determined using numerical matrix multiplication (NMM).[26] The NMM calculation has been performed on an equal spaced quadrature grid with the coordinate x ranging from -2.0 to 2.0 Bohr and a grid separation of 0.05 Bohr. For convergence 1024 iterations have been included. The free energy is then determined using trapezoid rule quadrature from the same grid points as generated from the NMM calculation. Using the same grid the primitive Fourier and cumulant values of the density matrix elements have been determined using Monte Carlo by generating Gaussian random noise for the Fourier coefficients. As with NMM the free energy is determined using trapezoid rule quadrature. The circles in Fig. 1 represent the primitive Fourier data, the squares represent the first-order cumulant data (partial averaging) and the triangles represents the data generated asymptotically through third cumulant order using Eqs. (15) and (16). The error bars generated from the Monte Carlo simulations are smaller than the resolution of the graph. The dramatic rapid convergence of the third-order cumulant approach is evident.

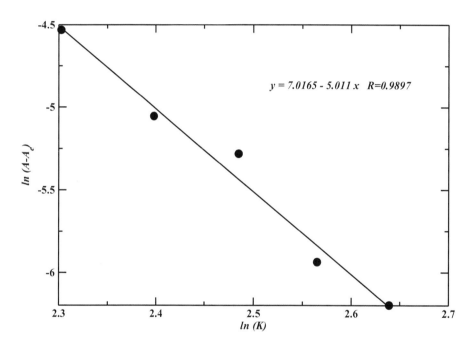

Fig. 2. The natural logarithm of the error in the free energy (error $= A - A_e$) computed with the asymptotic expression through third cumulant order as a function of the natural logarithm of the Fourier index K. The least-squares computed slope (shown in the form $y = b + mx$ with R the regression coefficient) is consistent with the formal asymptotic convergence rate of K^{-5}. The scatter of the plotted points reflects the statistical fluctuations of the Monte Carlo generated results.

Numerical investigations of the asymptotic convergence rates of the primitive Fourier and first-order cumulant methods have appeared in previous work.[22] To determine numerically the asymptotic convergence order of the asymptotic, third-order cumulant method, we assume the difference between the exact Helmholtz free energy A_e and the computed Helmholtz free energy A as a function K has the form

$$A - A_e = B/K^g \tag{20}$$

where B is a constant and g is the exponent defining the asymptotic behavior. Assuming K to be sufficiently large for asymptotic behavior, a graph of the natural logarithm of $A - A_e$ as a function of $\ln K$ should provide $-g$ as the slope. We have found the convergence of $A - A_e$ not to be monotonic until $K = 10$, and in Fig. 1 we present the previously discussed logarithmic plot for $K = 10$ through 14. For $K > 14$ the error in the free energy becomes too small to display meaningfully. The computed slope is consistent with an asymptotic convergence rate of K^{-5}.

4   Discussion

In a previous publication[17] we have formally explored the convergence characteristics of the cumulant expansion for Fourier path integrals. An important outcome of the formal work has been approximate expressions for the cumulants that retain many of the convergence characteristics of the full cumulant expansion with significant reduction in the amount of required numerical overhead. The potential for exploiting the results of Ref. [17] in real simulations appears high, but a full understanding of the potential of the cumulant-based methods requires numerical experimentation. In the current work we have begun the numerical investigation of the methods by illustrating some of the principal findings using a previously developed one-dimensional cage model of a quantum fluid. Using a Gaussian fit to a Lennard-Jones cage potential, we have compared the convergence characteristics of the cumulant expansion approximately through third-cumulant order. For the one-dimensional example we have numerically verified the derived asymptotic convergence rate, and we have demonstrated the significantly improved convergence characteristics for the free energy of the approximate third-order cumulant expressions compared to either the primitive Fourier method or the first-order cumulant (partial averaged) results. We have also verified the numerical tractability of the cumulant-based methods.

To further exploit these methods extensions to full many-particle example problems are required. In this regard we see that the products of derivatives appearing in Eqs. (15) and (16) for one-dimensional systems are replaced by traces of Hessian matrix products. Because only traces are required, the expected work should require acceptable computational resources. A detailed exploration of the cumulant-based methods for a realistic many-particle system will appear separately.

Acknowledgements.   JDD gratefully acknowledges grant support of this research through the DOE Multiscale and Optimization for Complex Systems program No. DE-SC0002413 and through DOE departmental program No. DE-00015561.

References

1. Berne, B.J., Thirumalai, D.: On the simulation of quantum systems: Path integral methods. Ann. Rev. Phys. Chem. 37, 401 (1986)
2. Bogojević, A., Balaž, A., Belić, A.: Asymptotic properties of path integral ideals. Phys. Rev. E 72(3), 036128 (2005),
   http://link.aps.org/doi/10.1103/PhysRevE.72.036128
3. Bogojević, A., Balaž, A., Belić, A.: Systematically accelerated convergence of path integrals. Phys. Rev. Lett. 94(18), 180403 (2005),
   http://link.aps.org/doi/10.1103/PhysRevLett.94.180403
4. Ceperley, D.M.: Path integrals in the theory of condensed helium. Rev. Mod. Phys. 67, 279 (1995)
5. Chandler, D., Wolynes, P.G.: Exploiting the isomorphism between quantum theory and classical statistical mechanics of polyatomic fluids. J. Chem. Phys. 74, 4078 (1981)
6. Chin, S.A.: Quantum statistical calculations and symplectic corrector algorithms. Phys. Rev. E 69(4), 046118 (2004),
   http://link.aps.org/doi/10.1103/PhysRevE.69.046118
7. Chin, S.A., Chen, C.R.: Gradient symplectic algorithms for solving the Schrödinger equation with time-dependent potentials. J. Chem. Phys. 117(4), 1409 (2002),
   http://link.aip.org/link/JCPSA6/v117/i4/p1409/s1&Agg=doi
8. Coalson, R.D., Freeman, D.L., Doll, J.D.: Partial averaging approach to Fourier coefficient path integration. J. Chem. Phys. 85, 4567 (1986)
9. Coalson, R.D., Freeman, D.L., Doll, J.D.: Cumulant methods and short time propagators. J. Chem. Phys. 91(7), 4242 (1989)
10. Doll, J.D., Freeman, D.L., Beck, T.L.: Equilibrium and dynamical Fourier path integral methods. Adv. Chem. Phys. 78, 61 (1990)
11. Doll, J., Coalson, R.D., Freeman, D.L.: Fourier path-integral Monte Carlo methods: Partial averaging. Phys. Rev. Lett. 55, 1 (1985)
12. Eleftheriou, M., Doll, J., Curotto, E., Freeman, D.L.: Asymptotic convergence rates of Fourier path integral methods. J. Chem. Phys. 110, 6657 (1999)
13. Feynman, R., Hibbs, A.: Quantum mechanics and path integrals. McGraw-Hill, New York (1965)
14. Freeman, D., Coalson, R., Doll, J.: Fourier path integral methods: A model study of simple fluids. J. Stat. Phys. 43, 931 (1986)
15. Jang, S., Jang, S., Voth, G.A.: Applications of higher order composite factorization schemes in imaginary time path integral simulations. J. Chem. Phys. 115(17), 7832 (2001), http://link.aip.org/link/JCPSA6/v115/i17/p7832/s1&Agg=doi
16. Kleinert, H.: Path integrals in quantum mechanics, statistics and polymer physics. World Scientific, Singapore (1995)
17. Kunikeev, S., Freeman, D.L., Doll, J.D.: Convergence characteristics of the cumulant expansion for Fourier path integrals. Phys. Rev. E 81, 066707 (2010)
18. Kunikeev, S., Freeman, D.L., Doll, J.: A numerical study of the asymptotic convergence characteristics of partial averaged and reweighted Fourier path integral methods. Int. J. Quant. Chem. 109, 2916 (2009)
19. Makri, N., Miller, W.H.: Exponential power series expansion for the quantum time evolution operator. J. Chem. Phys. 90(2), 904 (1989),
   http://link.aip.org/link/JCPSA6/v90/i2/p904/s1&Agg=doi
20. Miller, W.H.: Path integral representation of the reaction rate constant in quantum mechanical transition state theory. J. Chem. Phys. 63(3), 1166 (1975),
   http://link.aip.org/link/JCPSA6/v63/i3/p1166/s1&Agg=doi

21. Predescu, C., Doll, J., Freeman, D.L.: Asymptotic convergence of the partial averaging technique. arXiv:cond-mat/0301525 (2003)
22. Predescu, C., Sabo, D., Doll, J.D.: Numerical implementation of some reweighted path integral methods. J. Chem. Phys. 119, 4641 (2003)
23. Sakkos, K., Casulleras, J., Boronat, J.: High order chin actions in path integral monte carlo. J. Chem. Phys. 130(20), 204109 (2009),
    http://link.aip.org/link/JCPSA6/v130/i20/p204109/s1&Agg=doi
24. Schweizer, K.S., Stratt, R.M., Chandler, D., Wolynes, P.G.: Convenient and accurate discretized path integral methods for equilibrium quantum mechanical calculations. J. Chem. Phys. 75, 1347 (1981)
25. Suzuki, M.: General theory of fractal path integrals with applications to many-body theories and statistical physics. J. Math. Phys. 32, 400 (1991)
26. Thirumalai, D., Bruskin, E.J., Berne, B.J.: An iterative scheme for the evaluation of discretized path integrals. J. Chem. Phys. 79, 5063 (1983)
27. Trotter, H.F.: On the product of semi-groups of operators. Proc. Am. Math. Soc. 10, 545 (1959)

Optimization of Functionals of Orthonormal Functions in the Absence of Unitary Invariance

Peter Klüpfel[1], Simon Klüpfel[1],
Kiril Tsemekhman[2], and Hannes Jónsson[1,2]

[1] Science Institute, VR-III, University of Iceland, 107 Reykjavík, Iceland
[2] Chemistry Department, University of Washington, Seattle, USA

Abstract. We discuss the optimization of a functional with respect to sets of orthonormal functions where unitary invariance does not apply. This problem arises, for example, when density functionals with explicit self-interaction correction are used for systems of electrons. There, unitary invariance cannot be used to reformulate the minimization of the energy with respect to each of the functions as an eigenvalue problem as can be done for the commonly used GGA-DFT and Hartree-Fock theory. By including optimization with respect to unitary transformations as an explicit step in the iterative minimization procedure, fast convergence can, nevertheless, be obtained. Furthermore, by working with two sets of orthonormal functions, the optimal functions and a set of eigenfunctions, the implementation of the extended functional form in existing software becomes easier. The additional computations arising from the lack of unitary invariance can largely be carried out in parallel.

Keywords: functional optimization, orthonormal functions, electrons.

1 Introduction

The task of optimizing the value of a functional of orthonormal functions arises in many contexts in engineering, physics and chemistry [1]. One example is the description of many-electron systems using density functional theory (DFT) which has become a widely used tool in calculations of the basic properties of solids, liquids and molecules [2]. Various approximations to the exact but unknown energy functional are used, but those that are commonly used and can be applied to large systems have several limitations in terms of the accuracy of the results, as described below. In this article, we discuss a possible extension of the form of energy functionals and the corresponding modifications in the minimization procedure. The extended functional form calls for new numerical methods and software implementations for solving the resulting equations.

In Kohn-Sham (KS) DFT [3] using local (LDA) or semi-local (GGA) functionals, the energy due to Coulomb interaction between the electrons

$$E_{\mathrm{H}}[\rho] = \frac{1}{2} \int d^3\mathbf{r} d^3\mathbf{r}' \, \frac{\rho(\mathbf{r})\rho(\mathbf{r}')}{|\mathbf{r} - \mathbf{r}'|} \tag{1}$$

K. Jónasson (Ed.): PARA 2010, Part II, LNCS 7134, pp. 23–33, 2012.
© Springer-Verlag Berlin Heidelberg 2012

and the energy due to an external potential $v_{ext}(\mathbf{r})$

$$E_{ext}[\rho] = \int d^3\mathbf{r}\, v_{ext}(\mathbf{r})\rho(\mathbf{r}) \tag{2}$$

are evaluated directly from the total electron density, $\rho(\mathbf{r})$, rather than the much more complicated many-electron wave function. But, in order to get a good enough estimate of the kinetic energy, a set of orthonormal functions $\varphi^N = \{\varphi_1,\ldots,\varphi_N\}$

$$\int d^3\mathbf{r}\, \varphi_i^*(\mathbf{r})\varphi_j(\mathbf{r}) = \delta_{ij} \tag{3}$$

each depending on the coordinates of just one electron (single-particle functions), are introduced and the kinetic energy minimized with respect to all sets φ^N consistent with the total electron density $\rho(\mathbf{r}) = \sum_i^N \rho_i(\mathbf{r})$ where $\rho_i(\mathbf{r}) = |\varphi_i(\mathbf{r})|^2$

$$T^{KS}[\rho] = \min_{\varphi^N} \sum_i^N \int d^3\mathbf{r}\, \varphi_i^*(\mathbf{r})\left(-\frac{1}{2}\nabla^2\right)\varphi_i(\mathbf{r}) . \tag{4}$$

The remaining contributions to the energy, which include the quantum mechanical exchange and correlation energy as well as correction to the above estimate of the kinetic energy, are denoted by $E_{xc}[\rho]$. They are estimated by comparison with numerically exact calculations of the homogeneous electron gas (when using the LDA approximation) or - as in most calculations today - also include estimates of the effect of local variations by including dependence on the gradient of the density (the GGA approximation) [2]

$$E_{xc}^{KS}[\rho] = \int d^3\mathbf{r}\, \epsilon_{xc}(\rho, \boldsymbol{\nabla}\rho) . \tag{5}$$

The notation here ignores spin for simplicity. The ground state energy of the system is then obtained by variational minimization of the energy with respect to all electron density distributions, ρ, integrating to N electrons

$$E^{KS}[\rho] = T^{KS}[\rho] + E_H[\rho] + E_{xc}^{KS}[\rho] + \int d^3\mathbf{r}\, v_{ext}(\mathbf{r})\rho(\mathbf{r}). \tag{6}$$

Using Lagrange's method, the orthonormality constraints are incorporated into the objective functional by

$$S^{KS}[\rho] = E^{KS}[\rho] - \sum_{i,j}\lambda_{ji}\left[\int d^3\mathbf{r}\varphi_i^*(\mathbf{r})\varphi_j(\mathbf{r}) - \delta_{ij}\right] \tag{7}$$

where $\boldsymbol{\Lambda} = \{\lambda_{ij}\}$ is a matrix of Lagrange multipliers. The variational optimization of S^{KS} with respect to the orthonormal, single-particle functions [4] gives

$$\hat{H}\varphi_i(\mathbf{r}) = \sum_{j=1}^N \lambda_{ji}\varphi_j(\mathbf{r}) \tag{8}$$

where $\hat{H}$ is an operator (the Hamiltonian) defined as

$$\hat{H}\varphi_i(\mathbf{r}) = \frac{\delta E}{\delta \varphi_i^*(\mathbf{r})} \tag{9}$$

and turns out to be the same for all the functions. The functional is invariant under unitary transformations of the functions and variation with respect to φ_i rather than φ_i^* gives the same result. One can choose the particular set of functions for which Λ is diagonal (see section 3). The set of coupled equations for the φ^N functions then reduces to a set of eigenvalue problems

$$\hat{H}\varphi_i(\mathbf{r}) = \epsilon_i \varphi_i(\mathbf{r}) \tag{10}$$

which, however, are still coupled through the total electron density. A solution can be obtained using an iterative procedure starting with a guess and eventually obtaining self-consistency.

Functionals of this type are widely used in the modeling of solids and liquids. Various semi-local approximations to $E_{xc}[\rho]$ have been proposed and powerful software packages have been developed utilizing highly efficient optimization algorithms to solve the fundamental minimization problem [5]. However, several limitations of these functional approximations have also become apparent: (a) The predicted total energy is generally not accurate enough. Useful estimates of energy differences can still be obtained in many cases because of cancellation of error, but this is problematic when the two systems being compared are qualitatively different. For example, the energy of transition states compared with energy of stable states (i.e. reaction barriers) are typically underestimated [6,7]. (b) Electronic defect states tend to be overly delocalized and even unstable [8]. (c) Neither the functions φ_i nor energy eigenvalues ϵ_i have any known, directly observable meaning (but the ϵ_i are sometimes used as estimates of ionization energy or band gap, giving generally poor approximations). This list is far from being complete, but illustrates that the deficiencies of GGA functionals are significant. For a more complete discussion, see [9].

One approach to improve the semi-local approximation is to mix in some 'exact exchange' in so-called hybrid functionals [10,11] through a linear combination with LDA and GGA. Hybrid functionals can cure some of the deficiencies mentioned above for example improved bond energy and bond length [6,12]. The optimal linear combination coefficients, i.e. mixing parameters, are, however, not the same for all types of systems (for example molecules vs. solids) and this approach should be regarded as semi-empirical and relies on tuned cancellation of errors of different origin. For metallic systems, hybrid functionals in fact give poorer predictions than GGA. Although hybrid functionals are available in most major DFT software packages today, their application to systems with appreciable numbers of electrons is, furthermore, hampered by the expensive evaluation of the non-local, exact exchange. The computational effort scales as N^4 rather than the N^3 scaling for LDA and GGA.

2 Orbital Density Dependent Functionals

While the single-particle functions, which frequently are referred to as 'orbitals', are in GGA-DFT simply mathematical constructs that represent the total electron density and improve the estimate of the kinetic energy beyond what has been possible from the total electron density alone, these functions can in principle be interpreted as meaningful representations of the electrons and the corresponding probability density, ρ_i, represent the probability distribution of an electron. This is an assumption, consistent with intuition that is often invoked, but no proof of this has been presented. The form of the energy functional then should include explicit dependence on the orbital densities. We will refer to such an extended functional form as orbital density dependent (ODD) functionals. This can lead to much improved estimates of various properties, but the mathematical task of finding the optimal set of orbitals becomes more challenging.

The ODD functional form can, in particular, be used to correct for the so-called self-interaction error in GGA functionals. The evaluation of the Coulomb energy from the total electron density as in (1) includes interaction of the electrons with themselves, a self-interaction energy. Ideally, the E_{xc} correction term should remove this error, but in practice the approximations used for E_{xc}, such as PBE, only partly cancel it out. A better estimate of the Coulomb interaction is the orbital density dependent expression

$$E_{\mathrm{H}}^{\mathrm{ODD}}[\rho^N] = \frac{1}{2} \sum_{i \neq j} \int d^3\mathbf{r}\,d^3\mathbf{r}'\, \frac{\rho_i(\mathbf{r})\rho_j(\mathbf{r}')}{|\mathbf{r}-\mathbf{r}'|} = E_{\mathrm{H}}[\rho] - \frac{1}{2}\sum_{i=1}^{N} \int d^3\mathbf{r}\,d^3\mathbf{r}'\, \frac{\rho_i(\mathbf{r})\rho_i(\mathbf{r}')}{|\mathbf{r}-\mathbf{r}'|} \quad (11)$$

where the $i = j$ terms representing self-interaction are excluded. Here, ρ^N denotes the set of N orbital densities, $\rho^N = \{\rho_1 \ldots \rho_N\}$. Revised exchange-correlation functionals are necessary in order to account for this modification in the Coulomb term. The evaluation of this expression for the Coulomb energy requires $N+1$ solutions of the Poisson equation and thus is computationally much less demanding than the exact exchange of hybrid functionals and scales as one lower power in N. Furthermore, the $N + 1$ Poisson equations can be solved simultaneously on N nodes or sets of nodes, making parallel implementation easy and efficient. Parallel implementation of hybrid functionals is more difficult [13].

Perdew and Zunger [14] proposed an estimate of the total self-interaction energy for each orbital as

$$E^{\mathrm{SI}}[\rho_i] = \frac{1}{2} \int d^3\mathbf{r}'\, \frac{\rho_i(\mathbf{r})\,\rho_i(\mathbf{r}')}{|\mathbf{r}-\mathbf{r}'|} d^3\mathbf{r} - E_{\mathrm{xc}}[\rho_i] \quad (12)$$

and an improved estimate of the energy by explicit subtraction

$$E^{\mathrm{KS\text{-}SIC}}[\rho^N] = E^{\mathrm{KS}}[\rho] - \sum_i E^{\mathrm{SI}}[\rho_i]. \quad (13)$$

The energy is no longer invariant under unitary transformations of the orbitals. For example, if delocalized Bloch functions are used in a calculation of a crystal, the self-interaction energy is small or even zero, but if localized orbitals - which can be formed by a unitary transformation of the Bloch functions - are used, then E^{SI} is finite and can be significant.

The problem is now to optimize S^{ODD} with respect to the orbitals, where

$$S^{\text{ODD}}[\rho^N] = E^{\text{ODD}}[\rho^N] - \sum_{i,j} \lambda_{ji} \left[\int d^3\mathbf{r} \varphi_i^*(\mathbf{r})\varphi_j - \delta_{ij} \right]. \tag{14}$$

The orbitals φ^N are in general complex functions, yielding two equations for the extremum

$$\left.\begin{array}{l} \dfrac{\delta S^{\text{ODD}}}{\delta \varphi_i^*(\mathbf{r})} = 0 \\[2mm] \dfrac{\delta S^{\text{ODD}}}{\delta \varphi_i(\mathbf{r})} = 0 \end{array}\right\} \implies \left\{\begin{array}{l} \hat{H}_i\varphi_i(\mathbf{r}) = \displaystyle\sum_{j=1}^{N} \lambda_{ji}\varphi_j(\mathbf{r}) \\[2mm] \hat{H}_i\varphi_i(\mathbf{r}) = \displaystyle\sum_{j=1}^{N} \lambda_{ij}^*\varphi_j(\mathbf{r}) \end{array}\right. \tag{15}$$

where

$$\hat{H}_i\varphi_i(\mathbf{r}) = \frac{\delta E}{\delta \varphi_i^*(\mathbf{r})} \tag{16}$$

is the functional derivative of the energy with respect to the conjugate orbital. Note that both equations in (15) coincide whenever the matrix Λ is Hermitian. Thus, an alternative set of conditions for the optimal set of orbitals is given by

$$\hat{H}_i\varphi_i(\mathbf{r}) = \sum_{j=1}^{N} \lambda_{ji}\varphi_j(\mathbf{r}) \tag{17a}$$

$$\Lambda = \Lambda^\dagger. \tag{17b}$$

For GGA, Λ is guaranteed to be Hermitian, but not for ODD functionals. The orbitals φ^N obtained from an ODD functional are not arbitrary since the energy is not invariant under unitary transformations.

While there have been few self-consistent calculations using this functional form in the 30 years since the publication of the article by Perdew and Zunger (compared with the very large number of GGA calculations), see references in [9], it is clear that this functional form introduces flexibility that can be used to remove several deficiencies of the semi-local GGA functionals.

In the next section we will give a brief review of the minimization approaches that have previously been used for PZ-SIC. We then present our approach and compare the performance and reliability of various minimization schemes. We emphasize that the algorithms discussed here may be utilized for any functional of the ODD form, the PZ-SIC is used here only as an example. Development of an improved functional of the ODD form is currently ongoing.

3   Minimization of Energy Functionals

3.1   Minimization of Unitary Invariant Functionals

For GGA functionals and hybrid functionals, the functional derivative of the energy can be expressed by a single opeartor, $\hat{H}$, the same for all orbitals

$$\hat{H} = \hat{T} + v_{\text{ext}}(\mathbf{r}) + v_{\text{H}}(\mathbf{r}) + \hat{v}_{\text{xc}}(\mathbf{r}) \ . \tag{18}$$

For GGA, $\hat{v}_{\text{xc}}$ is a local multiplicative potential while hybrid functionals yield a non-local potential. Projection of (17a) yields an expression for the Lagrange multipliers

$$\lambda_{ij} = \lambda_{ji}^* = \int d^3\mathbf{r} \ \varphi_i^*(\mathbf{r})\hat{H}\varphi_j(\mathbf{r}) \ . \tag{19}$$

which always fulfills also eqn. (17b). As the constraint matrix $\mathbf{\Lambda}$ is Hermitian, it can be diagonalized using a unitary transformation W giving real eigenvalues ϵ_i

$$\lambda_{ij} = \sum_{k=1}^{N} \epsilon_k W_{ki}^* W_{kj} \tag{20}$$

and eigenfunctions $\psi^N = \{\psi_1, \ldots, \psi_N\}$

$$\psi_i(\mathbf{r}) = \sum_{k=1}^{N} W_{ik}^* \varphi_k(\mathbf{r}) \ . \tag{21}$$

The total density and energy do not change when the transformation is applied. The functions ψ^N are commonly taken to represent pseudo-particles of the non-interacting electron reference system. They span the total density ρ and make it possible to get a good estimate of the kinetic energy. One may be tempted to go beyond this and interpret these orbitals in terms of electrons since the defining equations are structurally equivalent to non-interacting Schrödinger equations. Any unitary transformation of the orbitals is, however, equally justified, but can typically give a range from highly localized to delocalized functions. The introduction of hybrid orbitals, for example sp^2 and sp^3 atomic orbitals, is an example of this flexibility in choosing the unitary transformation.

3.2   Minimization of ODD Functionals

The algorithm for the minimization of an ODD functional with respect to the orbitals needs to be substantially different from the one used to minimize GGA functionals since the functional derivatives give a different operator, $\hat{H}_i$ for each orbital

$$\hat{H}_i\varphi_i(\mathbf{r}) = \frac{\delta E}{\delta \varphi_i^*(\mathbf{r})} \ , \qquad \hat{H}_i = \hat{H}_0 + \hat{V}_i \tag{22}$$

where $\hat{H}_0$ is the unitary-invariant part of the operator and $\hat{V}_i$ an orbital dependent part. From eqn. (17), a projection can be used to evaluate the Lagrange
multipliers

$$\lambda_{ij} = \int d^3\mathbf{r}\; \varphi_i^*(\mathbf{r})\hat{H}_j\varphi_j(\mathbf{r})\;. \tag{23}$$

Note that in contrast to eqn. (19), the constraint matrix is not Hermitian. An
asymmetry is introduced by the orbital dependence. The second condition (17b)
should be enforced explicitly. This has consequences for the S^{ODD} objective
functional: Its imaginary part is related to the anti-Hermitian part of $\boldsymbol{\Lambda}$ and the
deviation from orthonormality.

$$\mathrm{Im}\{S^{\mathrm{ODD}}\} = \sum_{i,j=1}^{N} \frac{\lambda_{ij}^* - \lambda_{ji}}{2i} \left[\int d^3\mathbf{r}\; \varphi_i^*(\mathbf{r})\varphi_j(\mathbf{r}) - \delta_{ij} \right] \tag{24}$$

While any solution of the extremum condition (15) yields a Hermitian $\boldsymbol{\Lambda}$ matrix,
intermediate stages of an iterative procedure typically lead to asymmetric matrices and there is not a unique way to define a proper set of Lagrange multipliers.
Various possible choices to deal with this problem are discussed below.

Heaton, Harrison and Lin [16,17] presented an approach which corresponds
to solving equations (16) and (23) without considering the symmetry condition
for the Lagrange multipliers. An optimization with respect to unitary transformations of the orbitals was not included. Goedecker and Umrigar pointed out
problems with this approach [18].

Asymmetric and symmetric constraint An alternative approach is to use
one of the equations (15) to define the Lagrange multipliers and the other one
to define the orbitals, for example

$$\hat{H}_i\varphi_i(\mathbf{r}) = \sum_{j=1}^{N} \lambda_{ji}^{\mathrm{a}}\varphi_j(\mathbf{r})\;, \quad \lambda_{ij}^{\mathrm{a}} = \int d^3\mathbf{r}\; \varphi_i^*(\mathbf{r})\hat{H}_i\varphi_j(\mathbf{r}). \tag{25}$$

The constraint matrix $\boldsymbol{\Lambda}^{\mathrm{a}}$ is not necessarily Hermitian, but converges to a Hermitian matrix at the end of the iterative procedure.

Goedecker and Umrigar (GU) [18] used the Hermitian average, $\boldsymbol{\Lambda}^{\mathrm{s}}$, of the two
possible choices for the definition of the Lagrange multipliers

$$\hat{H}_i\varphi_i(\mathbf{r}) = \sum_{j=1}^{N} \lambda_{ji}^{\mathrm{s}}\varphi_j(\mathbf{r}), \quad \lambda_{ij}^{\mathrm{s}} = \int d^3\mathbf{r}\; \varphi_i^*(\mathbf{r})\frac{\hat{H}_i + \hat{H}_j}{2}\varphi_j(\mathbf{r}). \tag{26}$$

It can be shown that for a given set of orbitals, φ^N, this choice of the constraint
yields the direction of steepest descent for a correction to the orbitals.

Unitary Optimization Neither the symmetrized $\boldsymbol{\Lambda}^{\mathrm{s}}$ nor the asymmetric $\boldsymbol{\Lambda}^{\mathrm{a}}$
make explicit use of the necessary requirement for a Hermitian constraint matrix.
Defining $\boldsymbol{\kappa}$ as

$$\boldsymbol{\kappa} = (\boldsymbol{\Lambda}^{\mathrm{a}} - \boldsymbol{\Lambda}^{\mathrm{a\dagger}})/2 \tag{27}$$

yields an anti-Hermitian matrix with elements

$$\kappa_{ij} = \int d^3\mathbf{r} \; \varphi_i^*(\mathbf{r}) \frac{\hat{H}_j - \hat{H}_i}{2} \varphi_j(\mathbf{r}) \tag{28}$$

$$= \int d^3\mathbf{r} \; \varphi_i^*(\mathbf{r}) \frac{\hat{V}_j - \hat{V}_i}{2} \varphi_i(\mathbf{r}). \tag{29}$$

When the two orbital dependent operators are subtracted, the unitary invariant part is cancelled out, leaving only the difference of the orbital dependent parts.

Expressing explicitly the dependency of κ on unitary transformations amongst the orbitals

$$\varphi_i[\mathbf{U}](\mathbf{r}) = \sum_{j=1}^{N} U_{ji}\varphi_j(\mathbf{r}) \;, \qquad \rho_i[\mathbf{U}] = |\varphi_i[\mathbf{U}](\mathbf{r})|^2 \tag{30}$$

the equation defining the optimal unitary transformation $\mathbf{U}$ is

$$\kappa[\mathbf{U}] = 0 \;. \tag{31}$$

This is referred to as the "localization condition" [19]. The efficiency of unitary variant minimization algorithms depends on fast and reliable solution of this equation. Lagrange multipliers that are evaluated from orbitals satisfying the localization condition are guaranteed to fulfill the symmetry requirement (17b). Furthermore, using unitary optimization as a preconditioner for the constraint matrix, unifies the previously reported constraint definitions.

Reintroduction of Eigenfunctions Although the minimization can be carried out by solving the coupled set of equations (15), this is cumbersome and makes it difficult to incorporate ODD functionals into existing software which typically relies on the formulation of the optimization problem as an eigenvalue problem. This has led to the idea of using two sets of functions.

The first set, referred to as the optimal basis, is given by the functions φ^N which should converge to the solutions of the optimization problem (15). The second set, the 'canonical orbitals' ψ^N, is introduced to decouple the equations into single-particle eigenvalue equations analogous to the ones obtained for unitary invariant functionals. Both sets of functions span the same total density ρ and are related to each other by a unitary transformation $\mathbf{W}$

$$\varphi_i(\mathbf{r}) = \sum_{k=1}^{N} W_{ki}\psi_k(\mathbf{r}) \;, \qquad \psi_i(\mathbf{r}) = \sum_{k=1}^{N} W_{ik}^*\varphi_k(\mathbf{r}) \;. \tag{32}$$

In order to decouple (15) into single-particle equations, $\mathbf{W}$ has to diagonalize the constraint matrix

$$\lambda_{ij} = \sum_k \epsilon_k W_{ki}^* W_{kj} \;, \qquad \delta_{ij} = \sum_k W_{ki}^* W_{kj}. \tag{33}$$

This step requires a Hermitian constraint matrix, which is provided either by the symmetrized eqn. (26), or by unitary optimization at each iteration.

Expressing the condition for minimal energy (15) in terms of the ψ^N yields

$$\sum_{k'=1}^{N} W_{k'i} \hat{H}_0 \psi_{k'}(\mathbf{r}) + \sum_{k'=1}^{N} W_{k'i} \hat{V}_i \psi_{k'}(\mathbf{r}) = \sum_{jk'=1}^{N} \lambda_{ji} W_{k'j} \psi_{k'}(\mathbf{r}). \tag{34}$$

and the equations can be decoupled by forming the linear-combination $\sum_{i=1}^{N} W_{ki}^*$

$$\hat{H}_0 \psi_k(\mathbf{r}) + \sum_{ij=1}^{N} W_{ki}^* \hat{V}_i W_{ji} \psi_j(\mathbf{r}) = \varepsilon_k \psi_k(\mathbf{r}). \tag{35}$$

The resulting operator is still orbital dependent but now with respect to the canonical orbitals rather than the optimal ones

$$\left(\hat{H}_0 + \hat{V}_k^c \right) \psi_k(\mathbf{r}) = \epsilon_k \psi_k(\mathbf{r}) \tag{36}$$

where $\hat{V}_k^c$ is given by

$$\hat{V}_k^c f(\mathbf{r}) = \sum_{i=1}^{N} W_{ki}^* \hat{V}_i \varphi_i(\mathbf{r}) \int d^3\mathbf{r}' \ \psi_k(\mathbf{r}') f(\mathbf{r}'). \tag{37}$$

$\hat{V}_k^c$ is structurally simpler than the previous $\hat{V}_k$ in eqn. (22). It is invariant under unitary transformations of the functions ψ^N in a subtle way: The unitary transformation is simply compensated by an inverse change to $\mathbf{W}$ in eqn. (32) maintaining the same φ^N. The canonical orbitals ψ^N turn out to converge faster than the φ^N which improve mainly through the unitary optimization. Numerically, the separation into basis set optimization and unitary optimization is advantageous in the electronic structure problem as different energy scales are separated, i.e. the relatively small contribution from $\hat{V}_k$ is separated from the large contribution from $\hat{H}_0$. A similar procedure has been used in time-dependent DFT [20].

4   Performance

The performance of the minimization was benchmarked in all electron Gaussian type orbital based calculations of the N_2 molecule. The convergence was measured in terms of the residual, R,

$$R = \left[\sum_{i=1}^{N} \int d^3\mathbf{r} \left| \hat{H}_i \varphi_i(\mathbf{r}) - \sum_{j=1}^{N} \lambda_{ji} \varphi_j(\mathbf{r}) \right|^2 \right]^{1/2} \tag{38}$$

and a measure, K, of the error in the localization condition

$$K = \|\boldsymbol{\kappa}\| = \left[\sum_{ij=1}^{N} |\kappa_{ij}|^2 \right]^{1/2}. \tag{39}$$

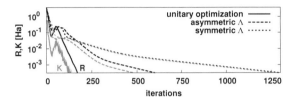

Fig. 1. Convergence of steepest descent minimization for a N_2 molecule starting from same initial orbitals. The residual R (black) and K (grey) are shown for different methods of evaluating the Λ matrix: symmetric, asymmetric and including unitary optimization until $K < 0.1R$.

Figure 1 compares different choices for dealing with the Λ matrix. The energy was minimized using the steepest descent method which allows for direct comparison of different functionals and algorithms. Both the symmetric definition (26) and the asymmetric one (25) result in slow convergence rate in the later stage of the minimization. However, the origin of the slow convergence is different in the two cases. For the symmetric definition, the convergence of R and K is roughly equally slow, but for the asymmetric definition, which gives faster convergence, R is slower. In the unitary optimization, which converges much faster, K is reduced to less than 10% of R by an intermediate unitary optimization, followed by the use of the symmetric constraint.

The effort involved in minimizing the ODD type LDA-SIC functional is compared with LDA, GGA (using PBE) and Hartree-Fock in Fig. 2. The ODD calculation required similar number of iterations as the LDA and GGA calculations. Hartree-Fock requires many more iterations. The CPU time needed to reach convergence is also shown. The ODD calculation turns out to be faster in this case than the PBE calculation because the gradient dependent terms, which are absent in LDA, involve significant computational effort. The Hartree-Fock calculation is much faster than the others despite the large number of iterations because the integrals can for this small system be stored in memory.

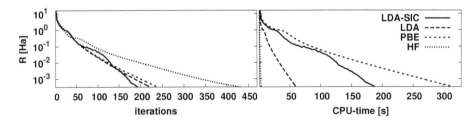

Fig. 2. Convergence of a steepest descent minimization of the energy of N_2 using various functionals: LDA, PBE (a GGA functional), LDA-SIC (an ODD functional) and Hartree-Fock (HF).

5 Parallelization

The main computational effort when ODD functionals are used is related to the evaluation of the N orbital dependent potentials. Each of them is as expensive as the GGA potential for the total electron density. But, since the calculations for the N orbitals are indendent, they can readily be distributed over N processors without the need for significant communication. The real time of an ODD calculation would then be similar to a regular GGA calculation even for a large system. Our results obtained so far indicate that performance similar to GGA functionals can be accomplished for this more general functional form.

Acknowledgements. This work was supported by the Icelandic Research Fund, N-Inner Nordic fund, EC Marie Curie network 'Hydrogen' and the US National Science Foundation.

References

1. Edelman, A., Arias, T.A., Smith, S.T.: SIAM J. Matrix Anal. Appl. 20, 303 (1998)
2. Kohn, W.: Rev. of Mod. Phys. 71, 1253 (1998)
3. Kohn, W., Sham, L.J.: Phys. Rev. 140, A1133 (1965)
4. Traditionally, the variation is carried out with respect to the total density, but for GGA functionals variation with respect to the $\varphi_i(\mathbf{r})$ gives the same results
5. Davidson, E.R.: Comp. Phys. Commun. 53, 49 (1989)
6. Nachtigal, P., Jordan, K.D., Smith, A., Jónsson, H.: J. Chem. Phys. 104, 148 (1996)
7. Patchkovskii, S., Ziegler, T.: J. Chem. Phys. 116, 7806 (2002)
8. Nuttall, R.H.D., Weil, J.A.: Can. J. Phys. 59, 1696 (1981) Pacchioni, G., Frigoli, F., Ricci, D., Weil, J.A.: Phys. Rev. B 63, 054102 (2000)
9. Kümmel, S., Kronik, L.: Reviews of Modern Physics 80, 3 (2008)
10. Becke, A.D.: J. Chem. Phys. 98, 5648 (1993); 104, 1040 (1996); 88, 1053 (1988)
11. Perdew, J.P., Ernzerhof, M., Burke, K.: J. Chem. Phys. 105, 9982 (1996)
12. Kurth, S., Perdew, J.P., Blaha, P.: Int. J. of Quantum Chem. 75, 889 (1999)
13. Bylaska et a, E., Tsemekhman, K., Baden, S.B., Weare, J.H., Jónsson, H.: J. Comp. Chem. 32, 5469 (2011)
14. Perdew, J.P., Zunger, A.: Phys. Rev. B 23, 5048 (1981)
15. Perdew, J.P., Burke, K., Ernzerhof, M.: Phys. Rev. Lett. 77, 3865 (1996)
16. Heaton, R.A., Harrison, J.G., Lin, C.C.: Phys. Rev. B 28, 5992 (1983)
17. Harrison, J.G., Heaton, R.A., Lin, C.C.: J. Phys. B 16, 2079 (1983)
18. Goedecker, S., Umrigar, C.J.: Phys. Rev. A 55, 1765 (1997)
19. Pederson, M.R., Heaton, R., Lin, C.C.: J. Chem. Phys. 80, 1972 (1984)
20. Messud, J., Dinh, P.M., Reinhard, P.-G., Suraud, E.: Phys. Rev. Lett. 101, 096404 (2008)

Simulated Annealing with Coarse Graining and Distributed Computing

Andreas Pedersen, Jean-Claude Berthet, and Hannes Jónsson

Science Institute, VR-III, University of Iceland, 107 Reykjavík, Iceland

Abstract. EON is a software package that uses distributed computing, systematic coarse graining and bookkeeping of minima and first order saddle points too speed up adaptive kinetic Monte Carlo simulations. It can be used to optimize continuously differentiable functions of a large number of variables. The approach is based on finding minima of the cost function by traversing low-lying, first-order saddle points from one minimum to another. A sequence of minima is thus generated in a path through regions of low values of the cost function with the possibility of 'temperature' controlled acceptance of higher lying saddle points. Searches of first order saddle points are carried out using distributed computing and the minimum-mode following method. Coarse graining which involves merging local minima into composite states and the recognition of previous search paths and saddle points are used to accelerate the exploration of the cost function. In addition to obtaining an estimate of the global minimum, a simulation using this approach gives information about the shape of the cost function in the regions explored. Example applications to the simulated annealing of a cluster of water molecules on a platinum metal surface and grain boundary in copper are presented.

Keywords: optimization, distributed computing, coarse graining, water cluster, grain boundary.

1 Introduction

Annealing has for a long time been used to improve the atomic scale structure of materials by eliminating strain and defects. Glassblowers, for example, use annealing to prevent new glass structures from cracking. When a material is annealed, its temperature is brought to an elevated level to accelerate thermally activated mobility of defects. However, the elevated temperature can also increase the population of defects. By cooling slowly enough from the elevated temperature, the defects become less stable while still being mobile enough to get annihilated. The result of this treatment is a lowering of the energy as the arrangement of the atoms is optimized. If the cooling is slow enough, all defects will be eliminated and the global energy minimum reached. This, however, may require impossibly long time.

Optimization of functions of many variables is often carried out using computer simulated annealing algorithms that mimmic roughly the annealing of

K. Jónasson (Ed.): PARA 2010, Part II, LNCS 7134, pp. 34–44, 2012.

materials. The 1983 article by Kirkpatrick, Gelatt and Vecchi [1] where such an approach was promoted and applied to circuit design now has over 12000 citations. The cost function to be minimized is taken to give the 'energy' of the system. A Monte Carlo algorithm based on random numbers is used to simulate an annealing process were changes in the arguments of the cost function are accepted or rejected in accordance with a fictitious 'temperature'. The reason for introducing temperature is to introduce and control the probability of accepting increases in the cost function since they may be an essential intermediate step to ultimately reach lower values.

The Adaptive kinetic Monte Carlo algorithm (AKMC) [2] can be used to accelerate simulations of the time evolution in materials without the need for a priori information about possible transitions in the system. This is unlike the regular KMC algorithm where the mechanism and rate of possible transitions is needed as input [3]. It is also different from the Metropolis Monte Carlo algorithm [4] in that the changes made to the variables represent likely transition mechanisms rather than just random moves and one can estimate the 'time' evolved at each iteration. The AKMC method has been applied successfully to several different problems, for example the annealing of grain boundaries in metals [5], reactions at surfaces [6] and crystal surface annealing during growth [7]. In AKMC, the possible transitions are found by locating first-order saddle points on the energy surface that are in the vicinity of currently known local minima. The probability of the possible transitions decreases exponentially as a function of the height of the saddle point over the current minimum. A random number is used to pick the next transition according to the relative rates of possible transitions. Several additional features in the implementation have been developed to speed up the simulation such as (1) distributed computing using the BOINC communication framework [18] were hundreds of computers connected by internet can be used simultaneously for the saddle point searches; (2) systematic coarse graining were local minima separated by low-lying saddle points are grouped together to form composite states; (3) bookkeeping of previously found saddle points and search paths to suggest new, short searches and terminate searches that are likely to converge to known saddle points. The software package EON [8] has been written to carry out such simulations. While it has so far been applied to studies of atomic scale systems, it can in principle be used for optimization of any cost function where first derivatives are continuous and can be evaluated readily. Since the algorithm follows regions where the cost function is relatively small, this approach may be particularly useful when the calculation of high values of the cost function can be problematic because of, for example, convergence problems.

In this article we describe briefly the algorithms on which the EON software is based and discuss two applications: (1) ordering of water molecules in a cluster on a Pt surface, and (2) atomic ordering at a grain boundary in Cu. We conclude with some remarks on the applicability of this software to other, more general optimization problems.

2   The Cost Function

The cost function is assumed to be a continuously differentiable function of N variables

$$f : \mathbb{R}^N \to \mathbb{R} \qquad (1)$$

In typical applications of AKMC such as the ones described below, N is on the order of 10^3. The cost function defines a surface in high dimensional space and the goal is to find the global minimum, or a representative minimum if there are several roughly equally low minima. When navigating on the surface, the extremal points where $\nabla f = 0$ and the function value is low are of particular interest, namely local minima and first order saddle points. At a minimum, the Hessian matrix, $H_{ij} = \frac{\partial^2 f}{\partial x_i \partial x_j}$, has only positive eigenvalues but at a first order saddle point it has one negative eigenvalue. Hereafter, a saddle point will be taken to mean *first order* saddle point. An *optimal path* is such that at each point on the path the gradient is pointing along the path [9,10]

$$\nabla f - \nabla f \cdot \hat{\tau} \, \hat{\tau} \; = \; 0 \qquad (2)$$

were $\hat{\tau}$ is the normalized tangent vector for the path at that point [11]. A first order saddle point is a maximum along an optimal path between two minima. We assume that the gradient ∇f of the cost function can be evaluated readily (recent developments in automatic differentiation [12] could prove valuable in this context), but second derivatives are not needed. The method used to find first order saddle points involves a minimization using a transformed gradient where the component along the minimum mode of the Hessian is reversed

$$\nabla f^{eff} = \nabla f - 2(\nabla f \cdot \hat{v}_\lambda)\hat{v}_\lambda \qquad (3)$$

Here, $\hat{v}_\lambda$ is a normalized eigenvector corresponding to the minimum eigenvalue, λ, of the Hessian. This projection locally transforms the gradient in the vicinity of a first order saddle point to a gradient characteristic of the vicinity of a minimum and the conjugate gradient method (without line search [13,14]), for example, can be used to converge on the first order saddle point. The minimum mode vector is found using the dimer method [13] without having to evaluate the Hessian or any of the second derivatives. Given a local minimum, a small random change of the variables away from the minimum is first made and the minimum mode following (MMF) method is then used to climb up the cost function surface and converge on to a saddle point. For more detail about this method and its performance, see refs. [14].

Once a search trajectory exits the region in the neighborhood of a minimum where all eigenvalues are positive, the MMF search path is stable and deterministic, i.e. a given point outside the positive region will converge onto a certain saddle point. A basin of attraction can be defined as illustrated in Fig. 1.

3   Adaptive Kinetic Monte Carlo Algorithm

The kinetic Monte Carlo method (KMC) [3] is an iterative algorithm where a Markov chain of states is generated using a predefined rule for transitions and

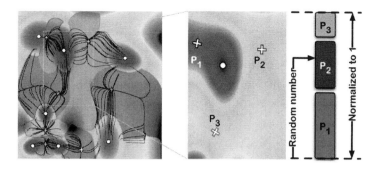

Fig. 1. Left panel: A model cost function and paths from multiple saddle point searches carried out with the MMF method. Minima are denoted by a white dot and saddle points by a white + with arms pointing along the eigenvectors of the Hessian. Shaded regions are basins of attraction of the saddle points. For each local minimum, several paths (solid lines) leading to saddle points are generated using the MMF method. The paths tend to merge as they approach a saddle point. Searches that lead to saddle points outside the figure are not shown. Right panel: Schematic illustration of an iteration in the AKMC method. For each saddle point, a rate can be evaluated from Eqn. 6. The normalized rate for each of the three possible transitions gives a transition probability, represented by the colored regions in the column, and a random number can be used to pick the next transition.

their rate constants

$$C^{KMC} = \{f_0, f_1, f_2, \ldots f_\eta\}. \tag{4}$$

The mechanism and probability of transitions needs to be read in as input before the simulation starts. In the adaptive kinetic Monte Carlo method (AKMC) [2] this input is not required but at each state, which is a minimum of the cost function, multiple low lying saddle points are found by MMF searches starting from slightly different starting points (as illustrated in Fig. 1) where the displacements are drawn from a Gaussian distribution . Each saddle point or rather the optimal path going through the saddle point, represents a transition mechanism and a possible new state in the Markov chain. After several low lying saddle points have been found, a random number in the interval $[0, 1]$ is used to pick one of the possible transitions according to the relative rates. By repeating this procedure, a chain of minima and associated saddle points is generated

$$C^{AKMC} = \left\{ f_0^{min}, \{f_1^{sp}, f_1^{min}\}, \{f_2^{sp}, f_2^{min}\}, \ldots \{f_\eta^{sp}, f_\eta^{min}\} \right\}. \tag{5}$$

The rate constant for a transition from $i-1$ to i via saddle point i is an exponential function of the value of the cost function at the saddle point, consistent with Boltzmann distribution of thermal energy

$$k_i = A \exp\left[-\frac{f_i^{sp} - f_{i-1}^{min}}{T} \right] \tag{6}$$

where T is the temperature and A is some prefactor which can in the simplest approach be taken to be the same for all transitions. In transition state theory

(TST) of thermal transitions on an energy surface [15], the rate constant is estimated from the probability of a thermal fluctuation that brings the system to a transition state separating the initial state from other states multiplied by the rate of crossing the transition state. After making second order Taylor expansions of the cost function about the saddle point and the minimum, the expression for the prefactor becomes [16]

$$A^{HTST} = \frac{\prod_j^D \nu_j^{min}}{\prod_j^{D-1} \nu_j^{sp}} \qquad (7)$$

where ν_j^{min} is the frequency associated with eigenmode j at the minimum and ν_j^{sp} at the saddle point. The mode with negative eigenvalue at the saddle point is not included. This expression does require the evaluation of the Hessian at the minimum and at the saddle point. It assigns higher probability for transitions where the 'valley' widens when going from the minimum to the saddle point and lower probability if the valley narrows. But, typically the value is similar for transitions in a given system and AKMC simulations are sometimes carried out using just a fixed prefactor, thereby avoiding the evaluation of second derivatives all together. The saddle point searches are continued until a predefined criterion has been reached, for example that the lowest saddle point has been found a certain number of times, or that no new saddle point in the relevant range (defined by the temperature) has been found in the last n searches where n is some predefined number.

From all the successful saddle point searches from the current minimum, i, in the Markov chain (some searches may not converge, and some may converge to a saddle point that is not directly connected by an optimal path to the current minimum), a table of possible transitions and their normalized probability, P_i is constructed. This is illustrated in Fig. 1. A random number is then used to pick one of the possible transitions and the system advanced to a new state, found by sliding down forward along the optimal path from the saddle point. From the Poisson distribution of residence time, the expectation value of the time evolved in each iteration can be estimated to be $\tau = 1/\sum_i k_i$ where the sum extends over all possible transitions from this state. Random sampling from this distribution is made by picking another random number, χ, in the interval $[0, 1]$ and estimating the time increment as $\tau = -\log \chi / \sum_i k_i$.

3.1   Distributed Computing

The most computationally demanding part of an AKMC simulation is the search for saddle points. But, each search is independent of the others, requires only the location of the current local minimum and an initial search direction as input. For large systems each search can take several minutes. The saddle point searches can, therefore, be distributed to several computers connected only by internet. The EON software is based on such a distributed computing approach [8]. In the first version, the communication middleware Mithral was used [17] but a more recent version of EON is based on BOINC. In the later implementation, the client software is executed as a remote procedure and the communication uses *http*. BOINC offers stability of the central server as the core parts are a SQL database and an Apache server. The client and server are decoupled which is particularly useful when debugging since the client calculation can be run as stand-alone calculation and ordinary debugging tools applied.

Even though BOINC supports distributed computing using public computers, we typically only run EON clients on local resources in clusters dedicated to research where EON-clients get executed when nodes become idle. Currently, EON is being extended to work also with the ARC-middleware [19] which will enable execution of EON on the Nordic data grid [20] and it is, furthermore, being extended to run on Amazon's 'elastic cloud' [21].

A rough estimate of the performance of a distributed AKMC simulation using EON carried out on a cluster of 2.8 GHz Intel Pentium 4 computers with GB internet connection shows that if a calculation on the client takes ca. 5 min. then the server can on average keep 100 clients busy. When saddle point searches are too fast for the server to keep up, two or more searches can be sent to the client at a time, thus saving on the overhead in the distribution of the tasks.

3.2   Coarse Graining

Kinetic Monte Carlo simulations, including AKMC, often get stuck in a small subset of states. This happens, for instance, when two states are connected by a transition that is much faster than any transition away from the pair. Essentially, the problem arises when there is large disparity in the rate constants. To address this problem, a coarse graining algorithm has been developed and implemented in EON. The problem is two-fold: (1) to automatically identify such sets of states and group them into a *composite state*, and (2) devise an algorithm for escaping the composite state and estimating the time that would have been spent there in the absence of coarse graining.

The first issue is addressed by defining a reference value of the cost function for each state, f_i^{cg} which starts out being equal to the minimum value $f_i^{cg} = f_i^{min}$. Each time this minimum is entered and added to the Markov chain, the reference value is incremented by a small amount, ϵ

$$\epsilon = \epsilon_0 \frac{f_i^{sp} - f_i^{min}}{f_i^{sp} - f_{low}^{min}} \tag{8}$$

where f_{low}^{min} is the lowest minimum found in the simulation so far and ϵ_0 is some constant chosen for the simulation (the results of the simulation are unaffected by the choice of ϵ_0 over several orders of magnitude as illustrated in a test case below). When a transition occurs with $f_i^{sp} < f_{i-1}^{cg}$, then state i is merged with state $i-1$ to form a new composite state. If either of the states i or $i-1$ or both already belong to a composite state, then the composite states are extended by an additional state or two composite states are merged depending on the situation.

This process of merging states into composite states is illustrated in Fig. 2. The simulation starts at the state corresponding to the second lowest minimum and there is a high saddle point separating it from the global minimum. Only after all the minima close to the initial minimum have been grouped into one large composite state, can the high saddle point be overcome and the global minimum is then found after only a few iterations. This example illustrates how EON could be used to find the global minimum of a cost function where multiple local minima are present and one or more high saddle point separate groups of minima. When a composite state has been formed from a set of minima, the time evolution within that set of states becomes irrelevant and only two things need to be determined: (1) From which one of the local minima will the system escape the composite state? (2) How long time will the system spend in the composite state before escaping. These questions can be answered rigorously using absorbing Markov chain theory [22].

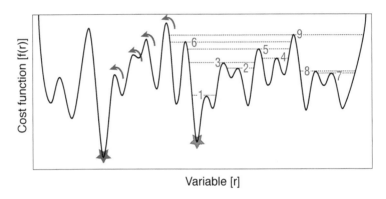

Fig. 2. Example of a cost function of one variable where the annealing simulation starts in a basin consisting of several local minima but is separated by a large barrier from the global minimum. Without coarse graining the simulation would be stuck for many iterations between the initial state (green star) and the adjacent shoulder state. These two are the first to get merged into a composite state (1). Then a new composite state (2) is formed, also by merging two minima. In the third merger the two composite states are combined into a composite state with four minima (3). Eventually, the large barrier is overcome when all 10 minima have been merged into one composite state (9). Then, the global minimum (red star) is quickly reached.

Test: Annealing in One-dimension. A problem where the coarse graining algorithm was tested on a cost function of one variable is illustrated in Figure 3. An AKMC simulation both with and without coarse graining was used to estimate the time it takes the system to go from a high initial state minimum to the global minimum, traversing a high barrier. The results are presented on Table 1 and show that the coarse graining does not affect the estimated time, even as the basic increment, ϵ_0, is varied over several orders of magnitude.

This simple test illustrates how the coarse graining speeds up the annealing simulation without affecting the results, in particular the estimated time. In this case, the acceleration is up to 100 fold. But, more importantly, the coarse graining can make a simulation doable while it is impossible without the coarse graining (for example, this same objective function but a significantly lower temperature). Fig. 3 shows how the number of iterations needed in a KMC simulation increases for the simple one-dimensional test case as the temperature is lowered. Without the coarse graining, the number of iterations needed to reach the global minimum is about three orders of magnitude larger at a temperature of 0.007 than at 0.1, but with the coarse graining there is only less than a factor of 2 increase.

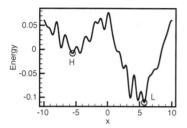 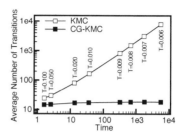

Fig. 3. A model cost function of one variable used to test the coarse graining algorithm. Starting in state H, the AKMC simulation with and without coarse graining was used to estimate the time needed to reach the global minimum, L. The results are given in Table 1 and show that the coarse graining does not affect the estimated time, even as the basic increment, ϵ_0, is varied over many orders of magnitude. Right panel: Average number of transitions (iterations) needed in the simulations versus the estimated mean time needed to go from point H to point L for different 'temperatures'. While the regular KMC needs many more iterations at low temperature, the coarse grained simulation only requires slightly more.

Table 1. The elapsed time and number of transitions required to move from local minimum H to global minimum L in Fig. 3, averaged over 1000 runs at a 'temperature' of 0.09, with no coarse graining (no CG) and with using various values of the basic increment, ϵ_0 in the coarse graining algorithm. The prefactor was set to $A = 10$.

$\log \epsilon_0$	Time	Ave. Trans.
no CG	708 ± 23	1547 ± 51
-5	670 ± 20	1287 ± 31
-4	684 ± 21	405 ± 6
-3	683 ± 18	71 ± 1
-2	698 ± 13	15 ± 0

3.3   Other Tricks to Improve Efficiency

Each minimum visited is stored, i.e. the value of the variables and the value of the cost function. When a minimum is revisited, new saddle point searches are not carried out unless the reference energy has increased beyond the region where good sampling of saddle points has been obtained previously. Also, since search paths tend to merge together close to the saddle points, as shown in Fig. 1 intermediate points along search paths in the basin of attraction (where one eigenvalue is negative) are stored and used to terminate later searches that come close to a previous search path. Only the values of variables that change most along the path are stored. We refer to this as the 'skipping-path' method [23]. Around each of the stored intermediate points, a spherical region with small radius is defined and any later search that comes within one of these regions gets terminated and the intermediate points of that search added to the list. When thorough sampling of saddle points is carried out, this can save substantial computational effort [23].

In some systems the transitions are 'local' in that they only involve appreciable change in some small subset of the variables. Then, one transition may not affect strongly transitions involving another set of variables. An example of that from the atomic scale simulations is a rearrangement of atoms in one part of the simulated system affecting only insignificantly atoms in another part of the system. The stored saddle points can, therefore, be good guesses for new saddle point searches where only a few iterations using the MMF method are needed to re-converge on saddle points. This can save substantial amount of computational effort [24].

4   Applications

Two applications of simulations with EON are presented here briefly, both involving the optimization of the structure of atomic scale systems. In the first, a cluster consisting of 7 water molecules adsorbed on a Pt(100) surface is annealed to find the optimal structure. The interaction potential function was developed by Zhu and Philpott [25] but the interaction between the water molecules and the surface has been scaled down to match recent density functional theory calculations [26]. The initial configuration is generated by adding two molecules to the optimal configuration of a cluster consisting of five molecules, which is a pentagonal ring. The structure is shown in Fig. 4. A shallow minimum slightly higher in energy has one of the molecules rotated in such a way that a hydrogen atom gets pointed away from the surface. The AKMC simulation jumps between those two states several times until eventually the two are joined in a composite state. A significant barrier is involved in breaking up the pentagonal ring to form a rectangular core of 6 molecules with one molecule weakly bound to the edge, but this leads to substantial lowering of the energy. After that, the AKMC simulation flips the under-coordinated edge molecule and this is repeated until the two configurations get merged in a composite state.

The second example comes from a study of a twist and tilt grain boundary in copper. The results of an extensive study have been presented elsewhere [5] but here we present for the first time the improvement obtained by using coarse graining. A potential function of the EMT form is used here to describe the intermolecular interactions [27]. The simulated system consists of 1309 atoms but the two outermost layers parallel to the grain boundary are kept frozen so the cost function (the energy of the system) is a function of 3567 variables, three coordinates for each movable atom. The goal is to gain information about the atomic arrangement at the grain boundary. Away from the grain boundary the atoms are in a FCC crystal arrangement. One of the questions addressed in these studies is how wide a region around the grain boundary atoms are anomalously coordinated and what arrangement of atoms is most characteristic for the grain boundary. Since the atoms that are not in an FCC crystal arrangement are the focus here and are most likely to be involved in transitions, the initial random displacement is generated in a spherical region including about 30 atoms centered on a randomly picked, non-FCC atom. Typically, 1 to 10 atoms are displaced by more than half an Ångström in a single transition [5].

The excess energy due to the non-FCC arrangement of atoms at the grain boundary, the interface energy, is shown as a function of iterations in the simulation for a temperature of 300 K in Fig. 4. At first, the coarse graining was turned off and while some transitions occurred and annealing took place, the simulation was not making any progress for a long time. The energy in the last 200 iterations is shown far to the left in the figure. The system is going back and forth between a few states with interface energy between 81.8 and 82.0 meV/Å^2. Then, coarse graining is turned on (iteration labeled '0') and a composite state is formed which enables the simulation to get to higher energy states, up to 82.4

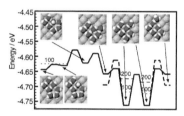
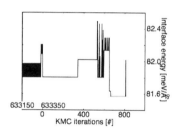

Fig. 4. Left: Simulation of a cluster of seven water molecules on a Pt(100) surface. The horizontal dashed lines show the reference level, f^{cg}, after 100 and 200 transitions. First, the two states on the far left are merged into a composite state. Then, the pentagonal ring breaks up and a rectangular hexamer core is formed with one water molecule at the edge. The difference between the two deep minima to the right is just the orientation of the low coordinated molecule. This kind of rotational flipping which does not represent a change in the energy or the structure is a significant problem in many simulations when coarse graining is not used. Right: Interface energy in a twist and tilt grain boundary between copper crystal grains as a function of iterations in an AKMC simulation. After 633350 iterations without coarse graining where the simulation was caught in a subset of 11 states and little progress was made (marked as '0' on the horizontal axis), the coarse graining algorithm is turned on. Then, a composite state of these minima was formed and the simulation reached higher energy states which eventually lead to a significant annealing event. By applying coarse graining, only 700 additional iterations were required to reach this annealing event while it took 300.000 additional iterations without coarse graining.

meV/Å^2 and eventually find a transition that reduces the energy significantly, an annealing event. It took less than 700 iterations from the time the coarse graining was turned on until the annealing event was observed. The same kind of annealing event was also observed by continuing the simulation without coarse graining, but then 300.000 iterations were needed.

Acknowledgments. The EON software is being developed in collaboration with the research group of Prof. G. Henkelman at the University of Texas at Austin. This work was supported by the Icelandic Research Fund, the University of Iceland research fund, and EC Marie Curie network 'Hydrogen'.

References

1. Kirkpatrick, S., Gelatt Jr., C.D., Vecchi, M.P.: Science 220, 671 (1983)
2. Henkelman, G., Jónsson, H.: J. Chem. Phys. 115, 9657 (2001)
3. Glauber, R.J.: J. Math. Phys. 4, 294 (1963); Martin, P.A.: J. Stat. Phys. 16, 149 (1977); Bortz, A.B., et al.: J. Comput. Phys. 17, 10 (1975)
4. Metropolis, N., Rosenbluth, A.W., Rosenbluth, M.N., Teller, A.H., Teller, E.: J. Chem. Phys. 21, 1087 (1953)
5. Pedersen, A., Henkelman, G., Schiøtz, J., Jónsson, H.: New Journal of Physics 11, 073034 (2009)
6. Xu, L., Henkelman, G.: J. Chem. Phys. 131, 244520 (2009)
7. Henkelman, G., Jónsson, H.: Physical Review Letters 90, 116101 (2003)
8. Pedersen, A., Jónsson, H.: Math. and Comput. in Simulations 80, 1487 (2010)
9. Jónsson, H., Mills, G., Jacobsen, K.W.: Classical and Quantum Dynamics in Condensed Phase Simulations. In: Berne, B.J., Ciccotti, G., Coker, D.F. (eds.), ch.16, p. 385. World Scientific (1998)
10. Henkelman, G., Uberuaga, B., Jónsson, H.: J. Chem. Phys. 113, 9901 (2000)
11. Henkelman, G., Jónsson, H.: J. Chem. Phys. 113, 9978 (2000)
12. http://www.autodiff.org
13. Henkelman, G., Jónsson, H.: J. Chem. Phys. 111, 7010 (1999)
14. Olsen, R.A., Kroes, G.J., Henkelman, G., Arnaldsson, A., Jónsson, H.: J. Chem. Phys. 121, 9776 (2004)
15. Wigner, E.: Tr. Far. Soc. 34, 29 (1938); Eyring, H.: J. Chem. Phys. 3, 107 (1935)
16. Vineyard, G.H.: J. Phys. Chem. Solids 3, 121 (1957)
17. http://www.mithral.com/projects/3rdparty/
18. http://boinc.berkeley.edu/
19. http://www.nordugrid.org/
20. http://www.ndgf.org/
21. http://aws.amazon.com/ec2/
22. Novotny, M.A.: A tutorial on advanced dynamic Monte Carlo methods for systems with discrete state spaces (2001), URL arXiv.org:cond-mat/0109182
23. Pedersen, A., Hafstein, S.F., Jónsson, H.: SIAM Journal of Scientific Computing 33, 633 (2011)
24. Xu, L., Henkelman, G.: J. Chem. Phys. 129, 114104 (2008)
25. Zhu, S.-B., Philpott, M.R.: J. Chem. Phys. 100, 6961 (1994)
26. Árnadóttir, L., Stuve, E., Jónsson, H.: Surf. Sci. 604, 1978 (2010); Surf. Sci. (in press, 2011)
27. Jacobsen, K.W., Stoltze, P., Nørskov, J.K.: Surf. Sci. 366, 394 (1996)

Path Optimization
with Application to Tunneling

Dóróthea M. Einarsdóttir, Andri Arnaldsson,
Finnbogi Óskarsson, and Hannes Jónsson

Science Institute, VR-III, University of Iceland, 107 Reykjavík, Iceland

Abstract. A method is presented for optimizing paths on high dimensional surfaces, i.e. scalar functions of many variables. The method involves optimizing simultaneously the end points and several intermediate points along the path and thus lends itself well to parallel computing. This is an extension of the nudged elastic band method (NEB) which is frequently used to find minimum energy paths on energy surfaces of atomic scale systems, often with several thousand variables. The method is illustrated using 2-dimensional systems and various choices of the object function, in particular (1) path length, (2) iso-contour and (3) quantum mechanical tunneling rate. The use of the tunneling paths to estimate tunneling rates within the instanton approximation is also sketched and illustrated with an application to associative desorption of hydrogen molecule from a copper surface, a system involving several hundred degrees of freedom.

Keywords: optimal paths, transitions, instantons.

1 Introduction

There can be several reasons for wanting to find a path on a surface that is optimal in some sense. Our motivation comes mainly from the need to find minimum energy paths (MEPs) on energy surfaces to estimate rates of transitions due to thermally activated, classical trajectories [1], or - as is the focus here - quantum mechanical tunneling through energy barriers [2]. The method used for the path optimization is, however, quite general and can be used in various contexts.

The surface is described by a continuously differentiable function, V, of N variables

$$V : \mathbb{R}^N \to \mathbb{R} \qquad (1)$$

In typical applications to transition rates in atomic scale systems, N is on the order of 10^3. We assume that the gradient ∇V of the object function can be evaluated readily, but second derivatives are not needed. The goal is to find a finite path on the surface that is optimal in some sense. For example, the MEP on an energy surface can be of interest since the point of highest energy on the path, a first order saddle point, gives the activation energy barrier for going from one local minimum to another and, thereby, determines the exponential dependence of the rate on temperature [3,4]. At every point on a MEP

K. Jónasson (Ed.): PARA 2010, Part II, LNCS 7134, pp. 45–55, 2012.
© Springer-Verlag Berlin Heidelberg 2012

$$\nabla V - \nabla V \cdot \hat{\tau}\,\hat{\tau} = 0 \tag{2}$$

were $\hat{\tau}$ is the unit tangent vector for the path at that point [5]. Furthermore, the curvature for all modes perpendicular to the path must be positive. The NEB is frequently used to find MEPs for estimating rates of thermal transitions in atomic scale systems where the atoms are described by classical dynamics [5,6]. Some systems have even included over a hundred thousand coordinate variables [7]. The path optimization method presented here is a generalization of the NEB method and can be used, for example, to calculate rates of thermal transitions in quantum mechanical systems were tunneling takes place.

Let $\mathbf{R}$ denote a vector of N variables and $V(\mathbf{R})$ the surface. The object function, $\tilde{S}$, can be defined as a functional of the path, $\mathbf{R}(s)$ where $s \in [0,1]$, that is $\tilde{S} = \tilde{S}[\mathbf{R}(s)]$. The object function can, for example, involve an integral over the path

$$\tilde{S}[\mathbf{R}(s)] = \int_{\mathbf{R}_0}^{\mathbf{R}_n} f(V(\mathbf{R}))d\mathbf{R} \tag{3}$$

where f is some function. The path will be represented by a set of discrete points along the path $\{\mathbf{R}_0, \mathbf{R}_1, \dots \mathbf{R}_n\}$ and the integral approximated using, for example, the trapezoidal rule. The task is then to find the values of the vectors $\mathbf{R}_i$ that minimize the object function for discretized paths

$$\tilde{S}[\mathbf{R}(s)] \approx S(\mathbf{R}_0, \dots \mathbf{R}_n) =$$
$$= \frac{1}{2} \sum_{i=1}^{n} (f(V(\mathbf{R}_i)) + f(V(\mathbf{R}_{i-1}))) |\mathbf{R}_i - \mathbf{R}_{i-1}| \tag{4}$$

There are $n-1$ discretization points representing the path between the two end points, $\mathbf{R}_0$ and $\mathbf{R}_n$, which can be constrained to have some predetermined values of V, i.e. $V(\mathbf{R}_0) = v_a$ and $V(\mathbf{R}_n) = v_b$. In the NEB, end points of the path are fixed (usually at minima), but in this more general formulation the position of the end points is adjusted during the optimization along the iso-contours corresponding to v_a and v_b.

2   Path Optimization

The optimization is started by specifying some trial set of discretization points $\{\mathbf{R}_0^0, \mathbf{R}_1^0, \dots \mathbf{R}_n^0\}$ and then iterating until $S(\mathbf{R}_0, \mathbf{R}_1, \dots \mathbf{R}_n)$ has reached a minimum value. Let the negative gradient of the functional, S, with respect to the discretization point, $\mathbf{R}_j$, be denoted by

$$\mathbf{g}_j = -\nabla_j S \tag{5}$$

This represents the direction of steepest descent for each one of the discretization points and can be used in a minimization algorithm to find the set of vectors $\{\mathbf{R}_0, \mathbf{R}_1, \dots \mathbf{R}_n\}$ that minimize S. But, only the component of $\mathbf{g}_j$ that is perpendicular to the path should be included in the optimization [1,6]. The distribution of the discretization points along the path is controlled separately and should not be affected by S. This projection is referred to as 'nudging'. The negative gradient, $\mathbf{g}_j$, is projected along the path

$$\mathbf{g}_j^{\|} = (\mathbf{g}_j \cdot \hat{\tau}_j)\hat{\tau}_j \tag{6}$$

and the rest of the vector is the perpendicular component

$$\mathbf{g}_j^{\perp} = \mathbf{g}_j - \mathbf{g}_j^{\|} . \tag{7}$$

The discretization points can be distributed along the path in various ways, for example by using a restraint method where a 'spring' acting between adjacent discretization points is added, Again, a projection is used to make sure this does not affect the location of the converged path. For the discretization points that are not at the ends, $\{\mathbf{R}_1, \mathbf{R}_2, \ldots \mathbf{R}_{n-1}\}$, the component of $\mathbf{g}_j$ parallel to the path is replaced by

$$\mathbf{g}_j^{sp} = k \left(|\mathbf{R}_{j+1} - \mathbf{R}_j| - |\mathbf{R}_j - \mathbf{R}_{j-1}| \right) \hat{\tau}_j \tag{8}$$

where k is analogous to a spring constant. A wide range of values can be chosen for k without affecting the results, but the convergence rate is in general faster if the $\mathbf{g}_j^{sp}$ are roughly of the same magnitude as the $\mathbf{g}_j$. The total $\mathbf{g}$ that is used in the optimization is then given by the vector sum

$$\mathbf{g}_j^{\mathrm{opt}} = \mathbf{g}_j^{\perp} + \mathbf{g}_j^{sp} \tag{9}$$

for $j = 1, \ldots, n-1$. In a steepest descent algorithm, all the discretization points $\mathbf{R}_j$ will be displaced in the direction of $\mathbf{g}_j^{\mathrm{opt}}$ at each iteration. A more efficient approach is discussed in section 2.1. If the spring constant, k, is the same for all pairs of adjacent discretization points, then the points will be equally spaced along the path when convergence has been reached. If a different distribution is desired, the values of k for each adjacent pair of discretization points can be chosen accordingly.

The steepest descent direction for the end points is defined differently since they should only move along the iso-contours corresponding to v_a or v_b. The component of $\mathbf{g}^{sp}$ parallel to the gradient of V needs to be zeroed so the end points only get displaced along the iso-contour. Furthermore, a restraint is added to pull the end points towards the iso-contour if curvature has resulted in a drift away from the iso-contour. Denoting the unit vector in the opposite direction of the gradient of V as

$$\hat{\mathbf{F}} = -\nabla V / |\nabla V| \tag{10}$$

the steepest descent direction for end point $\mathbf{R}_0$ can be written as

$$\mathbf{g}_0^{\mathrm{opt}} = \mathbf{g}_0^{sp} - \left(\mathbf{g}_0^{sp} \cdot \hat{\mathbf{F}}_0 - V(\mathbf{R}_0) + v_a \right) \hat{\mathbf{F}}_0 \tag{11}$$

where

$$\mathbf{g}_0^{sp} = k \left(\mathbf{R}_1 - \mathbf{R}_0 - \ell \, \hat{\mathbf{F}}_0 \right) \tag{12}$$

and $\widehat{\mathbf{F}}_0 = \widehat{\mathbf{F}}(\mathbf{R}_0)$. Here, the parameter ℓ has been introduced to make it possible to adjust the length of the path when the endpoints are not constrained (for example, in the iso-contour example below). If ℓ is chosen to be $\ell = L/n$ the path will have length L when the endpoints are free to move during the optimization. An analogous expression holds for the other end point, $\mathbf{R}_n$

$$\mathbf{g}_n^{\mathrm{opt}} = \mathbf{g}_n^{sp} - \left(\mathbf{g}_n^{sp} \cdot \widehat{\mathbf{F}}_n - V(\mathbf{R}_n) + v_b \right) \widehat{\mathbf{F}}_n \tag{13}$$

where

$$\mathbf{g}_n^{\mathrm{sp}} = k \ (\mathbf{R}_{n-1} - \mathbf{R}_n - \ell \ \widehat{\mathbf{F}}_n). \tag{14}$$

and $\widehat{\mathbf{F}}_n = \widehat{\mathbf{F}}(\mathbf{R}_n)$. $\mathbf{g}_0^{\mathrm{opt}}$ and $\mathbf{g}_n^{\mathrm{opt}}$ give the steepest descent direction for the two end points used in the iterative optimization while equation (9) applies to the intermediate discretization points.

2.1   Optimization of the Path

While the location of the discretization points of the path can be optimized by steepest descent displacements in the direction of $\mathbf{g}^{\mathrm{opt}}$, this tends to have slow convergence and various more efficient minimization algorithms can be employed. We have found it useful to divide this numerical optimization into two phases: an initial phase with a rather conservative algorithm and then a final phase with a quadratically convergent algorithm. In the case of atomic scale systems, such as the H_2/Cu system discussed in section 4, the transition is made when the RMS force has dropped to below 0.5 $eV/\text{Å}$.

In the beginning, the system can be far from the optimal path. Often, a good guess for the optimal path is not available. A method we have found to be robust and convenient to implement is based on modified classical dynamics where the effective mass associated with each degree of freedom is arbitrarily set to unity and the force is taken to be the steepest descent vector. By introducing a certain damping in the dynamics, convergence to a minimum is obtained. The damping involves zeroing the velocity from previous iteration, except for the component in the direction of the force in the current iteration when the projection of the velocity on the force is positive. This algorithm is explained in ref. [1]. In the second phase, a quadratically convergent algorithm such as conjugate gradients or BFGS is more efficient. Some modifications of these algorithms have to be made, though, because an object function corresponding to the steepest descent direction $\mathbf{g}^{opt}$ is not known. The projection (nudging) and addition of the springs modifies the steepest descent direction in such a way that it no longer corresponds to the gradient of S. A review of several minimization methods proposed for the optimization of elastic bands in the context of minimum energy paths and comparison of their efficiency has recently been published [8]. We expect similar performance for the elastic bands presented here, but systematic performance analysis have not yet been made.

Since the optimization of the paths is carried out by adjusting the location of each one of the discretization points simultaneously, and the calculation of the steepest descent direction only depends on coordinates of each point and its two nearest neighbors, this algorithm for path optimization lends itself well to parallel computing.

3   Examples

3.1   Example I: Shortest Path between Iso-contours

A simple illustration of the method described above is a search for the shortest path between two iso-contours of a given value $\nu_a = \nu_b = \nu$. Here, f can be chosen to be a constant, $f(V) = 1$, and $\ell = 0$. The object functional is simply

$$S^l(\mathbf{R}_0, \ldots, \mathbf{R}_n) = \sum_{i=1}^{n} |\mathbf{R}_i - \mathbf{R}_{i-1}| \tag{15}$$

Differentiation of S gives

$$\mathbf{g}_j = -\nabla_j S^l = -\frac{\mathbf{R}_j - \mathbf{R}_{j-1}}{|\mathbf{R}_j - \mathbf{R}_{j-1}|} + \frac{\mathbf{R}_{j+1} - \mathbf{R}_j}{|\mathbf{R}_{j+1} - \mathbf{R}_j|} \tag{16}$$

Using the equations (9), (11) and (13) in an iterative optimization scheme, gives points along the path which has the shortest distance between two iso-contours, as illustrated in figure 1.

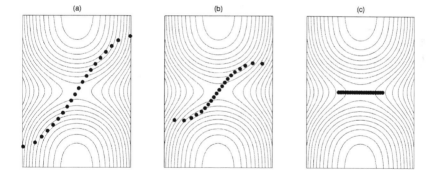

Fig. 1. Path optimization where the object function is the length of the path and the end points are confined to a contour on the surface. (a) Initial path with end points at arbitrary locations on two separate segments the contour; (b) intermediate configuration of the path during the optimization; (c) the final, converged, shortest path between the two contour lines.

3.2   Example II: Tracing Out an Iso-contour

Another example is a path that lies along an iso-contour, say $V = v_c$. Here, f is chosen to be $f(V) = (V - v_c)^2/2$ and $\ell = L/n$ where L is the desired length of the path. The object function becomes

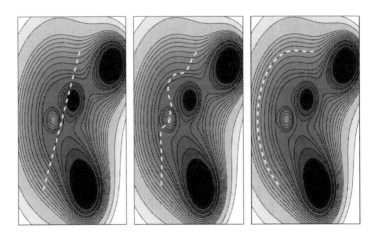

Fig. 2. Path optimization where the object function includes the squared deviation from the contour value. (left) An arbitrary initial path; (middle) intermediate configuration of the path during the optimization; (right) the final, converged path tracing the contour.

$$S^c(\mathbf{R}_0, \ldots \mathbf{R}_n) = \frac{1}{2} \sum_{i=0}^{n} (V(\mathbf{R}_i) - v_c)^2 \, |\mathbf{R}_i - \mathbf{R}_{i-1}| \tag{17}$$

Optimization using equations (9), (11) and (13), gives discretization points $\{\mathbf{R_i}\}$ for a path that lies along the v_c iso-contour as is shown in figure 2.

3.3   Example III: Tunneling Path

The optimization procedure described in the previous section can be used to find optimal, quantum mechanical tunneling paths. The function V then represents potential energy of the system and the vector $\mathbf{R}$ consists of the coordinates of all the particles in the system, some of which may undergo a tunneling transition from one position to another. In the path optimization, all particles in the system are allowed to move, unless boundary conditions restricting their movement are applied. In the JWKB method [2], the tunneling path for energy E_c is the path between classical turning points $V(\mathbf{R}_0) = V(\mathbf{R}_n) = E_c$ where the action, S^t, is minimized

$$\tilde{S}^t[\mathbf{R}(s)] = \frac{1}{\hbar} \int_{\mathbf{R}_0}^{\mathbf{R}_n} \sqrt{2\mu(V(\mathbf{R}) - E_c)} d\mathbf{R} \tag{18}$$

Here, μ is the effective mass which is conveniently taken into account by using mass weighted coordinates and forces [9]. The optimization yields a path corresponding to the lowest value of the integral and gives the highest JWKB estimate of the tunneling probability [2,10,11,12]. A reasonable initial guess for the path could be a straight line interpolation between the minima of the initial and final states, but any guess where the end points are placed on different sides of a saddle point higher than E_c will give a tunneling path.

After discretizing the integral, equation (18) becomes

$$S^t(\mathbf{R}_0, \ldots \mathbf{R}_n) = \frac{1}{2\hbar} \sum_{i=1}^{n} \left(\sqrt{2\mu(V(\mathbf{R}_i) - E_c)} \right.$$
$$\left. + \sqrt{2\mu(V(\mathbf{R}_{i-1}) - E_c)} \right) |\mathbf{R}_i - \mathbf{R}_{i-1}| \tag{19}$$

To simplify the notation, it is convenient to define a new function

$$\xi_i = \frac{1}{\hbar} \sqrt{2\mu(V(\mathbf{R}_i) - E_c)} \tag{20}$$

and rewrite the action integral as

$$S^t(\mathbf{R}_0, \ldots \mathbf{R}_n) = \frac{1}{2} \sum_{i=1}^{n} (\xi_i + \xi_{i-1}) |\mathbf{R}_i - \mathbf{R}_{i-1}| \tag{21}$$

Differentiating this expression with respect to the position of the intermediate discretization points $j = 1, \ldots, n-1$, gives

$$\mathbf{g}_j = -\nabla_j S^t = -\frac{1}{2} \left(\frac{\mu}{\hbar \xi_j} (d_j + d_{j+1}) |\nabla V(\mathbf{R}_j)| \widehat{\mathbf{F}}_j - \right. \tag{22}$$
$$\left. -(\xi_j + \xi_{j-1}) \widehat{\mathbf{d}}_j + (\xi_{j+1} + \xi_j) \widehat{\mathbf{d}}_{j+1} \right).$$

where $\widehat{\mathbf{F}}_j$ is again given by equation (10) and the d_j and $\widehat{\mathbf{d}}_j$ are defined as

$$d_j = |\mathbf{R}_j - \mathbf{R}_{j-1}| \tag{23}$$
$$\widehat{\mathbf{d}}_j = (\mathbf{R}_j - \mathbf{R}_{j-1})/d_j \tag{24}$$

The steepest descent direction, $\mathbf{g}^{\text{opt}}$, is given by equation (9) for the intermediate discretization points, but for the end points equation (11) and (13) are used. By iteratively moving the discretization points, the optimal tunneling path can be found, i.e. the values of $\{\mathbf{R}_0, \ldots \mathbf{R}_n\}$ that minimize S^t.

To illustrate how the method works, the tunneling path of a particle subject to the 2-dimensional potential function used in example II was found, starting initially with an arbitrary, straight path. Various stages of the optimization of the path are shown in figure 3: the initial guess, two intermediate paths during the optimization process and the final, optimal tunneling path. Note that the surface has extra minima and maxima which makes the problem somewhat challenging even though only 2 degrees of freedom are included.

Since the wave function decays exponentially in the classically forbidden region, the tunneling path can be displaced from the MEP into a region of higher potential energy if this leads to significant shortening of the path. This 'corner-cutting' can be seen form the converged path in figure 3.

The corner-cutting becomes stronger as the temperature is lowered, as can be seen from another 2-D model calculation shown in figure 4. The lower the temperature is, the further the path moves away from the MEP. This effect is particularly strong when the MEP has large curvature.

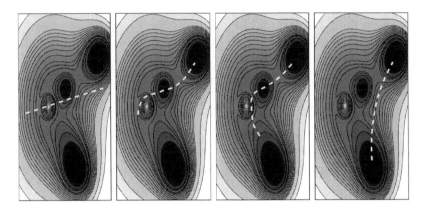

Fig. 3. Path optimization on a surface with an intermediate local minimum and a maximum, starting from an arbitrary straight line path, to illustrate the robustness of the method. Initial path (far left); intermediate paths during the optimization (middle two figures); and converged, optimal JWKB tunneling path (far right).

4   Application: Calculation of Tunneling Rates

The optimal, JWKB tunneling paths discussed in example III above can be used to estimate the tunneling rate at a given temperature rather than at given total energy. The theory is essentially a harmonic quantum transition state theory and is often referred to as 'instanton' theory [10,11]. The path that minimizes the object functional given by equation (18) turns out to be the same as a classical periodic orbit for the inverted potential energy surface , $-V(\mathbf{R})$, and is referred to as the instanton [10]. This is a closed Feynman path and it gives maximum tunneling probability at a temperature which can be related to the period, τ, of the periodic orbit through the relation $T = \hbar/k_B\tau$. The calculation of the period and location of discretization points in the statistical Feynman path corresponding to the optimized JWKB path can be obtained in a rather straight forward way by interpolation between the discretization points. As in harmonic transition state theory, where the reaction rate is estimated by approximating the potential energy surface around the classical saddle point by a quadratic expansion, the quantum mechanical rate can be obtained by expanding the effective quantum mechanical potential energy surface around the instanton to second order [12]. The instanton rate constant, k_{ins}, is given by

$$Q_R \; k_{\mathrm{ins}} = \sqrt{\frac{S_0}{2\pi\hbar}} \; \frac{k_B T P}{\hbar |\prod'_j \lambda_j|} \; e^{-V_{\mathrm{eff}}^{ins}/k_B T} \tag{25}$$

where Q_R is the partition function of the initial state, V_{eff}^{ins} is the value of the effective potential

$$V_{\mathrm{eff}}(\mathbf{R}_0, \dots \mathbf{R}_n) = \sum_{i=0}^{P} \left[\frac{1}{2} k_{\mathrm{sp}} \left| \mathbf{R}_{i+1} - \mathbf{R}_i \right|^2 + \frac{V(\mathbf{R}_i)}{P} \right] \tag{26}$$

evaluated at the instanton. Here, P is the number of discretization points in the Feynman path ($\mathbf{R}_{P+1}$ is set equal to $\mathbf{R}_P$) and k_{sp} is the temperature dependent spring constant

$$k_{sp}(T) = \mu P \left(\frac{k_B T}{\hbar} \right)^2 \qquad (27)$$

The λ_j in equation (25) are the frequencies of the normal modes of vibration of the instanton. One vibrational mode has zero eigenvalue, namely the one corresponding to displacement of the images along the path. This mode gives rise to S_0 which is twice the instanton action due to the (imaginary-time) kinetic energy

$$S_0 = \frac{\mu P k_B T}{\hbar} \sum_{j=1}^{P} |\mathbf{R}_j - \mathbf{R}_{j-1}|^2 \qquad (28)$$

The prime on the product sign in equation (25) denotes the absence of the zero-mode, since it cannot be treated with a quadratic approximation.

This procedure for estimating the rate constant from JWKB tunneling paths has been tested both on model 2-dimensional systems and for a large system involving several hundred degrees of freedom, the associative desorption of H_2 molecule from a Cu(110) surface. The desorption has been studied by several different methods in the past, including a full free energy method based on Feynman path integrals, the so-called RAW-QTST method [14]. Here, the JWKB paths were used to estimate the desorption rate as a function of temperature using the instanton approximation as described above. The calculation involved 432 degrees of freedom, the coordinates of the two hydrogen atoms and four layers of Cu atoms in a slab subject to periodic boundary conditions. The bottom two layers of Cu atoms in the slab were held fixed. The results are shown in figure 4. The agreement with the full free energy calculation is surprisingly good considering the fact that a gas phase molecule is being formed and that harmonic approximation, which the instanton approach is based on, applies mostly to systems where the effective range of the variables is limited, as is the case for atom coordinates in solids. The instanton approximation involves much less computational effort than RAW-QTST, by about a factor of 10^4, and with the path optimization method presented here, it can be used with atomic forces obtained from electronic structure calculations where each force evaluation can easily take tens of minutes of CPU time. For example, the rate constant for hydrogen atom tunneling in solids has been carried out using the method presented above coupled with density functional theory evaluation of the atomic forces, but those results will be presented elsewhere.

5   Discussion

A general method for finding optimal paths on a multidimensional surface has been presented here. Several two-dimensional problems have been used to illustrate the method, but the strength of the approach is its applicability to

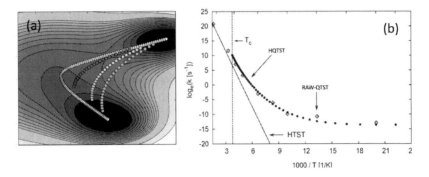

Fig. 4. (a) Model 2-D energy surface with minimum energy path (far left), and converged JWKB tunneling paths for high, intermediate and low (far right) energy. The end points move to the specified system energy. The lower the energy, the more the tunneling path 'cuts the corner', as it moves to a region of higher energy and becomes shorter. (b) Calculation of the rate of H_2 molecule desorption from a Cu(110) surface, including 432 atom coordinates as degrees of freedom. The number of discretization points was $n = 10$ for the highest energy, and $n = 40$ for the lowest energy. The temperature dependence of the rate constant for desorption shows an onset of tunneling at around 250 K. RAW-QTST labels the results from full quantum TST calculation [14]. HQTST labels the results obtained from the instanton approximation, which performs remarkably well here, especially considering that a gas phase molecule is formed.

problems where many, even thousands, of degrees of freedom need to be included. One example of a large system was presented in connection with the calculation of tunneling rate in an atomic scale system. There, the path optimization method provides an efficient way of finding the tunneling path within the so-called instanton approximation. The computational effort is similar to the widely used NEB method for finding minimum energy paths in classical systems where atomic forces from *ab initio* and density functional theory treatments of the electronic degrees of freedom are used as input. Calculations of tunneling rates using such atomic forces are not significantly harder.

An alternative approach to the implementation of the instanton approximation is to use the fact that the instanton path is a first order saddle point on the effective potential surface, V_{eff} given by equation (26), for closed Feynman paths [14]. Methods converging to first order saddle points, such as the minimum mode following method [15,16], can then be used to find tunneling paths. This approach has been used in refs. [17,18]. But, the approach presented here has several advantages over this methods. One is that the distribution of discretization points in the optimization of the JWKB tunneling path can be controlled and they can, for example, be chosen to be equally distributed while the replicas in the Feynman paths tend to cluster in the neighborhood of the end points. Also, the convergence to the saddle point has to be very tight in order to get just one negative eigenvalue. This is particularly problematic when atomic forces from electronic structure calculations are used as input. The method presented here has similar convergence properties and computational effort as the NEB method, which is now widely used for finding classical transition paths. Finally, multiple maxima and minima on the energy surface can make the convergence to the right saddle point problematic while the calculations of the tunneling path is more robust.

It is also possible to find the optimal tunneling path by constructing an elastic band of Feynman paths, forming the so-called minimum action path, using the NEB method as was done in [14]. This, however, involves much more computation as the total number of degrees of freedom in the optimization becomes $NP(n-1)$ where P is the number of discretization points in the Feynman paths, $(n+1)$ the number of discretization points in the minimum action path and N is the number of variables in the system.

Acknowledgements. We would like to thank Judith Rommel and Johannes Kaestner for useful comments on this manuscript. This work was supported by the Icelandic Research Fund, University of Iceland research fund and EC integrated project NESSHy.

References

1. Jónsson, H., Mills, G., Jacobsen, K.W.: Classical and Quantum Dynamics in Condensed Phase Simulations. In: Berne, B.J., Ciccotti, G., Coker, D.F. (eds.), ch. 16, p. 385. World Scientific (1998)
2. Razavy, M.: Quantum Theory of Tunneling. World Scientific Publishing (2003)
3. Wigner, E.: Trans. Faraday Soc. 34, 29 (1938); Eyring, H.: J. Chem. Phys. 3, 107 (1935)
4. Vineyard, G.H.: J. Phys. Chem. Solids 3, 121 (1957)
5. Henkelman, G., Jónsson, H.: J. Chem. Phys. 113, 9978 (2000)
6. Henkelman, G., Uberuaga, B., Jónsson, H.: J. Chem. Phys. 113, 9901 (2000)
7. Rasmussen, T., Jacobsen, K.W., Leffers, T., Pedersen, O.B., Srinivasan, S.G., Jónsson, H.: Physical Review Letters 79, 3676 (1997)
8. Sheppard, D., Terrell, R., Henkelman, G.: J. Chem. Phys. 128, 134106 (2008)
9. Wilson, E.B., Decius, J.C., Cross, P.C.: Molecular Vibrations: The Theory of Infrared and Raman Vibrational Spectra. Dover (1980)
10. Callan, C.G., Coleman, S.: Phys. Rev. D 16, 1762 (1977)
11. Miller, W.H.: J. Phys. Chem. 62, 1899 (1975)
12. Messina, M., Schenter, G., Garrett, B.C.: J. Chem. Phys. 103, 3430 (1995)
13. Skodje, R.T., Truhlar, D.G.: J. Chem. Phys. 77, 5955 (1982)
14. Mills, G., Schenter, G.K., Makarov, D., Jónsson, H.: Chem. Phys. Lett 278, 91 (1997); RAW Quantum Transition State Theory. In: Berne, B.J., et al. (eds.) Classical and Quantum Dynamics in Condensed Phase Simulations, page 405. World Scientific (1998)
15. Henkelman, G., Jónsson, H.: J. Chem. Phys. 111, 7010 (1999)
16. Kaestner, J., Sherwood, P.: J. Chem. Phys. 128, 014106 (2008)
17. Arnaldsson, A.: Ph.D. thesis, University of Washington, Seattle, WA, USA (2007)
18. Andersson, S., Nyman, G., Arnaldsson, A., Manthe, U., Jónsson, H.: J. Phys. Chem. A 113, 4468 (2009)

Shallow Water Simulations on Multiple GPUs

Martin Lilleeng Sætra[1] and André Rigland Brodtkorb[2]

[1] Center of Mathematics for Applications, University of Oslo,
P.O. Box 1053 Blindern, NO-0316 Oslo, Norway
m.l.satra@cma.uio.no
[2] SINTEF, Dept. Appl. Math., P.O. Box 124, Blindern, NO-0314 Oslo, Norway
Andre.Brodtkorb@sintef.no

Abstract. We present a state-of-the-art shallow water simulator running on multiple GPUs. Our implementation is based on an explicit high-resolution finite volume scheme suitable for modeling dam breaks and flooding. We use row domain decomposition to enable multi-GPU computations, and perform traditional CUDA block decomposition within each GPU for further parallelism. Our implementation shows near perfect weak and strong scaling, and enables simulation of domains consisting of up-to 235 million cells at a rate of over 1.2 gigacells per second using four Fermi-generation GPUs. The code is thoroughly benchmarked using three different systems, both high-performance and commodity-level systems.

1 Introduction

Predictions of floods and dam breaks require accurate simulations with rapid results. Faster than real-time performance is of the utmost importance when simulating these events, and traditional CPU-based solutions often fall short of this goal. We address the performance of shallow water simulations in this paper through the use of multiple graphics processing units (GPUs), and present a state-of-the-art implementation of a second-order accurate explicit high-resolution finite volume scheme.

There has been a dramatic shift in commodity-level computer architecture over the last five years. The steady increase in performance does no longer come from higher clock frequencies, but from parallelism through more arithmetic units: The newest CPU from Intel, for example, contains 24 single precision arithmetic units (Core i7-980X). The GPU takes this parallelism even further with up-to 512 single precision arithmetic units (GeForce GTX 580). While the GPU originally was designed to offload a predetermined set of demanding graphics operations from the CPU, modern GPUs are now fully programmable. This makes them suitable for general purpose computations, and the use of GPUs has shown large speed-ups over the CPU in many application areas [1,2]. The GPU is connected to the rest of the computer through the PCI Express bus, and commodity-level computers can have up-to two GPUs connected at full data speed. Such solutions offer the compute performance comparable to a small CPU cluster, and this motivates the use of multiple GPUs. In fact, three of the five fastest supercomputers use GPUs as a major source of computational power [3]. However, the extra floating-point performance comes at a price, as it is nontrivial to develop

K. Jónasson (Ed.): PARA 2010, Part II, LNCS 7134, pp. 56–66, 2012.

efficient algorithms for GPUs, especially when targeting multiple GPUs. It requires both different programming models and different optimization techniques compared to traditional CPUs.

Related Work. The shallow water equations belong to a wider class of problems known as hyperbolic conservation laws, and many papers have been published on GPU-acceleration of both conservation and balance laws [4,5,6,7,8,9,10]. There have been multiple publications on the shallow water equations as well [11,12,13,14,15,16], illustrating that these problems can be efficiently mapped to modern graphics hardware. The use of multiple GPUs has also become a subject of active research. Micikevicius [17] describes some of the benefits of using multiple GPUs for explicit finite-difference simulation of 3D reverse time-migration (the linear wave equation), and reports super-linear speedup when using four GPUs. Overlapping computation and communication for explicit stencil computations has also been presented for both single nodes [18] and clusters [19] with near-perfect weak scaling. Perfect weak scaling was shown by Acuña and Aoki [20] for shallow water simulations on a cluster of 32 GPU nodes, by overlapping computations and communication. Rostrup and De Sterck [21] further present detailed optimization and benchmarking of shallow water simulations on clusters of multi-core CPUs, the Cell processor, and GPUs. Comparing the three, the GPUs offer the highest performance.

In this work, we focus on single-node systems with multiple GPUs. By utilizing more than one GPU it becomes feasible to run simulations with significantly larger domains, or to increase the spatial resolution. Our target architecture is both commodity-level computers with up-to two GPUs, as well as high-end and server solutions with up-to four GPUs at full data speed per node. We present a multi-GPU implementation of a second-order well-balanced positivity preserving central-upwind scheme [22]. Furthermore, we offer detailed performance benchmarks on three different machine setups, tests of a latency-hiding technique called ghost cell expansion, and analyzes of benchmark results.

2   Mathematical Model and Discretization

In this section, we give a brief outline of the major parts of the implemented numerical scheme. For a detailed overview of the scheme, we refer the reader to [22,23]. The shallow water equations are derived by depth-integrating the Navier-Stokes equations, and describe fluid motion under a pressure surface where the governing flow is horizontal. To correctly model phenomena such as tsunamis, dam breaks, and flooding over realistic terrain, we need to include source terms for bed slope and friction:

$$\begin{bmatrix} h \\ hu \\ hv \end{bmatrix}_t + \begin{bmatrix} hu \\ hu^2 + \frac{1}{2}gh^2 \\ huv \end{bmatrix}_x + \begin{bmatrix} hv \\ huv \\ hv^2 + \frac{1}{2}gh^2 \end{bmatrix}_y = \begin{bmatrix} 0 \\ -ghB_x \\ -ghB_y \end{bmatrix} + \begin{bmatrix} 0 \\ -gu\sqrt{u^2+v^2}/C_z^2 \\ -gv\sqrt{u^2+v^2}/C_z^2 \end{bmatrix}. \quad (1)$$

Here h is the water depth and u and v are velocities along the abscissa and ordinate, respectively. Furthermore, g is the gravitational constant, B is the bottom topography, and C_z is the Chézy friction coefficient.

To be able to simulate dam breaks and flooding, we require that our numerical scheme handles wetting and drying of cells, a numerically challenging task. However, we also want other properties, such as well-balancedness, accurate shock-capturing without oscillations, at least second order accurate flux calculations, and that the computations map well to the architecture of the GPU. A scheme that fits well with the above criteria is the explicit Kurganov-Petrova scheme [22], which is based on a standard finite volume grid. In this scheme, the physical variables are given as cell averages, the bathymetry as a piecewise bilinear function (represented by the values at the cell corners), and fluxes are computed across cell interfaces (see also Figure 1). Using vectorized notation, in which $Q = [h, hu, hv]^T$ is the vector of conserved variables, the spatial discretization can be written,

$$
\begin{aligned}
\frac{dQ_{ij}}{dt} &= H_f(Q_{ij}) + H_B(Q_{ij}, \nabla B) - \left[F(Q_{i+1/2,j}) - F(Q_{i-1/2,j})\right] \\
&\qquad - \left[G(Q_{i,j+1/2}) - G(Q_{i,j-1/2})\right] \\
&= H_f(Q_{ij}) + R(Q)_{ij}.
\end{aligned}
\tag{2}
$$

Here $H_f(Q_{ij})$ is the friction source term, $H_B(Q_{ij}, \nabla B)$ is the bed slope source term, and F and G are the fluxes across interfaces along the abscissa and ordinate, respectively. We first calculate $R(Q)_{ij}$ in (2) explicitly, and as in [23], we use a semi-implicit discretization of the friction source term,

$$
\tilde{H}_f(Q_{ij}^k) = \begin{bmatrix} 0 \\ -g\sqrt{u_{ij}^{k\,2} + v_{ij}^{k\,2}}/h_{ij}^k C_{z\,ij}^2 \\ -g\sqrt{u_{ij}^{k\,2} + v_{ij}^{k\,2}}/h_{ij}^k C_{z\,ij}^2 \end{bmatrix}.
\tag{3}
$$

This yields one ordinary differential equation in time per cell, which is then solved using a standard second-order accurate total variation diminishing Runge-Kutta scheme [24],

$$
\begin{aligned}
Q_{ij}^* &= \left[Q_{ij}^n + \Delta t R(Q^n)_{ij}\right]/\left[1 + \Delta t \tilde{H}_f(Q_{ij}^n)\right] \\
Q_{ij}^{n+1} &= \left[\tfrac{1}{2}Q_{ij}^n + \tfrac{1}{2}\left[Q_{ij}^* + \Delta t R(Q^*)_{ij}\right]\right]/\left[1 + \tfrac{1}{2}\Delta t \tilde{H}_f(Q_{ij}^*)\right],
\end{aligned}
\tag{4}
$$

or a first-order accurate Euler scheme, which simply amounts to setting $Q^{n+1} = Q^*$. The timestep, Δt, is limited by a CFL condition,

$$
\Delta t \le \tfrac{1}{4}\min_\Omega\left\{\left|\Delta x/\lambda_x\right|, \left|\Delta y/\lambda_y\right|\right\}, \qquad
\begin{aligned}
\lambda_x &= u \pm \sqrt{gh}, \\
\lambda_y &= v \pm \sqrt{gh}
\end{aligned}
\tag{5}
$$

that ensures that the fastest numerical propagation speed is at most one quarter grid cell per timestep.

In summary, the scheme consists of three parts: First fluxes and explicit source terms are calculated in (2), before we calculate the maximum timestep according to the CFL condition, and finally evolve the solution in time using (4). The second-order accurate Runge-Kutta scheme for the time integration is a two-step process, where we first perform the above operations to compute Q^*, and then repeat the process to compute Q^{n+1}.

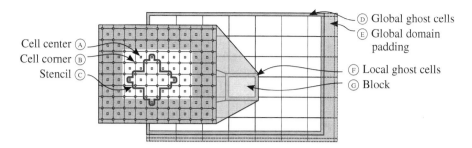

Fig. 1. Domain decomposition and variable locations for the single-GPU simulator. The global domain is padded ⓔ to fit an integer number of CUDA blocks, and global ghost cells ⓓ are used for boundary conditions. Each block ⓖ has local ghost cells ⓕ that overlap with other blocks to satisfy the data dependencies dictated by the stencil ⓒ. Our data variables Q, R, H_B, and H_f are given at grid cell centers ⓐ, and B is given at grid cell corners ⓑ.

3 Implementation

Solving partial differential equations using explicit schemes implies the use of stencil computations. Stencil computations are embarrassingly parallel and therefore ideal for the parallel execution model of GPUs. Herein, the core idea is to use more than one GPU to allow faster simulation, or simulations with larger domains or higher resolution. Our simulator runs on a single node, enabling the use of multithreading, and we use one global *control* thread in addition to one *worker* thread per GPU. The control thread manages the worker threads and facilitates domain decomposition, synchronization, and communication. Each worker thread uses a modified version of our previously presented single-GPU simulator [23] to compute on its part of the domain.

Single-GPU Simulator. The single-GPU simulator implements the Kurganov-Petrova scheme on a single GPU using CUDA [25], and the following gives a brief overview of its implementation. The simulator first allocates and initializes data according to the initial conditions of the problem. After initialization, we repeatedly call a step function to advance the solution in time. The step function executes four CUDA *kernels* in order, that together implement the numerical scheme. The first kernel computes the fluxes across all interfaces, and is essentially a complex stencil computation. This kernel reads four values from global memory, performs hundreds of floating point operations, and writes out three values to global memory again. It is also the most time consuming kernel, with over 87% of the runtime. The next kernel finds the maximum wave speed in the domain, and then computes the timestep size according to the CFL condition. The third kernel simply solves the ordinary differential equations in time to evolve the solution. Finally, the fourth kernel applies boundary conditions by setting the values of global *ghost cells* (see Figure 1).

Threaded Multi-GPU Framework. When initializing our simulator, the control thread starts by partitioning the global domain, and continues by initializing one worker thread

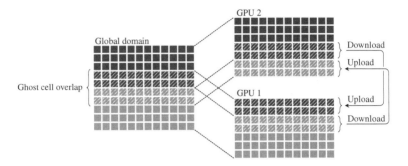

Fig. 2. Row decomposition and exchange of two rows of ghost cells. The shaded cells are a part of the overlapping ghost cell region between subdomains.

per subdomain, which attaches to a separate GPU. We can then perform simulation steps, where the control thread manages synchronization and communication between GPUs. An important thing to note about this strategy is that the control thread handles all multi-GPU aspects, and that each GPU is oblivious to other GPUs, running the simulation on its subdomain similar to a single-GPU simulation.

We use a row domain decomposition, in which each subdomain consists of several rows of cells (see Figure 2). The subdomains form overlapping regions, called ghost cells, which function as boundary conditions that connect the neighbouring subdomains. By exchanging the overlapping cells before every timestep, we ensure that the solution can propagate properly between subdomains. There are several benefits to the row decomposition strategy. First of all, it enables the transfer of continuous parts of memory between GPUs, thus maximizing bandwidth utilization. A second benefit is that we can minimize the number of data transfers, as each subdomain has at most two neighbours. To correctly exchange ghost cells, the control thread starts by instructing each GPU to download its ghost cells to *pinned* CPU memory, as direct GPU to GPU-transfers are currently not possible. The size of the ghost cell overlap is dictated by the stencil, which in our case uses two values in each direction (see Figure 1). This means that we need to have an overlap of four rows of cells, two from each of the subdomains. After having downloaded the ghost cells to the CPU, we need to synchronize to guarantee that all downloads have completed, before each GPU can continue by uploading the ghost cells coming from neighbouring subdomains. Note that for the second-order accurate Runge-Kutta time integration scheme, we have to perform the ghost cell exchange both when computing Q^* and when computing Q^{n+1}, thus two times per full timestep.

The multi-GPU simulator is based on our existing single-GPU simulator, which made certain assumptions that made it unsafe to execute from separate threads. This required us to redesign parts of the code to guarantee thread safety. A further difficulty related to multi-GPU simulation is that the computed timestep, Δt, will typically differ between subdomains. There are two main strategies to handle this problem, and we have investigated both. The simplest is to use a globally fixed timestep throughout the simulation. This, however, requires that the timestep is less than or equal to the smallest timestep allowed by the CFL condition for the full simulation period, which again implies that our simulation will not propagate as fast as it could have. The second strategy is to synchronize the timestep between subdomains for each timestep, and choose the smallest. This strategy requires that we split the step function into two parts, where the first computes fluxes and the maximum timestep, and the second performs time integration and applies boundary conditions. Inbetween these two substeps we can find the smallest global timestep, and redistribute it to all GPUs. This strategy ensures that the simulation propagates at the fastest possible rate, but at the expense of potentially expensive synchronization and more complex code.

Ghost Cell Expansion. Synchronization and overheads related to data transfer can often be a bottleneck when dealing with distributed memory systems, and a lot of research has been invested in, e.g., latency hiding techniques. In our work, we have implemented a technique called *ghost cell expansion* (GCE), which has yielded a significant performance increase for cluster simulations [26,27]. The main idea of GCE is to trade more computation for smaller overheads by increasing the level of overlap between subdomains, so that they may run more than one timestep per ghost cell exchange. For example, by extending the region of overlap from four to eight cells, we can run two timesteps before having to exchange data. When exchanging ghost cells for every timestep, we can write the time it takes to perform one timestep as

$$w_1 = T(m) + c_T + C(m, n) + c,$$

in which m and n are the domain dimensions, $T(m)$ is the ghost cell transfer time, c_T represents transfer overheads, $C(m, n)$ is the time it takes to compute on the subdomain, and c represents other overheads. Using GCE to exchange ghost cells only every kth timestep, the average time per timestep becomes

$$w_k = T(m) + c_T/k + C(m, n + \mathcal{O}(k)) + c,$$

in which we divide the transfer overheads by k, but increase the overlap, and thus the size of each subdomain. This means that each worker thread computes on a slightly larger domain, and we have larger but fewer data transfers.

4   Results and Analysis

To validate our implementation, we have compared the multi-GPU results with the original single-GPU simulator [23], which has been both verified against analytical solutions and validated against experiments. Our multi-GPU results are identical to those produced by the single-GPU implementation, which means that the multi-GPU implementation is also capable of reproducing both analytical and real-world cases.

We have used three different systems for benchmarking our implementation. The first system is a Tesla S1070 GPU Computing Server consisting of four Tesla C1060 GPUs with 4 GiB memory each[1], connected to an IBM X3550 M2 server with two

---

[1] Connected through two PCIe ×16 slots.

2.0 GHz Intel Xeon X5550 CPUs and 32 GiB main memory (see Figure 3). The second system is a SuperMicro SuperServer consisting of four Tesla C2050 GPUs with 3 GiB memory each (2.6 available when ECC is enabled)[2], and two 2.53 GHz Intel Xeon E5630 CPUs with 32 GiB main memory. The third system is a standard desktop PC consisting of two GeForce 480 GTX cards with 1.5 GiB memory each and a 2.67 GHz Intel Core i7 CPU with 6 GiB main memory. The first two machine setups represent previous and current generation server GPU nodes, and the third machine represents a commodity-level desktop PC.

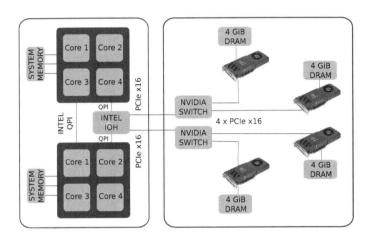

Fig. 3. Hardware setup of the Tesla S1070 GPU Computing Server with four Tesla C1060 GPUs (right) connected to an IBM X3550 M2 server (left).

As our performance benchmark, we have used a synthetic circular dam break over a flat bathymetry, consisting of a square 4000-by-4000 meters domain with a water column placed in the center. The water column is 250 meters high with a radius of 250 meters, and the water elevation in the rest of the domain is 50 meters. At time $t = 0$, the dam surrounding the water column is instantaneously removed, creating an outgoing circular wave. We have used the first-order accurate Euler time integrator in all benchmarks, and the friction coefficient C_z is set to zero. The bed slope and friction coefficient do not affect the performance in this benchmark.

Ghost Cell Expansion. We have implemented ghost cell expansion so that we can vary the level of overlap, and benchmarked three different domain sizes to determine the effect. Figure 4 shows that there is a very small overhead related to transferring data for sufficiently large domain sizes, and performing only one timestep before exchanging overlapping ghost cells actually yields the best overall results for the Tesla S1070 system. Expanding with more than eight cells, the performance of the simulator starts decreasing noticeably. From this, we reason that the overhead connected with data transfers between subdomains in these tests is negligible, compared to the transfer and computational time. Increasing the level of GCE only had a positive impact on the

---

[2] Connected through four PCIe ×16 slots.

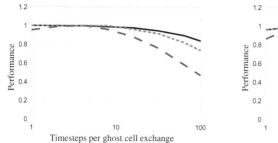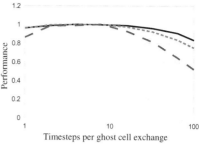

Fig. 4. Testing the impact of different numbers of ghost cell expansion rows with four GPUs. Tested on both The Tesla S1070 system (see Figure 3) (left), and the Tesla C2050-based system (right). The domains tested consists of 1024^2 (dashed), 4096^2 (densely dashed) or 8192^2 (solid) cells. The graphs have been normalized relative to the peak achieved performance for each domain size.

smallest domain for the Tesla S1070 system, where the transferred data volume is so small that the overheads become noticeable. On the Tesla C2050-based system, however, we see that the positive impact of GCE is more visible. We expect this is because this GPU is much faster, making the communication overheads relatively larger.

Our results show that ghost cell expansion had only a small impact on the shared-memory architectures we are targeting for reasonably sized grids, but gave a slight performance increase for the Tesla C2050 GPUs. This is due to the negligible transfer overheads. We thus expect GCE to have a greater effect when performing ghost cell exchange across multiple nodes, since the overheads here will be significantly larger, and we consider this a future research direction.

Since our results show that it is most efficient to have a small level of GCE for the Tesla S1070 system, we choose to exchange ghost cells after every timestep in all of our other benchmarks for this system. For the Tesla C2050-based system we exchange data after eight timesteps, as this gave the overall best results. Last, for the GeForce 480 GTX cards, which displayed equivalent behaviour to that of the Tesla C2050-based system, we also exchange ghost cells after performing eight timesteps.

Timestep Synchronization. We have implemented both the use of a global fixed timestep, as well as exchange of the minimum global timestep in our code, and benchmarked on our three test systems to determine the penalty of synchronization. In the tests we compared simulation runs with a fixed $\Delta t = 0.001$ in each subdomain, and runs with global synchronization of Δt. When looking at the results we see that the cost of synchronizing Δt globally has a negligible impact on the performance of the Tesla S1070 system, with an average 0.36% difference for domain sizes larger than five million cells on four GPUs. As expected, the cost is also roughly halved when synchronizing two GPUs compared to four (0.17%). For smaller domain sizes, however, the impact becomes noticeable, but these domains are typically not candidates for multi-GPU simulations. The Tesla C2050- and GeForce 480 GTX-based systems also display similar results, meaning that global synchronization of Δt is a viable strategy for reasonably sized domains.

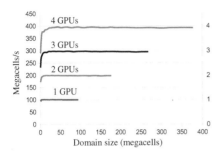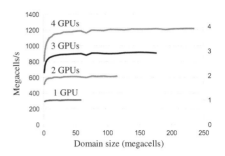

Fig. 5. (Left) Performance experiment on a Tesla S1070 system (see Figure 3) with up-to four GPUs. (Right) Performance experiment on a Tesla C2050-based system, using up-to four GPUs. The secondary y-axis on the right-hand side shows scaling relative to the peak achieved performance of a single GPU.

Weak and Strong Scaling. Weak and strong scaling are two important performance metrics that are used for parallel execution. While varying the number of GPUs, weak scaling keeps the domain size per GPU fixed, and strong scaling keeps the global domain size fixed. As we see from Figure 5, we have close to linear scaling from one to four GPUs. For domains larger than 25 million cells the simulator displays near perfect weak and strong scaling on all three systems. Running simulations on small domains is less efficient when using multiple GPUs for two reasons: First of all, as the global domain is partitioned between more GPUs, we get a smaller size of each subdomain. When these subdomains become sufficiently small, we are unable to fully occupy a single GPU, and thus do not reach peak performance. Second, we also experience larger effects of overheads. However, we quickly get close-to linear scaling as the domain size increases.

The Tesla C1060 GPUs have 4.0 GiB of memory each, which enables very large scale simulations: When using all four GPUs, domains can have up to 379 million cells, computing at 396 megacells per second. Because the most recent Tesla C2050 GPUs from NVIDIA have only 3.0 GiB memory per GPU, our maximum domain size is smaller (235 million cells), but our simulation speed is dramatically faster. Using four GPUs, we achieve over 1.2 gigacells per second. The fastest system per GPU, however, was the commodity-level desktop machine with two GeForce 480 GTX cards. These cards have the highest clock frequency, and we achieve over 400 megacells per second per GPU.

5   Summary and Future Work

We have presented an efficient multi-GPU implementation of a modern finite volume scheme for the shallow water equations. We have further presented detailed benchmarking of our implementation on three hardware setups, displaying near-perfect weak *and* strong scaling on all three. Our benchmarks also show that communication between GPUs within a single node is very efficient, which enables tight cooperation between subdomains.

A possible further research direction is to explore different strategies for domain decomposition, and especially to consider techniques for adaptive domain decompositions.

Acknowledgements. Part of this work is supported by the Research Council of Norway's project number 180023 (Parallel3D), and the authors are grateful for the continued support from NVIDIA. Thanks also to Ole W. Saastad for access to the Teflon GPU compute cluster at the University of Oslo.

References

1. Brodtkorb, A., Dyken, C., Hagen, T., Hjelmervik, J., Storaasli, O.: State-of-the-art in heterogeneous computing. Journal of Scientific Programming 18(1), 1–33 (2010)
2. Owens, J., Houston, M., Luebke, D., Green, S., Stone, J., Phillips, J.: GPU computing. Proceedings of the IEEE 96(5), 879–899 (2008)
3. Meuer, H., Strohmaier, E., Dongarra, J., Simon, H.: Top 500 supercomputer sites (November 2010), http://www.top500.org/
4. Hagen, T., Henriksen, M., Hjelmervik, J., Lie, K.A.: How to solve systems of conservation laws numerically using the graphics processor as a high-performance computational engine. In: Hasle, G., Lie, K.A., Quak, E. (eds.) Geometrical Modeling, Numerical Simulation, and Optimization: Industrial Mathematics at SINTEF, pp. 211–264. Springer, Heidelberg (2007)
5. Hagen, T.R., Lie, K.-A., Natvig, J.R.: Solving the Euler Equations on Graphics Processing Units. In: Alexandrov, V.N., van Albada, G.D., Sloot, P.M.A., Dongarra, J. (eds.) ICCS 2006. LNCS, vol. 3994, pp. 220–227. Springer, Heidelberg (2006)
6. Brandvik, T., Pullan, G.: Acceleration of a two-dimensional Euler flow solver using commodity graphics hardware. Journal of Mechanical Engineering Science 221(12), 1745–1748 (2007)
7. Brandvik, T., Pullan, G.: Acceleration of a 3D Euler solver using commodity graphics hardware. In: Proceedings of the 46th AIAA Aerospace Sciences Meeting. Number 2008-607 (2008)
8. Klöckner, A., Warburton, T., Bridge, J., Hesthaven, J.: Nodal discontinuous Galerkin methods on graphics processors. Journal of Computational Physics 228(21), 7863–7882 (2009)
9. Wang, P., Abel, T., Kaehler, R.: Adaptive mesh fluid simulations on GPU. New Astronomy 15(7), 581–589 (2010)
10. Antoniou, A., Karantasis, K., Polychronopoulos, E., Ekaterinaris, J.: Acceleration of a finite-difference weno scheme for large-scale simulations on many-core architectures. In: Proceedings of the 48th AIAA Aerospace Sciences Meeting (2010)
11. Hagen, T., Hjelmervik, J., Lie, K.A., Natvig, J., Henriksen, M.: Visual simulation of shallow-water waves. Simulation Modelling Practice and Theory 13(8), 716–726 (2005)
12. Liang, W.-Y., Hsieh, T.-J., Satria, M.T., Chang, Y.-L., Fang, J.-P., Chen, C.-C., Han, C.-C.: A GPU-Based Simulation of Tsunami Propagation and Inundation. In: Hua, A., Chang, S.-L. (eds.) ICA3PP 2009. LNCS, vol. 5574, pp. 593–603. Springer, Heidelberg (2009)
13. Lastra, M., Mantas, J., Ureña, C., Castro, M., García- Rodríguez, J.: Simulation of shallow-water systems using graphics processing units. Mathematics and Computers in Simulation 80(3), 598–618 (2009)
14. de la Asunción, M., Mantas, J., Castro, M.: Simulation of one-layer shallow water systems on multicore and CUDA architectures. The Journal of Supercomputing, 1–9 (2010) (published online)

15. de la Asunción, M., Mantas, J.M., Castro, M.J.: Programming CUDA-Based GPUs to Simulate Two-Layer Shallow Water Flows. In: D'Ambra, P., Guarracino, M., Talia, D. (eds.) Euro-Par 2010. LNCS, vol. 6272, pp. 353–364. Springer, Heidelberg (2010)
16. Brodtkorb, A., Hagen, T.R., Lie, K.A., Natvig, J.R.: Simulation and visualization of the Saint-Venant system using GPUs. Computing and Visualization in Science (2010) (forthcoming)
17. Micikevicius, P.: 3D finite difference computation on GPUs using CUDA. In: GPGPU-2: Proceedings of 2nd Workshop on General Purpose Processing on Graphics Processing Units, pp. 79–84. ACM, New York (2009)
18. Playne, D., Hawick, K.: Asynchronous communication schemes for finite difference methods on multiple GPUs. In: 10th IEEE/ACM International Conference on Cluster, Cloud and Grid Computing (CCGrid), pp. 763–768 (May 2010)
19. Komatitsch, D., Göddeke, D., Erlebacher, G., Michéa, D.: Modeling the propagation of elastic waves using spectral elements on a cluster of 192 GPUs. Computer Science - Research and Development 25, 75–82 (2010)
20. Acuña, M., Aoki, T.: Real-time tsunami simulation on multi-node GPU cluster. In: ACM/IEEE Conference on Supercomputing (2009) (poster)
21. Rostrup, S., De Sterck, H.: Parallel hyperbolic PDE simulation on clusters: Cell versus GPU. Computer Physics Communications 181(12), 2164–2179 (2010)
22. Kurganov, A., Petrova, G.: A second-order well-balanced positivity preserving central-upwind scheme for the Saint-Venant system. Communications in Mathematical Sciences 5, 133–160 (2007)
23. Brodtkorb, A., Sætra, M., Altinakar, M.: Efficient shallow water simulations on GPUs: Implementation, visualization, verification, and validation (preprint)
24. Shu, C.W.: Total-variation-diminishing time discretizations. SIAM Journal of Scientific and Statistical Computing 9(6), 1073–1084 (1988)
25. NVIDIA: NVIDIA CUDA reference manual 3.1 (2010)
26. Ding, C., He, Y.: A ghost cell expansion method for reducing communications in solving PDE problems. In: ACM/IEEE Conference on Supercomputing, pp. 50–50. IEEE Computer Society Press, Los Alamitos (2001)
27. Palmer, B., Nieplocha, J.: Efficient algorithms for ghost cell updates on two classes of MPP architectures. In: Akl, S., Gonzalez, T. (eds.) Proceedings of the 14th IASTED International Conference on Parallel and Distributed Computing and Systems, pp. 192–197. ACTA Press, Cambridge (2002)

High Performance Computing Techniques for Scaling Image Analysis Workflows

Patrick M. Widener[1], Tahsin Kurc[1], Wenjin Chen[2], Fusheng Wang[1], Lin Yang[2], Jun Hu[2], Vijay Kumar[3], Vicky Chu[2], Lee Cooper[1], Jun Kong[1], Ashish Sharma[1], Tony Pan[1], Joel H. Saltz[1], and David J. Foran[2]

[1] Center for Comprehensive Informatics, Emory University
patrick.widener@emory.edu
[2] Center for Biomedical Imaging & Informatics,
UMDNJ-Robert Wood Johnson Medical School
[3] Department of Computer Science and Engineering, Ohio State University

Abstract. Biomedical images are intrinsically complex with each domain and modality often requiring specialized knowledge to accurately render diagnosis and plan treatment. A general software framework that provides access to high-performance resources can make possible high-throughput investigations of micro-scale features as well as algorithm design, development and evaluation. In this paper we describe the requirements and challenges of supporting microscopy analyses of large datasets of high-resolution biomedical images. We present high-performance computing approaches for storage and retrieval of image data, image processing, and management of analysis results for additional explorations. Lastly, we describe issues surrounding the use of high performance computing for scaling image analysis workflows.

1 Introduction

High-resolution biomedical imaging provides a valuable tool for scientists to investigate the structure and function of biological systems at cellular, sub-cellular, and molecular levels. Information obtained at these scales can provide biomarkers for better prognostic accuracy and lead to new insights into the underlying mechanisms of disease progression. For example, current therapies and treatment regimens for breast cancer are based upon classification strategies which are limited in terms of their capacity to identify specific tumor groups exhibiting different clinical and biological profiles. Tumors can be analyzed using tissue microarrays (TMAs) to confirm clinico-pathologic correlations, which have been established with whole tissue sections[1]. The process could involve extracting and assimilating phenotypic and molecular features from images from multiple groups of patients, comparing and correlating this information with the information about the patient under study, and classifying the condition of the patient based on the analysis results.

The field of genomics research has been transformed with advances in high-throughput instruments which can generate large volumes of readings quickly. We are observing a similar trend in biomedical imaging. Advanced microscopy scanners are

K. Jónasson (Ed.): PARA 2010, Part II, LNCS 7134, pp. 67–77, 2012.

capable of rapidly (usually within several minutes) imaging glass slides in their entirety at high-resolution. The continuing increase in speed and resolution of imaging instruments will usher in high-throughput, high-resolution micro-scale feature analysis. Research studies will be able to collect large numbers of high-resolution images from TMAs and whole slides from cohorts of patients. In these studies, multiple images will be captured from the same tissue specimen using different stains and/or imaging modalities; images from the same patient will also be captured at multiple time points in a treatment or observation phase.

As advanced microscope scanners continue to gain favor in research and clinical settings, microscopy imaging holds great potential for highly detailed examination of disease morphology and for enhancing anatomic pathology. In order to realize this potential, researchers need high-performance systems to handle data and computation complexity of the high-throughput micro-scale feature analysis process. In addition, the systems should be able to leverage Grid computing both for taking advantage of distributed resources and for supporting sharing of analytical resources as well as image datasets and analysis results in collaborative studies.

In this paper we describe the requirements and challenges of supporting micro-scale analyses of large datasets of high-resolution biomedical images. We argue that an integrated software framework to address the requirements should provide support for researchers to efficiently store and retrieve large volumes of image data (image storage and management component), execute complex analyses on image datasets (analysis component), and manage, query, and integrate vast amounts of analysis results (results management component). We present high-performance computing approaches for these three components. We describe the implementation of a Grid-enabled analysis component for quantitative investigation of tissue microarrays and present empirical performance results.

2   Requirements and Challenges of Microscopy Image Analysis

The first challenge to enabling high-throughput, high-resolution analyses of micro-scale features is the fact that images obtained from contemporary scanners are very large. For example, scanning a 15 mm x 15 mm section of tissue results in a billion pixels, or 3 GB of RGB data, and image sizes of 10 GB or more are common for images obtained from a larger tissue section or scanned at higher resolutions. Large clinical studies may recruit hundreds of participants; studies on animal models of disease, such as those that make use of models based on mice, may have hundreds of specimens. These studies will generate thousands of images (e.g., studies on morphological changes in mouse placenta can generate up to thousand slides from a single placenta specimen to form a 3-dimensional representation of the placenta) over the course of the study, resulting in multiple terabytes of image data.

The second challenge is the computational complexity of analyzing images. Operations on image data may range from relatively simple intensity/color correction tasks to complex segmentation and feature extraction operations. A researcher may combine these operations into analysis workflows for characterization of micro-scale structures. In addition, multiple workflows composed of a number of interrelated algorithms may be needed to carry out segmentation and classification. One of the reasons for executing multiple workflows is to facilitate algorithm development and evaluation. The composition of analysis pipelines, the values of input parameters of analysis methods, and the characteristics of input datasets all affect the analysis results and the accuracy of the analysis outcome. Given the large volume of images and the vast number of features, it would not be feasible to manually inspect each image for every feature and fine-tune analysis pipelines. A workable approach is to apply a few hundred variations of analysis pipelines and input parameter values on a few hundred images. Systematic management, comparison, and analysis of the results from these experiments can weed out bad choices (for the intended study), reducing the number of potentially high-quality pipelines to 10-20. These pipelines are then executed on the whole collection of images. Besides increasing accuracy of and confidence in analysis results, an image dataset can be analyzed multiple times with different algorithms and pipelines to detect and extract different types of features.

Another challenge is the management of huge amounts of semantically complex analysis results. Image markups can be either geometric shapes or image masks; annotations can be calculations, observations, disease inferences or external annotations. Many of the analytical imaging results are anatomic objects such as lesions, cells, nuclei, blood vessels, etc. Features such as volume, area, elongation, are extracted from these objects, and the objects are classified (annotated) based on feature characteristics and domain knowledge. Annotations may draw from one or more domain ontologies, resulting in a semantically rich environment. An example query from one of our studies is Search for objects with an observation concept (astrocytoma), but also expand to include all its subclass concepts (gliosarcoma and giant cell glioblastoma). Spatial relationships among the objects are often important to understanding the biomedical characteristics of biology systems. Thus, additional annotations can be derived from existing annotations and spatial relationships among structures and features – common spatial relationships include containment, intersection or overlap, distance between objects, and adjacency relationships. Large image datasets and complex analyses result in large volumes of metadata about objects, markups, and features computed for each anatomic object, and semantic annotations (about cell types, genomic information associated with cells, etc). For instance, segmentation of whole slide images from brain tumor specimens can lead to 100,000 to 1,000,000 cells in each virtual slide. Classification categories include classes of brain tumor cells, normal brain cell categories, macrophages, endothelial cells, etc. Various markers can be used to identify possible cancer stem cells, mutations, along with markers designed to identify blood vessels. The process of classifying a given cell may involve 10-100 shape, texture, and stain quantification features. As a result, systematic analysis of a large dataset consisting of thousands of images can generate 10^{10} to 10^{13} features.

3   High Performance Computing Approaches

Distributed storage platforms can be leveraged to reduce I/O costs of storing and retrieving very large datasets of high-resolution images. To maximize the efficiency of parallel storage and data accesses for image data, data declustering and indexing techniques can be employed. In an earlier work[2], we evaluated several techniques for data distribution, indexing, and query processing of multi-resolution 3-dimensional image datasets. We implemented Hilbert-curve based, random, and round-robin distribution strategies for declustering of sub-image regions across storage nodes. A two-level R-tree based indexing scheme was employed. The two-level scheme consisted of an R-tree index for each mesh on the local chunks assigned to a node and another R-tree index on the bounding boxes of all the meshes in the dataset. We should note that storage and I/O costs can further be reduced by applying additional optimizations. These optimizations include incremental, adaptive declustering and partial replication. In the incremental, adaptive scheme, a dataset is initially declustered using a simple and inexpensive declustering algorithm (e.g., a round-robin assignment of image subregions to storage nodes) so that the data can be stored and made available for use quickly. The initial declustering can then be incrementally refined using information on data access patterns and a better, but potentially more expensive, declustering algorithm. Partial replication can be useful if there are multiple types of queries and/or if it is detected that certain regions of a dataset are accessed more frequently than others. In that case, instead of redistributing the entire dataset, the regions of the dataset can be replicated.

Processing of very large images and image datasets require careful coordination of data retrieval, distribution of data among processing nodes, and mapping of processing tasks to nodes. A combination of multiple parallelism approaches can be employed to quickly render results from a large dataset. Multiple images can be processed concurrently in a bag-of-tasks strategy, in which images are assigned to groups of processors in a demand-driven fashion. High-resolution images, however, may not fit in the main memory of a single processor. In addition, image analysis workflows may consist of operations that can process data in a pipelined, streaming manner. These characteristics of data and operations are suitable for combined use of task- and data-parallelism. We have developed a middleware system, referred to as out-of-core virtual microscope (OCVM)[3,4], based on the DataCutter infrastructure[5] in order to support multiple parallelism approaches. In this system, multiple instances of workflows can be created and executed with each instance processing a subset of images. Within each workflow instance, an image is partitioned into user-defined chunks (rectangular sub-regions) so that I/O operations can be coordinated by the runtime system rather than relying on the virtual memory. The processing operations constituting the workflow can be mapped to processors to reduce I/O and communication overheads. Multiple instances of an operation can be instantiated to allow for data-parallelism. In this setup, the retrieval, communication, and processing of chunks can be pipelined, wherever it is possible, and the chunks can be processed concurrently by multiple instances of an operation.

As we presented in Section 2, high-throughput, high-resolution analyses of micro-scale features will generate vast amounts of results. For example, in one of our projects, an analysis involving 213 whole-slide images segmented and annotated approximately 90 million nuclei. An XML results document for a single image, which included the boundaries of all segmented nuclei in the image along with 23 features computed for each nucleus, was close to 7GB in size. In order to scale to large volumes of data, databases of analysis results can be physically partitioned into multiple physical nodes on cluster based computing infrastructure. The distributed memory on a cluster system can also be leveraged to reduce I/O costs. We investigated the performance of different database configurations for spatial joins and cross-match operations[6]. The configurations included 1) a parallel database management system with active disk style execution support for some types of database operations, 2) a database system designed for high-availability and high-throughput (MySQL Cluster), and 3) a distributed collection of database management systems with data replication. Our experimental evaluation of crossmatch algorithms[7] shows that the choice of a database configuration can significantly impact the performance of the system. The configuration with distributed database management systems with partial replication provides a flexible environment, which can be adjusted to the data access patterns and dataset characteristics.

The other challenge associated with analysis results is the complexity of the results. As we presented in Section 2, semantic metadata is needed to describe analysis results (e.g., nuclear texture, blood vessel characteristics) and the context of the image analyses. An important aspect of semantic information systems is the fact that additional annotations/classifications (also referred to as implicit assertions) can be inferred from initial annotations (also called explicit assertions) based on the ontology and the semantics of the ontology language. Query execution and on-the-fly computation of assertions may take too long on a single processor machine. Pre-computation of inferred assertions, also referred to as the materialization process, can reduce the execution of subsequent queries. Execution strategies leveraging high-performance parallel and distributed machines can reduce execution times and speed up the materialization process[8]. One possible strategy is to employ data parallelism by partitioning the space in which the spatial objects are embedded. Another parallelization strategy is to partition the ontology axioms and rules, distributing the computation of axioms and rules to processors. This partitioning would enable processors to evaluate different axioms and rules in parallel. Inter-processor communication might be necessary to ensure correctness. This parallelization strategy attempts to leverage axiom-level parallelism. A third possible strategy is to combine the first two strategies with task-parallelism. In this strategy, N copies of the semantic store engine and M copies of the rule engine are instantiated on the parallel machine. The system coordinates the exchange of information and the partitioning of workload between the semantic store engine instances and the rule engine instances.

4   Scalable Image Analysis Workflows

In this section, we describe some of the relevant issues surrounding our construction and execution of scalable image analysis workflows.

4.1   Parallel and Grid-Enabled Analysis Techniques

We have developed a Grid service, based on the caGrid infrastructure[9], that encapsulates the computation of texton histograms given a set of TMA disc images. caGrid enables remote access to resources. Using the caGrid infrastructure, multiple algorithm developers can make their algorithms available using standard service interfaces and object models for method input and output. This in turn encourages sharing of new

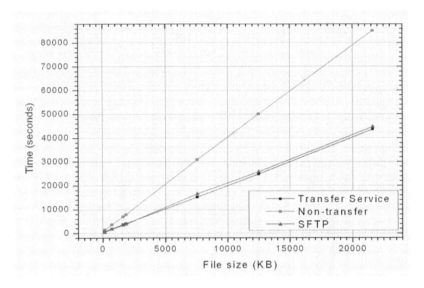

Fig. 1. Upload speed of a batch file with 1 to 100 TMA images

algorithms and tools among researchers and algorithm developers, because client programs can easily access the new resources without requiring resource specific modifications. Moreover, the service can be implemented in a scalable fashion on clustered computing resources to improve performance for increasing numbers of clients and sizes of data. In our case, the TMA analytical service is implemented using the Data-Cutter framework[5]. DataCutter is a stream-filter framework in which data is streamed through a network of interacting filters. Filters modify, shape, or annotate data as it passes through the network of streams, and can be grouped to achieve complex analysis tasks. DataCutter enables bag-of-tasks-style parallelism as well as combinations of task- and data-parallelism. Our current implementation uses the bag-of-tasks model. A single data processing operation or a group of interacting operations is treated as a single task. Multiple task instances can be instantiated on different computation nodes of a cluster. Images to be analyzed are distributed to and load-balanced among these instances using a demand-driven, master-slave strategy in which each slave node requests a new task from the master when it becomes available. The master node schedules upcoming tasks among the available slave nodes. Analysis results are returned to the client through the Grid.

We use the caGrid transfer service to exchange image data and analysis results efficiently between the caGrid service and clients. The caGrid transfer service allows us to upload or download large data sets without incurring the serialization-related costs associated with SOAP/XML messaging. Figure 1 compares the data upload speeds of different data transfer protocols for uploading a batch of files with one to one hundred TMA core images (size from 100KB to 20MB) over the Internet between a service

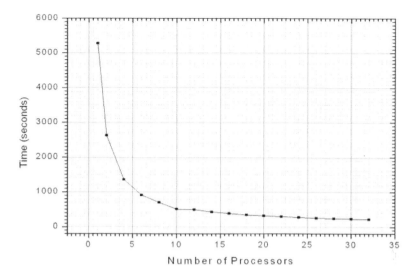

Fig. 2. Processing time of TMA disc images versus number of processors

and client. Our results demonstrated that the caGrid transfer service performs significantly faster than SOAP messaging, reflecting its avoidance of serialization costs, and performs comparably to the SFTP protocol.

Figure 2 shows that the processing time of TMA images versus the number of processors drops significantly as the number of CPUs increases. This result demonstrates the scalability advantages of structuring these types of analyses in a loosely-coupled, highly-parallel manner for execution on commodity cluster hardware. The experiment was performed on a cluster machine at Ohio State University, whose compute nodes are each equipped with AMD dual 250 Opteron CPUs, 8GB DDR400 RAM and 250GB SATA hard drive. The compute nodes are connected through dual GigE Ethernet and run CentOS 4.0.

4.2 Data Management and Query

We have implemented PAIS, a database infrastructure and accompanying data model, to manage microscopy analysis results. The database is designed to support both metadata and spatial queries. Examples of typical queries are: Find all cells in image A that are segmented by Algorithm B1 and are annotated as tumor cells or Find all nuclei segmented by Algorithm B1 that overlap nuclei segmented by Algorithm B2 in image A. The implementation uses the IBM DB2 Enterprise Edition 9.7.1 with DB2 Spatial Extender as the underlying database system. We chose DB2 since it is available free of charge for research and provides integrated support for spatial data types and queries through the spatial extender component.

To support efficient management and query of spatial information, we model and manage markup objects as spatial objects as supported by the spatial extension of DB2. We also employ in queries several spatial functions implemented in DB2 such as spatial relationship functions and functions that return information about properties and dimensions of geometries. Many of our spatial queries are different from traditional GIS queries. We have implemented additional optimizations to reduce query execution times. A preliminary study showed that the performance of spatial joins between two algorithms on the same image can be much improved by divide-and-conquer based approach. By dividing a region into four partitions, the cost of spatial overlap queries can be immediately reduced to less than half.

High-throughput, high-resolution analyses of micro-scale features require retrieval and processing of large image data and can benefit from parallel and distributed processing. We have developed workflows for image tiling, processing, and analysis for execution in high-performance cluster computing environments. Image tiling operations are performed using vendor software which integrates with our high-resolution microscopy image scanning hardware. Algorithm development and execution is performed using the MATLAB toolset. MATLAB provides a robust environment for prototyping, and gives us the ability to seamlessly move analysis tasks from desktop development to HPC cluster execution. These analysis tasks are executed on the Emory University EL-LIPSE cluster, a 1024-core distributed-memory cluster with 2GB memory available to each processor core. Our analysis workflows are executed in MATLAB on the ELLIPSE cluster.

We are also using our experience with this analysis workflow to inform our thoughts about how to handle analysis at larger scale. We have developed a middleware system, referred to as out-of-core virtual microscope (OCVM)[3,4], implemented using DataCutter, in order to support multiple parallelism approaches. OCVM supports processing of multiple images concurrently and of images which may be too big to fit in main memory. This support is provided through bag-of-tasks style parallelism and combined task-data parallelism. The bag-of-tasks support facilitates processing of multiple images or image regions concurrently, where each task (image or image region) is mapped to a processor using a demand-driven load distribution strategy. In the combined task-data parallelism mode, processing operations constituting a workflow can be mapped to processors to reduce I/O and communication overheads. Multiple instances of an operation can be instantiated to allow for data-parallelism. In this mode, the retrieval, communication, and processing of images (or image chunks, if an image is too big) can be pipelined, wherever it is possible, and the data elements can be processed in parallel by multiple instances of an operation. We have demonstrated the efficacy of OCVM on extremely large imaging data, ranging from several tens of gigabytes to terabytes.

5   Related Work

Several leading institutions have already undertaken ambitious projects directed toward digitally imaging, archiving, and sharing pathology specimens. One related software technology, referred to as Virtual Microscopy (VM), provides access to the resulting imaged specimens [10,11,12,13,14]. The Open Microscopy Environment project [15] develops a database-driven system for analysis of biological images. The system consists of a relational database that stores image data and metadata. Images in the database can be processed using a series of modular programs. These programs are connected to the database; a module in the processing sequence reads its input data from the database and writes its output back to the database so that the next module in the sequence can work on it. OME provides a data model of common specification for storing details of microscope set-up and image acquisition. CCDB/OpenCCDB [16] is a system and data model developed to ensure researchers can trace the provenance of data and understand the specimen preparation and imaging conditions that led to the data. CCDB implements an ontology link to support semantic queries and data sources federation. Our work, particularly in the context of the PAIS data model, seeks to provide data management along similar lines to these projects in a highly-scalable framework for parallel application development.

Several research projects have implemented techniques and tools for efficient management, query, and processing of scientific datasets. Manolakos and Funk[17] describe a Java-based tool for rapid prototyping of image processing operations. This tool uses a component-based framework, called JavaPorts, and implements a master-worker mechanism. Oberhuber[18] presents an infrastructure for remote execution of image processing applications using the SGI ImageVision library. Grid workflow management systems like Kepler[19] and Pegasus[20] seek to minimize the makespan by manipulating workflow-level parameters such as grouping and mapping of a workflows components. Glatard *et al.*[21] describe the combined use of data parallelism, services parallelism and job grouping for data-intensive application service-based workflows on the EGEE Grid. System-S[22] is a stream processing system developed at IBM. The system provides support for declaration and execution of system provided and user-defined operators on continuous streams of data on high-performance machines and in distributed environments. SciDB is a database management system under development, which is being designed to support very large scientific datasets[23]. SciDB is based on multi-dimensional array storage, rather than traditional relational tables, in order to reduce space and processing costs for scientific data. The MapReduce framework provides a programming model and runtime support for processing and generating large datasets on large cluster systems[24]. In MapReduce, two functions are provided by a user: A map function that processes (key,value) pairs and generates a set of (key,value) pairs; and a reduce function that merges and aggregates all the values with the same key.

While the projects described above address important pieces of the image analysis problem, we wanted a framework that had each of a set of important characteristics. For example, an open-source solution was desirable, as was one based in C/C++ for performance and scalability reasons. Programming models like Map/Reduce are not well-suited for more tightly-coupled analysis tasks that can be expressed using the DataCutter stream-filter model, and we wanted to remain compliant with service-style approaches provided by caGrid. We continue to investigate integration with complementary research such as workflow systems like Pegasus and data repositories like SciDB.

6   Conclusion

Microscopy imaging has been an underutilized tool in a researcher's arsenal of tools for basic and translational biomedical research. While advanced instruments for imaging tissues have been commercially available, the wider adoption of microscopy imaging in research and clinical settings has been hampered by the paucity of software tools for handling very large image datasets, complex analysis workflows, and managing huge volumes of analysis results. Use of parallel and distributed computing and storage environments can alleviate these challenges. It is possible to achieve good performance, but careful coordination and scheduling of I/O, communication, and computation operations in analysis workflows is necessary. Our work has showed that comprehensive systems for microscopy image analysis need to implement high-performance computing techniques throughout the system, including the storage and management of image data, execution of analysis algorithms, and management and exploration of analysis results. These systems should also leverage high performance computing technologies, including cluster and Grid systems, both for access to distributed computational and storage resources and for efficient sharing of data and tools in collaborative research efforts.

Acknowledgments. This research was funded, in part, by grants from the National Institutes of Health (National Institute of Biomedical Imaging and Bioengineering contract 5R01EB003587-03 and National Library of Medicine contract 1R01LM009239-01A1), PHS Grant UL1RR025008 from the CTSA program, by R24HL085343 from the NHLBI, by NCI Contract N01-CO-12400, 79077CBS10, 94995NBS23 and HHSN261200800001E, by NSF CNS-0615155 and CNS-0403342, and P20 EB000591 by the BISTI program. Additional funds were provided by the DoD via grant number W81XWH-06-1-0514 and by the U.S. Dept. of Energy under Contract DE-AC02-06CH11357.

References

1. Rimm, D.L., Camp, R.L., Charette, L.A., Costa, J., Olsen, D.A., Reiss, M.: Tissue microarray: A new technology for amplification of tissue resources. Cancer Journal 7(1), 24–31 (2001)
2. Zhang, X., Pan, T., Catalyurek, U., Kurc, T., Saltz, J.: Serving Queries to Multi-Resolution Datasets on Disk-based Storage Clusters. In: The Proceedings of 4th IEEE/ACM International Symposium on Cluster Computing and the Grid (CCGrid 2004), Chicago, IL (April 2004)
3. Kumar, V.S., Rutt, B., Kurc, T.M., Catalyurek, U.V., Pan, T.C., Chow, S., Lamont, S., Martone, M., Saltz, J.H.: Large-scale biomedical image analysis in grid environments. IEEE Transactions on Information Technology in Biomedicine 12(2), 154–161 (2008)
4. Kumar, V., Kurc, T., Ratnakar, V., Kim, J., Mehta, G., Vahi, K., Nelson, Y., Sadayappan, P., Deelman, E., Gil, Y., Hall, M., Saltz, J.: Parameterized specification, configuration and execution of data-intensive scientific workflows. Cluster Computing (April 2010)
5. Beynon, M., Chang, C., Catalyurek, U., Kurc, T., Sussman, A., Andrade, H., Ferreira, R., Saltz, J.: Processing large-scale multi-dimensional data in parallel and distributed environments. Parallel Comput. 28(5), 827–859 (2002)
6. Kumar, V., Kurc, T., Saltz, J., Abdulla, G., Kohn, S., Matarazzo, C.: Architectural Implications for Spatial Object Association Algorithms. In: The 23rd IEEE International Parallel and Distributed Processing Symposium (IPDPS 2009), Rome, Italy (May 2009)
7. Gray, J., Nieto-Santisteban, M.A., Szalay, A.S.: The zones algorithm for finding points-near-a point or cross-matching spatial datasets. CoRR, abs/cs/0701171 (2007)

8. Kurc, T., Hastings, S., Kumar, S., et al.: HPC and Grid Computing for Integrative Biomedical Research. International Journal of High Performance Computing Applications, Special Issue of the Workshop on Clusters and Computational Grids for Scientific Computing (August 2009)

9. Oster, S., Langella, S., Hastings, S., Ervin, D., Madduri, R., Phillips, J., Kurc, T., Siebenlist, F., Covitz, P., Shanbhag, K., Foster, I., Saltz, J.: caGrid 1.0: An Enterprise Grid Infrastructure for Biomedical Research. Journal of the American Medical Informatics Association (JAMIA) 15, 138–149 (2008)

10. Felten, C.L., Strauss, J.S., Okada, D.H., Marchevsky, A.: Virtual microscopy: high resolution digital photomicrography as a tool for light microscopy simulation. Hum. Pathol. 30(4), 477–483 (1999)

11. Singson, R.P., Natarajan, S., Greenson, J.K., Marchevsky, A.: Virtual microscopy and the Internet as telepathology consultation tools. A study of gastrointestinal biopsy specimens. Am J. Clin. Pathol. 111(6), 792–795 (1999)

12. Ramirez, N.C., Barr, T.J., Billiter, D.M.: Utilizing virtual microscopy for quality control review. Dis Markers 23(5-6), 459–466 (2007)

13. Okada, D.H., Binder, S.W., Felten, C.L., Strauss, J.S., Marchevsky, A.M.: Virtual microscopy and the internet as telepathology consultation tools: Diagnostic accuracy in evaluating melanocytic skin lesions. Am J. Dermatopholology 21(6), 525–531 (1999)

14. Afework, A., Beynon, M., Bustamante, F., et al.: Digital dynamic telepathology - the Virtual Microscope. In: The AMIA Annual Fall Symposium. American Medical Informatics Association (November 1998)

15. Goldberg, I., Allan, C., Burel, J.M., et al.: The open microscopy environment (OME) data model and xml file: Open tools for informatics and quantitative analysis in biological imaging. Genome Biol. 6(R47) (2005)

16. Martone, M., Tran, J., Wong, W., Sargis, J., Fong, L., Larson, S., Lamont, S., Gupta, A., Ellisman, M.: The cell centered database project: An update on building community resources for managing and sharing 3d imaging data. Journal of Structural Biology 161(3), 220–231 (2008)

17. Manolakos, E., Funk, A.: Rapid prototyping of component-based distributed image processing applications using JavaPorts. In: Workshop on Computer-Aided Medical Image Analysis, CenSSIS Research and Industrial Collaboration Conference (2002)

18. Oberhuber, M.: Distributed high-performance image processing on the internet, MS Thesis, Technische Universitat Graz (2002)

19. Ludascher, B., Altintas, I., et al.: Scientific workflow management and the kepler system. Research articles. Concurr. Comput.: Pract. Exper. 18(10), 1039–1065 (2006)

20. Deelman, E., Blythe, J., Gil, Y., Kesselman, C., Mehta, G., Patil, S., Su, M.-H., Vahi, K., Livny, M.: Pegasus: Mapping Scientific Workflows onto the Grid. In: Dikaiakos, M.D. (ed.) AxGrids 2004. LNCS, vol. 3165, pp. 11–20. Springer, Heidelberg (2004)

21. Glatard, T., Montagnat, J., Pennec, X.: Efficient services composition for grid-enabled data-intensive applications. In: Proceedings of the IEEE International Symposium on High Performance Distributed Computing (HPDC 2006), Paris, France, June 19 (2006)

22. Andrade, H., Gedik, B., Wu, K., Yu, P.: Scale-Up Strategies for Processing High-Rate Data Streams in System S. In: The 25th International Conference on Data Engineering (ICDE 2009), Shangai, China, pp. 1375–1378 (2009)

23. Cudre-Mauroux, P., Lim, H., Simakov, J., et al.: A Demonstration of SciDB: A Science-Oriented DBMS. In: 35th International Conference on Very Large Data Bases (VLDB 2009), Lyon, France (2009)

24. Dean, J., Ghemawat, S.: MapReduce: Simplified Data Processing on Large Clusters. In: 6th Symposium on Operating Systems Design and Implementation (OSDI 2004), San Francisco, CA (2004)

Parallel Computation of Bivariate Polynomial Resultants on Graphics Processing Units

Christian Stussak and Peter Schenzel

Martin Luther University Halle-Wittenberg, Institute of Computer Science,
Von-Seckendorff-Platz 1, D – 06120 Halle (Saale), Germany
{stussak,schenzel}@informatik.uni-halle.de

Abstract. Polynomial resultants are of fundamental importance in symbolic computations, especially in the field of quantifier elimination. In this paper we show how to compute the resultant $\operatorname{res}_y(f, g)$ of two bivariate polynomials $f, g \in \mathbb{Z}[x, y]$ on a CUDA-capable graphics processing unit (GPU). We achieve parallelization by mapping the bivariate integer resultant onto a sufficiently large number of univariate resultants over finite fields, which are then lifted back to the original domain. We point out, that the commonly proposed special treatment for so called unlucky homomorphisms is unnecessary and how this simplifies the parallel resultant algorithm. All steps of the algorithm are executed entirely on the GPU. Data transfer is only used for the input polynomials and the resultant. Experimental results show the considerable speedup of our implementation compared to host-based algorithms.

Keywords: polynomial resultants, modular algorithm, parallelization, GPU, CUDA, graphics hardware, symbolic computation.

1   Introduction

The resultant $\operatorname{res}(f, g)$ of two polynomials $f = \sum_{i=0}^{m} a_i x^i$ and $g = \sum_{i=0}^{n} b_i x^i$ is the determinant of their Sylvester matrix:

$$\operatorname{res}(f, g) = \det(\operatorname{Syl}(f, g)).$$

The $(m + n) \times (m + n)$ entries of the Sylvester matrix are determined by the coefficients a_i and b_i, which may be polynomials itself. Therefore, one might use it also in the case of multivariate polynomials, in our case for $f, g \in \mathbb{Z}[x, y]$.

The resultant is tightly coupled with the roots of polynomials. One of its main applications is the elimination of quantifiers in systems of algebraic equations. It is used for instance for solving systems of polynomial equations and in the analysis of implicit algebraic curves and surfaces. Unfortunately, computing resultants is a very time consuming task, especially in the multivariate case. Several algorithms have been proposed to reduce the complexity, respectively the running time in practice. The modular resultant algorithm of Collins [2] suits well for parallelization. The multivariate input polynomials over $\mathbb{Z}$ are mapped

K. Jónasson (Ed.): PARA 2010, Part II, LNCS 7134, pp. 78–87, 2012.

onto a certain number of univariate polynomials over prime fields. Resultants are calculated for the univariate case and then combined into the final resultant over the integers. The algorithm has been adapted for distributed systems [1] and for shared memory machines [5].

In this paper we show how to implement the polynomial resultant algorithm over $\mathbb{Z}[x, y]$ on a CUDA-capable graphics processing unit. We structure our paper according to the divide-conquer-combine strategy of the algorithm: In the first part we explain, how to apply parallel modular reduction and evaluation homomorphisms on the input polynomials. We show, that the notation of unlucky homomorphisms is unnecessary in the context of polynomial resultants, which also simplifies subsequent stages of the algorithm, and provide a fast implementation of GPU-based modular arithmetics. Then the parallel computation of univariate resultants using either global or shared memory of the GPU will be described and we present our approach in combining the intermediate results into a final resultant polynomial in $\mathbb{Z}[x]$. Finally we compare the described GPU parallelization with the sequential approach (see the benchmarks at the end).

2   Divide Phase: Applying Homomorphisms

According to Collins [2] we can use homomorphisms to split the resultant calculation into several modular tasks. A homomorphism $\varphi : R \to R'$ of commutative rings R and R' induces a homomorphism of $R[x]$ into $R'[x]$ defined by $\varphi(f(x)) = \varphi(\sum_i^m a_i x^i) = \sum_i^m \varphi(a_i) x^i$.

Lemma 1 (Collins). *If* $\deg(\varphi(f)) = \deg(f)$ *and* $\deg(\varphi(g)) = \deg(g) - k$, $0 \leq k \leq \deg(g)$, *then* $\varphi(\mathrm{res}(f, g)) = \varphi(\mathrm{lc}(f))^k \mathrm{res}(\varphi(f), \varphi(g))$.

A similar statement holds for $\deg(\varphi(f)) = \deg(f) - k$ and $\deg(\varphi(g)) = \deg(g)$. If both degrees decrease, the homomorphism is often considered to be *unlucky* as in [5], because the commonly used repeated division algorithm for computing the univariate resultant (see section 3) can not be applied. This is a drawback for a parallel algorithm, because unlucky homomorphisms have to be discarded. In the task-parallel algorithm of Hong and Loidl [5], unlucky homomorphisms are reported back to the main task. This can not trivially be done on a GPU. To sort out the bad cases, some kind of stream compaction would be necessary (e.g. by applying prefix sum techniques). We solve this problem using the following fact.

Lemma 2. *If* $\varphi(\mathrm{lc}(f)) = 0$ *and* $\varphi(\mathrm{lc}(g)) = 0$, *then* $\varphi(\mathrm{res}(f, g)) = 0$.

Proof. Because φ is a ring homomorphism, the computation of the determinant of a matrix commutes with the homomorphism. Thus $\varphi(\mathrm{res}(f, g)) = \varphi(\det(\mathrm{Syl}(f, g))) = \det(\varphi(\mathrm{Syl}(f, g)))$. The only nonzero elements in the first column of $\mathrm{Syl}(f, g)$ are $\mathrm{lc}(f)$ and $\mathrm{lc}(g)$, which are mapped onto zero by φ. Therefore, $\det(\varphi(\mathrm{Syl}(f, g))) = 0$. $\square$

According to lemma 2, the previously unlucky homomorphisms are actually lucky ones, because the value of the resultant is immediately known without further calculations.

Fig. 1. Speedup of our 32 bit modular multiplication compared to modular multiplication using 64 bit integer division on the NVIDIA GeForce GTX 260 and 480 GPUs.

2.1 Reduction Modulo a Prime Number

In the first step of the algorithm, we reduce the integer coefficients of f and g modulo pairwise different prime numbers p_i to coefficients in the prime field $\mathbb{Z}_{p_i}$, which completely eliminates the occurrence of expression swell in subsequent calculations. Several prime numbers are needed in order to reconstruct the integer coefficients of the resultant with the help of the Chinese Remainder Theorem (see section 4.2). To bound the number of primes needed, we use a bound on the size of the coefficients of the resultant, which is taken from [7, p. 97]. We refer there for details.

In our implementation the integer coefficients are represented in a number system with radix 2^{32}. Reduction modulo a prime number p is easily implemented by viewing the integer as a polynomial over $\mathbb{Z}_{2^{32}}$ and evaluating this polynomial in $\mathbb{Z}_p$ at $2^{32} \bmod p$ using the Horner scheme. According to lemma 2, all prime numbers are valid for the reduction. Therefore, we can reduce all coefficients of f and g modulo all prime numbers in parallel without having any dependencies.

Arithmetics in Prime Fields. In our current implementation each prime number p satisfies $2^{31} < p < 2^{32}$. Handling addition and subtraction in $\mathbb{Z}_p$ on 32 bit integers is trivial. But in general the result of a 32 bit multiplication $a \cdot b = hi \cdot 2^{32} + lo$ with $0 \leq hi, lo < 2^{32}$ has a size of 64 bit. Decomposing $a \cdot b = q \cdot p + r$, where $r = a \cdot b \bmod p$, by using 64 bit integer division to find q is very slow on current graphics processors. Instead we use a lower bound $\underline{q}$ on q to eliminate the hi part of $a \cdot b$ and reduce the residual 32 bit part modulo p afterwards. The exact value of q would be $\lfloor (hi \cdot 2^{32})/p \rfloor$, but this would also involve another division. The main ingredient of the fast modular multiplication is the precomputation of $2^{32}/p$. Due to the size of p it holds that $1 < 2^{32}/p < 2$, so the 0th binary digit of $2^{32}/p$ is implicitly known. The binary digits $-1, \ldots, -32$ are stored as a 32 bit integer u. Now $1 + u \cdot 2^{-32}$ is a good lower bound for $2^{32}/p$ with a maximum error smaller than 2^{-32}. The product $q' = \lfloor hi \cdot (1 + u \cdot 2^{-32}) \rfloor$ is now equal to $\underline{q}$ or $\underline{q} - 1$, depending on the round-off error. The true value of $\underline{q}$ is obtained by checking, if $hi \cdot 2^{32} - q' \cdot p > 2^{32}$. Because $(hi \cdot 2^{32} - \underline{q} \cdot p) + lo < 3p$, we have to subtract p at most two times to find the final remainder $r = a \cdot b \bmod p$. See Algorithm 2.1 for the pseudocode and Figure 1 for a runtime comparison of our algorithm and the modular multiplication using a 64 bit modulo operation.

Finally, modular inversion, the last operation in $\mathbb{Z}_p$, is implemented using the standard Euclidean algorithm.

Algorithm 2.1. Modular multiplication of 32 bit numbers $a, b \in \mathbb{Z}_p$ using only 32 bit integer addition and multiplication

Data: $a, b \in \mathbb{Z}_p$ with $2^{31} < p < 2^{32}$; $u = \lfloor (2^{32}/p - 1) \cdot 2^{32} \rfloor$; all variables are 32 bit integers

Result: $r = a \cdot b \bmod p$

```
1  begin
2  │   lo ← (a · b)lo                      ▷ least significant 32 bits of a · b
3  │   hi ← (a · b)hi                      ▷ most significant 32 bits of a · b
   │                                                    ▷ (__umulhi on the GPU)
4  │   q ← (hi · u)hi + hi         ▷ lower bound on hi div p with q ≥ hi div p − 1
5  │   lo ← (q · p)lo; hi ← (q · p)hi      ▷ parts of q · p used to eliminate hi
6  │   hi ← hi − hi                        ▷ try to eliminate hi
7  │   if lo ≠ 0 then
8  │   │   hi ← hi − 1          ▷ subtract carry arising from lower part of q · p
9  │   if hi = 1 then lo ← lo + p                     ▷ q was 1 to small
10 │   r ← lo − lo                                    ▷ 32 bit residue
11 │   if r < lo then r ← r − p                       ▷ cope with overflow
12 │   if r ≥ p then r ← r − p                        ▷ reduce modulo p
```

2.2 Evaluating Polynomials

Now let $f, g \in \mathbb{Z}_p[x, y]$. When calculating the resultant with respect to y, the evaluation homomorphism (see [2]) simply maps $\mathbb{Z}_p[x, y]$ onto $\mathbb{Z}_p[y]$ by evaluating $f(x, y)$ and $g(x, y)$ at $\deg_x(res_y(f, g)) + 1$ pairwise different positions $x = x_i \in \mathbb{Z}_p$. It follows from lemma 2, that the choice of the x_i is arbitrary as long as $\deg_x(res_y(f, g)) + 1 < p$. We choose $x_i = i$, which also simplifies the reconstruction phase (see 4.1). The required number of interpolation nodes is bounded by $\deg_x(res_y(f, g)) \leq \deg_y(f) \deg_x(g) + \deg_y(g) \deg_x(f)$ [7, p. 97].

The evaluation of the polynomials is again based on the Horner scheme. Due to the independence of the evaluation nodes and the coefficients of f and g, parallelization is easily achieved.

3 Conquer Phase: Calculating Resultants of Univariate Polynomials

We first state some properties of the resultant taken from [3, p. 408, p. 411], which we will use in our parallel algorithms.

Lemma 3. Let $f = \sum_{i=0}^{m} a_i y^i$ and $g = \sum_{i=0}^{n} b_i y^i$ be univariate polynomials over $\mathbb{Z}_p[y]$ of nonzero degree, $c \in \mathbb{Z}_p$ a nonzero constant and $f = p \cdot g + r$ with $\deg(r) = l$ a decomposition of f by g into quotient p and remainder r, then

$$res(c, g) = c^n, \quad res(f, g) = (-1)^{mn} res(g, f), \quad res(g, f) = b_n^{m-l} res(g, r),$$
$$res(cf, g) = c^n res(f, g), \quad res(y^k f, g) = b_0^k res(f, g), \quad k \geq 0.$$

The first line of equations in lemma 3 allows us to reduce the degrees of the involved polynomials by successive polynomial division until the remainder is constant. The repeated calculation of modular inversion during the polynomial division can be avoided by using polynomial pseudodivision instead. It computes the decomposition $b_n^{m-n+1} f = qg + r$ (see for example [6, p. 425ff]). Now we have

$$\text{res}(g, f) = b_n^{-n(m-n+1)} \, \text{res}(g, b_n^{m-n+1} f) = b_n^{(m-l)-n(m-n+1)} \, \text{res}(g, r). \quad (3.1)$$

If f and g are non constant, then $(m - l) - n(m - n + 1) \leq -l \leq 0$ and we can calculate $b_n^{n(m-n+1)-(m-l)}$. Suppose the product of all these factors of the resultant has been computed, then a single modular inverse yields the final result.

3.1   Parallelization on Global Memory

Due to our homomorphic mapping, the first idea to parallelize is to process all independent homomorphic images in parallel. Because it is currently not possible to store a sufficiently large number of images of even moderate degree within the shared memory of the GPU, the first version of our algorithm operates exclusively on global GPU memory. One thread calculates one resultant. If all the homomorphic images of f and g are stored within a 2D-Array, each polynomial occupying a single column, then coalesced memory access can be assured for the first polynomial division.

After the division the degree of the remainder will almost always be $\deg(g) - 1$, but it might be even lower for a few threads. This forces different threads to perform uncoalesced and therefore slow memory accesses during the next iteration. Although this case is rare, the uncoalesced access patterns are kept until the end of the resultant computation. To solve this issue we apply the last property of lemma 3 from right to left and calculate $\text{res}(g, y^k r)$ with $k = \deg(g) - 1 - \deg(r)$ instead of $\text{res}(g, r)$ in the next iteration. Thus, all threads carry on with a remainder of the same degree. But if the coefficient b_0 of g is zero, then $\text{res}(g, y^k r) = 0$, because of the newly introduced common factor y. In this case we proceed with $\text{res}(g+r, y^k r)$, because $g+r$ will only have a factor x, if also $y \mid g$ and $y \mid r$. We also treat the special case $r = 0$ resulting in $\text{res}(f, g) = 0$.

The number of accesses to global memory can be reduced further. After the initial polynomial division, we have $\deg(f) = \deg(g) + 1$. The next division gives a remainder of degree $\deg(f) - 2$. Thus, we have to calculate $cf - q_1 yg - q_0 g$, with $q_0, q_1 \in \mathbb{Z}_p$, to obtain the remainder. Both subtractions are done within a single loop, by reusing already loaded coefficients of g. Intermediate results are stored locally, thus saving another global memory store and load, improving performance on architectures without a memory cache.

3.2   Parallelization on Shared Memory

We now restrict the size of f and g to fit into the shared memory of the GPU. Without loss of generality we assume $\deg(f) \geq \deg(g)$ and use a thread block

---

Algorithm 3.1. Parallel computation of univariate polynomial resultant on shared memory.

Data: $f, g \in \mathbb{Z}_p[y]$ with $\deg(f) \geq \deg(g)$; t is the thread index
Result: $res = \mathrm{res}(f, g)$

1 **begin**
2     $k \leftarrow 0$ ▷ used to index factors of resultant
3   **if** $t = 0$ **then** $s \leftarrow 1$ ▷ sign of the resultant
4   **while** $\deg(g) > 0$ **do**
5        $(q, r) \leftarrow \mathrm{parallelPseudoDivide}(f, g)$ ▷ decomposition $cf = qg + r$
6   **if** $t = 0$ **then**
7           $m \rightarrow \deg(f); n \rightarrow \deg(g); l \rightarrow \deg(r)$
8           $\beta_k \leftarrow \mathrm{lc}(g)$ ▷ base of factor
9           $e_k \leftarrow n(m - n + 1) - (m - l)$ ▷ exponent of factor
10          $s \leftarrow s \cdot (-1)^{\deg(f)\deg(g)}$ ▷ update sign of resultant
11        $k \leftarrow k + 1$ $(f, g) \leftarrow (g, r)$
12   **if** $t = 0$ **then** $\beta_k \leftarrow \mathrm{lc}(g); e_k \leftarrow \deg f$ ▷ base case with $\deg(g) = 0$
13   **if** $t \leq k$ **then** $\beta_t \leftarrow \beta_t^{e_t}$ ▷ exponentiations of first k factors
14   reduce $\beta_0 = \prod_{i=0}^{k-1} \beta_i$ in parallel ▷ parallel reduction in $\mathcal{O}(\log k)$ steps
15   **if** $t = 0$ **then**
16        $res \leftarrow \beta_0^{-1}\beta_k s$ ▷ single modular inverse; base case factor; sign

---

with deg g threads to calculate a single resultant. Parallelization is achieved in the inner loop of the polynomial pseudodivision. Each thread is attached to a single coefficient during the computation. Additionally, we delay the exponentiation of the b_n factors (see equation (3.1)) until all of them are known. Then the exponentiation of all factors is done in parallel and their product is calculated using parallel reduction. The pseudocode is shown in Algorithm 3.1.

Although the thread utilization is reduced after each iteration, this algorithm is quite fast on GPUs, that do not have a cache for global memory. A runtime comparison of both algorithms is shown in Table 1.

4   Combine Phase: Reconstructing the Integer Resultant

We now reconstruct the final integer resultant from the homomorphic images of the resultant by polynomial interpolation and the Chinese Reminder Theorem.

4.1   Polynomial Interpolation

The interpolation is performed on the points (i, r_i), where r_i is the resultant at $x = i$. We apply the classical Newton interpolation, which represents a polynomial $p(x)$ in the form $p(x) = \sum_{k=0}^{n} v_k N_k(x)$ with Newton basis polynomials $N_k(x)$. The coefficients v_k are computed efficiently by the scheme of divided differences.

Newton Basis Polynomials. Due to lemma 2 we chose the interpolation nodes $0, 1, 2, \ldots, n$. For these nodes the Newton basis polynomials have the structure

$$N_n(x) = \prod_{k=0}^{n-1} (x - k) = \sum_{k=0}^{n} \begin{bmatrix} n \\ k \end{bmatrix} (-1)^{n-k} x^k, \tag{4.1}$$

where $\begin{bmatrix} n \\ k \end{bmatrix}$ are the Stirling numbers of the first kind. They are recursively given by

$$\begin{bmatrix} n \\ k \end{bmatrix} = \begin{bmatrix} n-1 \\ k-1 \end{bmatrix} + (n-1) \begin{bmatrix} n-1 \\ k \end{bmatrix} \tag{4.2}$$

with base cases $\begin{bmatrix} n \\ 0 \end{bmatrix} = \begin{bmatrix} n \\ n \end{bmatrix} = 1$ and $\begin{bmatrix} 0 \\ 0 \end{bmatrix} = \begin{bmatrix} 0 \\ k \end{bmatrix} = 0$ (see [4, p. 243ff]). In our implementation n threads[1] calculate the coefficients of $N_n(x)$ in parallel by applying equation (4.2) on the coefficients of $N_{n-1}(x)$, which are stored in shared memory. It is also possible to precompute the $N_n(x)$, but one has to keep the quadratic space requirements in mind.

Divided Differences. The definition of the divided differences supplied by [3, p. 188] is suitable for our application. With our special interpolation nodes it simplifies to

$$v_k = (((\cdots((r_k - v_0)k^{-1}) - v_1)(k-1)^{-1} - \cdots v_{k-2})2^{-1} - v_{k-1})1^{-1} \tag{4.3}$$

We start by computing the modular inverses $1^{-1}, 2^{-1}, \ldots, n^{-1}$ in parallel by n threads. Then thread k calculates v_k by applying equation (4.3) step by step. The v_k, which is subtracted by each thread in iteration $k + 1$, is available since iteration k and distributed among the threads via shared memory.

Overlapping Computations. A closer look at the previous two algorithms reveals another possibility for optimization. In iteration k we need k threads to compute the Newton basis polynomials and $n - k$ threads to work on the remaining divided differences. Thus, we can overlap both computations in a single kernel, keeping all n threads occupied during the whole interpolation process. Even the sum $p(x) = \sum_{k=0}^{n} v_k N_k(x)$ is computed along with the Newton polynomials. For iteration k of the sum, only v_k and $N_k(x)$ are needed. Thus, v_{k-1} and $N_{k-1}(x)$ are overwritten, as soon as they are no longer needed. Additionally, we store the growing number of coefficients for the $N_k(x)$ and the shrinking number of modular inverses needed by the v_k within the same array. The intermediate coefficients of the interpolation polynomial are kept locally by each thread. This results in a memory efficient algorithm, using only n 32 bit shared memory cells.

4.2   Lifting from Prime Fields to Integers

The last task is to combine the corresponding coefficients of the interpolation polynomials from different prime fields into an integer number. To find a solution

---

[1] The leading coefficient is always equal to one. No extra thread is required here.

for the simultaneous congruences $u \equiv u_k \mod p_k$ of an integer u and its modular images u_k, we follow Garner's algorithm, which is presented well in the book of Knuth [6, p. 284ff]. The integer u is represented in mixed radix form

$$u = v_0 + p_0(v_1 + p_1(v_2 + \cdots + p_{n-1}v_n)\cdots) \qquad (4.4)$$

with mixed radix digits $v_k \in \mathbb{Z}_{p_i}$, that are defined as

$$v_k = (\cdots((u_k - v_0)p_0^{-1} - v_1)p_1^{-1} - \cdots - v_{k-1})p_{k-1}^{-1} \mod p_k. \qquad (4.5)$$

The similarity between equations (4.3) and (4.5) is obvious and we use the same scheme for parallelization here.

Converting from Mixed Radix to Fixed Radix Notation. The conversion process involves alternate multiplication of a prime number and addition of a mixed radix digit, both 32 bit integers, on the intermediate value of u, which is a multi-precision integer. The multiplication is done in parallel. Each thread multiplies its fixed radix digit with the prime number, followed by a phase of carry propagation. The addition of the mixed radix digit is easily integrated as the carry at the least significant digit. Each carry is added on the corresponding digit, possibly generating another carry. We use the warp vote functions available on the GPU to determine within two synchronization steps, if there is still a carry to propagate. The warp vote tells us, if there is a carry within a warp. This information is written to shared memory and accumulated into the global carry flag by another warp vote. Although the carry might ripple from the least significant digit to the most significant one, this rarely occurs in practice. We follow this approach for reasons of simplicity in our first GPU-based implementation.

5   Experimental Results

We provide two types of benchmarks, which are both based on a NVIDIA GeForce GTX 260 and 480 GPUs and a 2.8GHz Intel Xeon Prestonia CPU.

In Table 1 we list the computation times for all parts of the parallel resultant algorithm for 65535 instances of the respective task. The time needed for one instance is to short for accurate measurements. Modular reduction and evaluation of polynomials are very fast operations. Their complexity is linear in the input size and each task is processed by a single thread without any communication. The complexity of other parts of the algorithm is quadratic in the input size. On the GTX 260 the univariate resultant on global memory is relatively slow compared to the parallel variant, that operates on shared memory. Due the on-chip cache for global memory, the GTX 480 behaves very well on the resultant on global memory. Because it allows full thread utilization without any synchronization, it is sometimes even faster than the resultant on shared memory. Polynomial interpolation and the lifting process from prime fields to integer coefficients via the Chinese Remainder Theorem are very similar in its nature. Unfortunately the lifting to integers involves mixed precision arithmetics and

Table 1. Computing times for the different parts of the parallel resultant algorithm for two different NVIDIA GeForce GPUs in milliseconds. The timings reflect the runtime for 65535 instances of the respective task. The input size determines: size of the number to reduce in digits (modular reduction); number of coefficients of the input polynomials (polynomial evaluation, univariate resultant on global memory (gmem) and shared memory (smem), interpolation); prime numbers (Chinese Remainder Theorem (crt)).

		2	4	8	16	32	64	128	256	512
						input size				
GTX 260	modular reduction	0.1	0.1	0.2	0.4	0.8	1.5	3.1	6.2	12.3
	evaluation	0.1	0.1	0.2	0.4	0.8	1.6	3.2	6.3	12.5
	resultant gmem	0.6	4.8	20.3	74.9	280.9	1127.6	4409.5	17423.2	94637.4
	resultant smem	8.8	32.7	43.0	62.9	121.5	280.6	868.8	3251.0	13230.8
	interpolation	2.5	5.7	10.8	21.4	52.4	150.0	497.1	1790.6	7040.5
	crt	13.1	35.6	83.2	183.0	530.2	1550.1	4258.2	11790.4	39005.0
GTX 480	modular reduction	0.1	0.1	0.1	0.1	0.1	0.3	0.5	1.1	2.2
	evaluation	0.1	0.1	0.1	0.1	0.2	0.4	0.9	1.8	3.6
	resultant gmem	0.1	0.3	1.2	3.6	12.0	46.1	202.1	761.5	5194.8
	resultant smem	4.1	13.3	17.7	26.1	42.9	84.8	249.3	870.0	3254.6
	interpolation	1.4	2.5	4.5	8.4	16.1	35.0	110.5	398.2	1576.6
	crt	7.0	17.8	40.5	89.1	197.0	542.9	1400.9	3531.9	9062.6

Table 2. Benchmarks for resultant computations on random polynomials of various degrees. We compare the timings in seconds of our CUDA-based resultant algorithm to the time MATHEMATICA 6 takes for the sequential modular resultant. The last column graphically shows the percentage of the computation time consumed by each task (□ modular reduction, ▨ evaluation, ▤ resultant smem, ■ interpolation, ▪ crt). In the first test group, the degree of the resultant and the size of the coefficients in the resultant are held approximately constant. The second group shows several examples, where the size of the coefficients in the input is smaller than 2^{32}. For the third test group the size of the coefficients in the resultant is approximately constant again.

$\deg(res)$	#moduli	#limbs	$\deg_y(f)$	$\deg_x(f)$	$\deg_y(g)$	$\deg_x(g)$	MATHE-MATICA	GTX 260	speedup	GTX 480	speedup	time distri-bution
510	509	10	26	10	25	10	232.92	0.881	264×	0.258	904×	
510	511	6	43	6	42	6	259.47	1.107	234×	0.352	737×	
510	502	5	52	5	50	5	278.19	1.347	207×	0.388	717×	
510	511	3	85	3	85	3	371.36	2.250	165×	0.588	632×	
510	507	4	128	2	127	2	467.83	4.195	112×	1.134	413×	
200	46	1	20	5	20	5	3.95	0.038	104×	0.005	725×	
300	69	1	30	5	30	5	10.38	0.099	105×	0.017	613×	
400	93	1	40	5	40	5	26.98	0.147	184×	0.041	657×	
200	511	13	20	5	20	5	27.95	0.389	72×	0.790	355×	
300	510	9	30	5	30	5	79.02	0.685	115×	0.145	547×	
400	510	7	40	5	40	5	148.83	0.757	197×	0.253	589×	

the computation of many modular inverses. The former also leads to a high communication overhead within a thread block, which is reflected in the benchmark results.

In Table 2 we list computing times for several resultants of polynomials of various degrees and coefficient sizes. Our results are compared against the sequential modular resultant algorithm implemented in Mathematica 6. We obtain substantial speedups with our approach. From the last column we see, how much time is spent on which part of the parallel modular resultant algorithm. The application of the homomorphisms is almost negligible. Due to the large number of homomorphic images, most time is spent on the univariate resultant. The next part, the Newton interpolation, is quite fast. It also benefits from our choice of interpolation nodes. The final phase, the lifting to integer coefficients, currently involves mixed precision arithmetics and lots of communication, preventing the algorithm from being as efficients as the interpolation.

6 Conclusion

We have presented a complete implementation of a parallel bivariate polynomial resultant over the integers on a CUDA-capable graphics processor. All stages of the algorithm exhibit a large amount of parallelism and we achieve high speedups on the GPU with this approach. The parts of the parallel implementation can also be used separately, e.g. to compute resultants over $\mathbb{Z}_p[x, y]$. Further investigation will be done to expand our results on other symbolic computations.

References

1. Bubeck, T., Hiller, M., Küchlin, W., Rosenstiel, W.: Distributed Symbolic Computation with DTS. In: Ferreira, A., Rolim, J.D.P. (eds.) IRREGULAR 1995. LNCS, vol. 980, pp. 231–248. Springer, Heidelberg (1995)
2. Collins, G.E.: The calculation of multivariate polynomial resultants. In: SYMSAC 1971: Proceedings of the Second ACM Symposium on Symbolic and Algebraic Manipulation, pp. 212–222. ACM, New York (1971)
3. Czapor, S., Geddes, K., Labahn, G.: Algorithms for Computer Algebra. Kluwer Academic Publishers (1992)
4. Graham, R.L., Knuth, D.E., Patashnik, O.: Concrete Mathematics: A Foundation for Computer Science. Addison-Wesley Longman Publishing Co., Inc., Boston (1990)
5. Hong, H., Loidl, H.W.: Parallel Computation of Modular Multivariate Polynomial Resultants on a Shared Memory Machine. In: Buchberger, B., Volkert, J. (eds.) CONPAR 1994 and VAPP 1994. LNCS, vol. 854, pp. 325–336. Springer, Heidelberg (1994), http://www.springerlink.com/content/fp43323vv51104q6/
6. Knuth, D.E.: The Art of Computer Programming: Seminumerical Algorithms, 3rd edn., vol. 2. Addison-Wesley Professional (1997)
7. Winkler, F.: Polynomial Algorithms in Computer Algebra. Springer-Verlag New York, Inc., Secaucus (1996)

Accelerating Model Reduction of Large Linear Systems with Graphics Processors

Peter Benner[1], Pablo Ezzatti[2], Daniel Kressner[3], Enrique S. Quintana-Ortí[4], and Alfredo Remón[4]

[1] Max-Planck-Institute for Dynamics of Complex
Technical Systems, Magdeburg, Germany
benner@mpi-magdeburg.mpg.de
[2] Centro de Cálculo-Instituto de la Computación,
Universidad de la República, Montevideo, Uruguay
pezzatti@fing.edu.uy
[3] Seminar für Angewandte Mathematik, ETHZ, Zürich, Switzerland
daniel.kressner@sam.math.ethz.ch
[4] Depto. de Ingeniería y Ciencia de Computadores,
Universidad Jaume I, Castellón, Spain
{quintana,remon}@icc.uji.es

Abstract. Model order reduction of a dynamical linear time-invariant system appears in many applications from science and engineering. Numerically reliable SVD-based methods for this task require in general $\mathcal{O}(n^3)$ floating-point arithmetic operations, with n being in the range $10^3 - 10^5$ for many practical applications. In this paper we investigate the use of graphics processors (GPUs) to accelerate model reduction of large-scale linear systems by off-loading the computationally intensive tasks to this device. Experiments on a hybrid platform consisting of state-of-the-art general-purpose multi-core processors and a GPU illustrate the potential of this approach.

Keywords: model reduction, dynamical linear systems, Lyapunov equations, SVD-based methods, GPUs.

1 Introduction

Model order reduction is an important numerical tool to diminish the simulation time or the cost of designing optimal controllers in many industrial processes, with dynamics modeled by a linear time-invariant (LTI) system:

$$
\begin{aligned}
E\dot{x}(t) &= Ax(t) + Bu(t), \quad t > 0, \quad x(0) = x^0, \\
y(t) &= Cx(t) + Du(t), \quad t \geq 0.
\end{aligned}
\tag{1}
$$

Here, $x(t)$ contains the states of the system, with initial state $x^0 \in \mathbb{R}^n$, $u(t)$ and $y(t)$ contain the inputs and outputs, respectively, and $E, A \in \mathbb{R}^{n \times n}$, $B \in \mathbb{R}^{n \times m}$, $C \in \mathbb{R}^{p \times n}$, $D \in \mathbb{R}^{p \times m}$. The system in (1) can also be described by the associated

K. Jónasson (Ed.): PARA 2010, Part II, LNCS 7134, pp. 88–97, 2012.

transfer function matrix (TFM) $G(s) = C(sE - A)^{-1}B + D$. A particularly important property is that the number of states (also known as the state-space dimension or the order) of the system, n, is in general much larger than the number of inputs and outputs, m and p, respectively.

The goal of model reduction is to find a reduced-order LTI system,

$$\hat{E}\dot{\hat{x}}(t) = \hat{A}\hat{x}(t) + \hat{B}u(t), \quad t > 0, \quad \hat{x}(0) = \hat{x}^0,$$
$$\hat{y}(t) = \hat{C}\hat{x}(t) + \hat{D}u(t), \quad t \geq 0, \tag{2}$$

of order r, with $r \ll n$, and associated TFM $\hat{G}(s) = \hat{C}(s\hat{E} - \hat{A})^{-1}\hat{B} + \hat{D}$ which approximates the dynamics of the original system defined by $G(s)$. The reduced-order realization (2) can then replace the original high-order system in a simulation or the design of an optimal controller, thus simplifying such tasks considerably. Model reduction of large-scale systems appears, e.g., in thermal, thermo-mechanical, electro-mechanical and acoustic finite element models [1]. We consider a system to be large-scale if $n \sim \mathcal{O}(1,000) - \mathcal{O}(100,000)$; while, often, $m, p \sim \mathcal{O}(10) - \mathcal{O}(100)$.

The numerical method for model reduction considered in this paper is based on the so-called state-space truncation approach and requires, at an initial stage, the solution of two coupled generalized Lyapunov equations. The reduced-order system is then obtained using a variant of the balanced truncation (BT) method [2,3], which only requires a few dense linear algebra computations. Although there exist several other approaches for model reduction (see, e.g., [1,4] and the references therein), those are specialized for certain problem classes and often lack properties such as error bounds or preservation of stability, passivity, or phase information. A comparison of the numerical properties of SVD-based methods (as BT) and Krylov subspace methods can be found in [1,5,6,7].

The Lyapunov equations are solved in our method via the matrix sign function, which yields a computational cost for the global model reduction procedure of $\mathcal{O}(n^3)$ flops (floating-point arithmetic operations). This calls for the application of high performance computing in the reduction of models already with n in the order of thousands.

Recent work on the implementation of BLAS and the major factorization routines for the solution of linear systems [8,9,10,11] has demonstrated the potential of graphics processors (GPUs) to yield high performance on dense linear algebra operations which can be cast in terms of matrix-matrix products. In [12] we built upon these works to deal with the solution of the standard Lyapunov equation on a GPU. Here, we further extend this work by tackling the different stages in SVD-based methods for model reduction of generalized linear systems, namely, the solution of the coupled generalized Lyapunov equations, the computation of the SVD, and auxiliary computations. The target architecture is a hybrid platform consisting of a general-purpose multicore processor and a GPU. We exploit these two resources by designing a hybrid numerical algorithm for model reduction that performs fine-grained computations on the CPU while off-loading computationally intensive operations to the GPU. We also overlap computations in both architectures in order to improve the performance.

The rest of the paper is structured as follows. In Section 2 we briefly review the BT method for model reduction, including the sign function-based Lyapunov solver, and the remaining stages of the method. There we also describe the approach to computing all these operations on the hybrid platform. In Section 3 we present experimental results that illustrate the accuracy and parallelism attained by the numerical algorithms on a platform consisting of two Intel QuadCore processors connected to an NVIDIA Tesla C1060 GPU via a PCI-e bus. Finally, in Section 4 we provide a few concluding remarks.

2   SVD-Based Methods for Model Reduction

BT model reduction [13,14,15,16] belongs to the family of absolute error methods, which aim at minimizing

$$\|G - \hat{G}\|_\infty = \sup_{\omega \in \mathbb{R}} \sigma_{\max}(G(j\omega) - \hat{G}(j\omega)),$$

where $j := \sqrt{-1}$ and $\sigma_{\max}(M)$ stands for the largest singular value of a matrix M.

Model reduction via BT methods employs information about the controllability Gramian W_c and the observability Gramian W_o of the system (1), given by the solutions of the coupled generalized Lyapunov matrix equations

$$AW_cE^T + EW_cA^T + BB^T = 0, \tag{3}$$
$$A^T\tilde{W}_oE + E^T\tilde{W}_oA + C^TC = 0, \tag{4}$$

with $W_o = E^T\tilde{W}_oE$. In most model reduction applications, the matrix pair (A, E) is stable (i.e., all its generalized eigenvalues are in the open left complex plane), so that W_c and W_o are both positive semidefinite. Therefore, the solutions of the Gramians can be factored as $W_c = S^TS$ and $W_o = R^TR$. (Here, S and R are usually refereed to as the *Cholesky factors* of W_c and W_o, though they are not necessarily Cholesky factors in a strict sense.)

Consider now the singular value decomposition (SVD) of the product

$$SR^T = U\Sigma V^T = [U_1 \, U_2] \begin{bmatrix} \Sigma_1 & \\ & \Sigma_2 \end{bmatrix} [V_1 \, V_2]^T, \tag{5}$$

where U and V are orthogonal matrices, and $\Sigma = \mathrm{diag}\,(\sigma_1, \sigma_2, \ldots, \sigma_n)$ is a diagonal matrix containing the singular values of SR^T, also known as the *Hankel singular values* (HSV) of the system. Given a partitioning of Σ into $\Sigma_1 \in \mathbb{R}^{r \times r}$ and $\Sigma_2 \in \mathbb{R}^{(n-r) \times (n-r)}$, and a conformal partitioning of U and V in (5), the *square-root* (SR) version of BT determines a reduced-order model of order r as

$$\hat{E} = T_lET_r, \; \hat{A} = T_lAT_r,$$
$$\hat{B} = T_lB, \quad \hat{C} = CT_r, \quad \hat{D} = D, \tag{6}$$

with

$$T_l = \Sigma_1^{-1/2}V_1^TRE^{-1} \quad \text{and} \quad T_r = S^TU_1\Sigma_1^{-1/2}. \tag{7}$$

The state-space dimension r of the reduced-order model can be chosen adaptively as this method provides a realization $\hat{G}$ satisfying

$$\|G - \hat{G}\|_\infty \leq 2 \sum_{j=r+1}^{n} \sigma_j.$$

In the following subsection we revisit the sign function-based generalized Lyapunov solver introduced in [17]. The solver yields low-rank approximations to the Cholesky or full-rank factors of the solution matrices which can reliably substitute S and R in the computations in (5) and (7).

2.1   The Sign Function Method

The matrix sign function was introduced in [18] as an efficient tool to solve stable (standard) Lyapunov equations. The following variant of the Newton iteration for the matrix sign function [17] can be used for the solution of the generalized Lyapunov equations (3)-(4):

Algorithm CGCLNC:

$A_0 \leftarrow A$, $\tilde{S}_0 \leftarrow B^T$, $\tilde{R}_0 \leftarrow C$
$k \leftarrow 0$
repeat
$\qquad A_{k+1} \leftarrow \frac{1}{\sqrt{2}} \left(A_k/c_k + c_k(EA_k^{-1})E \right)$
$\qquad$ Compute the rank-revealing QR (RRQR) decomposition
$\qquad\qquad \frac{1}{\sqrt{2c_k}} \left[\tilde{S}_k, \ c_k \tilde{S}_k(EA_k^{-1})^T \right] = Q_s \begin{bmatrix} U_s \\ 0 \end{bmatrix} \Pi_s$
$\qquad \tilde{S}_{k+1} \leftarrow U_s \Pi_s$
$\qquad$ Compute the RRQR decomposition
$\qquad\qquad \frac{1}{\sqrt{2c_k}} \left[\tilde{R}_k, \ c_k(\tilde{R}_k A_k^{-1})E \right] = Q_r \begin{bmatrix} U_r \\ 0 \end{bmatrix} \Pi_r$
$\qquad \tilde{R}_{k+1} \leftarrow U_r \Pi_r$
$\qquad k \leftarrow k + 1$
until $\|A_k - E\|_1 < \tau \|A_k\|_1$

On convergence, after j iterations, $\tilde{S} = \frac{1}{\sqrt{2}} \tilde{S}_j E^{-T}$ and $\tilde{R} = \frac{1}{\sqrt{2}} \tilde{R}_j E^{-1}$ of dimensions $\tilde{k}_o \times n$ and $\tilde{k}_c \times n$ are, respectively, full (row-)rank approximations of S and R, so that $W_c = S^T S \approx \tilde{S}^T \tilde{S}$ and $W_o = R^T R \approx \tilde{R}^T \tilde{R}$.

The Newton iteration for the sign function usually presents a fast convergence rate, which is ultimately quadratic. Initial convergence can be accelerated using several techniques. In our case, we employ a scaling defined by the parameter

$$c_k = \sqrt{\|A\|_\infty / \|EA_k^{-1}E\|_\infty}.$$

In the convergence test, τ is a tolerance threshold for the iteration that is usually set as a function of the problem dimension and the machine precision ε. In particular, to avoid stagnation in the iteration, we set $\tau = n \cdot \sqrt{\varepsilon}$ and perform one or two additional iteration steps after the stopping criterion is satisfied. Due to the quadratic convergence of the Newton iteration, this is usually enough to reach the attainable accuracy. The RRQR decomposition can be obtained by means of the traditional QR factorization with column pivoting [19] plus a reliable rank estimator.

Each iteration of algorithm CGCLNC requires the following operations: the LU decomposition of A_k ($\frac{2}{3}n^3$ flops), followed by the system solve EA_k^{-1} and the matrix product $(EA_k^{-1})E$ ($2n^3 + 2n^3$ flops); let s_k and r_k be the number of columns of $\hat{S}_k$ and $\hat{R}_k$; then an $n \times n \times s_k$ matrix product is required to construct $\tilde{S}_k(EA_k^{-1})^T$ ($2n^2 s_k$ flops), a system solve with r_k right-hand sides to obtain $\tilde{R}_k A_k^{-1}$ ($2n^2 r_k$ flops), and an $n \times n \times r_k$ matrix product to build $(\tilde{R}_k A_k^{-1})E$ ($2n^2 r_k$ flops); finally, two QR factorizations with column pivoting complete the major computations in the algorithm ($2n(s_k^2 + r_k^2) - \frac{2}{3}(s_k^3 + r_k^3)$ flops). (The latter flop count assumes that S_k and R_k are full rank, so the actual cost is smaller than this.) Other minor operations, as norms, scalings, etc., contribute with negligible computational costs.

2.2   Hybrid Implementation of the Lyapunov Solver

The objective of the hybrid implementation is to reduce the computational time executing each operation on the most convenient architecture while, whenever possible, overlapping the execution of operations in both architectures. On the other hand, a careful scheduling of operations is necessary to minimize the communication overhead, amortizing the cost of transferring the data between the memory spaces of the GPU and the CPU.

The hybrid algorithm proceeds as follows. At the beginning of each iteration, the CPU transfers matrix A_k to the GPU. Then, the CPU and the GPU cooperate in the LU factorization of matrix A_k. The solution of the EA_k^{-1} system is also obtained on the GPU while the CPU solves the $\tilde{R}_k A_k^{-1}$ system. Then, the computation of the matrix product $(EA_k^{-1})E$ proceeds on the GPU while the CPU computes $\tilde{S}_{k+1}$ and $\tilde{R}_{k+1}$, (in particular, this will require the computation of the two RRQR decompositions and four matrix-matrix products involving relatively small matrices). Finally, the matrix and scalings to construct A_{k+1} as in algorithm CGCLNC are computed on the CPU.

Some other necessary secondary operations are performed on the CPU since they require a minor computational effort.

The use of both architectures requires some data transfers. To control and minimize the communication overhead, data transfers are only scheduled if there is an important gain associated with them. Specifically, the data transfers needed at each iteration are:

1. Send A_k from the CPU to the GPU to compute its LU decomposition.
2. Send the factors resulting from the LU decomposition of A_k from the GPU to the CPU.

3. Send the solution of EA_k^{-1} from the GPU to the CPU so that $\tilde{S}_{k+1}$ can be computed there.
4. Send the result of $(EA_k^{-1})E$, required to compute A_{k+1}, from the GPU to the CPU.

Besides these data transfers, there are some minor communications being performed in the algorithm, in particular, by the LU decomposition kernel.

In summary, the most remarkable strengths of this implementation are:

- The use of a hybrid kernel for the LU decomposition. In this kernel the GPU and the CPU cooperate for computing the decomposition [10].
- The new code generated for the solution of triangular systems on the GPU. A great effort has been conducted to speed-up the execution of this operation; several GPU-based variants were implemented, employing techniques like padding. The best variant obtained, employed in this work, is a blocked routine that casts most of the arithmetic operations in terms of matrix-matrix products. As a result, this new version outperforms notoriously the CUBLAS implementation (it is approximately a 30% and 70% faster for the examples STEEL_I and FLOW_METER considered in Section 3, respectively) and yields a significant acceleration of one of the most time-consuming stages in the model reduction procedure.
- The use of two levels of parallelism. At the inner level, operations are performed using multi-threaded implementations of BLAS. At the outer level, different operations are executed concurrently in the two available resources: CPU and GPU.
- The reduced overhead introduced by communications: only transfers that are amortized over a large number of flops are performed, so that it will be unlikely that communication produces a loss of efficiency. Note that whether data transfers are or not amortized depends on the problem dimension, which in our case, ranges in $10^3 - 10^5$.

2.3 Remaining Stages in Model Reduction BT

Once the Cholesky factors $\tilde{S}$ and $\tilde{R}$ have been computed, the remaining operations to obtain the reduced order model comprise a matrix product of moderate dimension ($\tilde{S}^T\tilde{R} \approx SR^T$); the SVD of the result, see (5); and a few more matrix-matrix operations and a system solve, see (6)–(7). All these computations require a reduced number of flops and, therefore, are performed on the CPU.

3   Numerical Experiments

In this section we evaluate the numerical accuracy and parallel performance of the BT model reduction method. The target platform consists of two INTEL Xeon QuadCore E5410 processors at 2.33GHz, connected to an NVIDIA Tesla C1060 via a PCI-e bus. We employed the multi-threaded implementation of BLAS in

MKL (version 10.2) for the general-purpose processor and NVIDIA CUBLAS (version 2.1) for the GPU. We set `OMP_NUM_THREADS=8` so that one thread is employed per core in the parallel execution of the MKL routines in the Intel Xeon QuadCore processors.

In the following experiments, we evaluate the performance using single precision arithmetic on two model reduction problems from the Oberwolfach benchmark collection at the University of Freiburg[1]:

- STEEL_I: This model arises in a manufacturing method for steel profiles. The goal is to design a control that yields moderate temperature gradients when the rail is cooled down. The mathematical model corresponds to the boundary control for a 2-D heat equation. A finite element discretization, followed by adaptive refinement of the mesh results in the example in this benchmark. The dimensions of this problem are $n = 5,177$, $m = 7$, $p = 6$.
- FLOW_METER: This 2-D model of an anemometer-like structure mainly consists of a tube and a small heat source. The model is given by a spatially semi-discretized instationary convection-diffusion equation with Dirichlet boundary conditions and a parabolic inflow profile. The reference temperature is set to 300 K, and Dirichlet boundary conditions as well as initial conditions are set to 0 with respect to the reference. The dimensions of this problem are $n = 9,669$, $m = 1$, $p = 5$.

Table 1 shows the results obtained with our hybrid CPU-GPU algorithm for the solution of the coupled generalized Lyapunov equations associated with these systems. Columns 2, 3, 4 and 5 of the table show the time (in seconds) for the LU factorization of A_k, the solution of the four triangular systems in the computations EA_k^{-1} and $\tilde{R}_k A_k^{-1}$, the matrix product $(EA_k^{-1})E$, and the updates of the factors $\tilde{S}$ and $\tilde{R}$, respectively (including the time for all the data transfers associated to each one of the operations). The rest of columns show the global time per iteration of the hybrid implementation, the time per iteration for the same algorithm implemented on the multicore CPU, and the convergence criterion.

Most of the iteration time is spent in the computation of the LU decomposition (column 2), the solution of the four triangular systems (column 3) and the matrix-matrix product (column 4). Those are the operations which, in part or completely, are performed on the GPU.

The number of columns of the factors $\tilde{S}$ and $\tilde{R}$ is doubled at each iteration and, in consequence, the cost associated to the update of the factors increases with the iteration count. To keep the number of columns in the factors under control, an RRQR factorization is computed at each step [19]. This approach yields important gains when the number of iterations that are required for convergence is large enough to increment notoriously the size of the factors, as is the case for the two problems considered in this section. The increment in the number of columns of $\tilde{S}$ and $\tilde{R}$ results in an increment of the time required for their update (column 5). This time becomes relevant after some iterations, as this is

---

[1] `http://www.imtek.de/simulation/benchmark/`

Table 1. Performance of the hybrid CPU+GPU implementation of the Newton iteration for the solution of the Lyapunov equation with factored right-hand side

#Iter k	Time $PA_k = LU$	Time EA_k^{-1}, $\tilde{R}_k A_k^{-1}$	Time $(EA_k^{-1})E$	Time $\tilde{S}_k(EA_k^{-1})$, $\tilde{R}_k(A_k^{-1}E)$, compress	Time iteration (Hybrid)	Time iteration (CPU)	Conv. criterion $\frac{\|A_k+E\|_F}{\|E\|_F}$
			STEEL_I				
1	0.698	1.041	0.807	0.121	2.958	5.337	2.732e+02
2	0.544	1.023	0.788	0.047	2.618	5.286	2.064e+01
3	0.544	1.023	0.788	0.079	2.650	5.354	3.698e+00
4	0.544	1.023	0.788	0.159	2.732	5.465	1.140e+00
5	0.543	1.023	0.789	0.381	2.955	5.638	3.644e−01
6	0.545	1.023	0.788	0.909	3.486	6.219	7.936e−02
7	0.546	1.022	0.789	1.366	3.946	6.553	8.546e−03
8	0.543	1.023	0.788	1.866	4.442	6.909	5.706e−04
9	0.544	1.184	0.788	2.093	4.670	7.105	1.257e−05
10	0.546	1.209	0.788	2.185	4.767	7.250	7.319e−07
				ACCUMULATED TIME	35.224	61.156	
			FLOW_METER				
1	3.380	7.741	5.183	0.289	17.359	31.516	6.884e+01
2	2.906	7.673	5.116	0.109	16.512	31.580	6.758e+00
3	2.918	7.673	5.116	0.137	16.553	31.725	1.585e+00
4	2.888	7.673	5.116	0.202	16.592	31.970	5.010e−01
5	3.007	7.673	5.115	0.359	16.871	32.126	1.580e−01
6	2.893	7.674	5.116	0.702	17.099	32.329	5.044e−02
7	2.886	7.673	5.116	0.971	17.365	32.525	1.241e−02
8	2.890	7.674	5.116	1.066	17.462	32.842	1.702e−03
9	2.893	7.673	5.117	1.191	17.591	32.896	1.156e−04
10	2.891	7.673	5.115	1.236	16.634	32.997	1.396e−06
11	2.891	7.673	5.116	1.248	17.994	32.881	2.389e−07
				ACCUMULATED TIME	188.032	355.387	

mostly a BLAS-2 based computation performed on the CPU, e.g., being nearly half of the total iteration time for the STEEL_I problem after 9 iterative steps. Executing these operations on the GPU, though possible, would require some extra CPU-GPU communications and would slow down the execution of the initial iterations.

Compared with the execution of the same algorithm on a CPU, the use of the GPU yields an important reduction of the execution times on the most computationally expensive operations which carries over to the global execution time per iteration (the LU factorization, the solution of triangular systems and the matrix-matrix product). Furthermore, while some computations are off-loaded to the GPU, others are performed concurrently on the CPU. This second level of parallelism further reduces the total execution time.

4 Concluding Remarks

We have presented a new parallel algorithm for model reduction of large linear systems on a hybrid CPU-GPU platform. Our algorithm exploits the capabilities of both architectures, the multi-core CPU and the many-core GPU, obtaining a high performance implementation of a BT model reduction technique. We use two levels of parallelism: at the inner level, multi-thread implementations of the BLAS library (MKL and CUBLAS) compute the most time-consuming linear algebra kernels. At the outer level, operations proceed concurrently in both architectures.

Results show that model reduction of large-scale linear systems can be tackled with this kind of platforms in a reasonable computational time.

Future research resulting from this experience will include:

- Use of multiple GPUs to further reduce the computational time and increase the dimension of the affordable problems. The computation of the matrix-matrix product in $(EA_k^{-1})E$, due to its strong scalability, can be accelerated using a multi-GPU implementation. Also the computation time for the LU factorization of A_k can be reduced (see the results reported in [20]) as well as the solution of the triangular system performed on the GPU (EA_k^{-1}), since most of the operations are cast in terms of matrix-matrix products.
- Use of double precision arithmetic. Performance of current GPUs in double precision is considerably lower than single precision, but the new generation of GPUs will drastically reduce this difference. As an alternative, we will investigate the use of iterative refinement which given a single precision solution, obtains the double precision solution at a reduced cost.

References

1. Antoulas, A.: Approximation of Large-Scale Dynamical Systems. SIAM Publications, Philadelphia (2005)
2. Benner, P., Quintana-Ortí, E., Quintana-Ortí, G.: State-space truncation methods for parallel model reduction of large-scale systems. Parallel Comput. 29, 1701–1722 (2003)
3. Penzl, T.: Algorithms for model reduction of large dynamical systems. Linear Algebra and its Applications 415(2-3), 322–343 (2006)
4. Freund, R.: Reduced-order modeling techniques based on Krylov subspaces and their use in circuit simulation. In: Datta, B. (ed.) Applied and Computational Control, Signals, and Circuits, vol. 1, ch. 9, pp. 435–498. Birkhäuser, Boston (1999)
5. Benner, P.: Numerical linear algebra for model reduction in control and simulation. GAMM-Mitteilungen 29(2), 275–296 (2006)
6. Benner, P., Mehrmann, V., Sorensen, D. (eds.): Dimension Reduction of Large-Scale Systems. Lecture Notes in Computational Science and Engineering, vol. 45. Springer, Heidelberg (1976)
7. Schilders, W., van der Vorst, H., Rommes, J. (eds.): Model Order Reduction: Theory, Research Aspects and Applications. Mathematics in Industry, vol. 13. Springer, Heidelberg (2008)

8. Volkov, V., Demmel, J.: LU, QR and Cholesky factorizations using vector capabilities of GPUs. EECS Department, University of California, Berkeley, Tech. Rep. UCB/EECS-2008-49 (May 2008),
   http://www.eecs.berkeley.edu/Pubs/TechRpts/2008/EECS-2008-49.html
9. Bientinesi, P., Igual, F.D., Kressner, D., Quintana-Ortí, E.S.: Reduction to Condensed Forms for Symmetric Eigenvalue Problems on Multi-Core Architectures. In: Wyrzykowski, R., Dongarra, J., Karczewski, K., Wasniewski, J. (eds.) PPAM 2009. LNCS, vol. 6067, pp. 387–395. Springer, Heidelberg (2010)
10. Barrachina, S., Castillo, M., Igual, F.D., Mayo, R., Quintana-Ortí, E.S., Quintana-Ortí, G.: Exploiting the capabilities of modern GPUs for dense matrix computations. Concurrency and Computation: Practice and Experience 21, 2457–2477 (2009)
11. Ltaif, H., Tomov, S., Nath, R., Du, P., Dongarra, J.: A scalable high performance cholesky factorization for multicore with GPU accelerators. University of Tennessee, LAPACK Working Note 223 (2009)
12. Benner, P., Ezzatti, P., Quintana-Ortí, E.S., Remón, A.: Using Hybrid CPU-GPU Platforms to Accelerate the Computation of the Matrix Sign Function. In: Lin, H.-X., Alexander, M., Forsell, M., Knüpfer, A., Prodan, R., Sousa, L., Streit, A. (eds.) Euro-Par 2009. LNCS, vol. 6043, pp. 132–139. Springer, Heidelberg (2010)
13. Moore, B.: Principal component analysis in linear systems: Controllability, observability, and model reduction. IEEE Trans. Automat. Control AC-26, 17–32 (1981)
14. Safonov, M., Chiang, R.: A Schur method for balanced-truncation model reduction. IEEE Trans. Automat. Control AC-34, 729–733 (1989)
15. Tombs, M., Postlethwaite, I.: Truncated balanced realization of a stable non-minimal state-space system. Internat. J. Control 46(4), 1319–1330 (1987)
16. Varga, A.: Efficient minimal realization procedure based on balancing. In: Prepr. of the IMACS Symp. on Modelling and Control of Technological Systems, vol. 2, pp. 42–47 (1991)
17. Benner, P., Quintana-Ortí, E., Quintana-Ortí, G.: Solving linear-quadratic optimal control problems on parallel computers. Optimization Methods Software 23(6), 879–909 (2008)
18. Roberts, J.: Linear model reduction and solution of the algebraic Riccati equation by use of the sign function. Internat. J. Control 32, 677–687 (1980) (Reprint of Technical Report No. TR-13, CUED/B-Control, Cambridge University, Engineering Department, 1971)
19. Golub, G., Van Loan, C.: Matrix Computations, 3rd edn. Johns Hopkins University Press, Baltimore (1996)
20. Volkov, V., Demmel, J.W.: LU, QR and Cholesky factorizations using vector capabilities of GPUs. University of California at Berkeley, LAPACK Working Note 202 (May 2008)

Fast GPU-Based Fluid Simulations Using SPH

Øystein E. Krog and Anne C. Elster

Dept. of Computer and Information Science
Norwegian University of Science and Technology (NTNU)
oystein.krog@gmail.com, elster@idi.ntnu.no

Abstract. Graphical Processing Units (GPUs) are massive floating-point stream processors, and through the recent development of tools such as CUDA and OpenCL it has become possible to fully utilize them for scientific computing. We have developed an open-source CUDA-based acceleration framework for 3D Computational Fluid Dynamics (CFD) using Smoothed Particle Hydrodynamics (SPH). This paper describes the methods used in our framework and compares the performance of the implementation to previous SPH implementations. We implement two different SPH models, a simplified model for Newtonian fluids, and a complex model for Non-Newtonian fluids, which we use for simulation of snow avalanches. Having implemented two different models, we investigate the performance characteristics of SPH simulations on the GPU and find that despite the larger bandwidth-requirements of the complex model the GPU scales well. Our simulations are rendered interactively and in real-time. Using an NVIDIA GeForce GTX 470 Fermi-based card we achieve 215.4, 122.2 and 64.9 FPS for the simple model and 69.6, 37.4 and 19.1 FPS for 64K, 128K and 256K particles respectively.

Keywords: GPU, CFD, SPH, GPGPU, CUDA, Fluid, Newtonian, Non-Newtonian.

1 Introduction

Simulating fluids is a computationally intensive problem, especially in 3D. Due to large computational demands most fluid simulations are not done in real-time. We have developed a new open-source[1] framework for 3D SPH calculations on the GPU, where we provide computational primitives that accelerate the building blocks of the SPH algorithm. By using our framework the number of particles modeled in the simulation can be increased considerably, and the overall simulation speed is greatly increased compared to CPU-based simulations. Our work differs from previous works in several ways. We provide a modularized framework that can be used for implementing different SPH models. We use an acceleration algorithm that is well-suited for the GPU and finally we compare two different SPH models implemented using the same framework, thus giving a measure of how well the GPU scales with more complex models.

---

[1] http://code.google.com/p/gpusphsim/

K. Jónasson (Ed.): PARA 2010, Part II, LNCS 7134, pp. 98–109, 2012.

2   Previous Work

Some of first implementations of SPH on the GPU were by Harada *et al.*[4] and Zhang *et al.*[15]. This was before the introduction of CUDA and it was thus done using OpenGL and Cg which imposed severe limits on the implementations. Since then there has been growing interest in the implementation of SPH on the GPU resulting in several implementations that take full or partial advantage of the GPU and CUDA [14][2][5].

3   Computational Fluid Dynamics

Fluid dynamics is described using the the Navier-Stokes equations, and in their Lagrangian form consist of mass and momentum conservation:

$$\frac{d\rho}{dt} = -\rho \nabla \cdot \mathbf{v} \tag{1}$$

$$\frac{d\mathbf{v}}{dt} = -\frac{1}{\rho}\nabla p + \frac{1}{\rho}\nabla \cdot \mathbf{S} + \mathbf{f} \tag{2}$$

Where $\mathbf{v}$ is the velocity field, ρ the density field, ∇p the pressure gradient field resulting from isotropic stress, $\nabla \cdot \mathbf{S}$ the stress tensor resulting from deviatoric stress and $\mathbf{f}$ an external force field such as gravity. For incompressible Newtonian fluids the momentum conservation reduces to:

$$\frac{d\mathbf{v}}{dt} = -\frac{1}{\rho}\nabla p + \frac{\mu}{\rho}\nabla^2 \mathbf{v} + \mathbf{f} \tag{3}$$

Where the term μ is the dynamic viscosity of the fluid.

3.1   Smoothed Particle Hydrodynamics

In SPH the different effects of Navier-Stokes are simulated by a set of forces that act on each particle. These forces are given by scalar quantities that are inter-polated at a position $\mathbf{r}$ by a weighted sum of contributions from all surrounding particles within a cutoff distance h in the space Ω. In integral form this can be expressed as follows [9]:

$$A_i(\mathbf{r}) = \int_\Omega A(\mathbf{r}')W(\mathbf{r} - \mathbf{r}', h)d\mathbf{r}' \tag{4}$$

The numerical equivalent is obtained by approximating the integral interpolant by a summation interpolant [9]:

$$A_i(\mathbf{r}_i) = \sum_j A_j \frac{m_j}{\rho_j} W(\mathbf{r}_{ij}, h) \tag{5}$$

where j iterates over all particles, m_j is the mass of particle j, $\mathbf{r}_{ij} = \mathbf{r}_i - \mathbf{r}_j$ where $\mathbf{r}$ is the position, ρ_j the density and A_j the scalar quantity at position $\mathbf{r}_j$.
   For a more comprehensive introduction to SPH, please refer to [9].

3.2   Snow Avalanche SPH

Snow avalanches wary greatly in behavior, from powder-snow avalanches to so called dense-flow, or flowing snow avalanches. Snow avalanches can often appear as a viscous flow down a slope, and it is this obvious property which has prompted the use of fluid dynamics in avalanche simulation [1]. Several viscosity models exist for modeling Non-Newtonian fluids, and rheological parameters have been collected for flowing snow [7]. Many SPH models exist for viscoelastic fluids, from melting objects [11] to lava flows [12] and generalized rheological models [6]. We implement an SPH simulation of a Non-Newtonian fluid with configurable support for multiple rheological models to approximate the behavior of a snow avalanche.

4   Methods and Implementation

The simulation framework uses CUDA, a parallel computing architecture developed by NVIDIA. We parallelize the calculation of SPH by assigning a thread to each particle in the simulation. Each thread is then responsible for calculating the SPH sums over the surrounding particles.

When accessing memory on the GPU *coalesced* (correctly structured) access is very important. Due to the nature of the acceleration algorithm we use, perfectly coalesced access is unfortunately not possible. By utilizing the texture cache on the GPU this problem is greatly alleviated.

4.1   Nearest-Neighbor Search

The summation term in the SPH-formulation is computationally heavy, it requires looking at many nearby particles and computing interactions between them. To avoid a naive brute-force $O(N^2)$ search for neighbors, a nearest-neighbor search algorithm is commonly used, such as a linked list or a uniform grid. Our framework uses the acceleration algorithm found in the NVIDIA CUDA "Particles" demo [3], which is better suited to the GPU than many previously used algorithms. It can be summarized as follows:

1. Divide the simulation domain into a uniform grid.
2. Use the spatial position of each particle to find the cell it belongs to.
3. Use the particle cell position as input to a hash function (a spatial hash)
4. Sort the particles according to their spatial hash.
5. Reorder the particles in a linear buffer according to their hash value.

Particles in the same cell can then appear ordered in the linear buffer, the specifics depending on the spatial hash function. Finding neighbors is thus just a matter of iterating over the correct indices in the buffer. To sort the particles on the GPU we used a radix sort [13].

4.2 Non-Newtonian Fluids

Non-Newtonian fluids differ from Newtonian fluids in that their viscosity is not constant. In a Newtonian fluid the relation between shear stress and the strain rate is linear, with the constant of proportionality being the viscosity. For a Non-Newtonian fluid the relation is nonlinear and can even be time-dependent. There exist many classes of Non-Newtonian fluids, and many types of models, of which we implement several. The complex SPH model differs primarily from the simple SPH model in that it includes the much more complex stress calculation presented in [6] and in that the viscosity parameter is not constant but modeled using a rheological viscosity model.

4.3 SPH Models

We have implemented two different SPH models, a simple and a complex model. The simple SPH model is a partial implementation (discarding surface tension) of a well-known SPH model designed for interactivity [10]. In the complex SPH model we combine some of the techniques used in [10] with models from [6] and [12]. Compared to the Simple model we use a more accurate smoothing kernel, we use a more accurate calculation of shear forces and we support a range of rheological models which enable us to simulate different Non-Newtonian fluids.

4.4 SPH Formulation

By using the SPH formulation the Navier-Stokes equations can be approximated:

$$\rho_i = \sum_j m_j W(\mathbf{r}_{ij}, h) \tag{6}$$

$$\mathbf{f}_i^{pressure} = -\frac{1}{\rho} \nabla p(\mathbf{r}_i) = \sum_{j \neq i} m_j \left(\frac{p_i}{\rho_i^2} + \frac{p_j}{\rho_j^2} \right) \nabla W(\mathbf{r}_{ij}, h) \tag{7}$$

The incompressible fluid is simulated as a weakly compressible fluid where the incompressibility constraint is applied to the pressure p by using an equation of state given by the ideal gas law with an added rest density[4]: $p = k(\rho - \rho_0)$

Finally the stress force is calculated:

$$\mathbf{f}_i^{stress} = \frac{1}{\rho} \nabla \cdot \mathbf{S}_i = \sum_{j \neq i} \frac{m_j}{\rho_i \rho_j} (\mathbf{S}_i + \mathbf{S}_j) \cdot \nabla W(\mathbf{r}_{ij}, h) \tag{8}$$

Where the Non-Newtonian fluid stress tensor $\mathbf{S}$ is calculated as [6]. Thus we have that the acceleration for a particle is given by:

$$\mathbf{a}_i = f_i^{pressure} + f_i^{stress} + f_i^{external} \tag{9}$$

4.5   SPH Algorithm

Due to the data dependencies between the various forces, some of them must be calculated separately. Each calculation step is essentially a summation over neighboring particles, and we combine the force calculations using loop fusion as far as it is possible. For the complex SPH model we end up with the following steps:

1. Update the hashed, radix sorted, uniform grid.
2. Calculate the SPH density
3. Calculate the SPH velocity tensor
4. Calculate the SPH pressure and SPH stress tensor
5. Apply external forces and integrate in time using Leap-Frog.

For the simplified SPH model, the stress tensor is replaced with a simplified viscosity approximation. Thus viscosity force can be computed together with the pressure in step 4 and as a result the simple SPH model can drop an entire SPH summation loop.

5   Results

Comparing and evaluating the performance of the simulations is difficult due to the large amount of parameters and their effect on performance. In addition it is hard to compare to other SPH implementations due to different SPH models and parameters. For the simple SPH model we compare against Mller and Harada which use a very similar SPH model, using rest density selected to simulate water, with the dynamic viscosity set to 1. For the complex SPH model we have not found comparable implementations so must compare against our implementation

Fig. 1. A screenshot of the simple SPH model with 256K particles interacting with a terrain model. Hue-based gradient shading for the velocity of the particles.

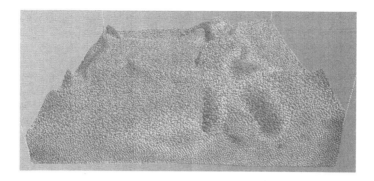

Fig. 2. A screenshot of the simple SPH model with 256K particles in a box of repulsive forces. Hue-based gradient shading for the velocity of the particles.

of the simple SPH model. To obtain absolute performance numbers we use a fairly simple simulation setup; a square simulation domain with simple repulsive forces as walls where a cubic volume of fluid is dropped into a shallow pool of water Figure 2. The performance numbers were measured when the fluid had reached a stable equilibrium to avoid errors introduced by fluctuations in simulation performance.

Hardware. For all our performance results we used a fairly high-end computer equipped with an Intel Core2 Quad Q9550 processor. We compare three different graphics cards, an NVIDIA GeForce GTX 260, a GeForce GTX 470 and a Tesla C2050, thus covering both the previous and the current generation of GPU processors.

Simulation Parameters. The simulation parameters were chosen for their stability and how realistic the resultant behavior fluid appeared. To improve the consistency and validity of the results we chose to use a simple Newtonian fluid rheology for performance measurements of the complex SPH model as well, which eliminates the complexity of measuring the performance of a model with a variable viscosity, while maintaining the computational complexity of the model.

We select a time step (dt) of 0.0005, and a rest density (ρ_0) of 1000. In addition we employ a simple aspring-like external boundary force with a stiffness and dampening coefficients of 20000 and 256 respectively. Finally we set a viscosity (μ) of 1.0. Our simulation is scaled with a factor of 0.0005.

Memory Usage. Since our implementations do not use constant or shared memory to any significant degree, the only memory usage that is of importance is the usage of the global memory on the device. Due to the hashed uniform grid structure the memory usage is highly efficient, since the particle data is stored in continuous memory buffers that are fully utilized. For the Simple SPH model the memory usage is $176N$ bytes where N is the number of particles. The memory usage of the complex SPH model is $240N$ bytes. This means that it is possible to simulate very large systems even on commodity hardware. Using the NVIDIA Tesla C2050, we have simulated up to 12 million particles with the Simple SPH model, thus using roughly 2GB of memory.

5.1   Performance Results

By manually optimizing register usage, reordering memory accesses and optimizing the block sizes for the CUDA code, performance gains as large as 40% over our earlier implementation were realized [8].

The performance of the Tesla C2050 is nearly identical to that of the GeForce GTX 470. The small difference in performance can most likely be attributed to the difference in memory speed (1500 MHz vs. 1674 MHz) and clock speed (1150 MHz vs. 1215 MHz). Our implementation is single precision only, so we can not utilize the greatest feature of the Tesla; greatly increased double precision performance.

It is interesting to note that though the Tesla has greater memory bandwidth due to the larger memory bus (384 bit vs. 320 bit), this does not seem to increase performance. This could be due to the slightly lower clocks of the Tesla or it may mean that the raw memory bandwidth is less important than the performance of the texture cache. Since the memory access pattern of the hashed uniform grid algorithm cannot be coalesced perfectly there will be uncoalesced memory access. In the context of this knowledge, the missing performance increase may mean that the performance is severely bottlenecked by these uncoalesced memory accesses, much in the same manner as a pipeline stall.

Access to the global memory on the GPU is not perfectly coalesced due to limitations of the GPU acceleration algorithm, however this limitation is greatly mitigated by use of the texture cache. The texture cache was found to increase performance between 55% and 200% [8].

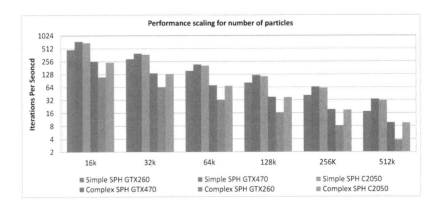

Fig. 3. Performance scaling for the simple SPH model

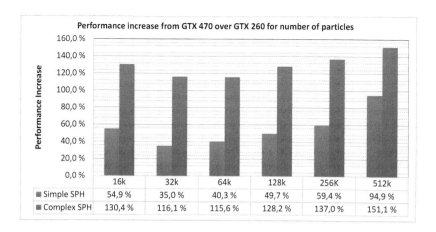

Fig. 4. Performance comparison between the GeForce GTX 470 and the GeForce GTX 260 for different amount of particles

5.2   NVIDIA Fermi Architecture

The new GF100/Fermi-architecture provides a large increase in performance compared to the GT200-architecture.

In Figure 4 we show the performance increase from using a GeForce GTX 470 over a GeForce GTX 260. These two GPUs have memory bandwidths of 111.9 GiB/s and 133.9 GiB/s, an increase of 20%, but the increase in measured performance is much larger, clearly indicating that the implementations are not completely bandwidth bound. The complex SPH model benefits the most, with improvements up to 150%, while the Simple SPH model sees improvements up to 95%. This large increase in performance can primarily be attributed to the new features of the Fermi architecture, which allows for greater utilization of the GPU resources (occupancy) and which minimizes the performance penalty of imperfectly coalesced memory access.

5.3   Kernels

We have measured the relative performance of the different kernels in our two SPH implementations. Our findings show that the most performance intensive parts are in the calculation of the SPH summation over neighboring particles. This is due to the large amount of computation and memory transfer (from global memory on the GPU) that occurs in these steps.

5.4   Rendering Overhead

We use direct rendering of the particle data using shaders on the GPU. This means that the rendering overhead is fairly small, though relatively large for small amounts of particles. With 16K particles the overhead is 95% and 60%

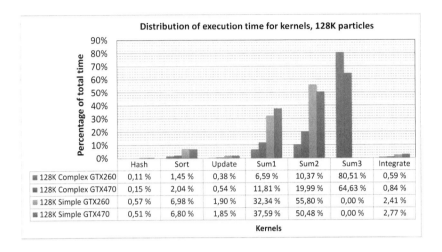

Fig. 5. Distribution of calculation time for the algorithm steps

for the simple model and 30% and 40% for the complex model, on the GT260 and the GTX470 respectively. For 256K particles this is reduced to 20% and 5% for the simple model and 8% and 3% for the complex model, on the GT260 and the GTX470 respectively.

5.5   Performance Review

We have compared our implementation performance for the Simple SPH model with that of other implementations. This algorithm has been widely implemented since it is very well suited for interactive or real-time simulation and as such it is possible to find comparable implementations. Unfortunately we have found that it is nearly impossible to do a review of earlier implementations that is both comprehensive and accurate since most authors do not specify all the parameters they use. In addition there are slight differences in the SPH models and finally also because of the different hardware used. Nonetheless we have attempted a comparison, if only to give a rough picture of the performance landscape.

We find that our GPU implementation is significantly faster than earlier GPU implementations, even for implementations using faster graphics cards. One such implementation[14] use a NVIDIA GTX 280 and get 66 iterations per second at 16K particles. Comparing their implementation against our implementation running on a GTX260 (without rendering) we see a 6x speedup. It is also interesting to note that our implementation seems to scale better, though the available data is not enough to draw any conclusions.

Harada *et al.*[4] achieves real-time performance at 17 FPS with 60000 particles on an NVIDIA GeForce 8800GTX and Zhang *et al.*[15] achieves 56 FPS with 60000 particles using the same GPU. Both these implementations use OpenGL and Cg and are thus very constrained compared to more recent implementations

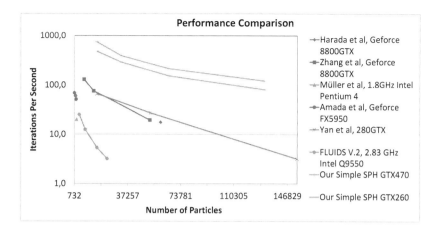

Fig. 6. Performance of our implementations compared to others

using CUDA. A very interesting comparison is with the FLUIDS V.2 software, which is a highly optimized SPH implementation for the CPU. Unfortunately FLUIDS can only use one of the cores in this CPU so it should be assumed that the performance could be almost quadrupled using all 4 cores. Comparing the FLUIDS software with our GPU implementation (with rendering), we see speedups of 91x for the GeForce GTX 470 and 49x for the GeForce GTX 260 with 16K particles.

5.6   Real-Time Appearance

By scaling the simulation domain, and relaxing the accuracy requirements by selecting large time step, the fluid simulations produce beliveable real-time fluid animations. Our complex SPH model is not as well suited to real-time simulation due to the necessity of a somewhat lower timestep in order to support higher viscosities, but by using 64K particles it is still possible to simulate avalanche-like animations in real-time. We found that using a Cross rheological model best captured the behavior of a flowing snow avalanche (Figure 7).

6   Conclusions

In this paper, we presented an implementation of Smoothing Particle Hydro-dynamics (SPH) on the GPU using. Our implementation achieves very good performance since we take advantage of the massive amount of parallelism available on modern GPUs, as well as use specialized acceleration data structures. As a result of the computational acceleration afforded by the use of GPUs our simulations can maintain very high performance with large problem sizes. This produces real-time simulations whose animation appears more correct and realistic than previously seen efforts.

Fig. 7. A screenshot of the complex SPH model with 64K particles. We use a cross rheological model to approximate the behavior of a flowing snow avalanche.

6.1   Current and Future Work

Simulations of snow can be used for everything from gaming to avalanche prediction. For games snow simulation can help create complex environments, which can lead to numerous possibilities for game-play mechanics. Predicting the behavior of snow avalanches can help prevent loss of both life and property. Our simulation framework can be used for more complex SPH models that can be used to produce qualitatively correct simulations.

The resource usage of our model has been investigated and it was found that it does not consume much memory but is very memory bandwidth intensive and suffers from imperfectly coalesced memory access, it would be interesting to research ways to improve the memory access pattern.

Finally, the visualization of the fluid model can be improved, at the moment a very simple but efficient and low-cost method of direct particle rendering is used. By using a surface reconstruction model such as Marching-Cubes a real surface can be rendered. It is also possible to use screen-space surface rendering techniques to approximate the fluid surface without the large computational cost associated with true surface reconstruction.

References

1. Bovet, E., Chiaia, B., Preziosi, L.: A new model for snow avalanche dynamics based on non-newtonian fluids. Meccanica (2008), http://www.springerlink.com/content/g82xk01766833788
2. Crespo, A.J.C.: Application of the Smoothed Particle Hydrodynamics model SPHysics to free-surface hydrodynamics. Ph.D. thesis, University of Vigo (2008)
3. Green, S., NVIDIA: CUDA Particles, Presentation slides. Tech. rep., NVIDIA (2008)

4. Harada, T., Koshizuka, S., Kawaguchi, Y.: Smoothed Particle Hydrodynamics on GPUs (2007), http://www.inf.ufrgs.br/cgi2007/cd_cgi/papers/harada.pdf

5. Herault, A., Bilotta, G., Dalrymple, R.A.: SPH on GPU with CUDA. Journal of Hydraulic Research 48(extra issue), 74–79 (2010)

6. Hosseini, S.M., Manzari, M.T., Hannani, S.K.: A fully explicit three-step SPH algorithm for simulation of non-Newtonian fluid flow. International Journal of Numerical Methods for Heat & Fluid Flow 17(7), 715–735 (2007), http://dx.doi.org/10.1108/09615530710777976

7. Kern, M.A., Tiefenbacher, F., McElwaine, J.N.: The rheology of snow in large chute flows. Cold Regions Science and Technology 39(2-3), 181–192 (2004), http://www.sciencedirect.com/science/article/B6V86-4CS4G7W-1/2/664993b41275bfb273c4b9b1d40cfd52

8. Krog, Ø.E.: GPU-based Real-Time Snow Avalanche Simulations. Master's thesis, NTNU (June 2010)

9. Liu, G.R., Liu, M.B.: Smoothed Particle Hydrodynamics: A Meshfree Particle Method. World Scientific Publishing Company (December 2003), http://amazon.com/o/ASIN/9812384561/

10. Müller, M., Charypar, D., Gross, M.: Particle-based fluid simulation for interactive applications. In: SCA 2003: Proceedings of the 2003 ACM SIGGRAPH/Eurographics Symposium on Computer Animation, pp. 154–159. Eurographics Association, Switzerland (2003)

11. Paiva, A., Petronetto, F., Lewiner, T., Tavares, G.: Particle-based non-newtonian fluid animation for melting objects. In: Brazilian Symposium on Computer Graphics and Image Processing, pp. 78–85 (2006)

12. Paiva, A., Petronetto, F., Lewiner, T., Tavares, G.: Particle-based viscoplastic fluid/solid simulation. Computer-Aided Design 41(4), 306–314 (2009), http://www.sciencedirect.com/science/article/B6TYR-4TTMNFW-1/2/3e798fdc322f7e878f386d435f80b01b, point-based Computational Techniques

13. Satish, N., Harris, M., Garland, M.: Designing efficient sorting algorithms for many-core GPUs. Tech. rep., NVIDIA Corporation, Los Alamitos, CA, USA (2009), http://dx.doi.org/10.1109/IPDPS.2009.5161005

14. Yan, H., Wang, Z., He, J., Chen, X., Wang, C., Peng, Q.: Real-time fluid simulation with adaptive SPH. Comput. Animat. Virtual Worlds 20, 417–426 (2009), http://portal.acm.org/citation.cfm?id=1568678.1568695

15. Zhang, Y., Solenthaler, B., Pajarola, R.: GPU accelerated SPH particle simulation and rendering. In: SIGGRAPH 2007: ACM SIGGRAPH 2007 Posters, p. 9. ACM, New York (2007)

Toward Techniques
for Auto-tuning GPU Algorithms

Andrew Davidson and John Owens

University of California, Davis

Abstract. We introduce a variety of techniques toward autotuning data-parallel algorithms on the GPU. Our techniques tune these algorithms independent of hardware architecture, and attempt to select near-optimum parameters. We work towards a general framework for creating auto-tuned data-parallel algorithms, using these techniques for common algorithms with varying characteristics. Our contributions include tuning a set of algorithms with a variety of computational patterns, with the goal in mind of building a general framework from these results. Our tuning strategy focuses first on identifying the computational patterns an algorithm shows, and then reducing our tuning model based on these observed patterns.

Keywords: GPU Computing, Auto-Tuning Algorithms, Data-Parallel Programming, CUDA.

1 Introduction

Given the complexity of emerging heterogeneous systems where small parameter changes lead to large variations in performance, hand-tuning algorithms have become impractical, and auto-tuning algorithms have gained popularity in recent years. When tuned, these algorithms select near-optimum parameters for any machine. We demonstrate a number of techniques which successfully auto-tune parameters for a variety of GPU algorithms. These algorithms are commonly used, were not synthetically created and were chosen to rely on different parameters and be performance bound in different ways.

Both work by Liu et al. [4] and Kerr et al. [2] use adaptive database methods for deriving relationships between input sets and their direct influence on the parameter space. Work by Li et al. [3] focuses on tuning one GEMM algorithm, while work by Ryoo et al. [5] mainly deals with kernel and compiler level optimizations for one machine (an 8800GTX). Our work differs from these in that we first try to identify computational patterns between algorithms and then tune their parameters using a number of different techniques.

Rather than use a single database method for all of our algorithms, which might be susceptible to non-linear relations, we try to identify relationships between each algorithm, possible input sets on any given machine; we then choose a tuning strategy accordingly. Since we now have these relationships, we can heavily prune the parameter search space before attempting to tune the algorithm,

K. Jónasson (Ed.): PARA 2010, Part II, LNCS 7134, pp. 110–119, 2012.

resulting in very fast tuning runs. For our set of algorithms, we show a quick tuning run (less than a few minutes for most machines) can generate the needed information to automatically choose near-optimum parameters for future runs. We will next discuss the general strategy (or philosophy) we used for developing each of our auto-tuned algorithms.

2   Algorithms

Algorithms designed in a data-parallel fashion are concerned with maintaining the highest throughput possible. This means parameters are often dependent not only on the machine, but also upon the workload being operated on. With this in mind, when attempting to auto-tune an algorithm we approach the problem using this philosophy:

- Identify the tunable parameters for the algorithm.
- Look for a relationship between the input space, and the parameter space.
- If such a relationship exists, we must build a model from the input space to parameter space, or model the parameter space solely on machine characteristics.

Therefore when considering the input parameters for each algorithm we identify two axes of tuning, *workload* specific and *machine* specific. The relationship between these two may vary widely between algorithms, but we concentrate on identifying common patterns between certain algorithms that can be used as a starting point for general tuning. As an example, in our N-Body algorithm we discovered a heavy dependence between our input parameters and the workload (input space). Yet for our reduction kernel, once the workload reaches a critical size, our tuning parameters rely solely on machine specific parameters. Therefore using micro-benchmarks we developed a model that estimates the correct parameters. Sections 2.2 covers this strategy in more depth. Next we will introduce our auto-tuning test suite, and the approaches we took to quickly tune each algorithm.

2.1   Test Suite

We studied four parallel algorithms extensively in our work. Reduction, a common parallel primitive which operates on a large set of values and reduces that set to one value. Next, scalar product, which sums up the products between two vectors. An N-Body algorithm which simulates the gravitational interactions between a set of objects. Finally, SGEMM, a fast matrix multiply. For the reduction, scalar product and N-Body algorithms we used the kernels found in the NVIDIA SDK as our base algorithms. These algorithms have already been highly optimized, and are considered standards for benchmarking. Since many of these algorithms have been hand-tuned for certain machines, we may receive only a marginal performance boost for certain devices and workloads. In particular, the NVIDIA SDK has changed a number of default parameters to obtain optimal performance on GT200 series cards. As a result, performance suffers on older cards which highlights the importance of auto-tuning algorithms which scale with architecture changes. For our SGEMM algorithm we used Chien's highly tuned and optimized kernel [1] which demonstrates higher performance than that released by NVIDIA's SGEMM. In our final auto-tuned algorithm we wish to trim as much of the parameter search space as possible, while maintaining an accurate model to select parameters.

Auto-tuning algorithms that use standard techniques such as model driven inputs, with interpolation between points, may result in sub-optimal parameters (one method does not fit all). This is where the major challenge lies, as great care must be taken to ensure your auto-tuning strategy is not susceptible to unpredictable non-linear effects.

2.2   Reduction

We made a few minor improvements to NVIDIA's SDK reduction kernel code. We chose this algorithm as it is obviously memory bound and each item can be accessed in any order (coalesced reads would obviously be preferred). Therefore tuning reduction would give insight into many other common parallel algorithms, such as scan and sort, which have similar computational patterns. For the reduction kernel, the parameters that require tuning are the number of threads per block, and the number of blocks. The optimum set of parameters may fluctuate with respect to the number of elements we reduce.

Through experiments and benchmarking, we were able to show a standard behavior for the optimum parameter set given a number of elements. Using the results from these tests, our auto-tuning method first searches for a thread cap for all elements, which is assumed to fully occupy the machine. The thread cap is defined as the maximum number of threads optimum to any valid input set (no optimum parameters will have more total threads than the thread cap).

Since our input set is a one-dimensional array in the reduction algorithm, it is easy to test what this thread cap is, and where it applies. All workloads greater than the number of elements where this thread cap applies, is also bound by the thread cap. Next we test for a lower switchpoint where the CPU outperforms the GPU, and the optimum parameters for that point. Using these two points, and their associated number of elements, we are able to select a number of total threads (threads per block and number of blocks) for any input set.

Therefore this algorithm is highly machine dependent, less workload dependent, and therefore much easier to tune. Highly workload dependent algorithms require micro-benchmarks in order to correlate the workload space with the parameter space. The next algorithms considered are cases where the workload has a more dominant effect on the parameter space.

2.3   Scalar Product

This algorithm is also memory bound and available in NVIDIA's SDK. However the parameters for tuning the Scalar Product kernel are more complex as the vector size and total number of vectors adds a dimension to the tuning process. Therefore while applying some of the same principles from our reduction method for a thread cap and associated work cap (the minimum workload at which our thread cap is applied), we also select a set of distributed points (vsize$_i$, n$_i$) such that vsize$_i$ < vsize$_m$ and n$_i$ < n$_m$, that are under the work cap and test for the optimum number of threads, these points will operate as a nearest neighbor spline.

Here our machine dependent parameters help prune our tuning space, as we generalize parameters for all input sets outside these limits. For all input sets under the thread cap we pre-tune each point from our distributed and select parameters for the knot closest to the input set. This works due to the fine grained corelation between the input set and the parameter set. We used a nearest neighbor approach rather than an interpolation approach, as our tests showed the closest knot approach generally performed better and performance was more stable (less variance).

Since we prune all points greater than (vsize$_m$, n$_m$), where vsize$_m$ and n$_m$ are dependent on machine parameters. We have therefore reduced the tuning space to a smaller subset of the original input space.

2.4   N-Body

We selected the N-Body algorithm as it has block-wise tunable parameters, is more arithmetically intense, and therefore not global memory bound. The tunable parameters in this case are two dimensions of block sizes that will be shared per block. Therefore tuning these parameters involves a tradeoff in register usage, and the amount of data sharing per block. These parameters are fairly fine-grained, and when paired with a fine-grained input set we find that there is a corelation between the input space and parameter space. In other words, the optimum parameter points does not vary widely between input set a$_i$ and a$_i$ + δx. Where δx is a small variation of the input set a$_i$.

This motivates us to use a nearest-neighbor spline technique and concentrate on intensely tuning a few distributed points. This technique allows us to greatly reduce the tuning search space to only a small subset, while maintaining near optimum performance for all inputs.

2.5   SGEMM

As mentioned previously we used Lung Sheng Chien's optimized code [1] (which is a further optimization of Vasily Volkov's code [6]) as the base kernel for our autotuning method. There are a number of variations to this set of code; we selected the one with best performance, that would not lead to out-of-bound problems (method8 in Chien's benchmark suite). Using templates, we created twenty-one valid versions of this method that relied on three input parameters. Due to Chien's reliance on precompiling kernels into cubin binaries before runs, we created twenty-one associated binaries for each version. Experimental results showed that a number of these valid versions were inherently inefficient, and were removed from the tuning suite.

We found that for larger matrices, one parameter set would deliver near optimum performance. Therefore, our strategy was to find the optimum parameter set (of the twenty-one available) for a given machine which would allow us to tune all input-sets larger than a certain matrix size. It was also noticed that in general, this preferred kernel had higher register usage per thread than other kernels. Whether or not this is a rule of thumb one can follow for quickly selecting an appropriate variation, requires more testing on a wider variety of devices.

For smaller matrices, our tests on a GTX260 and an 8600 GTS showed that there was more variety and workload dependence on the optimal kernel. Therefore, our complete tuning strategy for this algorithm is as follows:

- Find optimal kernel for large matrices. This relies on machine specific parameters.
- Find matrix size for which we switch strategies.
- Test a variety of candidates from a pre-tuning run, and finely tune all small matrices with these candidates.

Though this algorithm requires us to finely tune for smaller matrices, we gain an advantage by pruning all matrices greater than the switchpoint. Since smaller matrices are solved much faster, we are able to prune the most vital section of tuning space. Also, since we only select a few valid candidates for the smaller matrices, we further eliminate unnecessary kernel tests.

3   Results

We tested our methods, and ran a number of experiments on a higher end GTX 260 and GTX 280, medium-end 8800GT and 5600FX, and a low-end 8600 GTS. We had access to a Tesla C1060 for our reduction tests, however it was unavailable for test runs for our other algorithms.

In the next subsections, we will compare the performance for each of our auto-tuned algorithms versus the untuned default algorithms. For all auto-tuned algorithms, tuning runs are very short (around one to five minutes) due to our carefully pruning the tuning space so that we gather only necessary information for our tuning process.

3.1   Reduction

Figure 1 compares the performance of our auto-tuning method against that of the SDK's default parameters. The results show a speedup that brings memory performance close to bandwidth limit. Our auto-tuned method(blue plot) performs as well as a brute force check on all possible parameters (red dotted line), while only taking a minute or less to run a one-time tuning run. As a memory bound function, we found performance depended directly on the number of memory controllers available to the machine. Though this cannot be queried

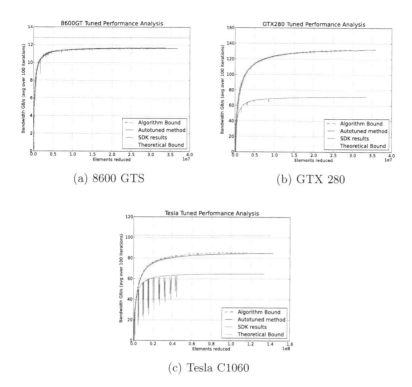

(a) 8600 GTS

(b) GTX 280

(c) Tesla C1060

Fig. 1. Auto-Tuned vs SDK Performance Comparison for Reduction. The theoretical bound is the reported maximum DRAM bandwidth. In all cases, the auto-tuned implementation performs as well as the algorithmic bound (brute force test of all implementations).

directly (one can query the number of memory controllers, but not the bandwidth at which each operates), some pretuning tests can supply this information, and be used to reduce the tuning space further.

The performance comparison in Figure 1 also shows this algorithm is dominated by machine dependent parameters. If there was a higher workload dependence, selecting optimal parameters would be more difficult, resulting in pockets of poorer performance. However, the auto-tuned reduction kernel consistently matches the optimum algorithmic bound curve. This algorithmic bound curve is a brute force performance check on every possible parameter combination.

3.2   Scalar Product

Though using knots adds complexity and possibility for non-optimal selections, our results still perform better than that of NVIDIA's SDK for most points. Since visualizing the performance of a multi-dimensional input space is difficult, we instead present our results in Table 1 as the relative performance from a set of random inputs. Our results in Table 1 show good speedups for almost all cases on the GTX 280 and 260 (achieving nearly 30 percent performance boost on the 280). While the performance gains on the 8800GT and 5600FX were not as drastic, we still were able to boost performance slightly.

Table 1. Performance speedups comparison of auto-tuned kernel vs default for a set of 1000 random selected points

Architecture (GPU)	Speedup
GTX 280	1.2817
GTX 260	1.1511
8800 GT	1.0201
5600 FX	1.0198

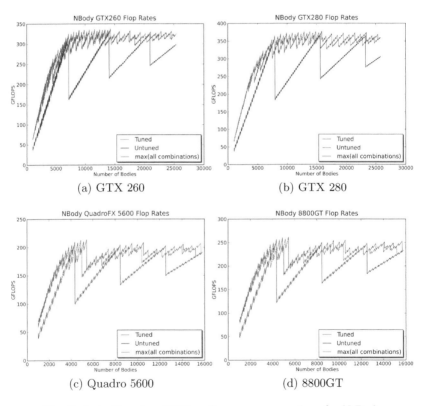

(a) GTX 260   (b) GTX 280

(c) Quadro 5600   (d) 8800GT

Fig. 2. Auto-Tuned vs SDK performance comparison for N-Body

3.3   N-Body

Figure 2 shows the performance comparison of our spline auto-tuned strategy versus the default parameters. The spikes in performance from the untuned implementation illustrate the workload dependence to the parameter space. As the workload changes, the untuned parameters fall in and out of sync with the optimum implementation.

Our spline strategy helps to minimize these spikes in performance, and maintain near-optimum performance by updating parameters as the workload changes. In some cases the speedups from these can be up to 2x that of the untuned performance. One can also vary the amount of tuning in order to get nearer and nearer to the optimum performance. The example in Figure 2 has twenty tuned points which are used as a reference.

3.4   SGEMM

As was illustrated in Section 2.4, small variations in parameters could lead to large variations in kernel performance. Therefore, we cannot use the same spline technique from our previous two algorithms. The default parameters that Chien [1] used were found to be the near-optimal for most cases on larger matrices on both the GTX260 and 8600GTS. For smaller size matrices ($N < M$) other kernels were preferred, as shown in Figure 3. More testing is needed to confirm this holds true for a variety of devices. However, this motivated our hybrid tuning technique where coarse tuning runs are used to select candidate parameters for smaller matrices, and then a dominating machine-dependent parameter set for all larger matrices.

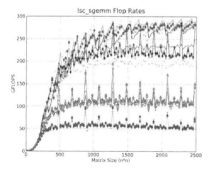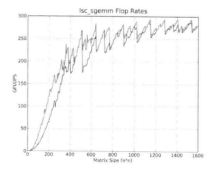

Fig. 3. Performance of the coarser grained parameter selection for our SGEMM. The right-hand figure demonstrates possible auto-tuned performance, versus the default kernel. For smaller matrices one could possibly achieve a 2x speedup(e.g. for matrices of 200×200, possible performance is about 130 GFlops/s versus 60 GFlops/sec).

3.5   General Auto-tuning Summary and Practices

This section serves as both a summary of our previous techniques, and to which types of algorithms each are applicable. Table 2 shows our tested algorithms, and how strongly their optimum parameters are machine and workload dependent. Once these are identified, one can begin developing an auto-tuning scheme for parameter selection.

Generally speaking, we find that for strongly workload dependent algorithms with a fine-grained input set and fine-grained parameter set, nearest neighbor spline strategies have returned good results. This was evident in both our N-Body and Scalar Product tuned algorithms. For our reduction kernel, we saw a strong dependence between our thread parameters and the device parameters (memory controllers). Therefore our strategy relies on this simple relationship. Finally, our SGEMM kernel displayed various levels of dependency, and we therefore developed a hybrid strategy for different workload ranges.

Table 2. Table summarizing each auto-tuned algorithm's dependencies on device specific parameters and workload specific parameters(input space)

Algorithm Name	Device Dependency	Workload Dependency
Reduction	Strong	Weak
Scalar Product	Medium	Strong
N-Body	Weak	Strong
SGEMM	Strong	Medium

4   Conclusion

We believe that hand-tuning algorithms for each machine will become an impractical method as systems become more diverse in capability, and algorithm bounds become more complex. Therefore, developing methods that either fully automate or assist the tuning process, could prove powerful tools for developers to boost utilization.

Future work is needed in developing firmer relationships between algorithms with *similar* computational patterns, and developing auto-tuning schemes between these algorithms. Testing on newer architectures, such as the recently released Fermi architecture is also needed. The Fermi 400x series cards contain a number of new features that would change tuning strategies for a number of algorithms. On top of faster global memory bandwidth, more shared memory within blocks, and compute power, these additions include faster atomic operations than previous cards, and more computational power for double precision operations.

Our work has shown a number of autotuning practices and methods which boost performance for a number of common algorithms. We believe this is an important stepping stone in developing a generalized tuning methodology for data parallel programs.

References

1. Chien, L.S.: Hand-Tuned SGEMM On GT200 GPU. Technical Report, Tsing Hua University (2010), http://oz.nthu.edu.tw/~d947207/NVIDIA/SGEMM/HandTuned Sgemm_2010_v1.1.pdf
2. Kerr, A., Diamos, G., Yalamanchili, S.: Modeling GPU-CPU Workloads and Systems. In: GPGPU 2010: Proceedings of the 3rd Workshop on General-Purpose Computation on Graphics Processing Units, pp. 31–42. ACM, New York (2010)
3. Li, Y., Dongarra, J., Tomov, S.: A Note on Auto-Tuning GEMM for GPUs. In: Allen, G., Nabrzyski, J., Seidel, E., van Albada, G.D., Dongarra, J., Sloot, P.M.A. (eds.) ICCS 2009. LNCS, vol. 5544, pp. 884–892. Springer, Heidelberg (2009)
4. Liu, Y., Zhang, E., Shen, X.: A Cross-Input Adaptive Framework for GPU Program Optimizations. In: IPDPS 2009: Proceedings of the 2009 IEEE International Symposium on Parallel and Distributed Processing, pp. 1–10. IEEE Computer Society Press, Washington, DC (2009)
5. Ryoo, S., Rodrigues, C.I., Stone, S.S., Baghsorkhi, S.S., Ueng, S.Z., Stratton, J.A., Hwu, W.W.: Program Optimization Space Pruning for a Multithreaded GPU. In: CGO 2008: Proceedings of the Sixth Annual IEEE/ACM International Symposium on Code Generation and Optimization, pp. 195–204 (April 2008)
6. Volkov, V., Demmel, J.W.: Benchmarking gpus to tune dense linear algebra. In: SC 2008: Proceedings of the 2008 ACM/IEEE Conference on Supercomputing, pp. 1–11. IEEE Press, Piscataway (2008)

An Interval Version of the Crank-Nicolson Method – The First Approach

Andrzej Marciniak

Institute of Computing Science, Poznan University of Technology
Piotrowo 2, 60-965 Poznan, Poland
andrzej.marciniak@put.poznan.pl

Abstract. To study the heat or diffusion equation, the Crank-Nicolson method is often used. This method is unconditionally stable and has the order of convergence $O(k^2 + h^2)$, where k and h are mesh constants. Using this method in conventional floating-point arithmetic, we get solutions including not only the method error, but also representation and rounding errors. Therefore, we propose an interval version of the Crank-Nicolson method from which we would like to obtain solutions including the discretization error. Applying such a method in interval floating-point arithmetic allows one to obtain solutions including all possible numerical errors. Unfortunately, for the proposed interval version of Crank-Nicolson method, we are not able to prove that the exact solution belongs to the interval solutions obtained. Thus, the presented method should be modified in the nearest future to fulfil this necessary condition. A numerical example is presented. Although in this example the exact solution belongs to the interval solutions obtained, but the so-called wrapping effect significantly increases the widths of these intervals.

Keywords: heat equation, Crank-Nicolson method, interval methods, floating-point interval arithmetic.

1 Introduction

In a number of our previous papers we developed interval methods for solving the initial value problem for ordinary differential equations (see e.g. [1] - [8]). These methods rely on on conventional Runge-Kutta and multistep methods. We have summarized our previous research in [9].

Now, our efforts are directed to construct similar methods for solving a variety of problems in partial-differential equations. It seems that it is possible to use the same technique as for ordinary differential equations. In [10], we have proposed an interval method for solving the Poisson equation. Here, we propose interval method based on the Crank-Nicolson scheme for solving the heat equation.

It should be mentioned that there are several other approaches to verify numerical solutions of different problems in partial-differential equations (see e.g. [11] – [15]).

K. Jónasson (Ed.): PARA 2010, Part II, LNCS 7134, pp. 120–126, 2012.
© Springer-Verlag Berlin Heidelberg 2012

2 Application of the Crank-Nicolson Method to Solve the Heat Equation

The parabolic partial-differential equation we consider in this paper is the heat or diffusion equation (one-dimensional in space)

$$\frac{\partial u}{\partial t}(x,t) = \alpha^2 \frac{\partial^2 u}{\partial x^2}(x,t), \quad a \le x \le b, \quad t > 0, \tag{1}$$

subject to the conditions

$$u(a,t) = 0, \quad u(b,t) = 0, \quad t > 0,$$

and

$$u(x,0) = f(t), \quad a \le x \le b.$$

The approach one uses to approximate the solution to this problem involves finite differences.

First, we select two mesh constants h and k with the stipulation that $m = (b-a)/h$ is an integer. The grid points are (x_i, t_j), where $x_i = a + ih$ for $i = 0, 1, \ldots, m$, and $t_j = jk$ for $j = 0, 1, 2, \ldots$. The implicit scheme, called the Crank-Nicolson method, is based on numerical approximations for solutions of the equation (1) at the points $(x_i, t_{j+1/2}) = (x_i, t_j + k/2)$ that lie between the rows in the grid.

Using the central-difference formula, we get

$$\frac{\partial u}{\partial t}(x_i, t_{j+1/2}) = \frac{u(x_i, t_{j+1}) - u(x_i, t_j)}{k} - \frac{k^2}{24}\frac{\partial^3 u}{\partial t^3}(x_i, \mu_j), \tag{2}$$

where $\mu_j \in (t_j, t_{j+1})$. The approximation used for

$$\frac{\partial^2 u}{\partial x^2}(x_i, t_{j+1/2})$$

is the average of the approximations for the terms

$$\frac{\partial^2 u}{\partial x^2}(x_i, t_{j+1}) \quad \text{and} \quad \frac{\partial^2 u}{\partial x^2}(x_i, t_j).$$

Formally, we have

$$\frac{\partial^2 u}{\partial x^2}(x_i, t_{j+1}) = \frac{\partial^2 u}{\partial x^2}(x_i, t_{j+1/2}) + \frac{k}{2}\frac{\partial^3 u}{\partial t \partial x^2}(x_i, t_{j+1/2}) + \frac{k^2}{8}\frac{\partial^4 u}{\partial t^2 \partial x^2}(x_i, \delta_j),$$

$$\frac{\partial^2 u}{\partial x^2}(x_i, t_j) = \frac{\partial^2 u}{\partial x^2}(x_i, t_{j+1/2}) - \frac{k}{2}\frac{\partial^3 u}{\partial t \partial x^2}(x_i, t_{j+1/2}) + \frac{k^2}{8}\frac{\partial^4 u}{\partial t^2 \partial x^2}(x_i, \tilde{\delta}_j),$$

where $\delta_j \in (t_{j+1/2}, t_{j+1})$ and $\tilde{\delta}_j \in (t_j, t_{j+1/2})$. Adding the above formulas, we get

$$\frac{\partial^2 u}{\partial x^2}(x_i, t_{j+1}) + \frac{\partial^2 u}{\partial x^2}(x_i, t_j) = 2\frac{\partial^2 u}{\partial x^2}(x_i, t_{j+1/2}) + \frac{k^2}{4}\frac{\partial^4 u}{\partial t^2 \partial x^2}(x_i, \tilde{\mu}_j),$$

where $\tilde{\mu}_j \in (t_j, t_j + 1)$, from which it follows that

$$\frac{\partial^2 u}{\partial x^2}(x_i, t_{j+1/2}) = \frac{1}{2}\left(\frac{\partial^2 u}{\partial x^2}(x_i, t_{j+1}) + \frac{\partial^2 u}{\partial x^2}(x_i, t_j)\right) - \frac{k^2}{8}\frac{\partial^4 u}{\partial t^2 \partial x^2}(x_i, \tilde{\mu}_j) . \quad (3)$$

Using the forward-difference method at the jth step in t and the backward-difference method at the $(j+1)$st step in t, we obtain

$$\frac{\partial^2 u}{\partial x^2}(x_i, t_j) = \frac{u(x_{i+1}, t_j) - 2u(x_i, t_j) + u(x_{i-1}, t_j)}{h^2}$$
$$- \frac{h^2}{12}\frac{\partial^4 u}{\partial x^4}(\xi_i, t_j) ,$$

$$\frac{\partial^2 u}{\partial x^2}(x_i, t_{j+1}) = \frac{u(x_{i+1}, t_{j+1}) - 2u(x_i, t_{j+1}) + u(x_{i-1}, t_{j+1})}{h^2}$$
$$- \frac{h^2}{12}\frac{\partial^4 u}{\partial x^4}(\tilde{\xi}_i, t_{j+1}) ,$$

$$(4)$$

where $\xi_i, \tilde{\xi}_i \in (x_{i-1}, x_{i+1})$. Neglecting the error terms in (2), (3) and (4), we get the following expression:

$$v_{i,j+1} - v_{ij} - \frac{\alpha^2 k}{2h^2}(v_{i+1,j} - 2v_{ij} + v_{i-1,j} + v_{i+1,j+1} - 2v_{i,j+1} + v_{i-1,j+1}) = 0 , \quad (5)$$

where v_{ij} approximates $u(x_i, y_j)$. This is the Crank-Nicolson method which has a local rounding error of order $O(k^2 + h^2)$ (see e.g. [16] – [18]), provided that the usual differentiability conditions are satisfied.

3   An Interval Crank-Nicolson Method

Taking the local rounding errors into consideration, the equation (5) can be written in the form

$$-\frac{\lambda}{2}v_{i-1,j+1} + (1 + \lambda)v_{i,j+1} - \frac{\lambda}{2}v_{i+1,j+1}$$

$$= \frac{\lambda}{2}v_{i-1,j} + (1 - \lambda)v_{ij} + \frac{\lambda}{2}v_{i+1,j} - \frac{\alpha^2 k h^2}{24}\left[\frac{\partial^4 u}{\partial x^4}(\xi_i, t_j) + \frac{\partial^4 u}{\partial x^4}(\tilde{\xi}_i, t_{j+1})\right] \quad (6)$$

$$- \frac{\alpha^2 k^3}{8}\frac{\partial^4 u}{\partial t^2 \partial x^2}(x_i, \tilde{\mu}_j) + \frac{k^3}{24}\frac{\partial^3 u}{\partial t^3}(x_i, \mu_j) ,$$

where $\lambda = \alpha^2 k/h^2$.

Let us assume that

$$\left|\frac{\partial^3 u}{\partial t^3}\right| \leq N , \left|\frac{\partial^3 u}{\partial t \partial x^2}\right| \leq M , \quad (7)$$

i.e.

$$\frac{\partial^3 u}{\partial t^3}(x, t) \in [-N, N] ,$$

and
$$\frac{\partial^3 u}{\partial t \partial x^2}(x,t) \in [-M, M] \,,$$

where $M, N = $ const. From (1) we have

$$\frac{\partial^3 u}{\partial t^3}(x,t) - \alpha^2 \frac{\partial^4 u}{\partial t^2 \partial x^2}(x,t) = 0 \,,$$

$$\frac{\partial^3 u}{\partial x^2 \partial t}(x,t) - \alpha^2 \frac{\partial^4 u}{\partial x^4}(x,t) = 0 \,.$$

Thus, from (7) it follows that

$$\left| \frac{\partial^4 u}{\partial t^2 \partial x^2}(x,t) \right| = \frac{1}{\alpha^2} \left| \frac{\partial^3 u}{\partial t^3}(x,t) \right| \le \frac{1}{\alpha^2} N \,,$$

$$\left| \frac{\partial^4 u}{\partial x^4}(x,t) \right| = \frac{1}{\alpha^2} \left| \frac{\partial^3 u}{\partial t \partial x^2}(x,t) \right| \le \frac{1}{\alpha^2} M \,,$$

and it means that

$$\frac{\partial^4 u}{\partial t^2 \partial x^2}(x,t) \in \frac{1}{\alpha^2}[-N, N] \,,$$

and

$$\frac{\partial^4 u}{\partial x^4}(x,t) \in \frac{1}{\alpha^2}[-M, M] \,.$$

Taking into account the above relations, we define an interval version of the Crank-Nicolson methods as follows:

$$-\frac{\lambda}{2} V_{i-1,j+1} + (1 + \lambda) V_{i,j+1} - \frac{\lambda}{2} V_{i+1,j+1} =$$

$$\tag{8}$$

$$= \frac{\lambda}{2} V_{i-1,j} + (1 - \lambda) V_{ij} + \frac{\lambda}{2} V_{i+1,j} + \frac{k}{6} \left(k^2[-N, N] - \frac{h}{2}[-M, M] \right) \,,$$

where $V_{ij} = [\underline{v}_{ij}, \overline{v}_{ij}]$.

The system of equations (8) is linear with a positive definite, symmetric, strictly diagonally dominant and tridiagonal matrix. It can be solved by an interval version of Crout reduction method.

In practice it can be difficult to determine the constants M and N since $u(x,t)$ is unknown. If it is impossible to determine M from any physical or other conditions of the problem considered, we propose to solve the problem by the conventional Crank-Nicolson method (5) and take

$$M \approx \frac{1.5}{kh^2} \max_{\substack{i=1,2,\ldots,m-1 \\ j=1,2,\ldots,n-1}} |v_{i+1,j} - v_{i+1,j-1} - 2(v_{ij} - v_{i,j-1}) + v_{i-1,j} - v_{i-1,j-1}| \,,$$

and

$$N \approx \frac{1.5}{k^3} \max_{\substack{i=0,1,\ldots,m \\ j=1,2,\ldots,n-2}} |v_{i,j+2} - 3v_{i,j+1} + 3v_{ij} - v_{i,j-1}| \,.$$

4   A Numerical Example

To have a view on the interval solutions obtained, let us consider a problem for which the exact solution is known. Let the method (8) be used to approximate the solution to the problem consisting of the equation

$$\frac{\partial u}{\partial t}(x,t) - \frac{\partial^2 u}{\partial x^2}(x,t) = 0 \ ,$$

subject to the conditions

$$u(x,0) = \cos\left(\frac{\pi x}{2}\right), \quad -1 \le x \le 1 \ ,$$

and

$$u(-1,t) = u(1,t) = 0, \quad t \ge 0 \ .$$

The exact solution of the above problem is as follows:

$$u(x,t) = \exp\left(-\frac{\pi^2 t}{4}\right)\cos\left(\frac{\pi x}{2}\right) \ . \tag{9}$$

The graph of this solution for $0 \le t \le 1$ is presented in Figure 1, and a particular value is

$$u(0,0.05) \approx 0.88393649689751144 \ .$$

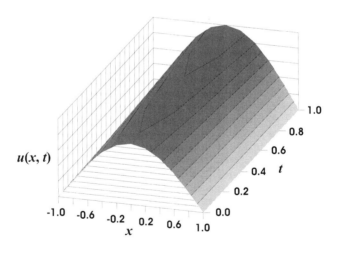

Fig. 1. The graph of the function (9)

Using the method (8) with $M = \pi^4/16$, $N = \pi^6/64$, $m = 20$, i.e. $h = 0.1$, and $k = 0.05$, and carried out all calculations in floating-point interval arithmetic (using the Delphi Pascal unit *IntervalArithmetic* described in [9]), we obtain

$$V(0,0.05) = [0.88374057912065346, 0.88430495470686242] \ .$$

The width of this interval is approximately equal to 5.6×10^{-4}. Unfortunately, for larger values of t we observe a sudden increase of the widths of interval solutions. This is caused by the so-called wrapping effect.

Let us note that the exact solution belongs to the interval solutions obtained. Although in many other numerical experiments carried out we have observed the same, it is not true in general. We have a number of examples in which the exact solution is outside interval solutions V_{ij} obtained by the method (8). It follows from the fact that it is impossible to prove that $u(x_i, t_j) \in V_{ij}$.

5   Conclusions and Further Studies

The interval method (8) based on the conventional Crank-Nicolson scheme is only a proposal for solving parabolic partial-differential equations such as the heat equation. Applying this method in floating-point interval arithmetic we can automatically include the representation and rounding errors into interval solutions.

Since we are not capable of proving that the exact solution belongs to the interval solutions obtained for the method (8), the presented method should be modified to fulfil this necessary condition. Moreover, the method should be also modified with respect to the increase of interval widths for the larger number of steps.

References

1. Gajda, K., Marciniak, A., Szyszka, B.: Three- and four-stage implicit interval methods of Runge-Kutta Type. Computational Methods in Science and Technology 6, 41–59 (2000)
2. Jankowska, M., Marciniak, A.: Implicit interval multistep methods for solving the initial value problem. Computational Methods in Science and Technology 8(1), 17–30 (2002)
3. Jankowska, M., Marciniak, A.: On explicit interval methods of Adams-Bashforth type. Computational Methods in Science and Technology 8(2), 46–47 (2002)
4. Jankowska, M., Marciniak, A.: On two families of implicit interval methods of Adams-Moulton type. Computational Methods in Science and Technology 12(2), 109–114 (2006)
5. Marciniak, A., Szyszka, B.: One- and two-stage implicit interval methods of Runge-Kutta type. Computational Methods in Science and Technology 5, 53–65 (1999)
6. Marciniak, A.: Implicit interval methods for solving the initial value problem. Numerical Algorithms 37, 241–251 (2004)
7. Marciniak, A.: Multistep interval methods of Nyström and Milne-Simpson types. Computational Methods in Science and Technology 13(1), 23–40 (2007)
8. Marciniak, A.: On multistep interval methods for solving the initial value problem. Journal of Computational and Applied Mathematics 199(2), 229–238 (2007)
9. Marciniak, A.: Selected Interval Methods for Solving the Initial Value Problem. Publishing House of Poznan University of Technology (2009)
10. Marciniak, A.: An interval difference method for solving the Poisson equation – the first approach. Pro Dialog 24, 49–61 (2008)

11. Nakao, M.T.: Solving nonlinear parabolic problems with result verification. Part I: one-space dimensional case. Journal of Computational and Applied Mathematics 38, 323–324 (1991)
12. Nagatou, K., Yamamoto, N., Nakao, M.T.: An approach to the numerical verification of solutions for nonlinear elliptic problem with local uniqueness. Numerical Functional Analysis and Optimization 20, 543–565 (1999)
13. Plum, M.: Safe numerical error bounds for solutions of nonlinear elliptic boundary value problems. In: Alefeld, G., Rohn, J., Rump, S. (eds.) Symbolic Algebraic Methods and Verification Methods, pp. 195–205. Springer, Heidelberg (2001)
14. Plum, M.: Computer-assisted enclosure methods for elliptic differential equations. Linear Algebra and its Applications 324, 147–187 (2001)
15. Nagatou, K., Hashimoto, K., Nakao, M.T.: Numerical verification of stationary solutions for Navier-Stokes problems. Journal of Computational and Applied Mathematics 199, 445–451 (2007)
16. Smith, G.D.: Numerical Solution of Partial Differential Equations: Finite Difference Methods. Oxford University Press (1995)
17. Lapidus, L., Pinder, G.F.: Numerical Solution of Partial Differential Equations in Science and Engineering. J. Wiley & Sons (1982)
18. Morton, K.W., Mayers, D.F.: Numerical Solution of Partial Differential Equations: An Introduction. Cambridge University Press (2005)

Parallel Detection of Interval Overlapping

Marco Nehmeier, Stefan Siegel, and Jürgen Wolff von Gudenberg

Institute of Computer Science, University of Würzburg
Am Hubland, D 97074 Würzburg, Germany
{nehmeier,st.siegel,wolff}@informatik.uni-wuerzburg.de

Abstract. In this paper we define the interval overlapping relation and develop a parallel hardware unit for its realization. As one application we consider the interval comparisons. It is shown that a detailed classification of the interval overlapping relation leads to a reduction of floating-point comparisons in common applications.

Keywords: interval arithmetic, interval relations, hardware unit.

1   Introduction

Detection of overlapping intervals is a problem that occurs in many application areas. The detection of overlapping boxes in computer graphics, overlapping time slots in scheduling problems, containment or membership tests, or enclosure tests in self-verifying scientific computing algorithms are some examples. In most of the applications, like in scheduling, the information whether 2 intervals overlap or not is not sufficient, but we also like to know how they overlap: completely contained in the interior vs. touching one bound, e.g. In this paper we, hence, define a general relation that describes the kind of overlapping between two one-dimensional intervals, and develop a hardware unit for its evaluation. As our intended first application of this unit we discuss the comparison relations in interval arithmetic.

An interval is a connected, closed, not necessarily bounded subset of the reals. It can be represented by its two bounds.

$$X := [\underline{x}, \overline{x}] = \{x \in \mathbb{R} \mid \underline{x} \le x \le \overline{x}\} \tag{1}$$

In this definition $\underline{x}$ can be $-\infty$, $\overline{x}$ can be $+\infty$, but the infinities never are members of an interval. The set of all intervals including the empty set is denoted as $\overline{\mathbb{IR}}$.

2   Definition

The interval overlapping relation is not a boolean relation but delivers 14 different states describing all the possible situations that occur when the relative positions of 2 intervals are regarded with respect to overlapping. Table 1 illustrates the meaning of the relation. Each row represents a different state. The

K. Jónasson (Ed.): PARA 2010, Part II, LNCS 7134, pp. 127–136, 2012.

columns 2 through 5 contain sketches of the scene with the interval A at the bottom and B on the top. Singleton or point intervals are denoted as dots where appropriate. As usual numbers grow from left to right. Let Q be the set of the 13 cases for non-empty intervals [1] listed in Tab. 1.

Definition 1 (Interval Overlapping). *The overlapping relation for two non-empty intervals is defined by the mapping*

$$\circledcirc_1 : (\overline{\mathbb{IR}} \setminus \emptyset) \times (\overline{\mathbb{IR}} \setminus \emptyset) \rightarrow Q \tag{2}$$

$$A \circledcirc_1 B \mapsto q_i \in Q , \quad i = 1 \ldots 13 \tag{3}$$

State 14, not in the table, characterises that one of the operands is empty. Then there is no overlapping at all.

Remark 1. Note that all possible situations are considered.

Table 1. The 13 different cases of the interval overlapping relation for non-empty intervals

$A \circledcirc_1 B$					$A \circledcirc B$	$A \subseteq B$	$A \supseteq B$	$A = B$	$A \cap B = \emptyset$
q_1					0001				•
q_2					0101				
q_3					1101				
q_4					0111	•			
q_5					1111	•			
q_6					1011	•			
q_7					1110				
q_8					1010				
q_9					0010				•
q_{10}					1001		•		
q_{11}					1100		•		
q_{12}					0110		•		
q_{13}					0011	•	•	•	

We represent the states by 4-bit strings resulting from specific comparisons of the bounds of the input intervals.

Definition 2 (Interval Overlapping Representation). *The interval over-lapping relation*

$$\infty : (\overline{\mathbb{IR}} \setminus \emptyset) \times (\overline{\mathbb{IR}} \setminus \emptyset) \rightarrow \{0,1\}^4 \tag{4}$$

$$A \infty B \mapsto (r_1, r_2, r_3, r_4) \tag{5}$$

for two non-empty intervals $A = [\underline{a}, \overline{a}], B = [\underline{b}, \overline{b}] \in \overline{\mathbb{IR}} \setminus \emptyset$ is defined by:

$$r_1 := ((\underline{a} \neq \underline{b}) \oplus (((\overline{a} \leq \underline{b}) \vee (\underline{a} > \overline{b}))$$
$$\wedge ((\underline{a} \neq \underline{b}) \wedge (\overline{a} \neq \overline{b})))) \tag{6}$$

$$r_2 := ((\overline{a} \neq \overline{b}) \oplus (((\overline{a} < \underline{b}) \vee (\underline{a} \geq \overline{b}))$$
$$\wedge ((\underline{a} \neq \underline{b}) \wedge (\overline{a} \neq \overline{b})))) \tag{7}$$

$$r_3 := (\underline{a} \geq \underline{b}) \tag{8}$$

$$r_4 := (\overline{a} \leq \overline{b}) \tag{9}$$

The function ∞ can be written as a composition of

$$\infty = \xi \circ \infty_1$$

where $\xi : Q \rightarrow \{0,1\}^4$ maps the state into a representation as defined in Tab. 1 columns 1 and 6. The 6th column of the table shows the state of overlapping coded into 4 bits.

In principle, each result bit refers to one comparison of the bounds.

$$r_1 := (\underline{a} \neq \underline{b})$$
$$r_2 := (\overline{a} \neq \overline{b})$$
$$r_3 := (\underline{a} \geq \underline{b})$$
$$r_4 := (\overline{a} \leq \overline{b})$$

With these simple definitions we could not separate the states q_1, q_2, q_3 or q_7, q_8, q_9, respectively. Therefore we developed the comparisons (6) and (7).

Corollary 1. *For two non-empty intervals $A = [\underline{a}, \overline{a}], B = [\underline{b}, \overline{b}] \in \overline{\mathbb{IR}} \setminus \emptyset$ the states with bitset "0000", "0100" or "1000" do not occur as a result of the interval overlapping relation $A \infty B$.*

Proof.

$$(A \neq \emptyset) \wedge (B \neq \emptyset) \wedge (\neg r_3 \wedge \neg r_4)$$
$$\overset{(8,9)}{\Rightarrow} (\underline{a} < \underline{b}) \wedge (\overline{a} > \overline{b})$$
$$\Rightarrow (\underline{a} \neq \underline{b}) \wedge (\overline{a} \neq \overline{b}) \wedge (\overline{a} > \underline{b}) \wedge (\underline{a} < \overline{b})$$
$$\overset{(6,7)}{\Rightarrow} r_1 \wedge r_2$$

$\square$

3   Comparisons in Interval Arithmetic

Interval arithmetic is currently being standardized. Our definition of intervals as connected, closed, not necessarily bounded subsets of the reals in section 1 follows the presumable standard P1788 [3]. Various competing sets of interval comparisons are under discussion. Nearly every combination of operator symbol and quantifier is proposed in the Vienna proposal [7]. That includes the so called "certainly" and "possibly" operations where the relation holds for some or all members of an interval, respectively. A smaller set of comparisons is given in the book [5].

There is, however, consensus that the subset-relation, either interior or proper or equal, the membership of a point in an interval, and the test for disjointness are mandatory.

In the following propositions we show that all these comparisons can easily be obtained from the interval overlapping relation.

Proposition 1 (Set Relations). *For two non-empty intervals* $A = [\underline{a}, \overline{a}]$, $B = [\underline{b}, \overline{b}] \in \overline{\mathbb{IR}} \setminus \emptyset$ *the three relations* $=, \subseteq$ *and* $\supseteq$ *as well the test for disjointness are implied by the interval overlapping relation* $A \otimes B$ *as follows:*

$$A \subseteq B \Leftrightarrow r_3 \wedge r_4 \tag{10}$$

$$A \supseteq B \Leftrightarrow (r_1 \oplus r_3) \wedge (r_2 \oplus r_4) \tag{11}$$

$$A = B \Leftrightarrow \neg r_1 \wedge \neg r_2 \wedge r_3 \wedge r_4 \tag{12}$$

$$A \cap B = \emptyset \Leftrightarrow \neg r_1 \wedge \neg r_2 \wedge \neg(r_3 \wedge r_4) \tag{13}$$

Proof.
   (10):

$$A \subseteq B \Leftrightarrow (\underline{b} \leq \underline{a}) \wedge (\overline{a} \leq \overline{b})$$

$$\overset{(8,9)}{\Leftrightarrow} r_3 \wedge r_4$$

(11):

$$A \supseteq B \Leftrightarrow (\underline{a} \leq \underline{b}) \wedge (\overline{b} \leq \overline{a})$$

$$\Leftrightarrow \neg(\underline{a} > \underline{b}) \wedge \neg(\overline{b} > \overline{a})$$

$$\Leftrightarrow ((\neg(\underline{a} \geq \underline{b}) \wedge (\underline{a} \neq \underline{b})) \vee ((\underline{a} \geq \underline{b}) \wedge \neg(\underline{a} \neq \underline{b})))$$

$$\wedge ((\neg(\overline{a} \leq \overline{b}) \wedge (\overline{a} \neq \overline{b})) \vee ((\overline{a} \leq \overline{b}) \wedge \neg(\overline{a} \neq \overline{b})))$$

$$\overset{\text{Def. } 2}{\Leftrightarrow} (r_1 \oplus r_3) \wedge (r_2 \oplus r_4)$$

(12):

$$A = B \Leftrightarrow (A \subseteq B) \wedge (A \supseteq B)$$

$$\overset{(10,11)}{\Leftrightarrow} r_3 \wedge r_4 \wedge (r_1 \oplus r_3) \wedge (r_2 \oplus r_4)$$

$$\Leftrightarrow \neg r_1 \wedge \neg r_2 \wedge r_3 \wedge r_4$$

(13):

$$A \cap B = \emptyset \Leftrightarrow (\overline{a} < \underline{b}) \vee (\underline{a} > \overline{b})$$
$$\Leftrightarrow (((\underline{a} \neq \underline{b}) \wedge (\overline{a} \neq \overline{b})) \wedge (\overline{a} \leq \underline{b}) \wedge (\overline{a} < \underline{b})$$
$$\wedge \neg(\underline{a} \geq \underline{b}) \wedge (\overline{a} \leq \overline{b}))$$
$$\vee (((\underline{a} \neq \underline{b}) \wedge (\overline{a} \neq \overline{b})) \wedge (\underline{a} \geq \overline{b}) \wedge (\underline{a} > \overline{b})$$
$$\wedge (\underline{a} \geq \underline{b}) \wedge \neg(\overline{a} \leq \overline{b}))$$
$$\overset{\text{Def. 2}}{\Leftrightarrow} \neg r_1 \wedge \neg r_2 \wedge (r_3 \oplus r_4)$$

□

Besides the formal proofs given in this section the formulas can be verified with the help of Tab. 1.

Corollary 2 (Set Membership). *The set membership* $a \in B$ *with* $a \in \mathbb{R}$, $B \in \overline{\mathbb{IR}} \setminus \emptyset$ *can be deduced to*

$$[a, a] \otimes B = (r_1, r_2, 1, 1) \tag{14}$$

with $r_1, r_2 \in \{0, 1\}$.

Proof.

$$a \in B \Leftrightarrow [a, a] \subseteq B \overset{(10)}{\Leftrightarrow} [a, a] \otimes B = (r_1, r_2, 1, 1)$$

□

Up to now we only have considered non-empty intervals. In many realizations of interval arithmetic the empty interval is represented as a pair of NaNs.

$$\emptyset := [\text{NaN}, \text{NaN}] \tag{15}$$

Corollary 3 (Empty Interval). *For two intervals* $A, B \in \overline{\mathbb{IR}}$ *where at least one of them is empty, the following equation*

$$A \otimes B = (0, 0, 0, 0) \tag{16}$$

holds if the empty interval is represented by two NaN*s.*

Proof. As defined in the IEEE standard for floating-point arithmetic [4], comparisons to NaN always return `false`. □

Hence, state 14 happens to be "0000" which was an unused bit combination so far.

Remark 2. With this definition of the empty set we can omit the assumption of non-empty intervals in Prop. 1 and Cor. 2.

Proposition 2 (Order Relations). *The order relations* $A \leq B, A \prec B,$ $A \geq B$ *and* $A \succ B$ *with* $A = [\underline{a}, \overline{a}], B = [\underline{b}, \overline{b}] \in \overline{\mathbb{IR}} \setminus \emptyset$ *follow from the interval overlapping relation* $A \varpi B$ *by*

$$A \leq B :\Leftrightarrow (\underline{a} \leq \underline{b}) \wedge (\overline{a} \leq \overline{b}) \Leftrightarrow (\neg r_1 \vee \neg r_3) \wedge r_4 \tag{17}$$

$$A \prec B :\Leftrightarrow (\overline{a} < \underline{b}) \qquad\qquad \Leftrightarrow \neg r_1 \wedge \neg r_2 \wedge \neg r_3 \wedge r_4 \tag{18}$$

$$A \geq B :\Leftrightarrow (\underline{a} \geq \underline{b}) \wedge (\overline{a} \geq \overline{b}) \Leftrightarrow (\neg r_2 \vee \neg r_4) \wedge r_3 \tag{19}$$

$$A \succ B :\Leftrightarrow (\underline{a} > \overline{b}) \qquad\qquad \Leftrightarrow \neg r_1 \wedge \neg r_2 \wedge r_3 \wedge \neg r_4 \tag{20}$$

Proof.
(17):

$$(\underline{a} \leq \underline{b}) \wedge (\overline{a} \leq \overline{b}) \Leftrightarrow ((\underline{a} = \underline{b}) \vee \neg(\underline{a} \geq \underline{b})) \wedge (\overline{a} \leq \overline{b})$$

$$\overset{(6,8,9)}{\Leftrightarrow} (\neg r_1 \vee \neg r_3) \wedge r_4$$

(18):

$$(\overline{a} < \underline{b}) \Leftrightarrow (\underline{a} \neq \underline{b}) \wedge (\overline{a} \neq \overline{b}) \wedge (\overline{a} < \underline{b}) \wedge \neg(\underline{a} \geq \underline{b}) \wedge (\overline{a} \leq \overline{b})$$

$$\overset{\text{Def. 2}}{\Leftrightarrow} \neg r_1 \wedge \neg r_2 \wedge \neg r_3 \wedge r_4$$

Proof of (19) and (20) analogous to (17) and (18). $\qquad\qquad\qquad\qquad\square$

Closely related with comparisons are the lattice operations like interval hull or intersection. They also can exploit the information obtained by one computation of the overlapping relation.

Proposition 3 (Intersection). *The intersection* $A \cap B$ *with* $A = [\underline{a}, \overline{a}],$ $B = [\underline{b}, \overline{b}] \in \overline{\mathbb{IR}}$ *follows from the interval overlapping relation* $A \varpi B$ *by*

$$A \cap B := \begin{cases} [\underline{a}, \overline{a}] & \text{if } (r_3 \wedge r_4) \\ \emptyset & \text{otherwise if } (\neg r_1 \wedge \neg r_2) \\ [\underline{a}, \overline{b}] & \text{otherwise if } (r_3) \\ [\underline{b}, \overline{a}] & \text{otherwise if } (r_4) \\ [\underline{b}, \overline{b}] & \text{otherwise} \end{cases} \tag{21}$$

Proof.
if $(r_3 \wedge r_4)$:

$$r_3 \wedge r_4 \overset{(10)}{\Rightarrow} A \subseteq B$$

$$\Rightarrow A \cap B = [\underline{a}, \overline{a}]$$

otherwise if $(\neg r_1 \wedge \neg r_2)$:

$$\neg r_1 \wedge \neg r_2 \wedge \neg(r_3 \wedge r_4) \overset{(13), \text{ Cor. 3}}{\Rightarrow} A \cap B = \emptyset$$

otherwise if (r_3):

$$(r_1 \vee r_2) \wedge r_3 \wedge \neg r_4 \overset{(8,9,13)}{\Rightarrow} (\underline{a} \geq \underline{b}) \wedge (\overline{a} > \overline{b}) \wedge (A \cap B \neq \emptyset)$$
$$\Rightarrow \quad A \cap B = [\underline{a}, \overline{b}]$$

otherwise if (r_4):

$$(r_1 \vee r_2) \wedge \neg r_3 \wedge r_4 \overset{(8,9,13)}{\Rightarrow} (\underline{a} < \underline{b}) \wedge (\overline{a} \leq \overline{b}) \wedge (A \cap B \neq \emptyset)$$
$$\Rightarrow \quad A \cap B = [\underline{b}, \overline{a}]$$

otherwise:

$$(r_1 \vee r_2) \wedge \neg r_3 \wedge \neg r_4 \overset{\text{Cor. 1}}{\Rightarrow} r_1 \wedge r_2 \wedge \neg r_3 \wedge \neg r_4$$
$$\overset{(11)}{\Rightarrow} A \supseteq B$$
$$\Rightarrow \quad A \cap B = [\underline{b}, \overline{b}]$$

□

4   Hardware Unit

Remark 3. We realize that 8 independent floating-point comparisons are needed. The hardware unit given in Fig. 1 thus consist of 8 comparators. They all work in parallel followed by at most 3 gates to obtain the result.

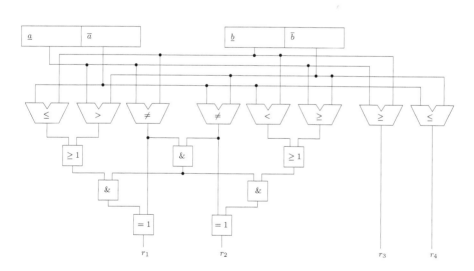

Fig. 1. Logic circuit of the interval overlapping relation

Algorithm 1. INewton (classical)

Input:
f : function
Y : interval
ϵ : epsilon
$yUnique$: flag
$Zero$: list of enclosing intervals
$Info$: flag vector
N : number
Output: $[Zero, Info, N]$
begin

  if $0 \notin f(Y)$ then
    | return *(Zero,Info,N)*
  $c \leftarrow \mu(Y)$;
  /* extended division   */
  $[Z_1, Z_2] \leftarrow f(c)/f'(Y)$;
  $[Z_1, Z_2] \leftarrow c - [Z_1, Z_2]$;
  $V_1 \leftarrow Y \cap Z_1$;
  $V_2 \leftarrow Y \cap Z_2$;
  if $V_1 = Y$ then
    $V_1 \leftarrow [\underline{y}, c]$;
    $V_2 \leftarrow [c, \overline{y}]$;

  if $V_1 \neq \emptyset$ and $V_2 = \emptyset$ then
    $yUnique \leftarrow yUnique$ or
    $V_1 \subset Y$;

  foreach $i = 1, 2$ do
    if $V_i = \emptyset$ then
      continue;

    if $drel(V_i) < \epsilon$ then
      $N = N + 1$;
      $Zero[N] = V_i$;
      $Info[N] = yUnique$;
    else
      $INewton(f, V_i, \epsilon,$
      $yUnique, Zero, Info, N)$;

  return *(Zero,Info,N)*;
end

Algorithm 2. INewtonRel (relational)

Input:
f : function
Y : interval
ϵ : epsilon
$yUnique$: flag
$Zero$: list of enclosing intervals
$Info$: flag vector
N : number
Output: $[Zero, Info, N]$
begin

  if $0 \notin f(Y)$ then
    | return *(Zero,Info,N)*
  $c \leftarrow \mu(Y)$;
  /* extended division   */
  $[Z_1, Z_2] \leftarrow f(c)/f'(Y)$;
  $[Z_1, Z_2] \leftarrow c - [Z_1, Z_2]$;
  $R_1 \leftarrow Y \oslash Z_1$;
  $R_2 \leftarrow Y \oslash Z_2$;
  if $R_1.subseteq()$ then
    $V_1 \leftarrow [\underline{y}, c]$;
    $V_2 \leftarrow [c, \overline{y}]$;
    $bisected \leftarrow true$;
  else if not $R_1.disjoint()$ and
  $R_2.disjoint()$ then
    $yUnique \leftarrow yUnique$ or
    $R_1.state() ==$ containedBy;
  foreach $i = 1, 2$ do
    if not $bisected$ then
      if $R_i.disjoint()$ then
      | continue;
      $R_i \leftarrow Y \oslash Z_i$;
      $V_i \leftarrow R_i.intersect()$;
    if $drel(V_i) < \epsilon$ then
      $N = N + 1$;
      $Zero[N] = V_i$;
      $Info[N] = yUnique$;
    else
      $INewtonRel(f, V_i, \epsilon,$
      $yUnique, Zero, Info, N)$;

  return *(Zero,Info,N)*;
end

In this paper we represent the states with 4 bits to have a compact numbering. We then need 8 comparisons to separate all states.

One may argue that specific relations like $A \subseteq B$ or $A = B$ can already be checked by 2 (parallel) floating-point comparisons. But the benefit of our general interval overlapping relation is that, if all 8 comparisons have been computed (in one parallel step), we have enough information to perform dependent interval comparisons only by bit-operations. The same holds for intersection that usually needs 3 floating-point comparisons. See the discussion of the interval Newton method in section 5.

5   Example

As an example for the use of the interval overlapping relation we discuss the extended interval Newton method [2]. In the usual formulation in Alg. 1 up to 6 interval comparisons and 2 intersections are used. We observe, however, that the same intervals are compared several times. Hence, the bitsets R_1 and R_2 gather all information with 2 calls of the interval overlapping relation in Alg. 2.

This reduction of the use of interval comparisons and intersections is done by replacing the 2 intersections by 2 calls of the interval overlapping relation storing the precise information about the relative positions of the interval operands. Then we can deduce all the necessery interval comparisons depending on the results and operands of the replaced intersections by applying the rules of Prop. 1 to the precomputed states R_1 and R_2. That means that the interval comparisons are replaced by bitset operations.

Additionally we introduce a flag *bisected* to determine, if a bisection of the input interval was performed. Otherwise we can use the stored information R_1 and R_2 to catch up the replaced intersection for an recursive call of the algorithm by applying the rules of Prop. 3.

6   Future Work

A companion paper [6] concerning interval comparisons has been submitted to the IEEE interval standard working group P1788. In this paper we emphasize the theoretical influence of the interval overlapping relation as a foundation for interval comparisons. An abstract datatype for the specification of the interval overlapping relation has been introduced. We further want to study its interface in an object oriented environment.

We plan to explore other applications in the area of computer graphics and time scheduling.

The hardware unit will be extended and optimized. Its collaboration with other hardware units for interval arithmetic will be discussed.

References

1. Allen, J.F.: Maintaining knowledge about temporal intervals. Communications of the ACM 26(11), 832–843 (1983)
2. Hammer, R., Hocks, M., Kulisch, U., Ratz, D.: C++ Toolbox for Verified Computing I: Basic Numerical Problems. Springer, Berlin (1995)
3. IEEE Interval Standard Working Group - P1788,
   http://grouper.ieee.org/groups/1788/
4. IEEE Task P754. IEEE 754-2008, Standard for Floating-Point Arithmetic. IEEE, New York (August 2008)
5. Kulisch, U.: Computer Arithmetic and Validity. Theory, Implementation, and Applications. de Gruyter Studies in Mathematics 33, Berlin (2008)
6. Nehmeier, M., Wolff von Gudenberg, J.: Interval Comparisons and Lattice Operations based on the Interval Overlapping Relation. Technical Report 476, Institute of Computer Science, University of Würzburg (September 2010)
7. Neumaier, A.: Vienna proposal for interval standardization, Version 3.0 (November 2008) in [3]

Using the Second-Order Information in Pareto-set Computations of a Multi-criteria Problem

Bartłomiej Jacek Kubica and Adam Woźniak

Institute of Control and Computation Engineering,
Warsaw University of Technology, Poland
bkubica@elka.pw.edu.pl, A.Wozniak@ia.pw.edu.pl

Abstract. The paper presents an extension of a previously developed interval method for solving multi-criteria problems [13]. The idea is to use second order information (i.e., Hesse matrices of criteria and constraints) in a way analogous to global optimization (see e.g. [6], [9]). Preliminary numerical results are presented and parallelization of the algorithm is considered.

Keywords: Pareto-front, Pareto-set, multi-criteria analysis, interval computations, second-order optimality conditions.

1   Introduction

We consider seeking the Pareto-set of the following problem:

$$\min_{x} q_k(x) \qquad k = 1, \dots, N \ , \tag{1}$$

$$\text{s.t.}$$

$$g_j(x) \leq 0 \qquad j = 1, \dots, m \ ,$$

$$x_i \in [\underline{x}_i, \overline{x}_i] \qquad i = 1, \dots, n \ .$$

Definition 1. *A feasible point x is* Pareto-optimal *(nondominated), if there exists no other feasible point x' such that:*

$$(\forall k) \quad q_k(y) \leq q_k(x) \ and$$
$$(\exists i) \quad q_i(y) < q_i(x) \ .$$

The set $P \subset \mathbb{R}^n$ of all Pareto-optimal points (Pareto-points) is called the Pareto-set.

Definition 2. *The Pareto-front is the image of the Pareto-set, i.e., the set of criteria values for all nondominated points.*

In the sequel one more definition will be needed.

Definition 3. *A point y dominates a set B, iff $D(y) \cap B = \emptyset$ and similarly a set B' dominates a set B, iff $(\forall y \in B')D(y) \cap B = \emptyset$.*

K. Jónasson (Ed.): PARA 2010, Part II, LNCS 7134, pp. 137–147, 2012.

The interpretation of the definitions is straightforward. A feasible point is Pareto-optimal if there is no other feasible point that would reduce some criterion without causing a simultaneous increase in at least one other criterion. Pareto-front is the image of Pareto-set in criterion space and D is the cone of domination in this space.

Interval methods allow to solve the problem of approximating the Pareto-set, using the branch-and-bound principle. Starting from the initial box and bisecting the boxes subsequently, we can quickly discard dominated boxes, enclosing the Pareto-set and Pareto-front with sets of boxes in decision and criteria spaces.

To discard or narrow boxes the algorithms use the following tools:

- checking if the box is non-dominated by other boxes (i.e., it may contain non-dominated points),
- set inversion of boxes from the criteria space to the decision space,
- the monotonicity test adapted to multi-criteria case (this test uses the first-order information).

No currently used interval algorithm ([3], [13], [16]) for computing the Pareto set uses the second-order information; gradients of the criteria and constraints are used, but not the Hesse matrices. The method of Toth and Fernandez [4] allows it by reducing the problem of Pareto-front seeking to repeated global optimization, but the approach can be applied to bi-criteria problems only.

Our idea is to extend the method proposed in [13] by using Hesse matrices of criteria and constraints in a way similar to well-known global optimization algorithms (see e.g. [6]), i.e., by solving the system of second order optimality conditions (in this case: Pareto-optimality conditions).

2   Generic Algorithm

In previous papers we developed an algorithm to seek the Pareto-set. It subdivides the criteria space in a branch-and-bound manner and inverts each of the obtained sets using a version of the SIVIA (Set Inversion Via Interval Analysis) procedure [8]. This version uses some additional tools (like the componentwise Newton operator) to speedup the computations.

The algorithm is expressed by the following pseudocode described with more details in previous papers ([13], [15]).

`compute_Pareto-set` $(\mathsf{q}(\cdot), \mathbf{x}^{(0)}, \varepsilon_y, \varepsilon_x)$
// $\mathsf{q}(\cdot)$ is the interval extension of the function
$\qquad q(\cdot) = (q_1, \ldots, q_N)(\cdot)$
// L is the list of quadruples $(\mathbf{y}, L_{\mathrm{in}}, L_{\mathrm{bound}}, L_{\mathrm{unchecked}})$
// for each quadruple: L_{in} is the list of interior boxes (in the decision space),
// L_{bound} – the list of boundary boxes and $L_{\mathrm{unchecked}}$ – of boxes to be checked yet
$\mathbf{y}^{(0)} = \mathsf{q}(\mathbf{x}^{(0)})$;
$L = \left\{ (\mathbf{y}^{(0)}, \{\}, \{\}, \{\mathbf{x}^{(0)}\}) \right\}$;
`while` (there is a quadruple in L, for which wid $\mathbf{y} \geq \varepsilon_y$)
$\qquad$ take this quadruple $(\mathbf{y}, L_{\mathrm{in}}, L_{\mathrm{bound}}, L_{\mathrm{unchecked}})$ from L;

```
        bisect y to y⁽¹⁾ and y⁽²⁾;
        for i = 1, 2
            apply SIVIA with accuracy εₓ to quadruple
                (y⁽ⁱ⁾, L_in, L_bound, L_unchecked);
            if (the resulting quadruple has a nonempty interior,
                i.e., L_in ≠ ∅)
                delete quadruples that are dominated by ȳ⁽ⁱ⁾;
            end if
            insert the quadruple to the end of L;
        end for
    end while
    // finish the Pareto-set computations
    for each quadruple in L do
        process boxes from L_unchecked until all of them get to L_in or L_bound;
    end do;
end compute_Pareto-set
```

Obviously, both loops in the above algorithm – the `while` loop and the `for each` loop can easily be parallelized.

3   Basic Idea

Let us formulate the set of Pareto optimality conditions; similar to the John conditions set [6]. For an unconstrained problem it has the following form (notation from [10] is used):

$$u_1 \cdot \frac{\partial q_1(x)}{\partial x_1} + \cdots + u_N \cdot \frac{\partial q_N(x)}{\partial x_1} = 0 , \qquad (2)$$

$$\cdots$$

$$u_1 \cdot \frac{\partial q_1(x)}{\partial x_n} + \cdots + u_N \cdot \frac{\partial q_N(x)}{\partial x_n} = 0 ,$$

$$u_1 + u_2 + \cdots + u_N = 1 ,$$

where $u_i \in [0, 1]$ $i = 1, \ldots, N$.

The above is a system of $(n+1)$ equations in $(n+N)$ variables. As the problem is supposed to have multiple criteria, clearly $N > 1$, which makes System (2) underdetermined. Solving underdetermined problems is less studied than well-determined ones (see paper [12] and references therein) and more difficult at the same time.

To consider a constrained multi-criteria problem, System (2) has to be extended slightly. In addition to multipliers u_i for all criteria $i = 1, \ldots, N$, we must have multipliers for all constraints: u_{N+j}, $j = 1, \ldots, m$.

The resulting system would take the following form:

$$u_1 \cdot \frac{\partial q_1(x)}{\partial x_1} + \cdots + u_N \cdot \frac{\partial q_N(x)}{\partial x_1} + \qquad (3)$$

$$+ u_{N+1} \cdot \frac{\partial g_1(x)}{\partial x_1} + \cdots + u_{N+m} \cdot \frac{\partial g_m(x)}{\partial x_1} = 0 ,$$

...

$$u_1 \cdot \frac{\partial q_1(x)}{\partial x_n} + \cdots + u_N \cdot \frac{\partial q_N(x)}{\partial x_n} +$$

$$+ u_{N+1} \cdot \frac{\partial g_1(x)}{\partial x_n} + \cdots + u_{N+m} \cdot \frac{\partial g_m(x)}{\partial x_n} = 0 ,$$

$$u_{N+1} \cdot g_1(x) = 0 ,$$

...

$$u_{N+m} \cdot g_m(x) = 0 ,$$

$$u_1 + u_2 + \cdots + u_N + u_{N+1} + \cdots + u_{N+m} = 1 ,$$

which is an underdetermined system of $(n + m + 1)$ equations in $(n + m + N)$ variables.

This system is used for narrowing the boxes by interval Newton operators in the SIVIA procedure. The procedure is similar to the one known from interval global optimization (see [6], [9]).

Previous experiments [12] with underdetermined equations systems, like (3) suggest that two methods are promising in solving them:

- the componentwise Newton operator [7],
- the Gauss-Seidel (GS) operator with rectangular matrix [9].

The first technique uses linearization of each equation with respect to only one of the variables at a time. Pairs equation-variable can be chosen using several heuristics. Our implementation uses the strategy of S. Herbort and D. Ratz [7] and tries to use all possible pairs subsequently.

The well-know GS operator is commonly used in interval algorithms. In our case it has to be used for a linear equations system with a rectangular matrix. This does not change much in the method: we choose one variable for reduction for each equation.

A slight adaptation of the classical GS procedure has to be done in preconditioning. We use the inverse-midpoint preconditioner, choosing a square submatrix with the Gauss elimination procedure, performed on the midpoint-matrix.

In current implementation of our algorithm we can use both versions of the Newton operator.

4   Implementation

Parallelization of the algorithm was done in a way described in [14] and [15]. This approach parallelizes the "outer loop" of the algorithm, i.e., operations on different boxes in the criteria space are done in parallel, but there is no nested parallelism on the SIVIA procedure applied to them. This allows larger grainsize, but makes us to execute costly operations on the list of sets in a critical section (deleting all dominated sets). Parallelization was obtained using POSIX threads [2] as in previous implementations [14], [15].

The program uses the C-XSC library [1] for interval operations.

5   Examples

We tested three versions of the algorithm:

- the version that uses 1st order information only; no Hesse matrices are computed nor considered,
- the version using the componentwise interval Newton operator with the S. Herbort and D. Ratz heuristic,
- the interval Gauss-Seidel operator with rectangular matrix.

Two problems to be solved were considered.

5.1   The Kim Problem

Our first example is a well-known hard problem for multi-criteria analysis [11]:

$$
\begin{aligned}
\min_{x_1,x_2} \Big(q_1(x_1,x_2) = &-\big(3(1-x_1)^2 \cdot \exp(-x_1^2 - (x_2+1)^2) + \\
&-10 \cdot \big(\frac{x_1}{5} - x_1^3 - x_2^5\big) \cdot \exp(-x_1^2 - x_2^2) + \\
&-3\exp(-(x_1+2)^2 - x_2^2) + 0.5 \cdot (2x_1 + x_2)\big), \\
q_2(x_1,x_2) = &-\big(3 \cdot (1+x_2)^2 \cdot \exp(-x_2^2 - (1-x_1)^2) + \\
&-10 \cdot \big(-\frac{x_2}{5} + x_2^3 + x_1^5\big) \cdot \exp(-x_2^2 - x_1^2) + \\
&-3\exp(-(2-x_2)^2 - x_1^2)\big)\Big),
\end{aligned}
\tag{4}
$$

$$
x_1, x_2 \in [-3,3] .
$$

The second example is related to a practical problem.

5.2   Tuning the PI Controller

A Proportional-Integral-Derivative (PID) controller can be found in virtually all kind of control equipments. In the so-called ideal non-interacting form it is described by the following transfer function:

$$
R(s) = k \cdot \big(1 + \frac{1}{Ts} + T_d s\big) ,
\tag{5}
$$

with parameters k, T and T_d. The selection of these parameters, i.e., the tuning of the PID actions, is the crucial issue in the control-loop design. A large number of tuning rules has been derived in the last seventy years starting with the well-known Ziegler-Nichols algorithm.

As many other engineering design problems PID tuning is generally a multi-objective one and can be solved using multi-objective optimization techniques. We are going to present an application of the derived algorithm for PID controller tuning for a non-minimum-phase (inverse response) plant. It is well known that in

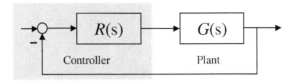

Fig. 1. Control system composed of a plant and a controller with closed-loop feedback

this case the PI controller, i.e., with proportional and integral terms, is adequate. Performance of the control loop can be measured in different way. In our example two objective functions were chosen: integrated square error (ISE) and some measure of overshoot. The ISE criterion minimized alone is insufficient, because it accepts oscillatory unit-step set-point response of the control-loop, so a second criterion has to be used. Controlled plant has inverse response, so we use as a measure of overshoot the span of closed-loop response to unit step in reference.

Applying an interval method to this problem was challenging as closed-form formulae are necessary to use them. Computing the formula for output signal of the system required computing the poles of the transition function, i.e., roots of its denominator, which is a quadratic function. The formulae are different for different signs of the discriminant Δ of this quadratic equation. As Δ is often an interval containing zero, one has to consider the interval hull of the results of all three formulae (for Δ positive, negative and equal to 0). As the automatic differentiation toolbox of C-XSC [1] does not have such operation, changes had to be done to the library code. Moreover the formulae are quite likely to result with improper operations, like division by 0 or computing the square root of an interval containing negative values. All such cases had to be carefully implemented.

Nevertheless, using our algorithm we are able to present the Pareto-front to the control-loop designer, so that they could consider conflicting criteria simultaneously and basing, e.g. on Haimes' multi-objective trade-off analysis [5] choose the controller parameters properly.

6   Computational Experiments

Numerical experiments were performed on a computer with 16 cores, i.e., 8 Dual-Core AMD Opterons 8218 with 2.6GHz clock. The machine ran under control of a Fedora 11 Linux operating system. The solver was implemented in C++, using C-XSC 2.4.0 library for interval computations. The GCC 4.4.4 compiler was used.

The following notation is used in the tables:

- "1st order" – the version of our algorithm that uses 1st order information only; no Hesse matrices are computed nor considered,
- "Ncmp" – the version using the componentwise interval Newton operator with the S. Herbort and D. Ratz heuristic,

- "GS" – the interval Gauss-Seidel operator with rectangular matrix,
- "high acc." at the version's name means results for smaller values of ε_y and ε_x.

For the Kim problem (4) we set computational accuracies in criteria space at $\varepsilon_y = 0.2$, and in decision space at $\varepsilon_x = 10^{-3}$. As high accuracy we used $\varepsilon_y = 0.1$, and $\varepsilon_x = 10^{-4}$, respectively.

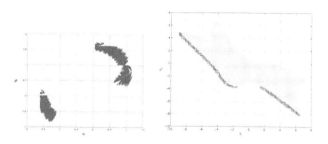

Fig. 2. Pareto-set in decision space and Pareto-front in criteria space computed for the Kim problem using 1st order information

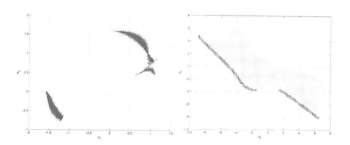

Fig. 3. Pareto-set and Pareto-front computed for the Kim problem using component-wise Newton version of the algorithm (Ncmp)

It is easy to observe by inspection that using 2nd order information results in dramatic improvement in accuracy of Pareto-set determination for the hard Kim problem.

For the PID tuning problem we used finer computational accuracies, i.e., in criteria space it was $\varepsilon_y = 0.02$, and in decision space – $\varepsilon_x = 10^{-5}$. As high accuracy we used $\varepsilon_y = 0.001$, and $\varepsilon_x = 10^{-6}$, respectively.

The obtained results allowed control designer to choose P and I parameters, i.e., k and T, resulting in very small values of rise time and inverse overshoot.

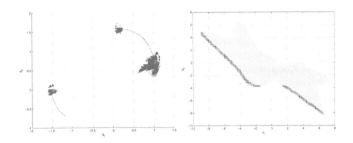

Fig. 4. Pareto-set and Pareto-front computed for the Kim problem using interval GS version of the algorithm (GS)

Table 1. Results for the Kim problem (4) and a single-threaded algorithm

	1st order	Ncmp	GS	GS (high acc.)
criteria evals.	10543390	4493434	2837044	26662794
criteria grad. evals	3906722	7642454	742694	8068624
criteria Hess. evals	0	1578964	1085072	11293078
bisecs.in crit.space	440	438	434	809
bisecs.in dec. space	956390	372062	253237	2796488
boxes deleted by monot.	17120	2234	6459	18411
boxes deleted by Newton	85716	128997	34390	274630
resulting quadruples	174	171	161	310
internal boxes	352922	101612	80782	1235138
boundary boxes	462448	94513	96066	1191551
Lebesgue measure crit.	4.84	4.76	4.48	2.16
Lebesgue measure dec.	0.76	0.24	0.19	0.13
time (sec.)	71	155	62	655

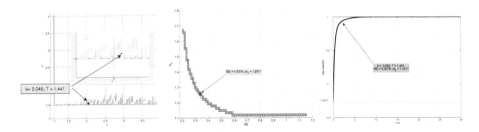

Fig. 5. Fragment of Pareto-set and Pareto-front computed for the PID tuning problem using interval GS version of the algorithm (GS) with the chosen PI controller settings and the response generated by the closed-loop system

Table 2. Results for the PID tuning problem and a single-threaded algorithm

	1st order	Ncmp	GS	Ncmp (high acc.)	GS (high acc.)
criteria evals.	396436222	228916	231500	4250430	4255408
criteria grad. evals	160959276	87394	23554	1606786	424160
criteria Hess. evals	0	154456	155306	2392690	2390932
bisecs.in crit.space	452	451	452	8261	8262
bisecs.in dec. space	40222521	20318	20532	445048	445494
boxes deleted by monot.	634	108	111	120	123
boxes deleted by Newton	170201	3819	519	54741	10159
resulting quadruples	147	145	145	3589	3589
internal boxes	23795930	9268	9273	223554	223575
boundary boxes	16250540	954	948	28910	28921
Lebesgue measure crit.	0.04	0.04	0.04	0.00192	0.00192
Lebesgue measure dec.	1.85	0.32	0.32	0.00515	0.00515
time (sec.)	4995	14	13	202	203

Table 3. Speedup for parallelized algorithms on the Kim problem

		1	2	4	6	8	10	12
1st order	time (sec.)	71	41	25	19	17	16	15
	speedup	1.0	1.73	2.84	3.74	4.18	4.44	4.73
Ncmp	time (sec.)	155	81	41	29	22	19	17
	speedup	1.0	1.91	3.78	5.34	7.05	8.16	9.12
GS	time (sec.)	62	32	16	11	9	11	10
	speedup	1.0	1.94	3.88	5.64	6.89	5.64	6.2
GS	time (sec.)	655	342	176	122	92	89	78
(high acc.)	speedup	1.0	1.92	3.72	5.37	7.12	7.36	8.40

Table 4. Speedup for parallelized algorithms on the PID tuning problem

		1	2	4	6	8	10	12
1st order	time (sec.)	4995	2575	1343	956	735	651	531
	speedup	1.0	1.94	3.72	5.22	6.80	7.67	9.41
Ncmp	time (sec.)	202	114	58	39	29	24	21
(high acc.)	speedup	1.0	1.77	3.48	5.18	6.97	8.42	9.62
GS	time (sec.)	203	103	53	35	27	22	18
(high acc.)	speedup	1.0	1.97	3.83	5.80	7.52	9.23	11.28

7 Results

It occurred that both version of the proposed modification of the algorithm allow to enclose the Pareto-set far more precisely. The difference for Pareto-fronts was only marginal (but not negligible), but for Pareto-sets (in the decision space) the paving generated by the algorithm using the 2nd order information is more than 3 times smaller (measuring with the Lebesgue measure) for the Kim problem and about 6 times smaller for the PID tuning problem.

The versions using the componentwise Newton operator is computationally intensive for the Kim problem. It seems to be caused by the necessity to compute Hesse matrices (in addition to gradients) using the automatic differentiation arithmetic.

The traditional interval Newton step, based on the Gauss-Seidel operator, requires only one computation of gradients of all functions to prepare to narrow the box. The componentwise operator has to recompute the gradient information in each step of the narrowing operator, i.e., to recompute all gradients each time.

This phenomenon does not affect efficiency for the PID tuning problem, which is probably caused by the following reason. All solutions for this problem lie on the boundaries. The version of the algorithm that uses 1st order derivatives only is not able to delete the interior boxes at the early stage and has to deal with them for several iterations. On the other hand both versions using the second order derivatives are able to delete the interior boxes relatively quickly and most of their work is analyzing the boundaries for which the Newton's method cannot be applied (in our implementation we do not use reduced gradients or Hesse matrices as in [9]) and consequently there is no difference between the different Newton operators used.

Parallelization, as it could be expected, improved the performance of the algorithm. The speedup is satisfactory for 4–6 threads (except the 1st order information only version for the Kim problem), but it usually scales worse further. This is probably caused by relatively time consuming critical sections (seeking for dominated quadruples in a linked list).

8 Conclusions

In the paper we tested two versions of the interval algorithm using the 2nd order information for seeking Pareto-set of a multi-criteria problem. We compared them with interval algorithm based on 1st order information only. It occurred that information enrichment – if processed properly – can improve both efficiency and accuracy of the algorithm several times.

Also, using the Gauss-Seidel interval operator with rectangular matrix seems to be a much better solution than using the componentwise Newton operator; at least for some problems (like the Kim problem).

9   Future Work

There is some more research to be done about the parallelization. It is going to be improved by reducing the critical sections. Instead of the synchronized linear search for quadruples to delete, we are going to store the information about obtained non-dominated points in a separate shared data structure.

Acknowledgments. The research has been supported by the Polish Ministry of Science and Higher Education under grant N N514 416934.

References

1. C-XSC interval library, http://www.xsc.de
2. POSIX Threads Programming, https://computing.llnl.gov/tutorials/pthreads
3. Barichard, V., Hao, J.K.: Population and Interval Constraint Propagation Algorithm. In: Fonseca, C.M., Fleming, P.J., Zitzler, E., Deb, K., Thiele, L. (eds.) EMO 2003. LNCS, vol. 2632, pp. 88–101. Springer, Heidelberg (2003)
4. Fernandez, J., Toth, B.: Obtaining an outer approximation of the efficient set of nonlinear biobjective problems. Journal of Global Optimization 38, 315–331 (2007)
5. Haimes, Y.Y.: Risk Modeling, Assessment, and Management. J. Wiley, New York (1998)
6. Hansen, E., Walster, W.: Global Optimization Using Interval Analysis. Marcel Dekker, New York (2004)
7. Herbort, S., Ratz, D.: Improving the efficiency of a nonlinear-system-solver using the componentwise Newton method, http://citeseer.ist.psu.edu/409594.html
8. Jaulin, L., Walter, E.: Set Inversion Via Interval Analysis for nonlinear bounded-error estimation. Automatica 29, 1053–1064 (1993)
9. Kearfott, R.B.: Rigorous Global Search: Continuous Problems. Kluwer, Dordrecht (1996)
10. Kearfott, R.B., Nakao, M.T., Neumaier, A., Rump, S.M., Shary, S.P., van Hentenryck, P.: Standardized notation in interval analysis (2002), http://www.mat.univie.ac.at/~neum/software/int/notation.ps.gz
11. Kim, I.Y., de Weck, O.L.: Adaptive weighted-sum method for bi-objective optimization: Pareto front generation. Structural and Multidisciplinary Optimization 29, 149–158 (2005)
12. Kubica, B.J.: Interval methods for solving underdetermined nonlinear equations systems. Presented at SCAN, Conference, El Paso, Texas, USA (2008)
13. Kubica, B.J., Woźniak, A.: Interval Methods for Computing the Pareto-Front of a Multicriterial Problem. In: Wyrzykowski, R., Dongarra, J., Karczewski, K., Wasniewski, J. (eds.) PPAM 2007. LNCS, vol. 4967, pp. 1382–1391. Springer, Heidelberg (2008)
14. Kubica, B. J., Woźniak, A.: A multi-threaded interval algorithm for the Pareto-front computation in a multi-core environment. Presented at PARA 2008 Conference, accepted for publication in LNCS 6126 (2010)
15. Kubica, B.J., Woźniak, A.: Optimization of the multi-threaded interval algorithm for the Pareto-set computation. Journal of Telecommunications and Information Technology 1, 70–75 (2010)
16. Ruetsch, G.R.: An interval algorithm for multi-objective optimization. Structural and Multidisciplinary Optimization 30, 27–37 (2005)

Comments on Fast and Exact Accumulation of Products*

Gerd Bohlender and Ulrich Kulisch

Institute for Applied and Numerical Mathematics 2,
Karlsruhe Institue of Technology, P.O. Box 6980, D-76049 Karlsruhe, Germany
{gerd.bohlender,ulrich.kulisch}@kit.edu
http://www.math.kit.edu/ianm2/

Abstract. A new IEEE arithmetic standard 1788 is currently being worked out. It will specify interval arithmetic and an exact dot product (EDP). In an EDP, an arbitrary finite number of products is accumulated without rounding errors.

These are essential tools for computations with reliable and accurate results. In high performance computing, it is necessary that implementations of interval arithmetic and the EDP must be as efficient as the ordinary floating-point arithmetic. In this paper, fast and accurate solutions for the EDP are presented.

Keywords: Interval arithmetic, accurate dot product, IEEE standard.

1 Introduction

In high performance computing, there is a growing need for reliability and high accuracy of results. Therefore, the IFIP Working Group on Numerical Software and other scientists repeatedly requested that a future arithmetic standard should consider and specify interval arithmetic and an exact dot product (EDP).

In contrast with floating-point arithmetic which only delivers approximations of mathematical results, interval arithmetic (when correctly applied) computes an enclosure of the corresponding exact mathematical results. This makes it possible to prove mathematical results in a rigorous way on the computer [11,20]. In an EDP, an arbitrary number of products may be accumulated without rounding errors. Combining these arithmetic tools, high accuracy can be achieved.

In 2008, an IEEE standardization group P1788 was set up for working out a standard on interval arithmetic. Since then, work is making progress and many details have been specified in "motions" [10]. On Nov. 18, 2009 the IEEE standards committee P1788 on interval arithmetic accepted a motion [14] for including the EDP into a future interval arithmetic standard. This justifies making the basic ideas for realizing the EDP known to a wider audience.

---

* Required by the IEEE Standards Committee P1788.

K. Jónasson (Ed.): PARA 2010, Part II, LNCS 7134, pp. 148–156, 2012.

2   Definition of Dot Products

Let $\mathbb{F} = \mathbb{F}(b, l, emin, emax)$ be a floating-point system with base b, l mantissa digits and an exponent range of $emin..emax$, e.g. for IEEE-arithmetic double precision we have:
$b = 2$; 64 bits word length; 1 bit sign; 11 bits exponent; $l = 53$ mantissa bits; $emin = -1022$, $emax = 1023$.

For all floating-point operands $a_i, b_i \in \mathbb{F}$, we are going to compute the two results (dot products)

$$s := \sum_{i=1}^{n} a_i \cdot b_i = a_1 \cdot b_1 + a_2 \cdot b_2 + \ldots + a_n \cdot b_n \qquad \textbf{(EDP)}$$

and

$$c := \square \sum_{i=1}^{n} a_i \cdot b_i = \square(a_1 \cdot b_1 + a_2 \cdot b_2 + \ldots + a_n \cdot b_n) = \square s \qquad \textbf{(CRDP)}$$

where all additions and multiplications are the operations for real numbers and $\square$ is a rounding symbol representing round to nearest, towards zero, upwards, or downwards. $s \in \mathbb{R}$ is a real number which we call the **Exact Dot Product (EDP)**, whereas $c \in \mathbb{F}$ is a floating-point number which we call the **Correctly Rounded Dot Product (CRDP)**.

This contrasts with the traditional computation of a dot product which involves a rounding in each multiplication and each addition ($2 \cdot n - 1$ roundings).

In Numerical Analysis the dot product is ubiquitous. It is not merely a fundamental operation in all vector and matrix spaces. It is the EDP which makes residual correction effective. This has a direct and positive influence on all iterative solvers of systems of equations [5,6].

The EDP is essential for fast long real and long interval arithmetic, as well as for assessing and managing uncertainty in computing. Using an EDP and such a long arithmetic, the result of every arithmetic expression can be guaranteed to a number of correct digits. By operator overloading variable precision interval arithmetic is very easy to use.

An EDP is mathematically precisely defined, leading to more portable programs.

Many algorithms and implementations have been developed for the dot product, as early as 1971 [16]. A survey of classical algorithms in hardware and software, as well as concepts for pipelined and parallel machines can be found in [3]. More recent highly optimized algorithms are presented e.g. in [17,19,21], their mathematical properties are carefully studied. An EDP has been available in all "XSC" programming languages developed in Karlsruhe since 1980 and maintained and further developed in Wuppertal since 1999. A description of the C++ library C-XSC can be found in [7,8]. The most recent version 2.4.0 of C-XSC, released December 18, 2009, provides several variants of dot product algorithms [22]; it may be downloaded from [9].

3   Realization of a CRDP by Addition with Remainder

Floating-point operations may be made "exact" by computing the remainder of each operation. Algorithms for these error free transformations were already developed in the 1960s. In particular, in the absence of overflow and underflow, the product and the sum of two floating-point numbers a and b may be exactly expressed as a pair of two floating numbers (s, r), where s is the rounded result and r is the exact remainder (see figures 1 and 2, which are updated versions of figures in [3]).

$$s = \Box(a + b), \quad r = (a + b) - s \qquad \textbf{(TwoSum)}$$
$$s = \Box(a \cdot b), \quad r = (a \cdot b) - s \qquad \textbf{(TwoProd)}$$

The quotient of two floating-point numbers may also be computed exactly – with an appropriate definition of the remainder. The semantics for such exact floating-point operations including the square root was studied in [4].

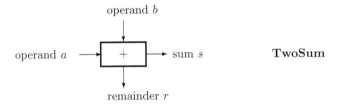

Fig. 1. Algorithm TwoSum

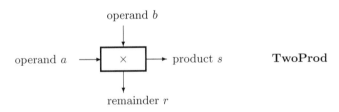

Fig. 2. Algorithm TwoProd

A first algorithm implementing a dot product with approximately k-fold accuracy (using older versions of such error free transformations) was presented already in 1972 [18]. With some additional error estimations and guard digits, this algorithm can be improved to compute an Exact Dot Product (EDP) [2].

Meanwhile, highly tuned versions of these error free transformations are commonly named *TwoProd* and *TwoSum*, resp. and the dot product algorithm based on these is named *DotK* because k iterations lead to k-fold accuracy of the result. Mathematical properties and implementation details are studied very closely in

[17]. Exactly speaking, this result is neither an EDP nor a CRDP, but a multi-precision interval enclosure of the exact result. But for large enough k, this is satisfactory for many applications.

Basing on these ideas, cascaded and pipelined versions of a dot product may be designed like in figures 3 and 4, which are essentially taken from [3].

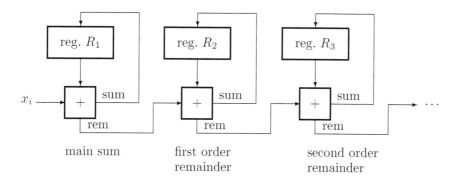

Fig. 3. Cascaded adders with remainder

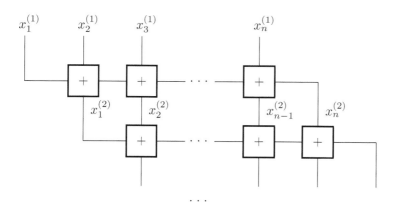

Fig. 4. Pipelined addition with remainder

4   Realization of an EDP by a Complete Register

Actually the simplest and fastest way for computing a dot product is to compute it exactly. Malcolm's algorithm was a first step in this direction [16]. By pipelining, the dot product can be computed in the time the processor needs to read the data, i.e., it comes with utmost speed. By a sample illustration we informally specify the implementation of the EDP on computers. While [14] defines **what** has to be provided, how to embed the EDP into the new standard IEEE 754, and how exceptions like NaN are to be dealt with, this section illustrates **how** the EDP can be implemented on computers. There is indeed no simpler way of accumulating a dot product. Any method that just computes an approximation also has to consider the relative values of the summands. This results in a more complicated method.

The hardware needed for the EDP is comparable to that for a fast multiplier by an adder tree, accepted years ago and now standard technology in every modern processor. The EDP brings the same speedup for accumulations at comparable costs. It exceeds any approximate software computation (including the CRDP) by several orders of magnitude.

Let $a = (a_i)$, $b = (b_i)$ be two vectors with n components which are floating-point numbers $a_i, b_i \in \mathbb{F}(b, l, emin, emax)$, for $i = 1(1)n$. We compute the sum

$$s := \sum_{i=1}^{n} a_i \cdot b_i \qquad \text{(EDP)},$$

where all additions and multiplications are the operations for real numbers.

All summands can be taken into a fixed-point register of length $2 \cdot emax + 2 \cdot l + 2 \cdot |emin|$ without loss of information.

If the register is built as an accumulator with an adder, all summands could be added in without loss of information. To accommodate possible overflows, it is convenient to provide a few, say g, guard digits of base b on the left. g can be chosen such that no overflows will occur in the lifetime of the computer.

For IEEE double precision format we have $l = 53$, $emin = -1022$, $emax = 1023$. Products may be represented exactly with $2 \cdot l = 106$ bit mantissas and an exponent range from $2 \cdot emin = -2044$ to $2 \cdot emax = 2046$. g may be chosen sufficiently large in such a way that the size of the fixed-point register is a multiple of 64. So, with $g = 92$ the entire unit consists of

$$L = g + 2 \cdot emax + 2 \cdot l + 2 \cdot |emin| = g + 4196 = 4288$$

bits. It can be represented by 67 words of 64 bits. L is independent of n for practical purposes (even the fastest current computer would have to work for nearly a million years, computing a single dot product, before an overflow may occur).

Figure 5 gives an informal description for realizing an EDP. The long register (here represented as a chest of drawers) is organized in words of 64 bits. The exponent of the products is obtained by adding the 11 bit exponents of the two factors. It consists of 12 bits. The leading 6 bits give the address of the three consecutive drawers to which the summand of 106 bits is added (The product of the two 53 bit mantissas consists of 106 bits). The low end 6 bits of the exponent are used for the correct positioning of the summand within the selected drawers.

A possible carry is absorbed by the next more significant word in which not all bits are 1. For fast detection of this word, a flag is attached to each word. It is set 1 if all bits of the word are 1. This means that a carry will propagate through

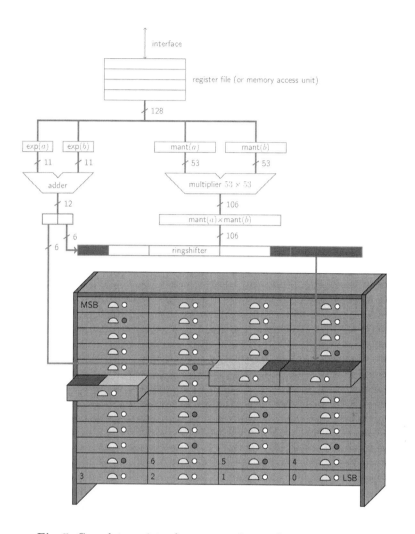

Fig. 5. Complete register for exact scalar product accumulation

the entire word. In the figure the flag is shown as a dark point. As soon as the exponent of the summand is available the flags allow selecting and incrementing the carry word. This can be done simultaneously with adding the summand into the selected drawers. Similar considerations may be applied to handle a possible borrow.

5   Circuitry for the Exact Dot Product

Basing on an architecture which is similar to figure 5, a coprocessor chip has been developed which demonstrates that an EDP may be implemented in hardware as efficiently as an ordinary dot product in floating-point arithmetic [12,13].

In figure 6 it can be seen that the EDP may be computed in a pipeline in the same time which is required to fetch the operands. That shows that no other implementation can be faster.

cycle	read	mult/shift	accumulate
——	read a_{i-1}		
——	read b_{i-1}		
——	read a_i	$c_{i-1} := a_{i-1} \cdot b_{i-1}$	
——	read b_i	$c_{i-1} := \text{shift}\,(c_{i-1})$	
——	read a_{i+1}	$c_i := a_i \cdot b_i$	address decoding / load
——	read b_{i+1}	$c_i := \text{shift}\,(c_i)$	add/sub c_{i-1} / store & store flags
——	read a_{i+2}	$c_{i+1} := a_{i+1} \cdot b_{i+1}$	address decoding / load
——	read b_{i+2}	$c_{i+1} := \text{shift}\,(c_{i+1})$	add/sub c_i / store & store flags
——	read a_{i+3}	$c_{i+2} := a_{i+2} \cdot b_{i+2}$	address decoding / load
——	read b_{i+3}	$c_{i+2} := \text{shift}\,(c_{i+2})$	add/sub c_{i+1} / store & store flags

Fig. 6. Pipeline for the accumulation of scalar products

6   Two Simple Applications

To be successful interval arithmetic has to be complemented by some easy way to use multiple or variable precision arithmetic. The fast and exact dot product is the tool to provide this very easily.

With the EDP quadruple or other multiple precision arithmetic can easily be provided on the computer [15,13]. This enables the use of higher precision operations in numerically critical parts of a computation. It helps to increase software reliability. A multiple precision number is represented as an array of floating-point numbers. The value of this number is the sum of its components. The number can be represented in the long register in the arithmetic unit.

Addition and subtraction of multiple precision numbers can easily be performed in this register. Multiplication of two such numbers is simply a sum of products. It can be computed easily and fast by means of the EDP. For instance, using fourfold precision the product of two such numbers $a = (a_1 + a_2 + a_3 + a_4)$ and $b = (b_1 + b_2 + b_3 + b_4)$ is obtained by

$$a \cdot b = (a_1 + a_2 + a_3 + a_4) \cdot (b_1 + b_2 + b_3 + b_4)$$
$$= \sum_{i=1}^{4} \sum_{j=1}^{4} a_i b_j. \qquad (1)$$

The result is a sum of products of floating-point numbers. It is independent of the sequence in which the summands are added.

A very impressive application is considered in [1], an iteration with the logistic equation (dynamical system)

$$x_{n+1} := 3.75 \cdot x_n \cdot (1 - x_n), \quad n \geq 0.$$

For the initial value $x_0 = 0.5$ the system shows chaotic behavior.

Double precision floating-point or interval arithmetic totally fail (no correct digit) after 30 iterations while long interval arithmetic still computes correct digits of a guaranteed enclosure after 2790 iterations using the EDP. It is the basic operation for long interval arithmetic. For definitions see [13], section 9.7.

7   Conclusions

Over the last decades, several algorithms and prototype implementations have been developed for an exact dot product (EDP). The future IEEE standard 1788 requires an EDP, improving the chances that chip makers will integrate it on their architectures. Work on details of the standard continues.

It can be shown that an EDP may be implemented as efficiently as ordinary floating-point arithmetic. High performance computing would benefit considerably from such an architecture.

References

1. Blomquist, F., Hofschuster, W., Krämer, W.: A Modified Staggered Correction Arithmetic with Enhanced Accuracy and Very Wide Exponent Range. In: Cuyt, A., Krämer, W., Luther, W., Markstein, P. (eds.) Numerical Validation. LNCS, vol. 5492, pp. 41–67. Springer, Heidelberg (2009)
2. Bohlender, G.: Genaue Berechnung mehrfacher Summen, Produkte und Wurzeln von Gleitkommazahlen und allgemeine Arithmetik in höheren Programmiersprachen. Dissertation, Universität Karlsruhe (1978)
3. Bohlender, G.: What Do We Need Beyond IEEE Arithmetic? In: Ullrich, C. (ed.) Computer Arithmetic and Self-Validating Numerical Methods, pp. 1–32. Academic Press, San Diego (1990)
4. Bohlender, G., Walter, W., Kornerup, P., Matula, D.W.: Semantics for Exact Floating Point Operations. In: Proceedings of 10th IEEE Symposium on Computer Arithmetic, June 26-28, pp. 22–26. IEEE (1991)
5. Bohlender, G., Kolberg, M., Cordeiro, D., Fernandes, G., Goldman, A.: A Multithreaded Verified Method for Solving Linear Systems in Dual-Core Processors. In: PARA – Workshop on State-of-the-Art in Scientific and Parallel Computing. Trondheim, 2008. Accepted for publication in LNCS. Springer, Heidelberg (2009)

6. Bohlender, G., Kolberg, M., Claudio, D.: Improving the Performance of a Verified Linear System Solver Using Optimized Libraries and Parallel Computation. In: Palma, J.M.L.M., Amestoy, P.R., Daydé, M., Mattoso, M., Lopes, J.C. (eds.) VECPAR 2008. LNCS, vol. 5336, pp. 13–26. Springer, Heidelberg (2008)
7. Hofschuster, W., Krämer, W.: C-XSC 2.0 – A C++ Library for Extended Scientific Computing. In: Alt, R., Frommer, A., Kearfott, R.B., Luther, W. (eds.) Num. Software with Result Verification. LNCS, vol. 2991, pp. 15–35. Springer, Heidelberg (2004)
8. Hofschuster, W., Krämer, W., Neher, M.: C-XSC and Closely Related Software Packages. In: Cuyt, A., et al. (eds.) Numerical Validation. LNCS, vol. 5492, pp. 68–102. Springer, Heidelberg (2009)
9. Hofschuster, W., Krämer, W.: C-XSC 2.4.0 – A C++ Class Library, www2.math.uni-wuppertal.de/~xsc/xsc/cxsc_new.html (accessed August 17, 2010)
10. IEEE Society IEEE Interval Standard Working Group - P1788, grouper.ieee.org/groups/1788/ (accessed August 17, 2010)
11. Kulisch, U., Lohner, R., Facius, A. (eds.): Perspectives on Enclosure Methods. Springer, Heidelberg (2001)
12. Kulisch, U.: Advanced Arithmetic for the Digital Computer – Design of Arithmetic Units. Springer, Heidelberg (2002)
13. Kulisch, U.: Computer Arithmetic and Validity – Theory, Implementation, and Applications. de Gruyter (2008)
14. Kulisch, U., Snyder, V.: The Exact Dot Product as Basic Tool for Long Interval Arithmetic. Computing, Wien (2010), dx.doi.org/10.1007/s00607-010-0127-7 (accessed December 10, 2010)
15. Lohner, R.: Interval Arithmetic in Staggered Correction Format. In: Adams, E., Kulisch, U. (eds.) Scientific Computing with Automatic Result Verification. Academic Press (1993)
16. Malcolm, M.A.: On Accurate Floating-Point Summation. Comm. ACM 14(11), 731–736 (1971)
17. Ogita, T., Rump, S.M., Oishi, S.: Accurate sum and dot product. SIAM Journal on Scientific Computing 26(6), 1955–1988 (2005)
18. Pichat, M.: Correction d'une somme en arithmétique à virgule flottante. Numerische Mathematik 19, 400–406 (1972)
19. Rump, S.M.: Ultimately Fast Accurate Summation. SIAM Journal on Scientific Computing 3(5), 3466–3502 (2009)
20. Rump, S.M.: Verification methods: Rigorous results using floating-point arithmetic. Acta Numerica, 287–449 (2010)
21. Yamanaka, N., Ogita, T., Rump, S.M., Oishi, S.: A Parallel Algorithm for Accurate Dot Product. Parallel Computing 34, 392–410 (2008)
22. Zimmer, M., Krämer, W., Bohlender, G., Hofschuster, W.: Extension of the C-XSC Library with Scalar Products with Selectable Accuracy. To Appear in Serdica Journal of Computing 4, 3 (2010)

An Interval Finite Difference Method of Crank-Nicolson Type for Solving the One-Dimensional Heat Conduction Equation with Mixed Boundary Conditions

Malgorzata A. Jankowska

Poznan University of Technology, Institute of Applied Mechanics
Piotrowo 3, 60-965 Poznan, Poland
malgorzata.jankowska@put.poznan.pl

Abstract. In the paper an interval method for solving the one-dimensional heat conduction equation with mixed boundary conditions is considered. The idea of the interval method is based on the finite difference scheme of the conventional Crank-Nicolson method adapted to the mixed boundary conditions. The interval method given in the form presented in the paper includes the error term of the conventional method.

Keywords: interval methods, finite difference methods, Crank-Nicolson method, partial differential equations, heat conduction equation with mixed boundary conditions.

1 Introduction

The boundary value problems for partial differential equations are important for a large class of problems in many scientific fields. Hence, some verified methods dealing with such problems have been developed. For example let us mention an interval approach to the Laplace equation [3], the Poisson equation [4] and the Navier-Stokes problems [7], [10]. An overview of some interval methods for the initial and boundary value problems for the ordinary and partial differential equations is also presented in [8].

Jankowska and Marciniak proposed an interval backward finite difference method for solving the one-dimensional heat conduction equation [1]. Then Marciniak made the first approach to an interval version of Crank-Nicolson method for solving the one-dimensional heat conduction equation with the Dirichlet boundary conditions [5].

Jankowska extended the research of Marciniak given in [5] and proposed an interval finite difference method of Crank-Nicolson type for solving the one-dimensional heat conduction equation with mixed boundary conditions. The interval method of Crank-Nicolson type shown in the paper includes the error term of the corresponding conventional method. Nevertheless, the method is verified purely theoretically at the moment.

K. Jónasson (Ed.): PARA 2010, Part II, LNCS 7134, pp. 157–167, 2012.

2 Heat Conduction Equation with Mixed Boundary Conditions and Conventional Crank-Nicolson Method

As an example of the parabolic partial differential equation we consider the one-dimensional heat conduction equation of the form

$$\frac{\partial u}{\partial t}(x,t) - \alpha^2 \frac{\partial^2 u}{\partial x^2}(x,t) = 0, \quad 0 < x < L, \ t > 0, \tag{1}$$

subject to the initial condition

$$u(x,0) = f(x), \quad 0 \le x \le L, \tag{2}$$

and the mixed boundary conditions

$$\frac{\partial u}{\partial x}(0,t) - Au(0,t) = \varphi_1(t), \quad t > 0, \tag{3}$$

$$\frac{\partial u}{\partial x}(L,t) + Bu(L,t) = \varphi_2(t), \quad t > 0. \tag{4}$$

Note that if $A = 0$ and $B = 0$ then the boundary conditions (3)-(4) are called the Neumann (or second-type) boundary conditions.

Let us set the maximum time $T_{\max}$ and choose two integers n and m. Then we find the mesh constants h and k such as $h = L/n$ and $k = T_{\max}/m$. Hence the grid points are (x_i, t_j), where $x_i = ih$ for $i = 0, 1, \ldots, n$ and $t_j = jk$ for $j = 0, 1, \ldots, m$.

Let us take the central difference formula for $\partial u/\partial t$, together with the local truncation error, at the point $(x_i, t_{j+1/2})$ of the form

$$\frac{\partial u}{\partial t}(x_i, t_{j+1/2}) = \frac{u(x_i, t_{j+1}) - u(x_i, t_j)}{k} - \frac{k^2}{24}\frac{\partial^3 u}{\partial t^3}\left(x_i, \eta_j^{(1)}\right), \quad \eta_j^{(1)} \in (t_j, t_{j+1}). \tag{5}$$

In order to find $\partial^2 u/\partial x^2 \left(x_i, t_{j+1/2}\right)$ we expand the derivative $\partial^2 u/\partial x^2 \left(x_i, t_j\right)$ in the Taylor series about $\left(x_i, t_{j+1/2}\right)$ and evaluate it at (x_i, t_j) and (x_i, t_{j+1}), respectively. We get

$$\frac{\partial^2 u}{\partial x^2}(x_i, t_j) = \frac{\partial^2 u}{\partial x^2}(x_i, t_{j+1/2}) - \frac{k}{2}\frac{\partial^3 u}{\partial t \partial x^2}(x_i, t_{j+1/2}) + \frac{k^2}{8}\frac{\partial^4 u}{\partial t^2 \partial x^2}\left(x_i, \widetilde{\eta}_j^{(2)}\right), \tag{6}$$

and

$$\frac{\partial^2 u}{\partial x^2}(x_i, t_{j+1}) = \frac{\partial^2 u}{\partial x^2}(x_i, t_{j+1/2}) + \frac{k}{2}\frac{\partial^3 u}{\partial t \partial x^2}(x_i, t_{j+1/2}) + \frac{k^2}{8}\frac{\partial^4 u}{\partial t^2 \partial x^2}\left(x_i, \overline{\eta}_j^{(2)}\right), \tag{7}$$

where $\widetilde{\eta}_j^{(2)} \in \left(t_j, t_{j+1/2}\right)$ and $\overline{\eta}_j^{(2)} \in \left(t_{j+1/2}, t_{j+1}\right)$. Now, let us add the formulas (6) and (7). We get

$$\frac{\partial^2 u}{\partial x^2}(x_i, t_{j+1/2}) = \frac{1}{2}\left(\frac{\partial^2 u}{\partial x^2}(x_i, t_j) + \frac{\partial^2 u}{\partial x^2}(x_i, t_{j+1})\right) \tag{8}$$

$$- \frac{k^2}{16}\left(\frac{\partial^4 u}{\partial t^2 \partial x^2}\left(x_i, \widetilde{\eta}_j^{(2)}\right) + \frac{\partial^4 u}{\partial t^2 \partial x^2}\left(x_i, \overline{\eta}_j^{(2)}\right)\right).$$

Furthermore, as a consequence of the Darboux's theorem we have

$$\frac{\partial^4 u}{\partial t^2 \partial x^2}\left(x_i, \eta_j^{(2)}\right) = \frac{1}{2}\left(\frac{\partial^4 u}{\partial t^2 \partial x^2}\left(x_i, \tilde{\eta}_j^{(2)}\right) + \frac{\partial^4 u}{\partial t^2 \partial x^2}\left(x_i, \overline{\eta}_j^{(2)}\right)\right), \qquad (9)$$

where $\eta_j^{(2)} \in (t_j, t_{j+1})$. Hence, the formula (8) with (9) gives

$$\frac{\partial^2 u}{\partial x^2}\left(x_i, t_{j+1/2}\right) = \frac{1}{2}\left(\frac{\partial^2 u}{\partial x^2}\left(x_i, t_j\right) + \frac{\partial^2 u}{\partial x^2}\left(x_i, t_{j+1}\right)\right) - \frac{k^2}{8}\frac{\partial^4 u}{\partial t^2 \partial x^2}\left(x_i, \eta_j^{(2)}\right). \qquad (10)$$

Finally, we take the central difference formula for $\partial^2 u / \partial x^2$, together with the local truncation error, at the points (x_i, t_j) and (x_i, t_{j+1}), respectively. We have

$$\frac{\partial^2 u}{\partial x^2}\left(x_i, t_j\right) = \frac{u\left(x_{i-1}, t_j\right) - 2u\left(x_i, t_j\right) + u\left(x_{i+1}, t_j\right)}{h^2} - \frac{h^2}{12}\frac{\partial^4 u}{\partial x^4}\left(\xi_i^{(1)}, t_j\right) \qquad (11)$$

and

$$\frac{\partial^2 u}{\partial x^2}\left(x_i, t_{j+1}\right) = \frac{u\left(x_{i-1}, t_{j+1}\right) - 2u\left(x_i, t_{j+1}\right) + u\left(x_{i+1}, t_{j+1}\right)}{h^2} - \frac{h^2}{12}\frac{\partial^4 u}{\partial x^4}\left(\xi_i^{(2)}, t_{j+1}\right), \qquad (12)$$

where $\xi_i^{(1)}, \xi_i^{(2)} \in (x_{i-1}, x_{i+1})$. Inserting (11) and (12) to (10) we obtain

$$\begin{aligned}
\frac{\partial^2 u}{\partial x^2}\left(x_i, t_{j+1/2}\right) = &\frac{1}{2h^2}\left(u\left(x_{i-1}, t_j\right) - 2u\left(x_i, t_j\right) + u\left(x_{i+1}, t_j\right)\right. \\
&\left. + u\left(x_{i-1}, t_{j+1}\right) - 2u\left(x_i, t_{j+1}\right) + u\left(x_{i+1}, t_{j+1}\right)\right) \\
&- \frac{h^2}{24}\left(\frac{\partial^4 u}{\partial x^4}\left(\xi_i^{(1)}, t_j\right) + \frac{\partial^4 u}{\partial x^4}\left(\xi_i^{(2)}, t_{j+1}\right)\right) - \frac{k^2}{8}\frac{\partial^4 u}{\partial t^2 \partial x^2}\left(x_i, \eta_j^{(2)}\right).
\end{aligned} \qquad (13)$$

Since we can express (1) at the grid points (x_i, t_j), $i = 0, 1, \ldots, n$, $j = 0, 1, \ldots, m$, inserting (5) and (13) in (1) yields

$$\begin{aligned}
&-\frac{\lambda}{2}u\left(x_{i-1}, t_{j+1}\right) + (1 + \lambda)\, u\left(x_i, t_{j+1}\right) - \frac{\lambda}{2}u\left(x_{i+1}, t_{j+1}\right) = \frac{\lambda}{2}u\left(x_{i-1}, t_j\right) \\
&\quad + (1 - \lambda)\, u\left(x_i, t_j\right) + \frac{\lambda}{2}u\left(x_{i+1}, t_j\right) + \widehat{R}_{i,j}, \\
&i = 0, 1, \ldots, n, \ j = 0, 1, \ldots, m-1,
\end{aligned} \qquad (14)$$

where $\lambda = \alpha^2\left(k/h^2\right)$, $\xi_i^{(1)}, \xi_i^{(2)} \in (x_{i-1}, x_{i+1})$, $\eta_j^{(1)}, \eta_j^{(2)} \in (t_j, t_{j+1})$ and $\widehat{R}_{i,j}$ are given as follows

$$\begin{aligned}
\widehat{R}_{i,j} = &\frac{k^3}{24}\frac{\partial^3 u}{\partial t^3}\left(x_i, \eta_j^{(1)}\right) - \frac{k^3 \alpha^2}{8}\frac{\partial^4 u}{\partial t^2 \partial x^2}\left(x_i, \eta_j^{(2)}\right) - \frac{h^2 k \alpha^2}{24}\left(\frac{\partial^4 u}{\partial x^4}\left(\xi_i^{(1)}, t_j\right)\right. \\
&\left. + \frac{\partial^4 u}{\partial x^4}\left(\xi_i^{(2)}, t_{j+1}\right)\right).
\end{aligned} \qquad (15)$$

Moreover, if we use the central difference formula for $\partial u / \partial x\, (0, t_j)$ and $\partial u / \partial x\, (L, t_j)$, and then we apply the boundary conditions (3)-(4) we get

$$u\left(x_{-1}, t_j\right) = u\left(x_1, t_j\right) - 2h\left(Au\left(x_0, t_j\right) + \varphi_1\left(t_j\right)\right) - \frac{h^3}{3}\frac{\partial^3 u}{\partial x^3}\left(\xi_j^{(L)}, t_j\right),$$

$$u\left(x_{n+1}, t_j\right) = u\left(x_{n-1}, t_j\right) - 2h\left(Bu\left(x_n, t_j\right) - \varphi_2\left(t_j\right)\right) + \frac{h^3}{3}\frac{\partial^3 u}{\partial x^3}\left(\xi_j^{(R)}, t_j\right),$$

$$j = 1, 2, \ldots, m, \tag{16}$$

where $\xi_j^{(L)} \in (x_{-1}, x_1)$, $\xi_j^{(R)} \in (x_{n-1}, x_{n+1})$. Let us note that $u\left(x_{-1}, t_j\right)$ and $u\left(x_{n+1}, t_j\right)$ are given only for $j = 1, 2, \ldots, m$. Hence we have to find the formulas for $u\left(x_{-1}, t_0\right)$ and $u\left(x_{n+1}, t_0\right)$ that are also required in (14).

First we take the central difference formula for $\partial u/\partial x$, together with the local truncation error, at the point (x_0, t_0) of the form

$$\frac{\partial u}{\partial x}(x_0, t_0) = \frac{u(x_{-1}, t_0) - u(x_1, t_0)}{2h} - \frac{h^2}{6}\frac{\partial^3 u}{\partial x^3}\left(\zeta^{(1)}, t_0\right), \quad \zeta^{(1)} \in (x_{-1}, x_1), \tag{17}$$

and the forward difference formula for $\partial u/\partial x$, together with the local truncation error, at the point (x_0, t_0) of the form

$$\frac{\partial u}{\partial x}(x_0, t_0) = \frac{-3u(x_0, t_0) + 4u(x_1, t_0) - u(x_2, t_0)}{2h} + \frac{h^2}{3}\frac{\partial^3 u}{\partial x^3}\left(\zeta^{(2)}, t_0\right),$$

$$\zeta^{(2)} \in (x_0, x_2). \tag{18}$$

Then, if we compare (17) with (18) we get

$$u\left(x_{-1}, t_0\right) = -3u\left(x_0, t_0\right) + 5u\left(x_1, t_0\right) - u\left(x_2, t_0\right) \tag{19}$$

$$+\frac{2}{3}h^3\left[\frac{1}{2}\frac{\partial^3 u}{\partial x^3}\left(\zeta^{(1)}, t_0\right) + \frac{\partial^3 u}{\partial x^3}\left(\zeta^{(2)}, t_0\right)\right].$$

Similarly, we take the central difference formula for $\partial u/\partial x$, together with the local truncation error, at the point (x_n, t_0) of the form

$$\frac{\partial u}{\partial x}(x_n, t_0) = \frac{u(x_{n-1}, t_0) - u(x_{n+1}, t_0)}{2h} - \frac{h^2}{6}\frac{\partial^3 u}{\partial x^3}\left(\zeta^{(3)}, t_n\right), \quad \zeta^{(3)} \in (x_{n-1}, x_{n+1}), \tag{20}$$

and the backward difference formula for $\partial u/\partial x$, together with the local truncation error, at the point (x_n, t_0) of the form

$$\frac{\partial u}{\partial x}(x_n, t_0) = \frac{3u(x_n, t_0) - 4u(x_{n-1}, t_0) + u(x_{n-2}, t_0)}{2h} + \frac{h^2}{3}\frac{\partial^3 u}{\partial x^3}\left(\zeta^{(4)}, t_0\right),$$

$$\zeta^{(4)} \in (x_{n-2}, x_n). \tag{21}$$

Then, if we compare (20) with (21) we get

$$u\left(x_{n+1}, t_0\right) = -3u\left(x_n, t_0\right) + 5u\left(x_{n-1}, t_0\right) - u\left(x_{n-2}, t_0\right) \tag{22}$$

$$-\frac{2}{3}h^3\left[\frac{1}{2}\frac{\partial^3 u}{\partial x^3}\left(\zeta^{(3)}, t_0\right) + \frac{\partial^3 u}{\partial x^3}\left(\zeta^{(4)}, t_0\right)\right].$$

Let $u_{i,j}$ approximate $u(x_i, t_j)$. If we omit the error term then from (14) we get the Crank-Nicolson method with the local truncation error $O(k^2 + h^2)$ of the form (see also [2], [9])

$$-\frac{\lambda}{2}u_{i-1,j+1} + (1+\lambda)u_{i,j+1} - \frac{\lambda}{2}u_{i+1,j+1} = \frac{\lambda}{2}u_{i-1,j} + (1-\lambda)u_{i,j} + \frac{\lambda}{2}u_{i+1,j},$$
$$i = 0, 1, \ldots, n, \quad j = 0, 1, \ldots, m-1, \tag{23}$$

where

$$u_{-1,0} = -3u_{0,0} + 5u_{1,0} - u_{2,0}, \quad u_{n+1,0} = -3u_{n,0} + 5u_{n-1,0} - u_{n-2,0},$$
$$u_{-1,j} = u_{1,j} - 2h(Au_{0,j} + \varphi_1(t_j)), \quad u_{n+1,j} = u_{n-1,j} - 2h(Bu_{n,j} - \varphi_2(t_j)),$$
$$j = 1, 2, \ldots, m, \tag{24}$$

with the initial condition given as follows

$$u_{i,0} = f(x_i), \quad i = 0, 1, \ldots, n. \tag{25}$$

3   Interval Finite Difference Method of Crank-Nicolson Type with Mixed Boundary Conditions

Let us first transform the exact formula (14) with (16), (19) and (22). We get as follows

$$(\lambda(1 + hA) + 1)u(x_0, t_1) - \lambda u(x_1, t_1) = \left(1 - \frac{5}{2}\lambda\right)u(x_0, t_0)$$
$$+3\lambda u(x_1, t_0) - \frac{\lambda}{2}u(x_2, t_0) - \frac{k}{h}\alpha^2 \varphi_1(t_1)$$
$$-\frac{\lambda}{6}h^3 \frac{\partial^3 u}{\partial x^3}\left(\xi_1^{(L)}, t_1\right) + \frac{\lambda}{3}h^3 \left[\frac{1}{2}\frac{\partial^3 u}{\partial x^3}\left(\zeta^{(1)}, t_0\right) + \frac{\partial^3 u}{\partial x^3}\left(\zeta^{(2)}, t_0\right)\right] + \widehat{R}_{0,0},$$
$$i = 0, \ j = 0, \tag{26}$$

$$(\lambda(1 + hA) + 1)u(x_0, t_{j+1}) - \lambda u(x_1, t_{j+1}) = (-\lambda(1 + hA) + 1)u(x_0, t_j)$$
$$+\lambda u(x_1, t_j) - \frac{k}{h}\alpha^2 (\varphi_1(t_j) + \varphi_1(t_{j+1}))$$
$$-\frac{\lambda}{6}h^3 \left(\frac{\partial^3 u}{\partial x^3}\left(\xi_j^{(L)}, t_j\right) + \frac{\partial^3 u}{\partial x^3}\left(\xi_{j+1}^{(L)}, t_{j+1}\right)\right) + \widehat{R}_{0,j},$$
$$i = 0, \ j = 1, 2, \ldots, m-1, \tag{27}$$

$$-\frac{\lambda}{2}u(x_{i-1}, t_{j+1}) + (1+\lambda)u(x_i, t_{j+1}) - \frac{\lambda}{2}u(x_{i+1}, t_{j+1}) = \frac{\lambda}{2}u(x_{i-1}, t_j)$$
$$+(1-\lambda)u(x_i, t_j) + \frac{\lambda}{2}u(x_{i+1}, t_j) + \widehat{R}_{i,j}$$
$$i = 1, 2, \ldots, n-1, \ j = 0, 1, \ldots, m-1, \tag{28}$$

$$-\lambda u\left(x_{n-1}, t_1\right) + \left(\lambda\left(1 + hB\right) + 1\right) u\left(x_n, t_1\right) = \left(1 - \frac{5}{2}\lambda\right) u\left(x_n, t_0\right)$$

$$+3\lambda u\left(x_{n-1}, t_0\right) - \frac{\lambda}{2} u\left(x_{n-2}, t_0\right) + \frac{k}{h}\alpha^2\varphi_2\left(t_1\right)$$

$$+\frac{\lambda}{6}h^3\frac{\partial^3 u}{\partial x^3}\left(\xi_1^{(R)}, t_1\right) - \frac{\lambda}{3}h^3\left[\frac{1}{2}\frac{\partial^3 u}{\partial x^3}\left(\zeta^{(3)}, t_0\right) + \frac{\partial^3 u}{\partial x^3}\left(\zeta^{(4)}, t_0\right)\right] + \widehat{R}_{n,0},$$

$$i = n, \ j = 0, \tag{29}$$

$$-\lambda u\left(x_{n-1}, t_{j+1}\right) + \left(\lambda\left(1 + hB\right) + 1\right) u\left(x_n, t_{j+1}\right) = \lambda u\left(x_{n-1}, t_j\right)$$

$$+\left(-\lambda\left(1 + hB\right) + 1\right) u\left(x_n, t_j\right) + \frac{k}{h}\alpha^2\left(\varphi_2\left(t_j\right) + \varphi_2\left(t_{j+1}\right)\right)$$

$$+\frac{\lambda}{6}h^3\left(\frac{\partial^3 u}{\partial x^3}\left(\xi_j^{(R)}, t_j\right) + \frac{\partial^3 u}{\partial x^3}\left(\xi_{j+1}^{(R)}, t_{j+1}\right)\right) + \widehat{R}_{n,j},$$

$$i = n, \ j = 1, 2, \ldots, m - 1. \tag{30}$$

The formulas (26)-(30) can be given in the following matrix representation:

$$Cu^{(j+1)} = D^{(j)}u^{(j)} + \widehat{E}^{(j)}, \quad j = 0, 1, \ldots, m - 1, \tag{31}$$

where $u^{(j)} = \left[u\left(x_0, t_j\right), \ u\left(x_1, t_j\right), \ \ldots, \ u\left(x_n, t_j\right)\right]^T$ and

$$C = \begin{bmatrix} \lambda\left(1 + hA\right) + 1 & -\lambda & 0 & \vdots & 0 & 0 & 0 \\ -\frac{\lambda}{2} & 1 + \lambda & -\frac{\lambda}{2} & \vdots & 0 & 0 & 0 \\ 0 & -\frac{\lambda}{2} & 1 + \lambda & \vdots & 0 & 0 & 0 \\ \cdots & \cdots & \cdots & \ddots & \cdots & \cdots & \cdots \\ 0 & 0 & 0 & \vdots & 1 + \lambda & -\frac{\lambda}{2} & 0 \\ 0 & 0 & 0 & \vdots & -\frac{\lambda}{2} & 1 + \lambda & -\frac{\lambda}{2} \\ 0 & 0 & 0 & \vdots & 0 & -\lambda & \lambda\left(1 + hB\right) + 1 \end{bmatrix}, \tag{32}$$

$$D^{(0)} = \begin{bmatrix} 1 - \frac{5}{2}\lambda & 3\lambda & -\frac{\lambda}{2} & \vdots & 0 & 0 & 0 \\ \frac{\lambda}{2} & 1 - \lambda & \frac{\lambda}{2} & \vdots & 0 & 0 & 0 \\ 0 & \frac{\lambda}{2} & 1 - \lambda & \vdots & 0 & 0 & 0 \\ \cdots & \cdots & \cdots & \ddots & \cdots & \cdots & \cdots \\ 0 & 0 & 0 & \vdots & 1 - \lambda & \frac{\lambda}{2} & 0 \\ 0 & 0 & 0 & \vdots & \frac{\lambda}{2} & 1 - \lambda & \frac{\lambda}{2} \\ 0 & 0 & 0 & \vdots & -\frac{\lambda}{2} & 3\lambda & 1 - \frac{5}{2}\lambda \end{bmatrix}, \tag{33}$$

$$D^{(j)} = \begin{bmatrix} -\lambda\left(1+hA\right)+1 & \lambda & 0 & \vdots & 0 & 0 & 0 \\ \frac{\lambda}{2} & 1-\lambda & \frac{\lambda}{2} & \vdots & 0 & 0 & 0 \\ 0 & \frac{\lambda}{2} & 1-\lambda & \vdots & 0 & 0 & 0 \\ \cdots & \cdots & \cdots & \ddots & \cdots & \cdots & \cdots \\ 0 & 0 & 0 & \vdots & 1-\lambda & \frac{\lambda}{2} & 0 \\ 0 & 0 & 0 & \vdots & \frac{\lambda}{2} & 1-\lambda & \frac{\lambda}{2} \\ 0 & 0 & 0 & \vdots & 0 & \lambda & -\lambda\left(1+hB\right)+1 \end{bmatrix} \tag{34}$$

$j = 1, 2, \ldots, m-1,$

$$\widehat{E}^{(0)} = \begin{bmatrix} -\frac{k}{h}\alpha^2\varphi_1\left(t_1\right) - \frac{\lambda}{6}h^3\frac{\partial^3 u}{\partial x^3}\left(\xi_1^{(L)}, t_1\right) + \\ +\frac{\lambda}{3}h^3\left[\frac{1}{2}\frac{\partial^3 u}{\partial x^3}\left(\zeta^{(1)}, t_0\right) + \frac{\partial^3 u}{\partial x^3}\left(\zeta^{(2)}, t_0\right)\right] + \widehat{R}_{0,0} \\ \widehat{R}_{1,0} \\ \cdots \\ \widehat{R}_{n-1,0} \\ \frac{k}{h}\alpha^2\varphi_2\left(t_1\right) + \frac{\lambda}{6}h^3\frac{\partial^3 u}{\partial x^3}\left(\xi_1^{(R)}, t_1\right) + \\ -\frac{\lambda}{3}h^3\left[\frac{1}{2}\frac{\partial^3 u}{\partial x^3}\left(\zeta^{(3)}, t_0\right) + \frac{\partial^3 u}{\partial x^3}\left(\zeta^{(4)}, t_0\right)\right] + \widehat{R}_{n,0} \end{bmatrix}, \tag{35}$$

$$\widehat{E}^{(j)} = \begin{bmatrix} -\frac{k}{h}\alpha^2\left(\varphi_1\left(t_j\right) + \varphi_1\left(t_{j+1}\right)\right) + \\ -\frac{\lambda}{6}h^3\left[\frac{\partial^3 u}{\partial x^3}\left(\xi_j^{(L)}, t_j\right) + \frac{\partial^3 u}{\partial x^3}\left(\xi_{j+1}^{(L)}, t_{j+1}\right)\right] + \widehat{R}_{0,j} \\ \widehat{R}_{1,j} \\ \cdots \\ \widehat{R}_{n-1,j} \\ \frac{k}{h}\alpha^2\left(\varphi_2\left(t_j\right) + \varphi_2\left(t_{j+1}\right)\right) + \\ +\frac{\lambda}{6}h^3\left[\frac{\partial^3 u}{\partial x^3}\left(\xi_j^{(R)}, t_j\right) + \frac{\partial^3 u}{\partial x^3}\left(\xi_{j+1}^{(R)}, t_{j+1}\right)\right] + \widehat{R}_{n,j} \end{bmatrix}, \tag{36}$$

$j = 1, 2, \ldots, m-1.$

Let us consider the finite difference scheme (26)-(30). The problem is how to find the intervals that contain the derivatives given in the error terms. From (1) we have

$$\frac{\partial^3 u}{\partial t^3}\left(x, t\right) = \alpha^2\frac{\partial^4 u}{\partial x^2\partial t^2}\left(x, t\right), \quad \frac{\partial^4 u}{\partial x^2\partial t^2}\left(x, t\right) = \frac{1}{\alpha^2}\frac{\partial^3 u}{\partial t^3}\left(x, t\right),$$

$$\frac{\partial^4 u}{\partial x^4}\left(x, t\right) = \frac{1}{\alpha^2}\frac{\partial^3 u}{\partial t\partial x^2}\left(x, t\right), \quad \frac{\partial^3 u}{\partial x^3}\left(x, t\right) = \frac{1}{\alpha^2}\frac{\partial^2 u}{\partial t\partial x}\left(x, t\right), \tag{37}$$

and we assume that

$$\frac{\partial^4 u}{\partial x^2\partial t^2}\left(x, t\right) = \frac{\partial^4 u}{\partial t^2\partial x^2}\left(x, t\right). \tag{38}$$

First, let us note that for $i = 0, 1, \ldots, n$, $j = 0, 1, \ldots, m-1$, we need

$$\frac{\partial^3 u}{\partial t^3}\left(x_i, \eta_j^{(1)}\right) = \alpha^2 \frac{\partial^4 u}{\partial x^2 \partial t^2}\left(x_i, \eta_j^{(1)}\right), \quad \frac{\partial^4 u}{\partial t^2 \partial x^2}\left(x_i, \eta_j^{(2)}\right) = \frac{1}{\alpha^2}\frac{\partial^3 u}{\partial t^3}\left(x_i, \eta_j^{(2)}\right), \tag{39}$$

where $\eta_j^{(1)}, \eta_j^{(2)} \in (t_j, t_{j+1})$. First, we assume that

$$\frac{\partial^3 u}{\partial t^3}\left(x_i, \eta_j^{(1)}\right) \in M_{i,j} = \left[\underline{M}_{i,j}, \overline{M}_{i,j}\right]. \tag{40}$$

Furthermore, since $\eta_j^{(2)}$ is such that $\eta_j^{(2)} \in (t_j, t_{j+1})$ and from (38)-(40) we have

$$\frac{\partial^4 u}{\partial t^2 \partial x^2}\left(x_i, \eta_j^{(2)}\right) \in \frac{1}{\alpha^2} M_{i,j} = \frac{1}{\alpha^2}\left[\underline{M}_{i,j}, \overline{M}_{i,j}\right]. \tag{41}$$

Next, we consider

$$\frac{\partial^4 u}{\partial x^4}\left(\xi_i^{(1)}, t_j\right) = \frac{1}{\alpha^2}\frac{\partial^3 u}{\partial t \partial x^2}\left(\xi_i^{(1)}, t_j\right), \quad \frac{\partial^4 u}{\partial x^4}\left(\xi_i^{(2)}, t_j\right) = \frac{1}{\alpha^2}\frac{\partial^3 u}{\partial t \partial x^2}\left(\xi_i^{(2)}, t_j\right). \tag{42}$$

Since, $\xi_i^{(1)}, \xi_i^{(2)} \in (x_{i-1}, x_{i+1})$, then for $i = 0, 1, \ldots, n$ we assume that

$$\frac{\partial^3 u}{\partial t \partial x^2}\left(\xi_i^{(1)}, t_j\right) \in Q_{i,j} = \left[\underline{Q}_{i,j}, \overline{Q}_{i,j}\right], \quad j = 0, 1, \ldots, m-1, \tag{43}$$

$$\frac{\partial^3 u}{\partial t \partial x^2}\left(\xi_i^{(2)}, t_j\right) \in Q_{i,j} = \left[\underline{Q}_{i,j}, \overline{Q}_{i,j}\right], \quad j = 1, 2, \ldots, m. \tag{44}$$

Then, for $j = 1, 2, \ldots, m$ we need to know

$$\frac{\partial^3 u}{\partial x^3}\left(\xi_j^{(L)}, t_j\right) = \frac{1}{\alpha^2}\frac{\partial^2 u}{\partial t \partial x}\left(\xi_j^{(L)}, t_j\right), \quad \frac{\partial^3 u}{\partial x^3}\left(\xi_j^{(R)}, t_j\right) = \frac{1}{\alpha^2}\frac{\partial^2 u}{\partial t \partial x}\left(\xi_j^{(R)}, t_j\right), \tag{45}$$

where $\xi_j^{(L)} \in (x_{-1}, x_1)$ and $\xi_j^{(R)} \in (x_{n-1}, x_{n+1})$. We assume that for $j = 1, 2, \ldots, m$, we have

$$\frac{\partial^2 u}{\partial t \partial x}\left(\xi_j^{(L)}, t_j\right) \in N_j^{(L)} = \left[\underline{N}_j^{(L)}, \overline{N}_j^{(L)}\right], \quad \frac{\partial^2 u}{\partial t \partial x}\left(\xi_j^{(R)}, t_j\right) \in N_j^{(R)} = \left[\underline{N}_j^{(R)}, \overline{N}_j^{(R)}\right]. \tag{46}$$

Finally, we also need

$$\frac{\partial^3 u}{\partial x^3}\left(\zeta^{(1)}, t_0\right) = \frac{1}{\alpha^2}\frac{\partial^2 u}{\partial t \partial x}\left(\zeta^{(1)}, t_0\right), \quad \frac{\partial^3 u}{\partial x^3}\left(\zeta^{(2)}, t_0\right) = \frac{1}{\alpha^2}\frac{\partial^2 u}{\partial t \partial x}\left(\zeta^{(2)}, t_0\right), \tag{47}$$

$$\frac{\partial^3 u}{\partial x^3}\left(\zeta^{(3)}, t_0\right) = \frac{1}{\alpha^2}\frac{\partial^2 u}{\partial t \partial x}\left(\zeta^{(3)}, t_0\right), \quad \frac{\partial^3 u}{\partial x^3}\left(\zeta^{(4)}, t_0\right) = \frac{1}{\alpha^2}\frac{\partial^2 u}{\partial t \partial x}\left(\zeta^{(4)}, t_0\right), \tag{48}$$

where $\zeta^{(1)} \in (x_{-1}, x_1)$, $\zeta^{(2)} \in (x_0, x_2)$, $\zeta^{(3)} \in (x_{n-1}, x_{n+1})$ and $\zeta^{(4)} \in (x_{n-2}, x_n)$. Similarly, we assume that for $j = 0$ we have

$$\frac{\partial^2 u}{\partial t \partial x}\left(\zeta^{(1)}, t_0\right) \in N_0^{(L)} = \left[\underline{N}_0^{(L)}, \overline{N}_0^{(L)}\right], \quad \frac{\partial^2 u}{\partial t \partial x}\left(\zeta^{(3)}, t_0\right) \in N_0^{(R)} = \left[\underline{N}_0^{(R)}, \overline{N}_0^{(R)}\right]. \tag{49}$$

Furthermore, we assume that we have such intervals $P^{(L)}$ and $P^{(R)}$ that

$$\frac{\partial^2 u}{\partial t \partial x}\left(\zeta^{(2)}, t_0\right) \in P^{(L)} = \left[\underline{P}^{(L)}, \overline{P}^{(L)}\right], \quad \frac{\partial^2 u}{\partial t \partial x}\left(\zeta^{(4)}, t_0\right) \in P^{(R)} = \left[\underline{P}^{(R)}, \overline{P}^{(R)}\right].$$

(50)

Now, substituting (40)-(41), (43)-(44), (46) and (49)-(50) to (26)-(30) we get an interval finite difference method of Crank-Nicolson type of the following form

$$(\lambda (1 + hA) + 1) U_{0,1} - \lambda U_{1,1} = \left(1 - \frac{5}{2}\lambda\right) U_{0,0} + 3\lambda U_{1,0} - \frac{\lambda}{2} U_{2,0}$$
$$- \frac{k}{h} \alpha^2 \Phi_1 (T_1) - \frac{\lambda}{6} \frac{h^3}{\alpha^2} N_1^{(L)} + \frac{\lambda}{3} \frac{h^3}{\alpha^2} \left(\frac{1}{2} N_0^{(L)} + P^{(L)}\right) + R_{0,0},$$
$$i = 0, \ j = 0,$$

(51)

$$(\lambda (1 + hA) + 1) U_{0,j+1} - \lambda U_{1,j+1} = (-\lambda (1 + hA) + 1) U_{0,j} + \lambda U_{1,j}$$
$$- \frac{k}{h} \alpha^2 (\Phi_1 (T_j) + \Phi_1 (T_{j+1})) - \frac{\lambda}{6} \frac{h^3}{\alpha^2} \left(N_j^{(L)} + N_{j+1}^{(L)}\right) + R_{n,j},$$
$$i = 0, \ j = 1, 2, \dots, m - 1,$$

(52)

$$-\frac{\lambda}{2} U_{i-1,j+1} + (1 + \lambda) U_{i,j+1} - \frac{\lambda}{2} U_{i+1,j+1} = \frac{\lambda}{2} U_{i-1,j}$$
$$+ (1 - \lambda) U_{i,j} + \frac{\lambda}{2} U_{i+1,j} + R_{i,j},$$
$$i = 1, \dots, n - 1, \quad j = 0, 1, \dots, m - 1,$$

(53)

$$-\lambda U_{n-1,1} + (\lambda (1 + hB) + 1) U_{n,1} = \left(1 - \frac{5}{2}\lambda\right) U_{n,0} + 3\lambda U_{n-1,0} - \frac{\lambda}{2} U_{n-2,0}$$
$$+ \frac{k}{h} \alpha^2 \Phi_2 (T_1) + \frac{\lambda}{6} \frac{h^3}{\alpha^2} N_1^{(R)} - \frac{\lambda}{3} \frac{h^3}{\alpha^2} \left(\frac{1}{2} N_0^{(R)} + P^{(R)}\right) + R_{n,0},$$
$$i = n, \ j = 0,$$

(54)

$$-\lambda U_{n-1,j+1} + (\lambda (1 + hB) + 1) U_{n,j+1} = \lambda U_{n-1,j} + (-\lambda (1 + hB) + 1) U_{n,j}$$
$$+ \frac{k}{h} \alpha^2 (\Phi_2 (T_j) + \Phi_2 (T_{j+1})) + \frac{\lambda}{6} \frac{h^3}{\alpha^2} \left(N_j^{(R)} + N_{j+1}^{(R)}\right) + R_{n,j},$$
$$i = n, \ j = 1, 2, \dots, m - 1,$$

(55)

where

$$R_{i,j} = \frac{k^3}{24} M_{i,j} - \frac{k^3}{8} M_{i,j} - \frac{kh^2}{24} (Q_{i,j} + Q_{i,j+1})$$

(56)

and

$$U_{i,0} = F(X_i), \quad i = 0, 1, \dots, n.$$

(57)

Note that X_i, $i = 0, 1, \ldots, n$, T_j, $j = 0, 1, \ldots, m$ are intervals such that $x_i \in X_i$ and $t_j \in T_j$. Furthermore, $F = F(X)$, $\Phi_1 = \Phi_1(T)$, $\Phi_2 = \Phi_2(T)$ denote interval extensions of the functions $f = f(x)$, $\varphi_1 = \varphi_1(t)$ and $\varphi_2 = \varphi_2(t)$, respectively.

The interval method (51)-(55) with (57) has also the matrix representation

$$CU^{(j+1)} = D^{(j)}U^{(j)} + E^{(j)}, \quad j = 0, 1, \ldots, m - 1, \tag{58}$$

where $U^{(j)} = [U_{0,j},\ U_{1,j},\ \ldots,\ U_{n,j}]^T$ and

$$E^{(0)} = \begin{bmatrix} -\frac{k}{h}\alpha^2\Phi_1(T_1) - \frac{\lambda}{6}\frac{h^3}{\alpha^2}N_1^{(L)} + \frac{\lambda}{3}\frac{h^3}{\alpha^2}\left(\frac{1}{2}N_0^{(L)} + P^{(L)}\right) + R_{0,0} \\ R_{1,0} \\ \cdots \\ R_{n-1,0} \\ \frac{k}{h}\alpha^2\Phi_2(T_1) + \frac{\lambda}{6}\frac{h^3}{\alpha^2}N_1^{(R)} - \frac{\lambda}{3}\frac{h^3}{\alpha^2}\left(\frac{1}{2}N_0^{(R)} + P^{(R)}\right) + R_{n,0} \end{bmatrix}, \tag{59}$$

$$E^{(j)} = \begin{bmatrix} -\frac{k}{h}\alpha^2\left(\Phi_1(T_j) + \Phi_1(T_{j+1})\right) - \frac{\lambda}{6}\frac{h^3}{\alpha^2}\left(N_j^{(L)} + N_{j+1}^{(L)}\right) + R_{0,j} \\ R_{1,j} \\ \cdots \\ R_{n-1,j} \\ \frac{k}{h}\alpha^2\left(\Phi_2(T_j) + \Phi_2(T_{j+1})\right) + \frac{\lambda}{6}\frac{h^3}{\alpha^2}\left(N_j^{(R)} + N_{j+1}^{(R)}\right) + R_{n,j} \end{bmatrix}, \tag{60}$$

$j = 1, 2, \ldots, m - 1$.

Theorem 1. *Let us assume that the local truncation error of the Crank-Nicolson scheme can be bounded by the appropriate intervals at each step. Moreover, let $F = F(X)$, $\Phi_1 = \Phi_1(T)$, $\Phi_2 = \Phi_2(T)$ denote interval extensions of the functions $f = f(x)$, $\varphi_1 = \varphi_1(t)$, $\varphi_2 = \varphi_2(t)$, given in the initial and boundary conditions (2)-(4) of the heat conduction equation (1). If $u(x_i, 0) \in U_{i,0}$, $i = 0, 1, \ldots, n$ and the linear system of equations (58) corresponding to the interval version of the Crank-Nicolson method (51)-(55) can be solved with some direct method, then for the interval solutions considered we have $u(x_i, t_j) \in U_{i,j}$, $i = 0, 1, \ldots, n$, $j = 1, 2, \ldots, m$.*

Remark 1. Taking into consideration the formulas (26)-(30) and (51)-(55) with their appropriate matrix representations (31) and (58), we conclude that the proof of the above theorem is a natural consequence of the thesis of Theorem 2.

Consider a finite system of linear algebraic equations of the form $Ax = b$, where A is an n-by-n matrix, b is an n-dimensional vector and the coefficients of A and b are real or interval values. The existence of the solution to $Ax = b$ is provided by Theorem 2 (see [6]).

Theorem 2. *If we can carry out all the steps of a direct method for solving $Ax = b$ in the interval arithmetic (if no attempted division by an interval containing zero occurs, nor any overflow or underflow), then the system has a unique solution for every real matrix in A and every real matrix in b, and the solution is contained in the resulting interval vector X.*

4   Conclusions

In the paper the interval version of Crank-Nicolson method for solving the one-dimensional heat conduction equation with the mixed boundary conditions is presented. In the form (51)-(55) it represents a scheme to compute a guaranteed result. However, we have to assume that the values of the derivatives in the error terms are enclosed by appropriate intervals.

   In practice it is not easy to find the endpoints of such intervals exactly. We can just approximate them with some possibly high order finite difference schemes. In this case, the interval method considered validates the conventional Crank-Nicolson method and we cannot guarantee that the exact solution belongs to the interval solution obtained. Although such approximation is not presented in the paper, it was developed and tested by the author on several problems. If we broaden the intervals of the error term by some experimentally chosen value we observe that the exact solution does belong to the interval solution. Such advantage of the proposed method over the conventional one seems to be a good starting point for further research.

References

1. Jankowska, M.A., Marciniak, A.: An Interval Finite Difference Method for Solving the One-Dimensional Heat Equation. LNCS (accepted)
2. Lapidus, L., Pinder, G.F.: Numerical Solution of Partial Differential Equations in Science and Engineering. J. Wiley & Sons (1982)
3. Manikonda, S., Berz, M., Makino, K.: High-order verified solutions of the 3D Laplace equation. WSEAS Transactions on Computers 4(11), 1604–1610 (2005)
4. Marciniak, A.: An Interval Difference Method for Solving the Poisson Equation - the First Approach. Pro Dialog 24, 49–61 (2008)
5. Marciniak, A.: An Interval Version of the Crank-Nicolson Method – The First Approach. In: Jónasson, K. (ed.) PARA 2010, Part II. LNCS, vol. 7134, pp. 120–126. Springer, Heidelberg (2012)
6. Moore, R.E., Kearfott, R.B., Cloud, M.J.: Introduction to Interval Analysis. SIAM, Philadelphia (2009)
7. Nagatou, K., Hashimoto, K., Nakao, M.T.: Numerical verification of stationary solutions for Navier-Stokes problems. Journal of Computational and Applied Mathematics 199(2), 445–451 (2007)
8. Nakao, M.T.: Numerical verification methods for solutions of ordinary and partial differential equations. Numerical Functional Analysis and Optimization 22(3-4), 321–356 (2001)
9. Smith, G.D.: Numerical Solution of Partial Differential Equations: Finite Difference Methods. Oxford University Press (1995)
10. Watanabe, Y., Yamamoto, N., Nakao, M.T.: A Numerical Verification Method of Solutions for the Navier-Stokes Equations. Reliable Computing 5(3), 347–357 (1999)

Using C-XSC
for High Performance Verified Computing

Walter Krämer, Michael Zimmer, and Werner Hofschuster

Bergische Universität Wuppertal, D-42119 Wuppertal, Germany
{kraemer,zimmer,hofschuster}@math.uni-wuppertal.de
http://math.uni-wuppertal.de/~xsc

Abstract. C-XSC is a C++ class library for scientific computing, with its main focus on reliable interval computations. Recently, several changes and new features have been implemented, making C-XSC much more suitable for tasks in high performance computing. However, these changes require that users take several factors into consideration when writing and compiling programs using C-XSC to get the best possible performance while still maintaining a sufficient level of numerical accuracy. This paper gives an overview of the most important points concerning these factors and tries to give background information and recommendations to the end user for the implementation of efficient C-XSC programs.

Remark: An accompanying extended version of this paper is available, see [10].

Keywords: C-XSC, high performance computing, compiler optimizations, dot product computation, error free transformation, BLAS, openMP, MPI.

1 Introduction

The C-XSC (XSC stands for eXtended Scientific Computing) library [8,7], derived from the earlier Pascal-XSC and several other XSC languages and libraries, was introduced in 1992 and is thus now nearly 20 years old. The main focus of the C-XSC development has always been to provide a rich feature set with high reliability and accuracy. One of the paradigms of its development was that every operation should be maximally accurate, i.e. at worst only one floating point rounding away from the exact result. The run time performance however, especially when dealing with larger, multidimensional problems, has not taken a big role in the development for a long time.

In recent years, the focus of the C-XSC development shifted into making C-XSC fit for applications in high performance computing, where large problems need to be solved effectively, while still preserving reliability and the features introduced so far. The goal is to make C-XSC a valid option for the end users who want to write such programs, whether they are interested in reliability and the possibilities of using interval computations provided by C-XSC, or just want to use a C++ library providing some basic data types and functionality.

K. Jónasson (Ed.): PARA 2010, Part II, LNCS 7134, pp. 168–178, 2012.

Parallel programming has become another point of interest in the development of C-XSC. Whether one wants to write programs for huge distributed memory clusters using MPI or on modern multicore systems using OpenMP or some thread library, the newer C-XSC versions now make it easier to do so and partially can take advantage of multicore machines themselves.

This paper discusses some of the most important factors to consider when writing programs with C-XSC that aim for high performance and efficiency. Recommendations for writing and compiling C-XSC programs as well as explanations of some possible pitfalls are given. The paper is structured as follows. Section 2 deals with the installation of C-XSC, while Section 3 is concerned with various aspects of writing C-XSC programs. The following Section 4 gives some advice on compiling C-XSC programs. Finally, the paper ends with a summarizing example in Section 5 and some final remarks in Section 6.

Some more detailed information, explanations, and source codes may be found in the accompanying extended version of this paper [10].

2   Installing an Efficient Version of the C-XSC Library

Many parts of the C-XSC library are defined directly in the header files to give the compiler the opportunity to inline calls to the functions defined there (for example all operators of the basic data types). This leads to a huge performance increase compared to compiling all code directly into the library itself. However, there are still big and vital portions of C-XSC that are compiled into the library and whose performance can be of high importance for the whole library. This section describes the issues to consider, when compiling C-XSC, and gives some recommendations for the best settings for most users.

The installation procedure of C-XSC is guided by the install script `install_cxsc`. It queries all important questions regarding compiler and configuration options to the user.

First, the user has to decide which compiler to use. If there are multiple compilers installed, the newest version should be used in most cases. The install script will then ask whether compiler optimizations should be used and, if this is the case, which flags are to be set. It is highly recommended to use at least the default optimization flags (`-O1 -finline-functions`), which should work correctly on every system.

If you use an x86 or x86-64 based system, you should activate the highest optimization level `-O3`. This might lead to reliability problems for parts of C-XSC that change the rounding mode, however, on these systems, assembler routines for the affected code can be activated, which will not suffer from these problems.

If you do not want to use the compiled C-XSC library on different systems (different processors), you should also tell the compiler to optimize specificially for your processor. The generated code will only run on systems with the same processor type, but with this option, the compiler will make use of the complete instruction set of the actual processor (for example SSE instructions). To activate this option, add `-march=native` for the gcc and `-xHost` for the icc to your optimization options.

When using icc, you should also activate so called interprocedural optimizations by adding `-ip` to your optimization options (for more detailed explanations of this option as well as the other optimization options, see [6]). The same applies for gcc [2].

If the FMA operation (Fused Multiply and Add) is available in hardware on the specific machine, C-XSC can make use of it. If you want to use it, add `-DCXSC_USE_FMA` to your optimization options. C-XSC will then call the standard library function `fma` to perform this operation at the appropriate places. If the FMA operation is supported in hardware and your standard library takes advantage of this (for example using gcc 4.4.2 on a system with a Nehalem processor should work), this will speed up some computations. If there is no hardware support, this option will also work, but will in fact slow down some computations, since the FMA operation is then simulated in software. Thus, this option should only be activated if hardware support is guaranteed.

After setting the compiler optimizations and when compiling on a 64 bit machine, the install script will ask whether it should generate 64 bit oder 32 bit code. In general it is recommended to compile 64 bit code. However, if you want to use the library later also on a 32 bit machine, you have to choose the option for 32 bit code generation.

The install script will then ask in which way C-XSC should set the rounding mode of the processor. When compiling on an x86 or x86-64 machine, the option to use assembler routines for this task will be available. It is strongly recommended to use these routines if available, since they are not only faster but also more reliable. If instead the hardware rounding option is chosen at this point, C-XSC will use the standard library function `fesetround`. This might lead to wrong numerical results when using compiler optimizations. The optimization level should be reduced in this case. The third option for the rounding mode is software rounding, which will simulate the correct rounding in software. This option should always work correctly, even with optimizations. However it will be very slow.

At the end of the installation procedure, the install script will finally ask whether it should start the code generation. At this point, all the needed makefiles and configuration files have been generated. You can now modify them by hand, if you want to add options or change anything that was not covered by the installation script.

One last thing to consider during the installation of C-XSC is the functionality of the error free transformations needed for the dot products in K-fold precision described in Section 3.1. The algorithms for the error free transformations will not work on processors that use 80-bit wide floating point registers (for example, the FPU of most x86 processors use such registers).

The best way on x86 machines is to use the SSE registers, which have no excess precision for floating point computations. With gcc, use the option `-mfpmath=sse` to enable this (at least SSE2 instructions have also to be activated with `-msse2`). The icc compiler will activate this by default on 32 bit x86 machines. If not, one has to use the option `-fp-model source`, which unfortunately can also slow down performance due to reduced optimizations on floating point operations.

3   Writing One's Own C-XSC Programs

This section shows how to write efficient C-XSC programs. The main focus of C-XSC has always been to achieve very high accuracy, with the paradigm (which still holds for the default settings of todays version) that the result of every operation should be maximally accurate, i.e. that the computed result is only one rounding away from the true, exact result.

In the newer C-XSC versions, a stronger focus on run time performance has been established. The changes described in the following subsections reflect this. It is however necessary to know how to use these new possiblities in order to write an efficient C-XSC program. Especially the changes described in Section 3.1 can have a major impact on the performance of a program when using matrix and vector operations.

3.1   Dot Products and BLAS Support

Since version 2.3.0, C-XSC includes the feature to switch the precision of all dot product computations at runtime. Before this version, every dot product, either explicitly computed oder implicitly computed in one of the operators (for example, when computing matrix*matrix), was computed in maximum precision using the so called long accumulator [11,12]. Since regrettably there is still no hardware support for the long accumulator (which would be faster than normal floating point dot products) available, it has to be simulated in software and thus is very slow.

There are two ways to compute dot products or dot product expressions in C-XSC. One is to use the operators of the appropriate data types. As an example, we want to compute the result of $z = b - Ax$, where b, x and z are real vectors (data type `rvector`) of length n and A is a real matrix (data type `rmatrix`) of dimension $n \times n$. The code for this example is given in Listing 1.1.

Listing 1.1. Computing $b - Ax$ using operators

```
rmatrix A(n,n);
rvector b(n), x(n), z(n);

//...fill A, b, x with values...

//set precision of operator dot products
opdotprec = 2;

//Compute expression
z = b - A*x;
```

When computing the matrix-vector product Ax via the appropriate operator, C-XSC has to compute n dot products. The precision used to compute these dot products can be set with the global C-XSC variable `opdotprec`. Setting this variable to some integer value $K \geq 0$ means that all following dot product computations in any operator are computed in K-fold double precision (thus if $K = 1$ normal floating point computations are used). If $K = 0$, the long accumulator is used as in previous C-XSC versions. This is also the default setting to preserve compatibility with older C-XSC programs. If $K \geq 2$, the DotK algorithm is used to simulate higher precision. More details can be found in [14,9].

A disadvantage when using operators is that while every dot product operation is computed in K-fold or even maximum precision, the result of every single operation is rounded into double precision so that the evaluation of a whole expression like $b - Ax$ might lead to results that are not very accurate due to the rounding errors of the intermediate values. To compute a complete expression in high precision, C-XSC provides the data type `dotprecision` (`idotprecision` for intervals, `cdotprecision` for complex values and `cidotprecision` for complex intervals), which represents a long accumulator. Listing 1.2 shows how to compute the above example $b - Ax$ that way.

Listing 1.2. Computing $b - Ax$ using the long accumulator

```
rmatrix A(n,n);
rvector b(n), x(n), z(n);

//...fill A, b, x with values...

//Initialize with 0
dotprecision dot(0.0);
//Set dot product precision
dot.set_dotprec(2);

for(int i=1 ; i<=n ; i++) {
   dot = b[i];
   accumulate(dot, -A[i], x);
   z[i] = rnd(dot);
}
```

First, a `dotprecision` variable `dot` is defined. The precision for all dot products computed with *this* variable can be set with the `set_dotprec` member function. To compute the expression, a loop over all elements of the result vector is used. In each iteration of the loop, the accumulator is initialised with the i-th element of b. Then, the dot product of the i-th row of A with the vector x is computed in the precision selected for `dot` (in the example above this is 2-fold precision) and then added to the accumulator. What is important here is that the added result with this method also has 2-fold double length precision and that it is added to the value b_i stored in the accumulator *with maximum precision*. Thus, computing a dot product expression in this way can drastically improve the accuracy of the result. The downside is that the computation will be slower (but often, depending on the number of additions, only slightly slower) and that the source code becomes more complex.

When writing a C-XSC program using matrix and vector computations, one should always ensure that the precision is not set higher than really needed. If pure floating point precision is sufficient for the specific application, setting the dot product precision to one will result in vastly improved runtime performance. Even if higher accuracy is needed, it will often be sufficient to use two fold precision in many practical cases, which will still be a lot faster than the default setting using the long accumulator.

Another option concerning dot products is the possibility to use optimized BLAS routines for matrix and vector computations. This feature has been introduced with C-XSC version 2.4.0. BLAS routines will be used for all matrix and vector operators explicitly or implicitly computing dot products (matrix*matrix, matrix*vector, vector*vector). To make use of the BLAS, opdotprec has to be set to one and your program must be compiled with the option -DCXSC_USE_BLAS. The program then also has to be linked against an appropriate BLAS and CBLAS library. For example, when using ATLAS BLAS (if compiled as a static library with the GNU Fortran compiler and installed in the directory /usr/local/atlas) you would have to add the following options to your compiler call:

```
-DCXSC_USE_BLAS -L/usr/local/atlas/lib -latlas -lcblas -lf77blas -latlas -lgfortran
```

Using BLAS routines is highly recommended, especially when computing matrix-matrix products, since this operation will normally see a drastic performance increase with these routines. Also, many BLAS libraries are already multithreaded and thus can take advantage of multicore or multiprocessor machines without any additional work on the side of the C-XSC programmer.

3.2   Using Data Types Efficiently

C-XSC provides a lot of different data types, most importantly the four basic types real, interval, complex and cinterval. Many other data types are based on these types, for example dense vectors (rvector, ivector, ...), dense matrices (rmatrix, imatrix, ...) and sparse matrices (srmatrix, simatrix, ...). This section contains some remarks for using these data types in an efficient way and choosing the most appropriate data type for a specific task.

The first important thing to note is that C-XSC uses the data type real instead of the built in data type double. A double can be converted into a real value by calling the corresponding constructor or by simply assigning the double value to the real variable. If one wants to convert from real to double, either a cast can be used or the function _double(real&) has to be called.

When using interval data types, one has always to keep in mind that reliable interval computations on a machine require every operation to be rounded outwards to guarantee an enclosure. To perform such an outward rounding, it is necessary to switch the rounding mode of the processor which is expensive in terms of run time performance. Because of this, one should always use the provided functions and operators and not perform operations like a dot product element wise.

When writing programs with matrices and vectors, it is important to distinguish dense matrices from sparse matrices (and vectors). If a matrix or vector consists mostly of zero entries, it is said to be sparse. To take advantage of the sparsity both in terms of performance as well as memory footprint, the sparse matrix and vector types should be used in this case. Listing 1.3 gives a small example.

Listing 1.3. Using dense and sparse matrices

```
rmatrix A(n,n); //dense

//A is diagonal matrix
for(int i=1 ; i<=n ; i++)
  A[i][i] = i;

//Create sparse matrix out of dense A
srmatrix B(A); //sparse

A = A*A; //dense times dense
A = B*A; //sparse times dense
B = B*B; //sparse time sparse
```

In this example, a diagonal matrix A is created and stored as a dense matrix. Then, a sparse matrix B is created out of A. A dense matrix-matrix product will be much slower than a sparse one, and still significantly slower than a sparse-dense product. Furthermore, everything said in Section 3.1 also applies to sparse data types.

When using sparse data types, some important facts about the underlying data structure (sparse matrices are stored using compressed column storage) have to be considered. Accessing single elements of a sparse matrix is quite expensive and should be avoided, if possible. Since the data structure is column based, accessing rows is also quite expensive, while accessing columns should be a lot faster. It might therefore be a good idea to first transpose a sparse matrix using the **transp** function and access the columns of the transposed matrix instead of the rows of the original matrix. As with dense matrices, it is also possible to access slices of a matrix in the form B(1,10,1,10) (this accesses the upper left 10×10 submatrix of B). This operation is quite expensive with sparse matrices and should not be overused. Further details about sparse matrices in C-XSC can be found in [16].

3.3   Using OpenMP/Multithreading

Systems with multiple cores and/or multiple processors have become mainstream in recent years. To make use of their possibilities, programs have to be multi-threaded. In general, the easiest and most convenient way to do this is using OpenMP, which in many cases allows one to make programs multithreaded by simply adding some compiler pragmas and without directly changing pre-existing serial code.

In older C-XSC versions (before version 2.3.0) one could easily run into problems when using multithreading, because in many calculations a global array of dotprecision objects was used internally. This was not transparent to the user, which made synchronisation between threads very hard. In newer C-XSC versions this has been changed, and implementing multithreaded C-XSC programs is now much easier. However, the user still has to take some global C-XSC configuration variables into account. First and foremost, the global variable opdotprec which, as described in Section 3.1, controls the precision of all dot product computations implied by corresponding operator calls. Changes to this variable in one thread will have an effect on all other threads. If this variable is changed during a dot product computation, this might even lead to undefined behaviour. It is therefore recommended not to set this variable within a parallel region.

Another global C-XSC variable you have to take care of is `stagprec`. When using staggered precision data types (`l_interval`, `lx_interval` etc.), this variable controls the length of the staggered representation and thus the accuracy of the computations. As with `opdotprec`, this variable should not be accessed inside of a parallel region.

3.4 Using MPI in Connection With C-XSC

When writing parallel programs for distributed memory machines, i.e. clusters, MPI has become a de-facto standard. In this form of parallelization, it is necessary to send data between the different running processes. MPI provides some built-in functions to do this for the standard C data types.

To also be able to send C-XSC data types via MPI, we provide an MPI-interface which gives the user access to communication functions that work the same way as the ordinary MPI routines. The recent changes in C-XSC (new dot products and additional sparse data types) made it necessary to update this interface appropriately. Older versions of the interface should be updated to the newest release available on our C-XSC website [19].

The MPI interface gives the user the possibility to write parallel C-XSC applications for large clusters which have the potential to solve huge problems. An example of this is the implementation of a verified solver for dense linear interval systems discussed e.g. in [9,17].

4 Compiling One's Own C-XSC Programs

This section deals with the compilation of C-XSC programs. Since huge portions of the C-XSC library code reside in the header files to allow the compiler to inline function calls, it is also very important for the performance of a C-XSC program, to compile the source code using appropriate compiler options.

First, we strongly recommend to update your compiler to the newest version. Newer versions can often give a significant performance boost "for free", especially if the compiler in use is already a few years old.

Inlining is of particular relevance to the performance of every C-XSC program. Since C-XSC uses object-oriented concepts, a huge number of function calls are necessary (for example, every use of an operator results in a function call). Since function calls are quite expensive, this can result in very bad run time performance, if the compiler is not forced to do inlining (function calls are replaced by direct copies of the function codes). Depending on the compiler, inlining may have already been activated by default. Nearly every compiler activates inlining on the highest optimization level -O3.

It is in general recommended to use the same optimization options as when installing C-XSC (see Section 2). This means

- for the GNU Compiler: `-O3 -march=native`
- for the Intel Compiler: `-O3 -xHost -ip`

Again, the options `-march=native` and `-xHost` shall only be used if the compiled program is not executed on different systems. Please refer to the manual of your compiler for more detailed explanations of the possible options.

It is in general a good idea to use BLAS routines for matrix and vector computations. As explained in Section 3.1, these can be activated by using the compiler switch `-DCXSC_USE_BLAS` and linking to a BLAS and a CBLAS library.

Some modules of the C-XSC Toolbox also can make use of optimized LAPACK routines. At the moment this only applies to the computation of an approximate inverse of a real matrix (header file `matinv_aprx.hpp`) and the solvers for dense linear systems (see Section 5). To make use of LAPACK the compiler switch `-DCXSC_USE_LAPACK` has to be used and the program must be linked against a LAPACK library. LAPACK support of C-XSC will be extended in the near future.

5   An Example: Runtime Comparison When Solving Linear Systems

In this very short section we discuss the possible improvement for applications using C-XSC when following the recommendations and suggestions given in this paper (more details and source code are presented in [10]). For this, we take solvers for dense linear systems as an example and compare the old Toolbox module LinSolve (see [3]) with the new optimized solvers, which were first implemented in a separate package [9] and are now also included directly in the C-XSC kernel. For both versions a C-XSC installation with full optimizations as explained in Section 2 is used.

The new solvers are compiled with all the recommended optimizations, use the new dot product algorithms as well as BLAS and LAPACK support, and also take advantage of multicore processors via OpenMP. We compare the time taken to solve a real linear point system of dimension $n = 1000$. (For the complete program listing please refer to [10].)

Using the Intel Compiler 11.1 with full optimizations and using the Intel MKL on a Nehalem based Intel Xeon 2.26 GHz computer with two processors with four cores each, the new solver takes about 0.34 seconds, while the old solver needs about 198 seconds. This is an extreme example (the improved solver is about 600 times faster than the traditional solver!), but it shows how much performance can be gained using and combining the new features of C-XSC in a sensible way.

6   Final Remarks and Future Prospect

In recent times the portable C++ class library C-XSC has made major advances to become a library that is not only providing a rich set of numerical features and that offers reliability and very high accuracy of numerical results, but is also very attractive for highly demanding applications from high performance computing (e.g. in solving parameter dependent linear systems [15]). The user always has the choice to use C-XSC in the traditional way with a focus on maximum accuracy, but now he can also adapt his programs for specific uses requiring higher performance. The following list summarizes some of the major advances:

- Choosable precision for dot products, ranging from dot products in pure floating point to exact dot products using a long accumulator.
- Support for BLAS and some LAPACK routines, allowing huge speed increases for some applications.
- Specific data types for sparse matrices and vectors that work much more efficiently than their dense counterparts.
- Easier usage of OpenMP with C-XSC.
- New multiple-precision data types `lx_real`, `lx_interval`, ... with corresponding (interval) operations and a rich set of elementary functions [1].

In the near future, C-XSC will offer even more functionality in terms of pure floating point algorithms by providing an interface to the routines from LA-PACK. Furthermore, parts of the C-XSC Toolbox will be optimized for multicore processors. With the current and future changes, C-XSC becomes more flexible and an interesting alternative for multiple numerical applications, especially for, but not restricted to the field of reliable computing.

This paper refers to C-XSC version 2.4.0 or higher. Note: C-XSC comes with a powerful installation script. This makes it easy to install C-XSC even for beginners by just accepting the default options proposed when executing the installation script. There is no need to first read this paper. But if you are interested in creating the C-XSC library for extensive numerical computations (typically done in high performance computing), you should know the possibilities to speed up computations using more specific options when installing C-XSC and when compiling your own C-XSC source code.

References

1. Blomquist, F., Hofschuster, W., Krämer, W.: A Modified Staggered Correction Arithmetic with Enhanced Accuracy and Very Wide Exponent Range. In: Cuyt, A., Krämer, W., Luther, W., Markstein, P. (eds.) Numerical Validation. LNCS, vol. 5492, pp. 41–67. Springer, Heidelberg (2009)
2. GCC online documentation, http://gcc.gnu.org/onlinedocs
3. Hammer, R., Hocks, M., Kulisch, U., Ratz, D.: Numerical Toolbox for Verified Computing I: Basic Numerical Problems. Springer, Heidelberg (1993)

4. ANSI/IEEE Std. 754-1985, A Standard for Binary Floating-Point Arithmetic. New York, 1985; reprinted in SIGPLAN 22(2), 9–25 (1987)
5. Forthcoming IEEE-1788 interval standard
6. Intel C++ Compiler User and Reference Guides. Available on the Intel Compiler Homepage, http://software.intel.com/en-us/intel-compilers
7. Hofschuster, W., Krämer, W.: C-XSC 2.0 – A C++ Library for Extended Scientific Computing. In: Alt, R., Frommer, A., Kearfott, R.B., Luther, W. (eds.) Num. Software with Result Verification. LNCS, vol. 2991, pp. 15–35. Springer, Heidelberg (2004)
8. Klatte, R., Kulisch, U., Wiethoff, A., Lawo, C., Rauch, M.: C-XSC - A C++ Class Library for Extended Scientific Computing. Springer, Heidelberg (1993)
9. Krämer, W., Zimmer, M.: Fast (Parallel) Dense Linear System Solvers in C-XSC Using Error Free Transformations and BLAS. In: Cuyt, A., Krämer, W., Luther, W., Markstein, P. (eds.) Numerical Validation. LNCS, vol. 5492, pp. 230–249. Springer, Heidelberg (2009)
10. Krämer, W., Zimmer, M., Hofschuster, W.: Using C-XSC for High Performance Verified Computing. Preprint BUW-WRSWT 2009/5, University of Wuppertal (2009)
11. Kulisch, U., Miranker, W.: The arithmetic of the digital computer: A new approach. SIAM Rev. 28(1), 1–40 (1986)
12. Kulisch, U.: Die fünfte Gleitkommaoperation für Top-Performance Computer. Berichte aus dem Forschungsschwerpunkt Computerarithmetik, Intervallrechnung und numerische Algorithmen mit Ergebnisverifikation, Univ. Karlsruhe (1997)
13. Kulisch, U.: Computer Arithmetic and Validity: Theory, Implementation, and Applications, de Gruyter Studies in Mathematics (2008)
14. Ogita, T., Rump, S.M., Oishi, S.: Accurate sum and dot product. SIAM Journal on Scientific Computing 26, 6 (2005)
15. Popova, E., Kolev, L., Krämer, W.: A Solver For Complex-Valued Parametric Linear Systems. Serdica Journal of Computing 4(1), 123–132 (2010)
16. Zimmer, M., Krämer, W., Hofschuster, W.: Sparse Matrices and Vectors in C-XSC, BUW-WRSWT 2009/7, Preprint 2009/7, Universität Wuppertal (2009)
17. Zimmer, M.: Laufzeiteffiziente, parallele Löser für lineare Intervallgleichungssysteme in C-XSC. Master Thesis, University of Wuppertal (2007)
18. Zimmer, M., Krämer, W., Bohlender, G., Hofschuster, W.: Extension of the C-XSC Library with Scalar Products with Selectable Accuracy. Serdica Journal of Computation 4(3), 349–370 (2010)
19. C-XSC Website, http://www.math.uni-wuppertal.de/wrswt/xsc/cxsc_new.html

Efficient Implementation
of Interval Matrix Multiplication

Hong Diep Nguyen

INRIA - LIP(UMR 5668 CNRS - ENS de Lyon - INRIA - UCBL),
Université de Lyon, France
hong.diep.nguyen@ens-lyon.fr

Abstract. The straightforward implementation of interval matrix product suffers from poor efficiency, far from the performances of highly optimized floating-point matrix products. In this paper, we show how to reduce the interval matrix multiplication to 9 floating-point matrix products - for performance issues - without sacrificing the quality of the result. Indeed, we show that, compared to the straightforward implementation, the overestimation factor is at most 1.18.

Keywords: interval arithmetic, interval matrix multiplication, efficiency.

1 Introduction

Interval arithmetic is a means of obtaining guaranteed results: computing enclosures of the exact results. Nonetheless, it suffers from lower performance than non-guaranteed floating-point computations. Theoretically, the factor of performance between interval arithmetic and floating-point arithmetic in theory is 4. It is even much worse in practice, especially when a large number of operations is involved.

In this paper we will study the case of matrix multiplication. Our implementation, based on the natural algorithm, gets good performances at the price of a slight overestimation of the result. The idea is to exploit existing libraries which are well optimized for floating-point matrix operations such as BLAS, ATLAS, etc. In the same vein, Rump proposed a fast algorithm which uses only four floating-point matrix products [9]. Rump's algorithm returns a result wider than the result obtained by replacing each floating-poing operation between two numbers by its interval counterpart: the factor of overestimation of this algorithm is 1.5 in the worst case. This paper proposes a new algorithm which costs 9 floating-point matrix products with a factor of overestimation in the worst case of 1.18.

This paper is organized as follows. Section 2 briefly presents interval arithmetic with some basic operations. A more thorough introduction to interval arithmetic as well as some applications of interval arithmetic can be found in [6,3,5]. These operations are extended to matrix operations which are studied in section 3. Section 4 reminds the principle of Rump algorithm. Section 5 is devoted to our proposed algorithm.

K. Jónasson (Ed.): PARA 2010, Part II, LNCS 7134, pp. 179–188, 2012.

Notations

In this paper, bold-face lower-case letters represent scalar intervals and bold-face upper-case letters represent interval matrices. Below are some additional notations:

$[\circ]$ is an expression computed by interval arithmetic,

$(\circ)_\downarrow$, $(\circ)_\uparrow$ are expressions computed by floating-point arithmetic with downward and upward rounding mode respectively,

$[i, s]$ is an interval whose endpoints are i and s,

$\{m, r\}$ is an interval whose mid-point is m and radius is r,

$\underline{a}$, $\overline{a}$ are the lower and upper bounds of a,

$\text{mag}(a)$ is the maximal magnitude of a: $\text{mag}(a) \overset{def}{=} \max\{|a|, a \in a\} = \max(|\underline{a}|, |\overline{a}|)$.

2   Interval Arithmetic

Intervals are used to represent connected closed sets of real values. Interval arithmetic defines operations between intervals. The result of an operation between intervals is also an interval containing all the possible results between all possible pairs of real values taken from input intervals: $r = a \circ b = \square\{a \circ b, a \in a, b \in b\}$, where $\square$ denotes the hull of a set, i.e. the smallest interval enclosing the set.

Due to the monotonicity property, the sum of two intervals $r = a + b$ can be computed via the sum of their two respective lower and upper bounds:

$$\underline{r} = \underline{a} + \underline{b} \qquad \overline{r} = \overline{a} + \overline{b}. \tag{1}$$

Interval multiplication $r = a * b$ is formed by taking the minimum and maximum value of the four products between two pairs of endpoints of the two input intervals.

$$\begin{cases} \underline{r} = \min(\underline{a} * \underline{b}, \underline{a} * \overline{b}, \overline{a} * \underline{b}, \overline{a} * \overline{b}) \\ \overline{r} = \max(\underline{a} * \underline{b}, \underline{a} * \overline{b}, \overline{a}` * \underline{b}, \overline{a} * \overline{b}). \end{cases} \tag{2}$$

Hence, in theory the factor, in terms of performance, between interval arithmetic and floating-point arithmetic is 4.

Implementation

Intervals are implemented on computers using floating-point numbers. To ensure the inclusion property of the results, rounding errors must be taken into account: the lower endpoint must be computed with downward rounding mode, and the upper endpoint with upward rounding mode.

For the interval addition $r = a + b$, this yields:

$$\underline{r} = (\underline{a} + \underline{b})_\downarrow \qquad \overline{r} = (\overline{a} + \overline{b})_\uparrow. \tag{3}$$

The interval product $r = a * b$ is computed following (2) by four floating-point products. Nevertheless, to accomodate rounding errors, each floating-point product must be computed twice, once with upward and once with downward rounding mode. Thus in total, it requires eight floating-point products.

$$\begin{cases} \underline{r} = \min((\underline{a} * \underline{b})_\downarrow, (\underline{a} * \overline{b})_\downarrow, (\overline{a} * \underline{b})_\downarrow, (\overline{a} * \overline{b})_\downarrow) \\ \overline{r} = \max((\underline{a} * \underline{b})_\uparrow, (\underline{a} * \overline{b})_\uparrow, (\overline{a} * \underline{b})_\uparrow, (\overline{a} * \overline{b})_\uparrow). \end{cases} \quad (4)$$

We can reduce the number of floating-point products by inspecting the sign of each component. But testing the sign and branching accordingly is costly also.

Particular Cases

If a and b are Centered in Zero , i.e., $\underline{a} = -\overline{a}$ and $\underline{b} = -\overline{b}$ then $\underline{a} + \underline{b} = -(\overline{a} + \overline{b})$. Hence, the sum $r = a + b$ can be computed by:

$$\overline{r} = (\overline{a} + \overline{b})_\uparrow, \qquad \underline{r} = -\overline{r}. \quad (5)$$

If b is Centered in Zero then $\max(x * \underline{b}, x * \overline{b}) = |x| * \overline{b}$ and $\min(x * \underline{b}, x * \overline{b}) = -|x| * \overline{b}$ for all $x \in \mathbb{R}$. Hence according to (2), $r = a * b$ can be computed by:

$$\overline{r} = \max(|\underline{a}|, |\overline{a}|) * \overline{b}$$
$$= \mathrm{mag}\,(a) * \overline{b}$$
$$\underline{r} = -\max(|\underline{a}|, |\overline{a}|) * \overline{b}$$
$$= -\overline{r}.$$

Using floating-point arithmetic, r can be computed by only one floating-point product in upward rounding mode:

$$\overline{r} = \left(\mathrm{mag}\,(a) * \overline{b}\right)_\uparrow, \qquad \underline{r} = -\overline{r}. \quad (6)$$

3 Interval Matrix Operations

Let's now study the case of matrix operations. Suppose that each interval matrix here is of dimension $n \times n$ and is represented by two floating-point matrices, one for the lower endpoints and the other for the upper endpoints.

The addition of two interval matrices $C = A + B$ can be computed by performing scalar additions between corresponding elements of the two matrices. Thus

$$\underline{C} = (\underline{A} + \underline{B})_\downarrow \qquad \overline{C} = (\overline{A} + \overline{B})_\uparrow.$$

For the case of interval matrix multiplication, each element of the result matrix is computed by:

$$C_{i,j} = \sum_k A_{i,k} * B_{k,j}. \quad (7)$$

Using this natural formula, the computation of each product element requires n interval multiplications and n interval additions, or, following (3) and (4), it requires $8n$ floating-point products and $2n$ floating-point additions. Hence, the overall cost in theory is $8n_*^3 + 2n_+^3$ floating-point operations. Nonetheless, this cost does take into account neither $\min, \max$ functions nor the cost of rounding mode changes.

Particular Cases

A is centered in Zero ,i.e., $A_{i,k}$ is centered in zero for all i, k. From (6) we get

$$A_{i,k} * B_{k,j} = \left[-\overline{A}_{i,k} * \text{mag}\,(B)_{k,j}\,, \overline{A}_{i,k} * \text{mag}\,(B)_{k,j} \right]$$

$$\Rightarrow C_{i,j} \quad = \left[-\sum_k \overline{A}_{i,k} * \text{mag}\,(B)_{k,j}\,, \sum_k \overline{A}_{i,k} * \text{mag}\,(B)_{k,j} \right]$$

$$\Rightarrow \overline{C} \quad = \left(\overline{A} * \text{mag}\,(B) \right)_\uparrow$$

$$\underline{C} \quad = -\overline{C}.$$

Hence $A * B$ can be computed by one floating-point matrix product with upward rounding mode.

A is Nonnegative. It means that $0 \le \underline{A}_{i,k} \le \overline{A}_{i,k}$ for all i, k. Hence

$$\begin{cases} \max(\underline{A}_{i,k} * \underline{B}_{k,j}, \underline{A}_{i,k} * \overline{B}_{k,j}) = \underline{A}_{i,k} * \overline{B}_{k,j} \\ \min(\underline{A}_{i,k} * \underline{B}_{k,j}, \underline{A}_{i,k} * \overline{B}_{k,j}) = \underline{A}_{i,k} * \underline{B}_{k,j} \\ \max(\overline{A}_{i,k} * \underline{B}_{k,j}, \overline{A}_{i,k} * \overline{B}_{k,j}) = \overline{A}_{i,k} * \overline{B}_{k,j} \\ \min(\overline{A}_{i,k} * \underline{B}_{k,j}, \overline{A}_{i,k} * \overline{B}_{k,j}) = \overline{A}_{i,k} * \underline{B}_{k,j}. \end{cases}$$

Denote by $x^{\text{pos}} = \max(x, 0)$ and $x^{\text{neg}} = \min(x, 0)$ then

$$\begin{cases} \max(\underline{A}_{i,k} * \overline{B}_{k,j}, \overline{A}_{i,k} * \overline{B}_{k,j}) = \overline{A}_{i,k} * \overline{B}_{k,j}^{\text{pos}} + \underline{A}_{i,k} * \overline{B}_{k,j}^{\text{neg}} \\ \min(\underline{A}_{i,k} * \underline{B}_{k,j}, \overline{A}_{i,k} * \underline{B}_{k,j}) = \underline{A}_{i,k} * \underline{B}_{k,j}^{\text{pos}} + \overline{A}_{i,k} * \underline{B}_{k,j}^{\text{neg}} \end{cases}$$

$$\Rightarrow \begin{cases} \overline{A_{i,k} * B_{k,j}} = \overline{A}_{i,k} * \overline{B}_{k,j}^{\text{pos}} + \underline{A}_{i,k} * \overline{B}_{k,j}^{\text{neg}} \\ \underline{A_{i,k} * B_{k,j}} = \underline{A}_{i,k} * \underline{B}_{k,j}^{\text{pos}} + \overline{A}_{i,k} * \underline{B}_{k,j}^{\text{neg}} \end{cases}$$

$$\Rightarrow \begin{cases} \overline{C} = \left(\overline{A} * \overline{B}^{\text{pos}} + \underline{A} * \overline{B}^{\text{neg}} \right)_\uparrow \\ \underline{C} = \left(\underline{A} * \underline{B}^{\text{pos}} + \overline{A} * \underline{B}^{\text{neg}} \right)_\downarrow. \end{cases}$$

In this case, an interval matrix product can be computed using 4 floating-point matrix products.

Similarly, if A is nonpositive, 4 floating-point matrix products suffice to compute an interval matrix product.

A does not Contain Zero in its Interior. We can then split A into positive and negative part by

$$A_{i,j}^{\text{pos}} = \begin{cases} \{a \in A_{i,j} \mid a \ge 0\} & \text{if} & \overline{A}_{i,j} \ge 0 \\ 0 & \text{if} & \overline{A}_{i,j} < 0 \end{cases}$$

$$A_{i,j}^{\text{neg}} = \begin{cases} \{a \in A_{i,j} \mid a \le 0\} & \text{if} & \underline{A}_{i,j} \le 0 \\ 0 & \text{if} & \underline{A}_{i,j} > 0. \end{cases}$$

A does not contain zero in its interior, thus either $A_{i,j}^{\text{pos}} = A_{i,j}$, and $A_{i,j}^{\text{neg}} = 0$ or $A_{i,j}^{\text{pos}} = 0$, and $A_{i,j}^{\text{neg}} = A_{i,j}$. Hence

$$A_{i,j} * B_{j,k} = A_{i,j} * B_{j,k}{}^{\text{pos}} + A_{i,j} * B_{j,k}{}^{\text{neg}}$$

$$\Rightarrow A * B \quad = A^{\text{pos}} * B + A^{\text{neg}} * B$$

In total, it costs eight floating-point matrix products.

For these three particular cases, if all operations are performed in infinite precision or if there is no rounding error, then the computed result is exact.

In general, when there is no assumption about the input intervals, then one could resort to the natural algorithm. However, it would not be efficient. A better solution is to exploit floating-point matrix multiplication because there are available libraries which are well optimised for floating-point matrix operations.

4   Rump's Algorithm

Rump proposes an algorithm which makes use of floating-point operations for speed [9,10]. This algorithm is based on the midpoint-radius representation of intervals.

Let $A = \{m_A, r_A\}$ and $B = \{m_B, r_B\}$ be two interval matrices, with m_A, m_B their midpoints and r_A, r_B their radius respectively. The product $A * B$ is enclosed by an interval matrix C whose midpoint and radius are computed by:

$$m_C = m_A * m_B$$
$$r_C = (|m_A| + r_A) * r_B + r_A * |m_B|.$$

In fact, because of rounding errors, $m_A * m_B$ cannot be computed exactly. Thus it must be computed with both upward and downward rounding mode to obtain en enclosure of the midpoint. The radius r_C must also be computed with upward rounding to ensure the inclusion property.

$$m_C = \left[(m_A * m_B)_\downarrow, (m_B * m_B)_\uparrow \right]$$
$$r_C = ((|m_A| + r_A) * r_B + r_A * |m_B|)_\uparrow.$$

Finally, the result can be easily converted back to endpoints representation by two floating-point additions.

$$\overline{C} = (\overline{m_C} + r_C)_\uparrow \qquad \underline{C} = (\underline{m_C} - r_C)_\downarrow.$$

In total, this algorithm uses **four** floating-point products. In practice, the interval matrix product implemented using Rump's algorithm exhibits a speedup by a factor 20 to 100 [9], in comparison with the BIAS's approach [4] which is based on the natural algorithm.

Over-estimation

Without taking into account the rounding errors, Rump's algorithm always provides over-estimated results, compared to the natural algorithm. The factor of over-estimation in the worst case is 1.5.

For example, by definition $[0, 2] * [0, 4] \overset{def}{=} [0, 8]$.

Meanwhile, Rump's algorithm gives

$$[0,2] = \{1,1\} \quad [0,4] = \{2,2\}$$
$$\{1,1\} * \{2,2\} = \{1*2, (1+1)*2+1*2\}$$
$$= \{2,6\}$$
$$= [-4,8].$$

The over-estimation factor in this case is $12/8 = 1.5$.

5 Proposition

Rump's algorithm can be considered as decomposing A and B into a sum of two components representing its midpoint and radius respectively. $A * B$ is then replaced by its development, which is a sum of four sub-products. Due to the sub-distributive property of interval product, the result yielded by this sum is an enclosure of the original product.

Our idea is to find another decomposition such that sub-products can be efficiently computed, and that the over-estimation is small.

As we can see in Section 3, an interval product can be efficiently computed when one of the two multipliers is centered in zero or does not contain zero.

Proposition 1. *Let A be an interval matrix. If A is decomposed into two interval matrices A^0 and A^* which satisfy:*

- $A_{i,j}^0 = 0, A_{i,j}^* = A_{i,j}$ *if* $\underline{A}_{i,j} * \overline{A}_{i,j} \geq 0$,
- $A_{i,j}^0 = \left[\underline{A}_{i,j}, -\underline{A}_{i,j}\right], A_{i,j}^* = \left[0, \underline{A}_{i,j} + \overline{A}_{i,j}\right]$ *if* $\underline{A}_{i,j} < 0 < |\underline{A}_{i,j}| \leq \overline{A}_{i,j}$,
- $A_{i,j}^0 = \left[-\overline{A}_{i,j}, \overline{A}_{i,j}\right], A_{i,j}^* = \left[\underline{A}_{i,j} + \overline{A}_{i,j}, 0\right]$ *if* $\underline{A}_{i,j} < 0 < \overline{A}_{i,j} < |\underline{A}_{i,j}|$

then

- A^0 *is centered in zero,*
- A^* *does not contain zero in its interior, and*
- $A^0 + A^* = A$.

Proof. Easily deduced from the formula.

Proposition 2. *Let A and B be two interval matrices and (A^0, A^*) a decomposition of A by Proposition 1. Let C be an interval matrix computed by*

$$C = A^0 * B + A^* * B \tag{8}$$

Then $A * B$ is contained in C. Denote $C \stackrel{def}{=} A \circledast B$.

Proof. Following Proposition 1: $A = A^0 + A^*$. Interval multiplication is sub-distributive, hence $A * B \subseteq A^0 * B + A^* * B$, or $A * B \subseteq C$.

A^0 is centered in zero and A^* does not contain zero, so according to Section 3, $A^0 * B$ and $A^* * B$ can be computed using 1 and 8 floating-point matrix products respectively. Hence, the overall cost is 9 floating-point matrix products. The following section will study the over-estimation factor of this operation in the worst case.

Over-estimation Factor

Let us first consider the product of two scalar intervals a and b with a being decomposed, following Proposition 1, into two parts: $a = a^0 + a^*$. Suppose that all calculations here are performed in infinite precision. It means that rounding errors will not be taken into account.

If a is Centered in Zero then $a^* = 0 \rightarrow a * b = a^0 * b$. Thus the result is exact.

If a does not Contain Zero in its Interior then $a^0 = 0 \rightarrow a * b = a^* * b$. Thus the result is exact too.

If a Contains Zero. Again, let us perform a case study (see details in [7]). Case $\underline{a} < 0 < |\underline{a}| < \overline{a}$. In this case $a^0 = [\underline{a}, -\underline{a}]$ and $a^* = [0, \overline{a} + \underline{a}]$. Proposition 2 yields that, in case b is either nonnegative or nonpositive, $a \circledast b = a * b$. In case b contains 0 in its interior and $\overline{b} > |\underline{b}|$, then let us denote by

$$\begin{cases} M = \max(|\underline{a}|/\overline{a}, |\underline{b}|/\overline{b}) \\ m = \min(|\underline{a}|/\overline{a}, |\underline{b}|/\overline{b}) \end{cases}$$

then

$$\begin{cases} 0 < m \leq M \leq 1 \\ \min(\underline{a}/\overline{a}, \underline{b}/\overline{b}) &= -M \\ \underline{a}/\overline{a} + \underline{b}/\overline{b} &= -m - M \\ \underline{a}/\overline{a} * \underline{b}/\overline{b} &= m * M. \end{cases}$$

Denote $\mathtt{diam}(x)$ the diameter of an interval x, then the factor of overestimation is computed by:

$$\begin{aligned} \frac{\mathtt{diam}(a \circledast b)}{\mathtt{diam}(a * b)} &= \frac{1 - \underline{a}/\overline{a} - \underline{b}/\overline{b} - \underline{a}/\overline{a} * \underline{b}/\overline{b}}{1 - \min(\underline{a}/\overline{a}, \underline{b}/\overline{b})} \\ &= \frac{1 + M + m - Mm}{1 + M} \\ &= \frac{1 + M + m(1 - M)}{1 + M} \\ &\leq \frac{1 + M + M(1 - M)}{1 + M}. \end{aligned}$$

Inspecting the last function of unknown M between 0 and 1, we have that its maximum is $4 - 2\sqrt{2} \approx 1.18$ and this maximum is reached for $m = M = \sqrt{2} - 1$, or $\underline{a}/\overline{a} = \underline{b}/\overline{b} = 1 - \sqrt{2}$.

The case where b contains 0 in its interior and $\overline{b} < |\underline{b}|$ is similar.

Case $\underline{a} < 0 < \overline{a} < |\underline{a}|$. Again, in the case where b is either nonnegative or nonpositive, $a \circledast b = a * b$. In the case b contains 0 in its interior and $\overline{b} > |\underline{b}|$, then the definition of M and m must be adapted to ensure that they are both less than 1:

$$\begin{cases} M = \max(\overline{a}/\,|\underline{a}|\,,|\underline{b}|\,/\overline{b}) \\ m = \min(\overline{a}/\,|\underline{a}|\,,|\underline{b}|\,/\overline{b}) \end{cases}$$

and the rest of the proof is very similar to the previous case.

Let's now extend to the case of matrix multiplication. Each element of the result matrix is computed by

$$C_{i,j} = \sum_k A^0_{i,k} * B_{k,j} + \sum_k A^*_{i,k} * B_{k,j}$$
$$= \sum_k (A^0_{i,k} * B_{k,j} + A^*_{i,k} * B_{k,j}).$$

Since $\mathtt{diam}(a + b) = \mathtt{diam}(a) + \mathtt{diam}(b)$, then

$$\mathtt{diam}(C_{i,j}) = \mathtt{diam}\left(\sum_k (A^0_{i,k} * B_{k,j} + A^*_{i,k} * B_{k,j})\right)$$
$$= \sum_k \mathtt{diam}(A^0_{i,k} * B_{k,j} + A^*_{i,k} * B_{k,j}).$$

With the assumption of no rounding error, the over-estimation factor in the worst case for this scalar product is $4 - 2\sqrt{2}$. Hence

$$\mathtt{diam}(C_{i,j}) \leq (4 - 2\sqrt{2}) * \sum_k \mathtt{diam}(A_{i,k} * B_{k,j})$$
$$\leq (4 - 2\sqrt{2}) * \mathtt{diam}\left(\sum_k A_{i,k} * B_{k,j}\right)$$
$$\Rightarrow \mathtt{diam}(C) \leq (4 - 2\sqrt{2}) * \mathtt{diam}(A * B).$$

Hence, the over-estimation factor in the worst case of interval matrix product is also $4 - 2\sqrt{2} \approx 1.18$.

6   Numerical Experiments

We implemented, in MatLab using the IntLab library [8], a function called `igemm` following Proposition 2. The `igemm` function uses in total 9 floating-point matrix products, along with some matrix operations of order $\mathcal{O}(n^2)$ to manipulate data. Hence, in theory, the factor in terms of execution time between `igemm` and a floating-point matrix product is 9.

In practice, when the matrix dimension is small, the difference between operations of order $\mathcal{O}(n^2)$ and operations of order $\mathcal{O}(n^3)$ is small. That is not only because the theoretical factor n is small, but also that operations of order $\mathcal{O}(n^3)$ better exploit memory locality and parallelism.

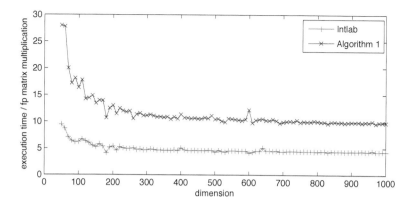

Fig. 1. Performance of `igemm`

As shown on Figure 1, when the matrix dimension is small, i.e smaller than 100, the execution time of `igemm` with respect to a floating-point matrix multiplication, marked by the × symbol, is high. Nevertheless, when the matrix dimension gets higher, this factor gets smaller and gets close to the theoretical factor.

The same phenomenon can be observed with Rump's algorithm, marked by the + symbol.

7   Conclusion

The algorithm presented in this paper implements the product of interval matrices using floating-point operations. It constitutes a trade-off between the performances of optimized floating-point libraries, and a slight overestimation of the result. We have proven that the width of computed result is always less than 1.18 times the width of exact result. In particular, if one of the two multipliers does not contain zero in its interior then the computed result is exact using exact arithmetic.

The performance of the algorithm given above relies entirely on the performance of the employed floating-point library. Such a library must support directed rounding modes to be of use. This requirement is not really an issue, since libraries such as LAPACK [1] or ATLAS [11], which are renowned for their performances, support directed roundings.

However, the use of directed rounding modes restricts the algorithm that can be used as a basic algorithm to implement the natural formula. Indeed, fast algorithms for matrix products [2], which have a complexity below $\mathcal{O}(n^3)$, do not preserve the monotonicity of operations in finite precision (they use substraction as well as addition and multiplication) and they cannot be used. Furthermore, fast algorithms rely on algebraic properties such as $(x + y) - y = x$, which do not hold in interval arithmetic.

References

1. Anderson, E., Bai, Z., Bischof, C., Blackford, S., Demmel, J., Dongarra, J., Du Croz, J., Greenbaum, A., Hammarling, S., McKenney, A., Sorensen, D.: LAPACK Users' Guide, 3rd edn. Society for Industrial and Applied Mathematics, Philadelphia (1999)
2. Ceberio, M., Kreinovich, V.: Fast Multiplication of Interval Matrices (Interval Version of Strassen's Algorithm). Reliable Computing 10(3), 241–243 (2004)
3. Jaulin, L., Kieffer, M., Didrit, O., Walter, E.: Applied Interval Analysis. Springer, Heidelberg (2001)
4. Knuppel, O.: PROFIL/BIAS–A fast interval library. Computing 53(3–4), 277–287 (2006)
5. Moore, R.E., Kearfott, R.B., Cloud, M.J.: Introduction to Interval Analysis. SIAM, Philadelphia (2009)
6. Neumaier, A.: Interval Methods for Systems of Equations. Cambridge University Press (1990)
7. Nguyen, H.D.: Efficient implementation of interval matrix multiplication (extended version). Research report, INRIA, 04 (2010)
8. Rump, S.M.: INTLAB - INTerval LABoratory, http://www.ti3.tu-hamburg.de/rump/intlab
9. Rump, S.M.: Fast and parallel interval arithmetic. BIT 39(3), 534–554 (1999)
10. Rump, S.M.: Computer-assisted proofs and self-validating methods. In: Einarsson, B. (ed.) Handbook on Accuracy and Reliability in Scientific Computation, ch. 10, pp. 195–240. SIAM (2005)
11. Whaley, R.C., Petitet, A.: Minimizing development and maintenance costs in supporting persistently optimized BLAS. Software: Practice and Experience 35(2), 101–121 (2005), http://www.cs.utsa.edu/

The Computing Framework
for Physics Analysis at LHCb

Markward Britsch

Max-Planck-Institut für Kernphysik,
Saupfercheckweg 1, 69117 Heidelberg, Germany
Markward.Britsch@mpi-hd.mpg.de

Abstract. The analysis of high energy physics experiment data is a good example for the need of real-time access and processing of large data sets. In this contribution we introduce the basic concepts and implementations of high energy data analysis on the example of the LHCb experiment at the LHC collider. Already now, but even more once running under nominal conditions it will produce unprecedented amounts of data for years. The contribution will also discuss the potential of parts of the existing implementations to be used in the analysis middleware for future dedicated projects on fast distributed analysis frameworks for LHC data.

Keywords: Particle physics, real-time access, large data sets.

1 Introduction

The analysis of high energy physics experiment data is a good example for the need of real-time access and processing of large data sets (see, *e.g.*, [1,2,3]). In this contribution we introduce the basic concepts of high energy physics data analysis on the example of the LHCb (LHC beauty) experiment [4]. LHCb is one of the four major experiments at the proton–proton collider LHC (Large Hadron Collider) at CERN near Geneva. It is build for precise measurements of CP violation [5] and rare decays. CP is the symmetry under parity transformation (P) followed by the charge conjugation transformation (C). About 730 physicists from 54 institutes are involved. The LHC started in late 2009 and is about to produce unprecedented amounts of data for the next years. LHCb alone is recording data with a rate in the order of 1 PB per year. This contribution will also discuss the software implementations used by LHCb and show why the existing solution does not satisfy all the physicist's needs. It is aimed at scientists of other fields, especially computer science. The reader is introduced to the state of the art of the existing solutions and learns about its shortcomings, helping him to participate in generating new solutions.

The remainder of this contribution is structured as follows. Chapter 2 describes how high energy physics data is taken and processed in modern experiments on the example of LHCb. Chapter 3 describes the current LHCb analysis

K. Jónasson (Ed.): PARA 2010, Part II, LNCS 7134, pp. 189–195, 2012.

software including hardware and software requirements. This is followed by comments on the relevance for a system for real-time access processing of large data sets in Chapter 4. The contribution closes with a summary and conclusions in Chapter 5.

2   High Energy Physics Experiment Data

For technical reasons, the proton beams in the LHC are not continuous but broken up into packages of protons, so called bunches. An event is then called the data collected corresponding to one crossing of bunches in the detector. In such a bunch crossing only a few out of the $\mathcal{O}(10^{11})$ protons collide. The read out of the detector for each event has to be triggered by hardware, called the level zero trigger. There are different triggers, including higher level (software) triggers, with different possible trigger conditions. The data at this stage is the so called raw data that is read out of the detector.

After the data is taken, it is processed in a so called reconstruction process in which single pieces of raw data are combined to more physical objects. As most high energy physics detectors, LHCb consists of different types of detectors as can be seen in a schematic view in Figure 1. For instance, the so called tracking detectors are position sensitive detectors for charged particles. The detection of a passage of such a particle is called a hit. These hits are combined into so called tracks, describing the trace of a charged particle in the detector as visualized in Figure 2. Note that a track is more than pure positional information, as it contains also, *e.g.*, the momentum of the particle obtained from the curvature of the charged particle's trajectory in the magnetic field of the experiment's magnet.

The data is organized in events, *i.e.*, the data corresponding to the same collision of proton bunches. Since the computer resources are limited, the physicists doing the analyses – of which there are several hundreds – can not have access to the full data set. Instead, in a centralized so called stripping process, a small subset of events suitable for further analysis is selected. The same procedure also categorizes the events into possible different physical processes, the so called stripping lines. The stripping is done about once every three months.

This needed (due to limited computing resources) but mostly unwanted filtering has some disadvantages as the physicists do not have direct access to the data. Firstly to define and check the selection criteria of the stripping, only a small amount of unfiltered data can be used. This can cause some bias or inefficiency or even lead to very rare but interesting events being missed altogether. Secondly the correction of bugs found after a stripping or missed deadlines for code submission can only be corrected for the next stripping which is done a few times per year. And lastly high statistics measurements that use more than a small fraction of all the events are not possible. These are the reasons why real time access and processing of the whole data set would be a great advantage.

After the reconstruction and stripping steps, the data is written to files in the so called DST format and made public within LHCb via the grid (see, *e.g.*, [7,8]). These files are the ones used for physics analysis.

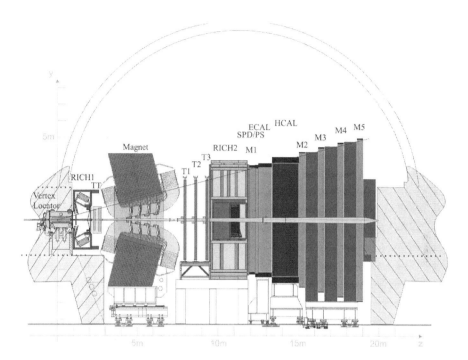

Fig. 1. A schematic few of the LHCb detector [6]. The proton beams are inside a evacuated pipe that is depicted here going from left to right through the whole detector. The point where the proton bunches cross, *i.e.*, where the protons collide, is at the very left of this view inside what is called Vertex Locator. The tracking detector stations are labeled TT, T1, T2 and T3.

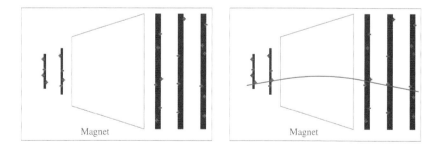

Fig. 2. Symbolic visualization of the reconstruction of tracks from hits. The blue areas represent the tracking detectors with the red and brown crosses as the hit positions. The volume containing the magnetic field is indicated in green. On the left the hits are shown only while on the right the red line marks the line fit to some of the hits, *i.e.*, a track. Note that the track is bend in the magnetic field (as charged particle trajectories are) from which the momentum of the particle can be deduced.

The grid is organized in tiers, where CERN is Tier-0 (see Figure 3). There are 6 Tier-1s major computing centers in different countries. Tier-2s are local computer cluster at institutes. The output files of the stripping are distributed to CERN and the Tier-1s, while the Tier-2s are mainly used for Monte-Carlo simulation production. Sending a job to the grid means sending a compiled library to one of the computing centers where the data in need is stored. For LHCb the software for distributed analysis is called DIRAC [9,10], while the user front-end is ganga [11,12]. DIRAC stands for Distributed Infrastructure with Remote Agent Control. It is based on the pilot job technique to ensure fitting local environments and load balancing. Ganga is the front-end for job definition and management, implemented in Python. It was developed to be used as a user interface to the grid for ATLAS and LHCb and allows easy switching between testing on the host machine, running on a local batch system and running on grid resources.

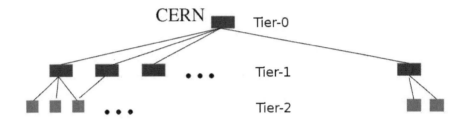

Fig. 3. The tier structure of the grid

3   The LHCb Analysis Software Framework

The LHCb software [13] is based on the open gaudi framework [15,16]. This framework is providing interfaces and services necessary for building high energy physics experiment frameworks for event data processing applications. It is experiment independent and used by different ones like ATLAS and LHCb. The LHCb software contains the DaVinci package [17] which is a framework adapted to the specific needs of LHCb analyses. It interfaces and uses the ROOT software [18,19] which is the standard framework for statistical analysis and visualization in high energy physics. DaVinci allows the user to access the information stored in the DST files and, for example, perform the analysis of a particular decay mode. Here the first step usually is a selection of candidates of the specific decay, *i.e.*, a set of tracks that could stem from the decay of a particle of interest and the properties of which are being studied. At this level, the data organization usually changes from event based to candidate based. The output typically consists of (ROOT) histograms or (ROOT) n-tuples. Histograms are used as estimators for the probability density function of characteristics, such as the momentum distribution, of all selected candidates. N-tuples on the other hand contain the candidates as list of variable – value pairs.

There are different ways the user can use `DaVinci`. One is by writing C++ code and compiling it as a library using the `DaVinci` framework. This library is then linked into the `DaVinci` framework. There is the possibility to use python scripts, to control the analysis (*i.e.*, setting parameters). There are also built-in tools that can be called from the same python script, so that a typical analysis becomes a hybrid between C++ and python code. Since there are many default tools ready to be used even some analyses without any user-written C++ code can be done.

A large step towards python is the `GaudiPython` project [20]. Using this software it is possible to access all the `ROOT` classes directly, *i.e.*, like in C++, including calling methods. In contrast to above, this is also done on full event and candidate level. In addition there is the software package `bender` [21,22], adding to `GaudiPython` the possibility to add your own classes in python, which makes a purely python analysis as flexible as a C++ based one, although with some loss in performance.

3.1   Hard- and Software Requirements

Here we give some information on the hardware and operating system requirements for the LHCb software any project on real-time access and processing of large data sets that aims at using LHCb data needs. The operating system in need is Linux SLC5 (64 bit) with compiler gcc 4.3.x (alternatively Linux SLC4 32/64 bit, gcc 3.4.x). Depending on the exact installation and the number of alternative versions of the different projects, the LHCb software needs $\mathcal{O}(10\text{ GB})$ on hard disk. For distributing this software to all nodes of a real-time access and processing of large data sets project, one would compress the software. In an example of a 3 GB installation the produced bzip2 file had a size of 552 MB.

The memory requirements for an analysis job is about 2 GB. Typical analysis times are $\mathcal{O}(10\text{ ms})$ per event, where an event has about 100 kB on average. But note that there are very different kinds of analyses so that the processing time can change dramatically between them.

4   Relevance for Projects on Fast Distributed Analysis Frameworks

As described above, using `GaudiPython` or `bender`, it is possible to do a complete LHCb analysis in python. An analysis that is purely written in a script language like python is of course an ideal candidate for being used in massively distributed analyses when the shipping of the whole analysis code from the computer of the user to all the worker nodes becomes prohibited by bandwidth. This will probably be the case in forthcoming projects dedicated to fast distributed analysis frameworks for LHC data. Thus one approach to the analysis framework of such a project could involve using these software frameworks and restricting analysis to be done in python only.

5   Summary and Conclusion

We have described the LHCb data analysis as an example for a typical high energy physics experiment. We have shown that this is an example for the need of a system for real-time access and processing of large data sets. The analysis chain in LHCb has been described, including the needed, but unwanted filtering step (stripping), which leads to the desire for such a system. There are some software implementations already in the LHCb software, with potential for use in dedicated future distributed analysis systems. These allow the physicist to do his analysis purely in the script language python which would be an advantage for a distributed system compared to shipping a large compiled code or library.

References

1. Schmelling, M., et al.: RAVEN-a Random Access, Visualization and Exploration Network for the Analysis of Petabyte Sized Data Sets. Poster at the 4th International Conference Distributed Computing and Grid-Technologies in Science and Education (2010)
2. Ballintijn, M., et al.: Parallel interactive data analysis with PROOF. Nuclear Instruments and Methods in Physics Research Section A: Accelerators, Spectrometers, Detectors and Associated Equipment 559, 13–16 (2006)
3. Dean, J., Ghemawat, S.: MapReduce: Simplified Data Processing on Large Clusters. Communications of the ACM 51(1), 107–113 (2008)
4. The LHCb Collaboration: The LHCb Detector at the LHC. Journal of Instrumentation 8, 8005+ (August 2008)
5. Beyer, M.: CP violation in particle, nuclear and astrophysics. Lecture Notes in Physics. Springer, Heidelberg (2002)
6. Lindner, R.: LHCb Collection, http://cdsweb.cern.ch/record/1087860
7. Lamanna, M.: The LHC computing grid project at CERN. Nuclear Instruments and Methods in Physics Reseacrch Section A: Accelerators, Spectrometers, Detectors and Associated Equipment 534, 1–6 (2004)
8. Bird, I., et al.: LHC computing Grid: Technical Design Report. LCG-TDR-001 CERN-LHCC-2005-024 (2005)
9. Bargiotti, M., Smith, A.C.: DIRAC data management: consistency, integrity and coherence of data. Journal of Physics: Conference Series 119, 062013 (2008), http://iopscience.iop.org/1742-6596/119/6/062013
10. The DIRAC team, DIRAC Website (2010), https://lhcb-comp.web.cern.ch/lhcb-comp/DIRAC/default.htm
11. Mościcki, J.T., et al.: Ganga: A tool for computational-task management and easy access to Grid resources. Computer Physics Communications 180, 2303–2316 (2009)
12. The ganga team, ganga (2010), http://ganga.web.cern.ch/ganga/index.php
13. de Olivia, E.C., et al.: LhcB Computing Technical Design Report, CERN/LHCC/205-019 (2005)

14. Koppenburg, P., The LHCb Collaboration: Reconstruction and analysis software environment of LHCb. Nuclear Physics B (Proceedings Supplements) 156, 213–216 (2006)
15. Barrand, G., et al.: GAUDI A software architecture and framework for building HEP data processing application. Computer Physics Communications 140, 45–55 (2001)
16. The Gaudi team, Gaudi (2010), http://proj-gaudi.web.cern.ch/proj-gaudi/
17. The DaVinci team, DaVinci (2010), http://lhcb-release-area.web.cern.ch/LHCb-release-area/DOC/davinci/
18. Antcheva, I., et al.: ROOT – A C++ framework for petabyte data storage, statistical analysis and visualization. Computer Physics Communications 180, 2499–2512 (2009)
19. The ROOT team, ROOT (2010), http://root.cern.ch
20. The GaudiPython team, GaudiPython (2010), https://twiki.cern.ch/twiki/bin/view/LHCb/GaudiPython
21. Barrand, G.B., et al.: Python-based physics analysis environment for LHCb. In: Proceedings of Computing in High Energy and Nuclear Physics (2004)
22. The Bender team, Bender (2010), https://lhcb-comp.web.cern.ch/lhcb-comp/Analysis/Bender/index.html

Taming the Raven – Testing the Random Access, Visualization and Exploration Network RAVEN

Helmut Neukirchen

Faculty of Industrial Engineering, Mechanical Engineering and Computer Science
University of Iceland, Dunhagi 5, 107 Reykjavík, Iceland
helmut@hi.is

Abstract. The *Random Access, Visualization and Exploration Network* (RAVEN) aims to allow for the storage, analysis and visualisation of peta-bytes of scientific data in (near) real-time. In essence, RAVEN is a huge distributed and parallel system.

While testing of distributed systems, such as huge telecommunication systems, is well understood and performed systematically, testing of parallel systems, in particular high-performance computing, is currently lagging behind and is mainly based on ad-hoc approaches.

This paper surveys the state of the art of software testing and investigates challenges of testing a distributed and parallel high-performance RAVEN system. While using the standardised *Testing and Test Control Notation* (TTCN-3) looks promising for testing networking and communication aspects of RAVEN, testing the visualisation and analysis aspects of RAVEN may open new frontiers.

Keywords: Testing, Distributed Systems, Parallel Systems, High-Performance Computing, TTCN-3.

1 Introduction

The RAVEN project aims to address the problem of the analysis and visualisation of inhomogeneous data as exemplified by the analysis of data recorded by a *Large Hadron Colider* (LHC) experiment at the *European Organization for Nuclear Research* (CERN). A novel distributed analysis infrastructure shall be developed which is scalable to allow (near) real-time random access and interaction with peta-bytes of data. The proposed hardware basis is a network of intelligent *Computing, data Storage and Routing* (CSR) units based on standard PC hardware. At the software level the project would develop efficient protocols for data distribution and information collection upon such a network, together with a middleware layer for data processing, client applications for data visualisation and an interface for the management of the system [1].

Testing of distributed systems, for example huge telecommunication systems, is mature and performed rigorously and systematically based on standards [2]. In contrast, a literature study on testing of parallel computing systems, such as high-performance cluster computing, reveals that testing is lagging behind

K. Jónasson (Ed.): PARA 2010, Part II, LNCS 7134, pp. 196–205, 2012.

in this domain. However, as computing clusters are just a special kind of distributed systems[1], it seems worthwhile to apply the industry-proven mature testing methods for distributed systems also for testing software of parallel systems. The future RAVEN system is an example for such a parallel system. As testing and testability should already be considered when designing a system [3], this paper investigates the state of the art and the challenges of testing a distributed and parallel high-performance RAVEN system, even though RAVEN is still at it's initial state of gathering requirements and neither a testable implementation nor a design is available, yet.

This paper is structured as follows: Subsequent to this introduction, Section 2 provides as foundation an overview on the state of the art of software testing. Section 3 gives a glimpse of the standardised *Testing and Test Control Notation* (TTCN-3) which is suitable for testing distributed systems. As the main contribution, this paper discusses, in Section 4, the challenges of testing RAVEN. Final conclusions are drawn in Section 5.

2   An Overview on Software Testing

Software testing is the most important means to give confidence that a system implementation meets its requirements with respect to functional and real-time behaviour. Even though testing is expensive[2], it pays off as it is able to reveal defects early and thus prevents them from manifesting during productive use.

In his seminal textbook on software testing [4], G. Myers defines testing as "[...] *the process of executing a program with the intent of finding errors*". However, software testing is no formal proof. Hence, E.W. Dijkstra remarked that testing can be used to show the presence of bugs, but never to show their absence [5].

While Myers refers in his above definition to a *program* which is tested, a more general term for the object of test is *item under test*. The item might range from a single software component (*unit test*) to a whole software system[3] (*system test* – the item under test is here typically called *system under test* (SUT)) via a composed set of components (*integration test*). The possible levels (sometimes called scopes) of testing are just one dimension of testing as shown in Fig. 1. The second dimension refers to the goal or type of testing: *structural testing* has the goal to cover the internal structure of an item under test, for example, the branches of control flow. To achieve this, knowledge of the internal structure (for example, conditional statements) is required (*glass-box test* [4]). The goal of *functional testing* is to assess an item under test with respect to the functionality it should fulfil with respect to it's specification disregarding internal implementation details (*black-box test* [6]). *Non-functional testing* aims at checking the fulfillment

---

[1] The main difference between distributed systems and parallel systems is probably that distributed systems focus on communication, whereas parallel system focus on computation.

[2] Up to 50% of the overall software development costs are incurred in testing [4].

[3] This can be even a large distributed or parallel system.

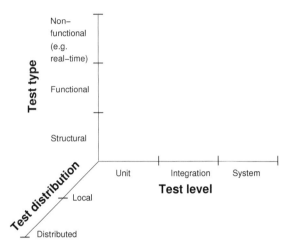

Fig. 1. Dimensions of testing

of non-functional requirements. A variety of different non-functional properties exist, for example, performance with respect to real-time requirements, scalability, security, or usability. The third dimension of testing comes from the fact that the *tester* (or *test system* as it is often called) may be *distributed* (that is: the test runs on multiple nodes) or *local* (that is: the test runs on just one single node). In particular if the item under test itself is distributed, a distributed tester may ease testing or enable certain tests in the first place. The three dimensions are independent from each other, thus the different test types can be performed at all levels and in a local or distributed fashion.

Testing, in particular functional testing, is typically performed by sending a *stimulus* (for example, a function call, a network message, or an input via a user interface) to the item under test and observing the *response* (for example, a return value, a network reply message or an output at the user interface). Based on the observation, a *test verdict* (for example, *pass* or *fail*) is assigned. Testing may be performed manually by running the item under test, providing input and observing the output. Distributed tests, that require co-ordinated test actions, and also real-time performance tests are preferably automated.

Due to the composition and interaction of components, testing at different levels is likely to reveal different defects [7,8]. Hence, testing at just one level is not considered sufficient, but performed at all levels. To reveal defects as soon as possible, a component is unit tested as soon as it is implemented. In case of a class, for example, a unit test would aim at covering all methods. Once multiple components are integrated, they are subject to integration testing. Here, a test would aim at covering the interface between the integrated components. Once the whole system is completed, a system test is performed. In this case, usage scenarios would be covered. All these tests are typically performed within an

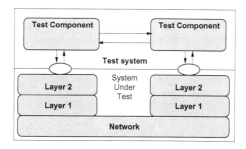

Fig. 2. System test

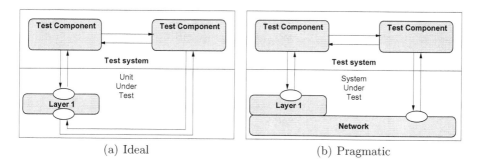

Fig. 3. Unit test

artificial *test environment*, often at special test lab. However, a special kind of system test is the *acceptance test* that is performed in the final productive environment.

Testing in artificial test environments allows test on single or integrated components to be performed in isolation by replacing components on which the item under test depends by special test components. This allows for better control of the item under test and makes sure that really the item under test is tested and not implicitly any of the other components. A layered architecture of a network protocol stack shall serve as an example for this technique: At the system test level, all the involved components (network layer and higher layers 1 and 2 of the protocol stack on both sides) are tested (Fig. 2). At the unit test level, a single unit (such as a class of an object-oriented implementation or an implementation of a network protocol layer as in the example in Fig. 3(a)) is tested. As shown in Fig. 3(a), the environment of that unit under test is replaced by a test environment consisting of test components that act at the interfaces of the unit under test. In practice, this ideal approach may not be possible (or is too expensive): in a layered architecture, higher layers may have hard-coded dependencies on lower layers (thus the lower layer cannot be replaced) or the lower layers provide quite complex functionality that cannot easily be replaced by a test component. The ISO/IEC standard 9646 *Conformance Testing Methodology and Framework* (CTMF) [2] suggests to circumvent this problem by testing first the lowest layer in an isolated unit test. Then, the next layer is tested together with the already tested underlying layer (Fig. 3(b)) and so on. As a result of this incremental approach, each layer (or unit) can be tested separately (assuming that the lower layers have been adequately tested) even if the ideal approach is not possible for pragmatic reasons.

3   Distributed Testing with TTCN-3

The *Testing and Test Control Notation* (TTCN-3) [9] is a language for specifying and implementing software tests and automating their execution. Due to the fact that TTCN-3 is standardised by the *European Telecommunications Standards Institute* (ETSI) and the *International Telecommunication Union* (ITU), several commercial tools and in-house solutions support editing test suites and compiling them into executable code. A vendor lock-in is avoided in contrast to other existing proprietary test solutions. Furthermore, tools allow to execute the tests, to manage the process of test execution, and to analyse the test results.

While TTCN-3 has its roots in functional black-box testing of telecommunication systems, it is nowadays also used for testing in other domains such as Internet protocols, automotive, aerospace, service-oriented architectures, or medical systems. TTCN-3 is not only applicable for specifying, implementing and executing functional tests, but also for other types of tests such as real-time performance, scalability, robustness, or stress tests of huge systems. Furthermore, all levels of testing are supported.

TTCN-3 has the look and feel of a typical general purpose programming language. Most of the concepts of general purpose programming languages can be found in TTCN-3 as well, for example, data types, variables, functions, parameters, visibility scopes, loops, and conditional statements. In addition, test and distribution related concepts are available to ease the specification of distributed tests.

As TTCN-3 is intended for black-box testing, testing a *system under test* (SUT)[4] takes place by sending a stimulus to the SUT and observing the response. In TTCN-3, communication with the SUT may be message-based (as, for example, in communication via low-level network messages) or procedure-based (as, for example, in communication via high-level procedure or method calls). Based on the observed responses, a TTCN-3 test case can decide whether an SUT has passed or failed a test.

In practise, testing a distributed system often requires that the test system itself is distributed as stimuli and observations need to be performed at different nodes. In contrast to other test solutions, TTCN-3 supports distributed testing – not only the SUT may be distributed or parallel, but also the test itself may consist of several *test components* that execute test behaviour in parallel.

---

[4] Or any other test item depending on the test level.

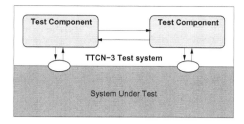

Fig. 4. A sample distributed TTCN-3 test system

The parallel test components may even communicate with each other to co-ordinate their actions or to come to a common test verdict. Figure 4 shows an example of a distributed TTCN-3 test system.

TTCN-3 test cases are abstract, this, for example, means they do not care on which concrete nodes the parallel test components are distributed. It is therefore the responsibility of the TTCN-3 test execution tool to perform the mapping of the abstract distributed test onto a concrete distributed test environment [10]. Thus, the abstract TTCN-3 test cases can be re-used in different distributed environments.

TTCN-3 and it's predecessors TTCN and TTCN-2 have been successfully applied by industry and standardisation for testing huge distributed systems (such as the GSM, 3G, and 3G LTE mobile telecommunication systems). In addition to pure functional tests, TTCN-3 has also been used for performance and load tests that involve testing millions of subscribers [11]. While these applications of TTCN-3 were mainly in the domain of testing "classical" distributed systems, only one work is known where TTCN-3 is used in the domain of "classical" parallel systems: Rings, Neukirchen, and Grabowski [12] investigate the applicability of TTCN-3 in the domain of Grid computing, in particular testing workflows of Grid applications.

More detailed information on TTCN-3 can be found in the TTCN-3 standard [9], in an introductory article [13], in a textbook [14], and on the official TTCN-3 website [15].

4   Testing RAVEN

The intended overall architecture of RAVEN is depicted in Fig. 5(a): the *Computing, data Storage and Routing* (CSR) nodes are connected via an underlying network. High-level data analyses are co-ordinated by an analysis layer that distributes the work load to the individual CSR nodes where the respective data to be analysed resides. The analysis results are then visualised by a corresponding visualisation layer that also provides the general graphical user interface. In addition, to support analysis by the human eye, a direct visualisation of the data is possible. To this aim, the visualisation layer accesses directly the CSR nodes.

For testing at the different test levels, the approaches described in Section 2 can be applied: In accordance to Fig. 3(b), Fig. 5(b) depicts how to perform a

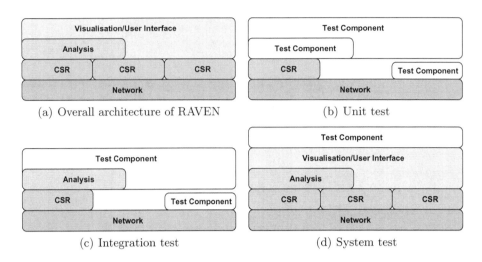

(a) Overall architecture of RAVEN   (b) Unit test

(c) Integration test   (d) System test

Fig. 5. RAVEN and different test levels

unit test of a CSR node. An integration test of a CSR node and the analysis component that have been integrated together is shown Fig. 5(c). For the system (and acceptance) test, the whole system is covered using representative usage scenarios, however the system under test is interfaced via the user interface only as depicted in Fig. 5(d).

Functional testing of RAVEN should be performed at all test levels. For functional testing of the networking and communication aspects, the proven TTCN-3-based standard approach from protocol testing for telecommunication and network systems [2] is applicable including distributed testing. For testing the user interface and visualisation aspects, the standard *capture/replay* testing approach[5] may not work here as RAVEN aims at providing completely new, yet unknown kinds of graphical user interfaces that are not supported by current capture/replay tools. However, testing of the user interface parts that are based on standard user interface technology should be possible. Testing of the analysis aspect should be feasible as long as small "toy" examples are used, where the result is known in advance.

4.1 Challenges of Testing RAVEN

While functional testing of RAVEN seems to be feasible as described above, the non-functional test types (performance test, scalability test, load test) that seem to be crucial for a system such as RAVEN that aims at providing (near) real-time responses and scalability can be expected to be a challenge. While

---

[5] In capture/replay testing, a correct interaction with the system under test via the user interface is recorded (user inputs as well as resulting system outputs). For testing, the recorded user inputs are replayed and the resulting actual system outputs are compared against the expected recorded system outputs.

these tests may be possible at unit level, performance results from unit level may not be extrapolated to the system level as, for example, scalability at unit level does not imply scalability of the system as a whole due to management and communication overheads. Thus, the non-functional tests need to be performed at the system level. However, for system level testing of a system of this size, probe effects [16] may occur at the software (and hardware) scale: by observing (= testing) a system, we unavoidably influence it. For example, the communication overhead to co-ordinate distributed test components reduces the network capacity that is available for the RAVEN system that is being tested. Similarly, the actual observation requires CPU time that is lacking in the RAVEN system under test.

A further challenge is how to test analysis algorithms working on peta-bytes of test data. Comparison of the observed output with an the expected output may be difficult if the expected output is not known in advance as it can only be calculated by the analysis algorithm under test itself[6]. Furthermore, these peta-bytes of test data need to be generated and stored. However, as RAVEN itself will be challenged by storing and processing peta-bytes of data, the test environment will be challenged as well to manage the test data. Thus, testing of RAVEN will only be possible to a certain limit within an artificial test environment. RAVEN will require a huge amount of hardware and it will not be economically feasible to use a comparable amount of hardware just for setting up an artificial test environment. This fact inevitably results in the conclusion that performance and scalability beyond a certain size will only testable by productive use in the real environment. As such, a system test within a test environment will not possible. Instead, RAVEN can only be tested immediately in it's real hardware environment – that is, only acceptance testing will be possible[7]. However, performance and scalability assessments of RAVEN beyond a certain size may be evaluated by simulation or analytical techniques based on performance models [18,19].

5   Conclusions

We have considered the state of the art in software testing and at the distributed test language TTCN-3, and we investigated possibilities and challenges of testing a RAVEN system.

It seems that the state of the art in software testing is in principle mostly mature enough for testing a RAVEN system. However, it is the fact that the large scale of RAVEN itself opens new frontiers that poses problems. As a result, a true system test in an artificial test environment will not be possible because

---

[6] This is a current research topic, see for example the "First International Workshop on Software Test Output Validation" 2010.

[7] The approach taken by others (for example the Apache Hadoop project for processing huge data sets [17]) confirms this: only small tests are performed in an artificial test environment – "big" tests (performance and scalability) involving huge amounts of data are essentially productive-use tests.

beyond a certain limit (in terms of size and complexity), only acceptance testing in the final environment will be possible. To some extent, the problems to be expected concerning the test of RAVEN are unavoidable due to testing overheads and probing effects on the software, hardware and network scale. The challenges of testing RAVEN should not lead to the conclusion to perform no testing at all, but to the contrary: to test where possible.

TTCN-3 is a test language that has concepts for testing distributed and parallel systems, mainly by supporting distributed testing based on concurrently running parallel test components. While these concepts are successfully applied for testing distributed systems, there is a striking lack of applying them for testing parallel systems. One reason might be that most software in parallel computing is from the domain of scientific computing. Typically, this software is written by the scientists themselves who are experts in their scientific domain, but typically not experts in software engineering thus lacking a background in software testing. Another reason is probably that distributed systems and testing them has a strong focus on message exchange and communication, while in parallel systems this is only a minor aspect as the main focus is on computation. However, both kinds of systems can be considered as similar when it comes to black-box testing them; thus, TTCN-3 should be applicable for testing software of parallel systems as well as it is for testing software of distributed systems. However, it still needs to be investigated whether a generic test solution like TTCN-3 (and the implementing TTCN-3 tools) is sufficient and in particular efficient enough, or if specifically tailored and hand tuned test solutions are required.

Finally, as RAVEN aims at not being just deployable on a single cluster, but to extend to external computing and storage resources in the Internet such as cloud computing, a research project has just been started by the author that investigates testability issues in cloud computing environments.

References

1. Schmelling, M., Britsch, M., Gagunashvili, N., Gudmundsson, H.K., Neukirchen, H., Whitehead, N.: RAVEN – Boosting Data Analysis for the LHC Experiments. In: Jónasson, K. (ed.) PARA 2010, Part II. LNCS, vol. 7134, pp. 206–214. Springer, Heidelberg (2012)
2. ISO/IEC: Information Technology – Open Systems Interconnection – Conformance testing methodology and framework. International ISO/IEC multipart standard No. 9646 (1994-1997)
3. Wallace, D.R., Fujii, R.U.: Software verification and validation: An overview. IEEE Software 6, 10–17 (1989)
4. Myers, G.: The Art of Software Testing. Wiley (1979)
5. Dijkstra, E.: Notes on Structured Programming. Technical Report 70-WSK-03, Technological University Eindhoven, Department of Mathematics (April 1970)
6. Beizer, B.: Black-Box Testing. Wiley (1995)
7. Weyuker, E.: Axiomatizing Software Test Data Adequacy. IEEE Transactions on Software Engineering 12(12) (December 1986)
8. Weyuker, E.: The Evaluation of Program-based Software Test Data Adequacy Criteria. Communications of the ACM 31(6) (June 1988), doi:10.1145/62959.62963

9. ETSI: ETSI Standard (ES) 201 873 V4.2.1: The Testing and Test Control Notation version 3; Parts 1–10. European Telecommunications Standards Institute (ETSI), Sophia-Antipolis, France (2010)

10. Din, G., Tolea, S., Schieferdecker, I.: Distributed Load Tests with TTCN-3. In: Uyar, M.Ü., Duale, A.Y., Fecko, M.A. (eds.) TestCom 2006. LNCS, vol. 3964, pp. 177–196. Springer, Heidelberg (2006), doi:10.1007/11754008_12

11. Din, G.: An IMS Performance Benchmark Implementation based on the TTCN-3 Language. International Journal on Software Tools for Technology Transfer (STTT) 10(4), 359–370 (2008), doi:10.1007/s10009-008-0078-x

12. Rings, T., Neukirchen, H., Grabowski, J.: Testing Grid Application Workflows Using TTCN-3. In: International Conference on Software Testing Verification and Validation (ICST), pp. 210–219. IEEE Computer Society (2008), doi:10.1109/ICST.2008.24

13. Grabowski, J., Hogrefe, D., Réthy, G., Schieferdecker, I., Wiles, A., Willcock, C.: An introduction to the testing and test control notation (TTCN-3). Computer Networks 42(3), 375–403 (2003), doi:10.1016/S1389-1286(03)00249-4

14. Willcock, C., Deiß, T., Tobies, S., Keil, S., Engler, F., Schulz, S.: An Introduction to TTCN-3. Wiley, New York (2005)

15. ETSI: TTCN-3, http://www.ttcn-3.org

16. Fidge, C.: Fundamentals of distributed system observation. IEEE Software 13, 77–83 (1996)

17. Apache Software Foundation: Apache Hadoop, http://hadoop.apache.org/

18. Law, A., Kelton, W.: Simulation Modeling and Analysis. McGraw-Hill (200)

19. Skadron, K., Martonosi, M., August, D., Hill, M., Lilja, D., Pai, V.: Challenges in Computer Architecture Evaluation. IEEE Computer 36(8), 30–36 (2003), doi:10.1109/MC.2003.1220579

RAVEN – Boosting Data Analysis for the LHC Experiments

Michael Schmelling[1,*], Markward Britsch[1], Nikolai Gagunashvili[2],
Hans Kristjan Gudmundsson[2], Helmut Neukirchen[3], and Nicola Whitehead[2]

[1] Max-Planck-Institute for Nuclear Physics, Heidelberg, Germany
{michael.schmelling,markward}@mpi-hd.mpg.de
[2] University of Akureyri, Akureyri, Iceland
{nikolai,hgk,nicolaw}@unak.is
[3] University of Iceland, Reykjavik, Iceland
helmut@hi.is

Abstract. The analysis and visualization of the LHC data is a good example of human interaction with petabytes of inhomogeneous data. After outlining the computational requirements for an efficient analysis of such data sets, a proposal, RAVEN – a Random Access, Visualization and Exploration Network for petabyte sized data sets, for a scalable architecture meeting these demands is presented. The proposed hardware basis is a network of "CSR"-units based on off-the-shelf components, which combine Computing, data Storage and Routing functionalities. At the software level efficient protocols for broadcasting information, data distribution and information collection are required, together with a middleware layer for data processing.

Keywords: LHC, Particle Physics, RAVEN, data analysis, CSR-unit.

1 Introduction

In particle physics the basic units which make up a data set are so-called "events". In former times an event did correspond to a photograph showing the interaction of a high energy particle with an atomic nucleus in a bubble chamber, at the LHC [1] it is the information recorded from a single bunch crossing of the two proton beams. A bubble chamber photograph is shown in Fig.1.

The bubble chamber is a detector device which allows to collect and to display the full information about a high energy particle physics interaction in a very intuitive form. Its main drawback is that it can only record events at a rate of a few Hz, which renders it unsuitable to look for really rare types of interactions. As a consequence, over the last 30 years they have been replaced by electronic detectors which nowadays are able to scrutinize high energy interactions with rates up to 40 MHz and to store information from potentially interesting events

---

* Corresponding author.

K. Jónasson (Ed.): PARA 2010, Part II, LNCS 7134, pp. 206–214, 2012.
© Springer-Verlag Berlin Heidelberg 2012

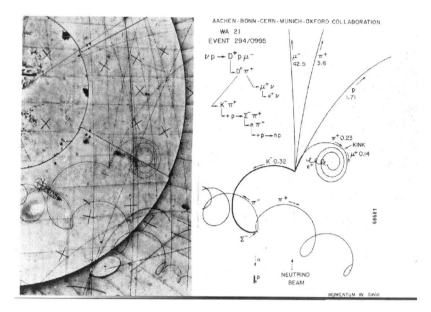

Fig. 1. Bubble chamber photograph of a high energy collision between elementary particles. Secondary particles are created from the available energy and travel before they decay or induce secondary interactions. The information about the reaction is contained in the momentum vectors of the final state particles, their charges and the points (vertices) of interactions. (Photo by CERN).

with a few kHz. For example at LHCb [2], one of the four large LHC experiments at CERN, the typical amount of information corresponds to 50 kB per event and events can be stored with a rate of 2 kHz. With an expected number of 2×10^{10} events per year, the annual data volume amounts to O(1) PB.

Individual events are reconstructed by means of sophisticated numerical algorithms. Those start from the raw information collected by the, depending on the specific experiment, 1 - 200 million readout channels of the detector. From those they extract the equivalent information one would have from a bubble chamber photograph, i.e. particle trajectories, vertices, decay chains etc. An example how the information from a modern electronic detector can be visualized is shown in Fig.2.

The final analysis of the reconstructed data is conceptually simple in the sense that all events are equivalent, i.e. at the event-level it parallelizes trivially. Also the information content of a single event has a relatively simple structure, consisting of lists of instances of a few basic elements such as "tracks" or "vertices" which contain the measured information about the final state particles created in a high energy collision. Different events will differ in the number of those objects and the relations between them.

In contrast to analysis tasks in other branches science where the main problem is accessing the relevant data items, data analysis in particle physics is

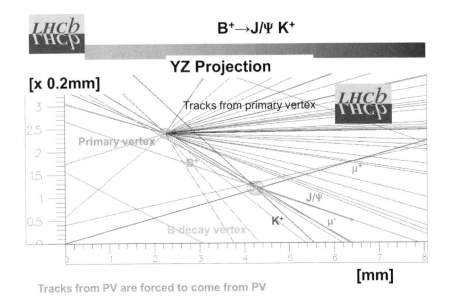

Fig. 2. Visualization of an interaction recorded by the LHCb experiment. Information from electronically read out detectors is used to reconstruct trajectories of particles created in a high-energy proton-proton collision. The computer generated image shows a zoom to the region of the primary vertex where two protons from the counter-rotating beams did collide. In a addition to a large number of particle created at the primary vertex, the reconstruction also shows the decay of a so-called B-meson which after creation at the primary vertex travels a few millimeters before decaying into three longer lived final state particles (K^+, μ^+, μ^-).

completely dominated by the processing of the event information. Compared to the processing step reading and decoding the data usually requires negligible CPU resources. To illustrate this, consider the problem of finding for example decays of so-called D^0-mesons into a pair of final state particles in LHCb. Finding such decays in an event requires checking all combinations of two tracks and to decide whether or not this pair is consistent with coming from a D^0-decay. In LHCb one has to check on average 73 combinations per event, with each single check requiring up to hundreds of floating point operations. In addition, D^0 decays into two final state particles, though still frequent compared to many other decay channels of interest, are already rather rare. Only about 1 percent of all events contains a D^0, and only about 4 percent of those decay into the specific two-particle final state. Already for this "easy" example the data analysis has to cope with a situation where the background is about 200,000 times larger that the signal.

The basic mode of data analysis in particle physics is characterized by two steps. In the first step the data set is scrutinized for events containing a specific signature. Events with this signature then are analyzed in detail, either by extracting some characteristic information or by iterating the selection process with additional criteria.

It is evident that depending on the selection criteria the size of the event sample used in a specific analysis can vary by many orders of magnitude. On the other hand, the maximum communication bandwidth available to return information back to the user will essentially by fixed, i.e. the interaction between user and the full data set must be such that the network traffic stays below a certain limit.

The quantities of interest in a typical particle physics analysis are probabilities or probability density functions for a certain process or configuration to occur. Numerical estimates are obtained by means of histograms, i.e. simple counters for how often a certain condition is observed. The analysis framework thus must be able to handle this kind of cumulative information, which even for very large event samples reduces to a limited set of numbers.

In addition to cumulative information from many or even all events, the system must be able to transmit some or all information from a few selected events. This is of particular relevance for very rare types for final states, such as for example events with a candidate Higgs decay or other exotic processes and which require an in depth analysis of single events.

The combination of the two access modes becomes particularly relevant in the context of interactive searches for special event types starting from the full data set. Here powerful visualization tools and user interfaces are required, which provide an intuitive representation of the properties of the event set, together with the possibility of interactive select-and-zoom schemes to focus on certain candidates.

2   Computing Requirements

During the construction of the CERN Large Hadron Collider (LHC) it was realized that the analysis of the data produced by the LHC experiments requires a computing infrastructure which at the time went beyond the capabilities of a single computing center, and which since then has been built up in the framework of the Worldwide LHC Computing Grid (WLCG) [3,4,5]. The design of the WLGC was driven by the requirement to allow a sharing of the effort between many partners and the ability to cope with future increases of the computing demands.

Despite the fact that many new concepts regarding data distribution and sharing of computing load have been implemented, the computing models for the analysis of the LHC data (see e.g. [6]) are still very close to the approach by earlier generation particle physics experiments. They focus on filtering the huge initial data sets to small samples relevant for particular physics question, which then are handled locally by the physicist doing the analysis.

While making efficient use of limited resources, this scheme has some obvious shortcomings.

- At a given time direct access is possible to only a small fraction of the total event sample. This reduced sample also has to serve to define and check the selection criteria for the selection jobs. As a consequence the selection may be biased or inefficient.
- The time constant for full access to the data is given by the frequency of the selection runs which go through the complete data set. Programming errors or missed deadlines for code submission can easily result in serious delays for the affected analyses.
- High statistics measurements, i.e. analysis which use information from more than a small fraction of all events, are not feasible. The same holds for finding exceptional rare events which are not caught by selection criteria based on prior expectations.

What is needed is a framework which allows random access on petabyte-size datasets. It should have a scalable architecture which allows to go to real time information retrieval from the entire data set. The initial use case of this infrastructure will be faster and more efficient access to the data for classical analysis scenarios. Beyond that, however, also novel ways of interacting with the data and new ways of data visualization will evolve.

3   Design Aspects

The requirements outlined above suggest a design similar to that of a biological brain: a dense network of many "simple" nodes combining data storage, processing and the routing of information flow. For the use case of particle physics, each node would store a small fraction of the total event sample, have the possibility to run an analysis task on those events and route information back to the user having submitted the analysis query. In the following these nodes will be referred to as Computing-Storage-Routing (CSR) units, which at the hardware level are standard commodity CPUs. With an appropriate middleware-layer a network of such CSR-units will then constitute a RAVEN system: a Random Access, Visualization and Exploration Network for petabyte sized data sets.

The key feature which guarantees exact scalability is a peer-to-peer architecture [7] where every node is able to perform every functionality required by the system. This departs significantly from the current Grid installation which is built around a system of services that are associated with distinct units, such as for example "worker nodes", "storage elements" or "work-load management systems". While the current Grid-approach is natural in the sense that different functionalities are identified and implemented separately, it results in a rather complex infrastructure with corresponding requirements in terms of maintenance. The RAVEN approach, in contrast, aims at defining a protocol or rule-set which allows the system to organize itself.

Another important aspect of RAVEN is redundant and possibly also encrypted data storage. While encryption should simply ensure the confidentiality of the data also in case that public computing resources are used, redundant storage assures that the entire data set can still be processed even if some nodes become

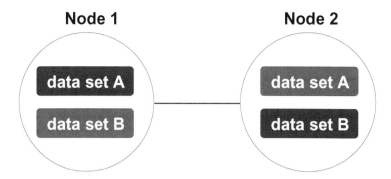

Fig. 3. Simple of example of redundant data storage on two nodes. If both nodes are present, the analysis starts in parallel on different subsets. Node 1 will start on data set A, node 2 will start with B. If one node is unavailable, either because it's down or busy with another task, then the other node will process the entire data set.

unavailable. A simple sketch how duplication of data between two nodes can serve these purposes is shown in Fig. 3. Although encryption adds to the computing costs, typical applications in particle physics analysis are such that the decoding step only adds a small overhead to the actual analysis.

For a particular analysis or visualization task, instructions would be broadcast to all CPUs. These instructions will then be executed on the local event samples, and the information retrieved from those events routed back to the user.

As discussed before, with respect to the information that is returned one has to distinguish between cumulative data, and per-event data. Since all data have to go back to a single node, per-event data should either be of only limited volume per event or should be transmitted for only a subset of all events. Cumulative data on the other hand, such as histograms, flowing back through the network can be accumulated on-the-fly such that the total amount of information transmitted over the network stays small even for very large event samples. Figure 4 illustrates the case.

4   Implementation Aspects

A central feature of the design of a RAVEN system is its scalability, which almost automatically comes from the fact the different events are independent and thus can be spread over as many CSR-units as are available. Scalability allows to develop RAVEN on a small test system and later expand the working system to the size required for a particular application, possibly also taking advantage of cloud-computing infrastructures.

A particular implementation dealing with 1 PB of data spread over 10^5 CSR-units would correspond to 10 GB per node. Assuming a processing speed of 10 MB/s, which seems possible today, the data set could be processed within a quarter of an hour. A test system should typically have one percent of the capacity of the 1 PB system.

One problem that has to be addressed for RAVEN is the creation of ad-hoc routing and communication topologies for a given analysis query to be used both to distribute the query to all nodes and to collect the results of the analysis. Here a big challenge arises from the fact that logically certain next-neighbor topologies may be required which then have to be mapped to actual routing schemes by taking into account the existing hardware capabilities and the data flow that needs to be handled. Furthermore, since many analyses will only access subsets of the full data set, the system should be able to process multiple queries simultaneously.

Another issue is the distribution of data, analysis code and actual query of a specific analysis. One big challenge is the distribution of the full data set. Here different data items have to go to different nodes, which in view of the total data volume that has to be distributed is a non-trivial task. Uniform distribution can be achieved by some hashing scheme, where a hash-code of every event or file determines on which node it will be stored and analyzed [8]. If the RAVEN system is able to autonomously distribute the address-space spanned by the hash code among its members, then an event entering the system via any node can be routed to its proper destination. It is also easy to check whether a particular event is already stored on the system. The data distribution scheme also should take care of the redundant storage scheme. Finally, the RAVEN system must be able to automatically detect new CSR-units joining the system and to migrate part of the data to the new resources.

Apart from data distribution also bookkeeping of available data has to be addressed. Although particle physics analyses can be performed on subsets of the total event sample, a proper interpretation of the results requires the knowledge about the actual events that have been processed. Even if the redundancy built into the system will normally guarantee access to the full data set, a monitoring of which events contribute to a particular result has to be foreseen.

While the distribution of the full data set will happen only rarely, updates of the analysis code will be more frequent, though still rare compared to analysis queries. The latter two can be distributed via a broadcast mechanism. The splitting into analysis code and query is motivated by the goal to minimize the network traffic. Instead of distributing the full analysis code, which for a typical LHC experiments amounts to $O(1)$ GB, with each query, a layered ("middleware") approach suggests itself. Here the (in general machine dependent) analysis code forms a software layer on top of the operating system. This "analysis middleware" then provides a machine independent high level language to perform the actual physics analysis.

While the mapping of the classical analysis models based on histograms or n-tuples on a RAVEN infrastructure is relatively straightforward, the system calls for novel approaches to exploit its real-time capabilities in new visualization tools for the interaction of a human being with petabytes of data.

The performance of the system can be optimized by making sure that events falling into the same class with respect to a specific selection are distributed as evenly as possible. An analysis query addressing only that subset then will

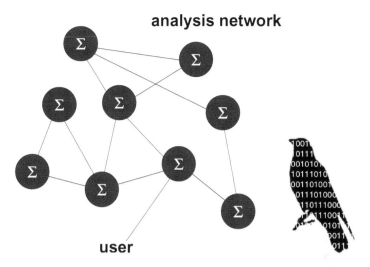

Fig. 4. Sketch of a RAVEN network. Histogram data, for example, produced by the analysis jobs on the different CSR-units are routed back to the node connected to the user, and updated on-the-fly on the way back. The links show which nodes are aware of their neighbors, i.e. the network topology routing and data distribution have to deal with.

harness a large number of CPU simultaneously and finish with minimal time. Providing analysis jobs with the possibility to tag events as belonging to a certain class should lead to a system which is able to automatically migrate data between nodes in order to minimize access times.

Another level of optimization would be to store event related information which is created by a specific analysis for further use. Information that should be kept in a persistent store can either be specified by the user, or selected automatically, e.g. storing by default all information that is determined with computational cost above a certain threshold.

5   Prior Work

Realization of the RAVEN project will benefit greatly from already existing knowledge in networking, middleware design, distributed data storage and computing. Projects which are in principle interesting from the point of view of RAVEN are for example **BitTorrent** [9,10] for broadcasting information over a network, the **Apache Hadoop** [11] project addressing MapReduce-based [12] scalable, distributed computing, the **BOINC** [13,14] framework for volunteer computing and grid computing, or the **xrootd** [15,16] server for low latency high bandwidth data access in the **root** [17,18] framework, which defines the de-facto standard for data analysis in particle physics. Additional input could come from the Grid-middleware developers e.g. **gLite** [19,20] or the **Linux** community [21].

6   Summary

Physics analysis of the data recorded by the LHC experiments calls for new computing architectures which ares scalable to allow fast parallel access to petabytes of data. One possible approach is the RAVEN system, featuring redundant storage, on-the-fly accumulation of results and a rigorous middleware-approach to the data analysis.

References

1. Evans, L., Bryant, P. (eds.): LHC Machine. JINST, vol. 3, p. S08001 (2008)
2. The LHCb Collaboration, Augusto Alves Jr., A., et al.: The LHCb Detector at the LHC. JINST 3, S08005 (2008)
3. The LHC Computing Grid: LCG, http://lcg.web.cern.ch/lcg/
4. Lamanna, M.: The LHC computing grid project at CERN. Nucl. Instrum. Meth. A 534, 1–6 (2004)
5. Eck, C., et al.: LHC computing Grid: Technical Design Report. Version 1.06. LCG-TDR-001 CERN-LHCC-2005-024
6. The LHCb Collaboration, Antunes Nobrega, R., et al.: LHCb Computing Technical Design Report. CERN/LHCC 2005-019
7. Oram, A.: Peer-to-Peer: Harnessing the Power of Disruptive Technologies. O'Reilly & Associates, Inc., Sebastopol (2001) ISBN 059600110X
8. Balakrishnan, H., Kaashoek, M.F., Karger, D., Morris, R., Stoica, I.: Looking up data in P2P systems. Commun. ACM 46, 43–48 (2003)
9. BitTorrent, Inc.: BitTorrent, http://www.bittorrent.com
10. Cohen, B.: Incentives Build Robustness in BitTorrent. In: 1st Workshop on Economics of Peer-to-Peer Systems, University of California, Berkeley, CA, USA (2003)
11. The Apache Software Foundation: Apache Hadoop, http://hadoop.apache.org/
12. Dean, J., Ghemawat, S.: MapReduce: simplified data processing on large clusters. Commun. ACM 51, 107–113 (2008)
13. BOINC project: BOINC, http://boinc.berkeley.edu/
14. Anderson, D.P.: BOINC: a system for public-resource computing and storage. In: Fifth IEEE/ACM International Workshop on Grid Computing (2004)
15. XRootD project: Scalla/XRootD, http://project-arda-dev.web.cern.ch/project-arda-dev/xrootd/site
16. Hanushevsky, A., Dorigo, A., Furano, F.: The Next Generation Root File Server. In: Proceedings of Computing in High Energy Physics (CHEP) 2004, Interlaken, Switzerland (2004)
17. The ROOT team: ROOT, http://root.cern.ch/
18. Antcheva, I., et al.: ROOT – A C++ framework for petabyte data storage, statistical analysis and visualization. Computer Physics Communications 180, 2499–2512 (2009)
19. gLite Open Collaboration: gLite, http://glite.web.cern.ch
20. Laure, E., et al.: Programming the Grid with gLite. Computational Methods in Science and Technology 12, 33–45 (2006)
21. The Linux Foundation: The Linux Foundation, http://www.linuxfoundation.org

Bridging HPC and Grid File I/O with IOFSL[*]

Jason Cope[1], Kamil Iskra[1], Dries Kimpe[2], and Robert Ross[1]

[1] Mathematics and Computer Science Division, Argonne National Laboratory
[2] Computation Institute, University of Chicago / Argonne National Laboratory
{copej,iskra,dkimpe,rross}@mcs.anl.gov

Abstract. Traditionally, little interaction has taken place between the Grid and high-performance computing (HPC) storage research communities. Grid research often focused on optimizing data accesses for high-latency, wide-area networks, while HPC research focused on optimizing data accesses for local, high-performance storage systems. Recent software and hardware trends are blurring the distinction between Grids and HPC. In this paper, we investigate the use of I/O forwarding — a well established technique in leadership-class HPC machines— in a Grid context. We show that the problems that triggered the introduction of I/O forwarding for HPC systems also apply to contemporary Grid computing environments. We present the design of our I/O forwarding infrastructure for Grid computing environments. Moreover, we discuss the advantages our infrastructure provides for Grids, such as simplified application data management in heterogeneous computing environments and support for multiple application I/O interfaces.

1 Introduction

Grid computing environments, such as the TeraGrid project, funded by the National Science Foundation (NSF) , have recently begun deploying massively-parallel computing platforms similar to those in traditional high-performance computing (HPC) centers. While these systems do not support distributed or multi resource MPI applications[8,2], they do support a variety of HPC applications well suited for tightly coupled resources, including high-throughput workloads [20] and massively parallel workloads [6]. To efficiently connect these resources, TeraGrid has has focused on enhancing Grid data services. This trend is evident in the goals for the emerging third phase of TeraGrid operations, known as TeraGrid "eXtreme Digital."

This shift in resource usage and deployments aligns Grids more closely with traditional HPC data centers, such as the DOE leadership computing

---

[*] The submitted manuscript has been created by UChicago Argonne, LLC, Operator of Argonne National Laboratory ("Argonne"). Argonne, a U.S. Department of Energy Office of Science laboratory, is operated under Contract No. DE-AC02-06CH11357. The U.S. Government retains for itself, and others acting on its behalf, a paid-up nonexclusive, irrevocable worldwide license in said article to reproduce, prepare derivative works, distribute copies to the public, and perform publicly and display publicly, by or on behalf of the Government.

K. Jónasson (Ed.): PARA 2010, Part II, LNCS 7134, pp. 215–225, 2012.
© Springer-Verlag Berlin Heidelberg 2012

facilities at Argonne National Laboratory and Oak Ridge National Laboratory. This realignment poses several data access challenges. One such challenge is enabling efficient, remote data access by Grid applications using large numbers of processing elements. Massively parallel applications can overwhelm file systems with large numbers of concurrent I/O requests. Leadership-class computing platforms face a similar data access problem for local data access to high-performance storage systems. Grid computing platforms experience similar problems for both local and remote data accesses. Existing Grid data management tools do not address the impact of increased concurrency on remote data access performance, nor do they account for the limited capacity of network and storage resources as application data continues to increase.

In this paper, we describe how I/O forwarding can improve the performance of Grid application data accesses to both local and remote storage systems. In the following sections, we present our I/O forwarding infrastructure for Grid computing environments and show how this infrastructure optimizes application remote data accesses in Grids. Section 2 presents I/O forwarding and its use in HPC. Section 3 describes typical I/O mechanisms used by Grid applications and how I/O forwarding integrates into Grids. Section 4 describes our experiments to evaluate the GridFTP driver compared to IOFSL and POSIX I/O. In Section 5, we describe related work and conclude this paper.

2   HPC I/O

In this section, we introduce the concept of I/O forwarding, followed by a description of our portable, open source implementation.

2.1   Revised I/O Software Stack

The current generation of leadership-class HPC machines, such as the IBM Blue Gene/P supercomputer at Argonne National Laboratory or the Roadrunner machine at Los Alamos National Laboratory, consists of a few hundred thousand processing elements. Future generations of supercomputers will incorporate millions of processing elements. This significant increase in scale is brought about by an addition in the number of nodes along with new multi-core architectures that can accommodate an increasing number of processing cores on a single chip.

While the computational power of supercomputers keeps increasing with every generation, the same is not true for their I/O subsystems. The data access rates of storage devices has not kept pace with the exponential growth in processing performance. In addition to the growing bandwidth gap, the increase in compute node concurrency has revealed another problem: the parallel file systems available on current leadership-class machines, such as PVFS2 [4], GPFS [15], Lustre [5] and PanFS [12] were designed for smaller systems with fewer filesystem clients. While some of these filesystems incorporate features for enhanced scalability, they are often not prepared to deal with the enormous increase in clients brought on by the increasing trend toward more concurrency.

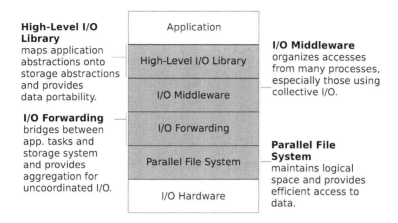

High-Level I/O Library maps application abstractions onto storage abstractions and provides data portability.

I/O Forwarding bridges between app. tasks and storage system and provides aggregation for uncoordinated I/O.

I/O Middleware organizes accesses from many processes, especially those using collective I/O.

Parallel File System maintains logical space and provides efficient access to data.

Fig. 1. I/O Forwarding in HPC systems

MPI-IO, distributed as part of the MPI library, is the standard parallel I/O API for HPC systems. In certain cases, by using *collective I/O*, the MPI-IO implementation is able to reduce the number of requests made to the filesystem. However, not all applications use the MPI-IO interface or are able to use collective I/O, so improvements made at the MPI-IO layer may not be available to the entire spectrum of scientific applications. Parallel high-level libraries such as Parallel-NetCDF [11] use MPI-IO thus face many of the same limitations outlined above. POSIX implementations and serial high-level libraries are an artifact from an earlier generation and are available only on current HPC systems to support legacy applications.

In order to address this I/O bottleneck, another layer needed to be introduced into the I/O software stack. Clients, instead of directly making requests to the parallel filesystem, forward their I/O operations to an I/O forwarder node, which performs the I/O operation on their behalf. One I/O node is typically responsible for 32 to 128 compute clients. Because of its position in the I/O path, the I/O forwarder is able to perform a wide range of optimizations that were not previously possible. For example, it can aggregate requests of unrelated software running on multiple compute nodes, thereby reducing both the number of requests and the number of clients visible to the parallel filesystem. Since the I/O forwarding software—running on the I/O node—does not share any resources (CPU or memory) with the compute clients, it is free to dedicate memory and compute power to optimizing I/O traffic without slowing down computation.

Another benefit of moving the actual I/O calls to the forwarder is that the compute client can be simplified. Instead of requiring a full I/O stack, it needs only to be able to send and receive requests to the I/O forwarder. The I/O forwarder then takes care of using the correct protocol to access the remote filesystem. Likewise, authentication (to the remote filesystem) can be handled by the I/O forwarder. This approach enables compute clients to use a simpler, local authentication scheme to authenticate to the I/O forwarder. Figure 1 shows the resulting I/O software stack.

2.2   I/O Forwarding Scalability Layer

In view of the importance of I/O forwarding in HPC systems, it is desirable to have a high-quality implementation capable of supporting multiple architectures, filesystems, and high-speed interconnects. While a few I/O forwarding solutions are available for the IBM Blue Gene and other leadership-class platforms, such as the Cray XT, they are each tightly coupled to one architecture [21,7]. The lack of an open-source, high-quality implementation capable of supporting multiple architectures, filesystems, and high-speed interconnects has hampered research and makes the deployment of novel I/O optimizations difficult.

To address this issue, we created a scalable, unified I/O forwarding framework for high-performance computing systems called the I/O Forwarding Scalability Layer (IOFSL) [1,14]. IOFSL includes features such as the coalescing of I/O calls on the I/O node, reducing the number of requests to the file system, and full MPI-IO integration, which translates into improved performance for MPI applications. IOFSL implements the ZOIDFS protocol for forwarding I/O requests between applications and IOFSL servers. The ZOIDFS protocol is stateless, uses portable file handles instead of file descriptors, and provides list I/O capabilities.

Ongoing work includes the integration of techniques for improving HPC I/O performance, such as data analytics [22] and checkpointing techniques [13] workloads.

3   Grid Data Access

Two approaches to application data accesses in Grids have emerged. They are described in Section 3.1. Section 3.2 describes how IOFSL can be used to improve the performance and enhance the usability of these approaches.

3.1   Traditional Grid I/O

The first approach stages data at the resource where the application executes or offloads data locally generated by an application to a remote storage system. This approach often uses GridFTP to perform bulk data transfers between the high-performance storage systems attached to Grid resources. While this approach offers good performance, since remote I/O is used only for staging files in and out the local storage, it has a number of drawbacks. For one, maintaining consistency between the local and remote copy is difficult. The second issue is related to the access granularity. Typically, the whole file needs to be transferred, reducing efficiency if the application requires only a subset of the file data.

The second approach is to host data on wide-area filesystems. These file-systems construct a distributed, shared storage space, which is mounted locally on each Grid resource to provide local application access. Examples of Grid-specific filesystems include Gfarm [17] and Chirp [18]. These filesystems typically do not provide traditional I/O semantics and are currently not well supported by parallel applications. For example, in Gfarm, files are basically write-once, and parallel read-write I/O has to be emulated through versioning and creating new files [16].

In addition to these Grid-specific filesystems, traditional HPC filesystems such as Lustre and GPFS have been adapted for Grid environments. While these do offer familiar parallel I/O facilities, the high latencies and large number of filesystem clients severely limits their performance and stability.

3.2   I/O Forwarding in a Grid Environment

When designing IOFSL, portability and modularity were important goals. IOFSL does not make any assumptions about operating system kernels, inter-connects, filesystems, or machine architectures. Hence, it can be easily retargeted to other environments, such as computational Grids.

In large HPC systems, I/O forwarding isolates local compute clients, con-nected by a high bandwidth, low latency interconnect from the more distant, higher latency parallel filesystem. At the same time, it protects the filesystem from being crippled by a storm of requests, by aggregating and merging requests before sending them to the filesystem. From the viewpoint of the remote filesystem, this reduces the number of visible clients and requests, hence increasing performance.

In a Grid environment, these optimizations are also applicable, albeit on a different scale. While latencies might be much higher, the same discontinuity exists when an application running on a local Grid resource needs to fetch data from a remote data store. As is the case in large HPC systems, a large number of simultaneous requests to a remote site might adversely affect the stability and throughput of the remote file server. This observation is valid both for data staging and for wide-area Grid filesystems.

Figure 2 shows the location of I/O forwarding in a Grid environment. Being located at the gateway between the local compute resources and the remote data, IOFSL acts as both a connection and a request aggregator: local applications can share the same set of outgoing connections, increasing efficiency and reducing the load on the remote filesystem. For example, if GridFTP is used as a data transport between the site where data is stored and the site where data is consumed or generated, when using IOFSL, the number GridFTP connections will not depend on the number of clients. Instead, each I/O forwarder can be configured to use an optimal number of GridFTP connections to obtain the data. Clients interacting with the I/O forwarder will transparently share existing connections when possible.

Another important advantage of deploying I/O forwarding in a Grid envi-ronment is that, to the client software, IOFSL can offer a more familiar access API. Currently, IOFSL implements two client-side APIs: POSIX and MPI-IO.

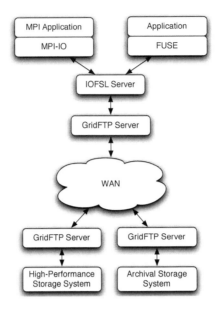

Fig. 2. I/O forwarding in a Grid environment

For POSIX, there are FUSE and SYSIO implementations. The former enables redirecting I/O accesses of unmodified binary applications. While the latter requires relinking applications with the SYSIO library, it provides support on platforms that do not support FUSE (for example, minimal operating system kernels such as Cray's Catamount kernel [9] or IBM BG/P's compute node kernel).

By directly supporting MPI-IO, the de facto I/O API for parallel MPI programs, IOFSL enables unmodified MPI applications (such as parallel analysis or visualization software) to transparently access remote data using GridFTP or other protocols not normally supported by HPC software. In this case, IOFSL effectively acts as a bridge between a local HPC program and remote Grid-style storage.

Dedicating some nodes as I/O forwarders also helps with high-latency network links, a typical problem when spanning multiple remote sites using a POSIX-like filesystem such as Lustre or GPFS. By using local system memory of the I/O forwarders for buffering read and write data, IOFSL is able to transform synchronous client accesses into asynchronous remote accesses, reducing the detrimental effects of high-latency links. Often requested data can be buffered locally, where it can be accessed over a low-latency, high-bandwidth network. As IOFSL transparently captures all file accesses, I/O requests from multiple programs can be optimized if they are requesting the same data. For example, several independent requests to the same data can be coalesced to a single request.

3.3   IOFSL GridFTP Module Implementation

IOFSL provides access to remote data using the GridFTP driver. This driver is similar to other IOFSL drivers because it bridges generalized application I/O requests to a specific I/O subsystem. This IOFSL driver maps the GridFTP API to the ZOIDFS protocol. The driver manages all GridFTP file handles in red–black trees and provides the application with portable, opaque IOFSL file handles. Using this driver, IOFSL servers can proxy application I/O requests to remote GridFTP servers when an application cannot directly access the data because of IP routing constraints or other connection limitations. Implementing this driver presented several challenges becuase it contains several features that existing IOFSL drivers do not have. We currently use the GridFTP 4.2 client library to provide GridFTP support.

In order to access remote data, the location of the data must be encoded into the I/O request. For other IOFSL drivers, such as the POSIX and PVFS2 drivers, the file path is sufficient for IOFSL to locate the data since those file systems are locally available to the nodes hosting IOFSL software. To access remote data with the IOFSL GridFTP driver, we require that applications prefix the file path with the remote access protocol to use, the remote host address, and the port the GridFTP server is using. For example, an application that requires access to the /etc/group file hosted on server 192.168.1.100 that hosts a GridFTP server listening on port 12345 using the ftp protocol will construct a file path /ftp/192.168.1.100/12345/etc/group.

Unlike other IOFSL drivers, the GridFTP client uses an asynchronous operation model. The existing IOFSL drivers use synchronous data management operations, which are easier to adapt to the synchronous ZOIDFS interface. To map the asynchronous GridFTP operations to the synchronous ZOIDFS interfaces, we developed a set of callbacks and monitors that poll the GridFTP client library for operation completion. Supporting these operations also required additional locking within the ZOIDFS driver operations to protect the GridFTP library from concurrent requests. Without additional optimizations, the additional locking within this driver can limit the performance of the IOFSL because of reductions in parallelism. Fortunately, higher-level IOFSL optimizations that can aggregate multiple operations into a single request will reduce the number of pending GridFTP operations and lock contention with the IOFSL GridFTP driver.

The GridFTP 4.2 client library used by the IOFSL driver did not fully support the ZOIDFS capabilities and interface. Several operations, including link and symlink, are not available through GridFTP, and IOFSL cannot support these operations for applications. GridFTP cannot provide all file attributes, including file access times and group identifiers. For attribute retrieval operations, the ZOIDFS GridFTP backend will fetch the available attributes and assumes that application is aware that other attributes are invalid. List I/O capabilities are supported for GridFTP write operations, but are not supported for GridFTP read operations. The IOFSL GridFTP driver must treat all read list I/O requests as individual requests, thus increasing the number of requests in flight that the server must manage.

4 Evaluation

We evaluated the IOFSL GridFTP driver to determine how well it performed and to demonstrate the new capabilities provided by this driver.

To demonstrate the basic capability of this driver, we performed several experiments to evaluate the GridFTP driver functionality and the baseline performance of the GridFTP driver compared to an existing IOFSL driver. We used the Argonne Leadership Computing Facility's Eureka Linux cluster. Eureka is a 100-node Linux cluster that uses a Myricom 10G interconnect. Each node in the cluster contains eight Intel-based cores and 16 GB of RAM. In these tests, the compute nodes of this cluster executed the application code, and the login nodes hosted our GridFTP and IOFSL servers. All network communication in these experiments use TCP/IP.

In the following tests, we evaluated the write performance of the GridFTP driver to a local file system (accessed through a GridFTP server) on the Eureka cluster login node. IOR was used to simulate an I/O bound application. We also collected data for these experiments using the POSIX IOFSL driver. The POSIX driver experiments accessed the data directly. When using the IOFSL GridFTP driver, application I/O requests are forwarded to the IOFSL server, and the IOFSL server delegates the application requests to the GridFTP server.

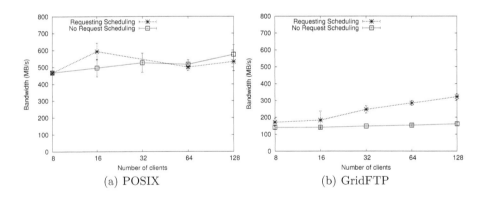

(a) POSIX   (b) GridFTP

Fig. 3. Effect of request scheduling on POSIX and GridFTP access methods

We observed that the request merging optimization increased the performance of the GridFTP driver as the number of application processes increased. Figure 3 illustrates the observed mean bandwidths and standard deviations of these experiments. Without request merging, the overhead for issuing each I/O request is large because of the GridFTP server overhead and the additional locking within the IOFSL GridFTP driver. The request merging optimization is able to aggregate multiple I/O requests into a single list I/O request to the GridFTP server. For high-latency network connections or I/O requests, this optimization can improve performance through the reduction of I/O requests. We observe this improvement when using the optimization with the GridFTP driver for IOFSL. For the POSIX access method, having a lower per request cost, the effect is less clear. It is difficult for the IOFSL to improve performance when using the POSIX access method for these tests because there is not enough filesystem contention. The additional processing overhead of the request merger hinders the performance of the IOFSL server.

Note that Figure 3 is not meant to compare the performance of the GridFTP access method with that of the POSIX access method. These methods each serve distinct purposes, and typically only one of them will be available for accessing a specific file. For example, while almost all compute nodes fully support POSIX I/O, GridFTP access from a compute node will rarely be available because of network limitations (the compute nodes do not have direct outside access) or software restrictions (microkernel operating systems limit software portability). The primary contribution of the IOFSL GridFTP module is that it provides a remote data access capability for systems that limit remote connectivity to compute nodes or other internal infrastructure.

5   Related Work and Conclusions

In this section, we present our future work related to the IOFSL GridFTP driver and present our conclusions from our initial experiments.

5.1   Related Work

In [3], a method is described to allow MPI-IO access to GridFTP stores. It differs from our work in that the MPI application itself makes the GridFTP connection, as opposed to the I/O forwarder node when IOFSL is used. This approach precludes optimizations such as request merging or link aggregation.

Stork [10] tries to improve I/O access time by explicitly scheduling data staging. While IOFSL will also buffer data using local temporary storage, it does this transparently—without explicit data staging—and on a sub file granularity.

Condor [19] enables remote I/O by shipping I/O operations back to the submission site. It requires application to re-link with the Condor library. While our approach also uses function call forwarding, the calls are not shipped to the remote site but to local aggregators.

5.2   Conclusions

In this paper, we provide an overview of the IOFSL project and how the I/O forwarding layer can be used to bridge HPC and Grid file I/O requests. We describe the concept of I/O forwarding in HPC systems and show how the same technique can be applied to Grid computing environments. We discuss its advantages and disadvantages and show how it enables connecting existing HPC and POSIX applications with Grid data stores.

We demonstrate how our work enables transparent GridFTP access. We evaluated our initial GridFTP driver using the IOR benchmark to simulate an I/O-bound application accessing remote data within a cluster. This driver demonstrates that we can effectively bridge HPC and Grid file I/O requests and service remote data requests of applications without modifications to the applications. Our current work includes improving the performance of the driver by reducing lock contention within the GridFTP driver and evaluating the use of this driver to proxy I/O requests from the compute nodes of an IBM Blue Gene/P system to remote data sources.

Acknowledgments. This work was supported by the Office of Advanced Scientific Computer Research, Office of Science, U.S. Dept. of Energy, under Contract DE-AC02-06CH11357. The IOFSL project is supported by the DOE Office of Science and National Nuclear Security Administration (NNSA). This research used resources of the Argonne Leadership Computing Facility at Argonne National Laboratory, which is supported by the Office of Science of the U.S. Department of Energy under contract DE-AC02-06CH11357.

References

1. Ali, N., Carns, P., Iskra, K., Kimpe, D., Lang, S., Latham, R., Ross, R., Ward, L., Sadayappan, P.: Scalable I/O forwarding framework for high-performance computing systems. In: IEEE Int'l Conference on Cluster Computing (Cluster 2009) (September 2009)
2. Allen, G., Dramlitsch, T., Foster, I., Karonis, N., Ripeanu, M., Seidel, E., Toonen, B.: Supporting efficient execution in heterogeneous distributed computing environments with cactus and globus. In: Proceedings of SC 2001, November 10-16 (2001)
3. Baer, T., Wyckoff, P.: A parallel I/O mechanism for distributed systems. In: IEE Cluster 2004, pp. 63–69. IEEE (2004)
4. Carns, P.H., Ligon III, W.B., Ross, R.B., Thakur, R.: PVFS: A parallel file system for Linux clusters. In: Proceedings of the 4th Annual Linux Showcase and Conference, pp. 317–327 (2000)
5. Cluster File Systems, Inc.: Lustre: A scalable high-performance file system. Tech. rep., Cluster File Systems (November 2002),
   http://www.lustre.org/docs/whitepaper.pdf
6. Grinberg, L., Karniadakis, G.: A scalable domain decomposition method for ultra-parallel arterial flow simulations. Communications in Computational Physics 4(5), 1151–1169 (2008)
7. Iskra, K., Romein, J.W., Yoshii, K., Beckman, P.: ZOID: I/O-forwarding infrastructure for petascale architectures. In: ACM SIGPLAN Symposium on Principles and Practice of Parallel Programming, Salt Lake City, UT, pp. 153–162 (2008)
8. Karonis, N., Toonen, B., Foster, I.: MPICH-G2: A Grid-enabled implementation of the Message Passing Interface. Journal of Parallel and Distributed Computing 63(5), 551–563 (2003)
9. Kelly, S., Brightwell, R.: Software architecture of the light weight kernel, Catamount. In: Proceedings of the 2005 Cray User Group Annual Technical Conference (2005)

10. Kosar, T., Balman, M.: A new paradigm: Data-aware scheduling in grid computing. Future Generation Computer Systems 25(4), 406–413 (2009)

11. Li, J., Liao, W., Choudhary, A., Ross, R., Thakur, R., Gropp, W., Latham, R., Siegel, A., Gallagher, B., Zingale, M.: Parallel netCDF: A high-performance scientific I/O interface. In: ACM/IEEE Conference on Supercomputing, Phoenix, AZ (November 2003)

12. Nagle, D., Serenyi, D., Matthews, A.: The Panasas ActiveScale storage cluster—delivering scalable high bandwidth storage. In: ACM/IEEE Conference on Supercomputing (November 2004)

13. Nowoczynski, P., Stone, N., Yanovich, J., Sommerfield, J.: Zest: Checkpoint storage system for large supercomputers. In: 3rd Petascale Data Storage Workshop, IEEE Supercomputing, pp. 1–5 (November 2008)

14. Ohta, K., Kimpe, D., Cope, J., Iskra, K., Ross, R., Ishikawa, Y.: Optimization Techniques at the I/O Forwarding Layer. In: IEEE Int'l Conference on Cluster Computing (Cluster 2010) (September 2010)

15. Schmuck, F., Haskin, R.: GPFS: A shared-disk file system for large computing clusters. In: USENIX Conference on File and Storage Technologies (2002)

16. Tatebe, O., Morita, Y., Matsuoka, S., Soda, N., Sekiguchi, S.: Grid datafarm architecture for petascale data intensive computing. In: CCGrid 2002, p. 102. IEEE Computer Society (2002)

17. Tatebe, O., Soda, N., Morita, Y., Matsuoka, S., Sekiguchi, S.: Gfarm v2: A Grid file system that supports high-performance distributed and parallel data computing. In: Proceedings of the Computing in High Energy and Nuclear Physics Conference, CHEP 2004 (2004)

18. Thain, D., Moretti, C., Hemmes, J.: Chirp: A practical global filesystem for cluster and Grid computing. Journal of Grid Computing 7(1), 51–72 (2009)

19. Thain, D., Tannenbaum, T., Livny, M.: Distributed computing in practice: The Condor experience. Concurrency and Computation Practice and Experience 17(2-4), 323–356 (2005)

20. Wilde, M., Ioan Raicu, I., Espinosa, A., Zhang, Z., Clifford, B., Hategan, M., Kenny, S., Iskra, K., Beckman, P., Foster, I.: Extreme-scale scripting: Opportunities for large task-parallel applications on petascale computers. Journal of Physics: Conference Series 180(1) (2009)

21. Yu, H., Sahoo, R.K., Howson, C., Almasi, G., Castanos, J.G., Gupta, M., Moreira, J.E., Parker, J.J., Engelsiepen, T.E., Ross, R., Thakur, R., Latham, R., Gropp, W.D.: High performance file I/O for the Blue Gene/L supercomputer. In: International Symposium on High-Performance Computer Architecture (February 2006)

22. Zheng, F., Abbasi, M., Docan, C., Lofstead, J., Liu, Q., Klasky, S., Prashar, M., Podhorszki, N., Schwan, K., Wolf, M.: Predata - preparatory data analytics on peta-scale machines. In: Proceedings of 24th IEEE International Parallel and Distributed Processing Symposium (2010)

Fine Granularity Sparse QR Factorization
for Multicore Based Systems

Alfredo Buttari

CNRS-IRIT,
118 route de Narbonne, F-31062 Toulouse, France
alfredo.buttari@irit.fr

Abstract. The advent of multicore processors represents a disruptive event in the history of computer science as conventional parallel programming paradigms are proving incapable of fully exploiting their potential for concurrent computations. The need for different or new programming models clearly arises from recent studies which identify fine-granularity and dynamic execution as the keys to achieve high efficiency on multicore systems. This work presents an implementation of the sparse, multifrontal QR factorization capable of achieving high efficiency on multicore systems through using a fine-grained, dataflow parallel programming model.

Keywords: Multifrontal, Sparse, QR, Least-Squares, Multicore.

1 Introduction

The QR factorization is the method of choice for the solution of least-squares problems arising from a vast field of applications including, for example, geodesy, photogrammetry and tomography (see [15,3] for an extensive list).

The cost of the QR factorization of a sparse matrix, as well as other factorizations such as Cholesky or LU, is strongly dependent on the fill-in generated, i.e., the number of nonzero coefficients introduced by the factorization. Although the QR factorization of a dense matrix can attain very high efficiency because of the use of Householder reflections (see [16]), early methods for the QR factorization of sparse matrices were based on Givens rotations with the objective of reducing the fill-in. One such method was proposed by Heath and George [10], where the fill-in is minimized by using Givens rotations with a row-sequential access of the input matrix. In order to exploit the sparsity of the matrix, such methods suffered a considerable lack of efficiency due to the poor utilization of the memory subsystem imposed by the data structures that are commonly employed to represent sparse matrices.

The *multifrontal method*, first developed for the Cholesky factorization of sparse matrices [8] and then extended to the QR factorization [12,9], quickly gained popularity over these approaches thanks to its capacity to achieve high performance on memory-hierarchy computers. In the multifrontal method, the factorization of a sparse matrix is cast in terms of operations on relatively smaller dense matrices (commonly referred to as *frontal matrices* or, simply, *fronts*)

K. Jónasson (Ed.): PARA 2010, Part II, LNCS 7134, pp. 226–236, 2012.

which gives a good exploitation of the memory subsystems and the possibility of using Householder reflections instead of Givens rotations while keeping the amount of fill-in under control. Moreover, the multifrontal method lends itself very naturally to parallelization because dependencies between computational tasks are captured by a tree-structured graph which can be used to identify independent operations that can be performed in parallel.

Several parallel implementations of the QR multifrontal method have been proposed for shared-memory computers [14,2,7]; all of them are based on the same approach to parallelization which suffers scalability limits on modern, multicore systems (see Section 3.1).

This work describes a new parallelization strategy for the multifrontal QR factorization that is capable of achieving very high efficiency and speedup on modern multicore computers. This method leverages a fine-grained partitioning of computational tasks and a dataflow execution model [17] which delivers a high degree of concurrency while keeping the number of thread synchronizations limited.

2   The Multifrontal QR Factorization

The multifrontal method was first introduced by Duff and Reid [8] as a method for the factorization of sparse, symmetric linear systems and, since then, has been the object of numerous studies and the method of choice for several, high-performance, software packages such as MUMPS [1] and UMFPACK [6].

At the heart of this method is the concept of an *elimination tree*, extensively studied and formalized later by Liu [13]. This tree graph describes the dependencies among computational tasks in the multifrontal factorization. The multifrontal method can be adapted to the QR factorization of a sparse matrix thanks to the equivalence of the R factor of a matrix A and the Cholesky factor of the normal equation matrix $A^T A$. Based on this equivalence, the elimination tree for the QR factorization of A is the same as that for the Cholesky factorization of $A^T A$.

In a basic multifrontal method, the elimination tree has n nodes, where n is the number of columns in the input matrix A, each node representing one pivotal step of the QR factorization of A. Every node of the tree is associated with a frontal matrix that contains all the coefficients affected by the elimination of the corresponding pivot. The whole QR factorization consists in a bottom-up traversal of the tree where, at each node, two operations are performed:

- **assembly:** a set of rows from the original matrix is assembled together with data produced by the processing of child nodes to form the frontal matrix;
- **factorization:** one Householder reflector is computed and applied to the whole frontal matrix in order to annihilate all the subdiagonal elements in the first column. This step produces one row of the R factor of the original matrix and a complement which corresponds to the data that will be later assembled into the parent node (commonly referred to as a *contribution block*). The Q factor is defined implicitly by means of the Householder vectors computed on each front.

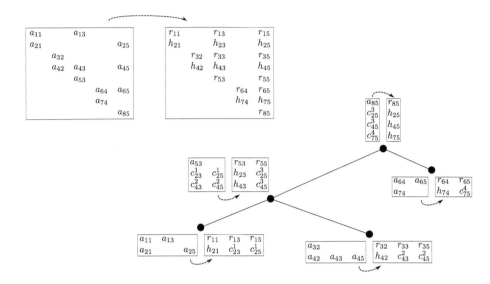

Fig. 1. Example of multifrontal QR factorization. The c_{ij} coefficients denote the contribution blocks.

Figure 1 shows how the QR factorization of the small 8×5 matrix in the top-left part can be achieved through the multifrontal method. The related elimination tree is depicted in the bottom-right part of the figure. Beside each node of the tree, the corresponding frontal matrix is shown after the assembly and after the factorization operations (the transition between these two states is illustrated by the dashed arrows).

In practical implementations of the multifrontal QR factorization, nodes of the elimination tree are amalgamated to form *supernodes*. The amalgamated pivots correspond to rows of R that have the same structure and can be eliminated at once within the same frontal matrix without producing any additional fill-in. This operation can be performed by means of efficient Level-3 BLAS routines. The amalgamated elimination tree is also commonly referred to as *assembly tree*.

In order to reduce the operation count of the multifrontal QR factorization, two optimizations are commonly applied:

1. once a frontal matrix is assembled, its rows are sorted in order of increasing index of the leftmost nonzero (Figure 2 (*middle*)). The number of operations can thus be reduced by ignoring the zeroes in the bottom-left part of the frontal matrix;
2. the frontal matrix is completely factorized (Figure 2 (*right*)). Despite the fact that more Householder vectors have to be computed for each frontal matrix, the overall number of floating point operations is lower since frontal matrices are smaller. This is due to the fact that contribution blocks resulting from the complete factorization of frontal matrices are smaller.

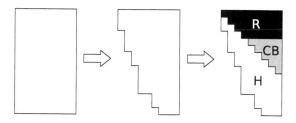

Fig. 2. Techniques to reduce the complexity of the multifrontal QR factorization

A detailed presentation of the multifrontal QR method, including the optimization techniques described above, can be found in Amestoy et al. [2].

The multifrontal method can achieve very high efficiency on modern computing systems because all the computations are arranged as operations on dense matrices; this reduces the use of indirect addressing and allows the use of efficient Level-3 BLAS routines which can achieve a considerable fraction of the peak performance of modern computing systems.

The factorization of a sparse matrix is commonly preceded by a preprocessing phase, commonly referred to as the *analysis phase*, where a number of (mostly symbolic) operations are performed on the matrix such as row and column permutations to reduce the amount of fill-in, the determination of the elimination tree or the symbolic factorization to estimate the amount of memory needed during the factorization phase.

In the rest of this paper, we assume that the analysis phase is already performed, and thus we only focus on the factorization; specifically, we assume that a fill-reducing permutation of the input matrix and the corresponding assembly tree have been computed.

3   Thread-Level Parallelism

Sparse computations are well known for being hard to parallelize on shared-memory, multicore systems. This is due to the fact that the efficiency of many sparse operations, such as the sparse matrix-vector product, is limited by the speed of the memory system. This is not the case for the multifrontal method; since computations are performed as operations on dense matrices, a *surface-to-volume* ratio between memory accesses and computations can be achieved which reduces the utilization of the memory system and opens opportunities for multithreaded, parallel execution.

In a multifrontal factorization, parallelism is exploited at two levels:

- tree-level parallelism: computations related to separate branches of the assembly tree are independent and can be executed in parallel;
- node-level parallelism: if the size of a frontal matrix is big enough, its partial factorization can be performed in parallel by multiple threads.

3.1   The Classical Approach

The classical approach to shared-memory parallelization of QR multifrontal solvers (see [14,2,7]) is based on a complete separation of the two sources of concurrency described above. The node parallelism is delegated to multithreaded BLAS libraries and only the tree parallelism is handled at the level of the multifrontal factorization. This is commonly achieved by means of a task queue where a task corresponds to the assembly and factorization of a front. A new task is pushed into the queue as soon as it is ready to be executed, i.e., as soon as all the tasks associated with its children have been treated. Threads keep polling the queue for tasks to perform until all the nodes of the tree have been processed.

Although this approach works reasonably well for a limited number of cores or processors, it suffers scalability problems mostly due to two factors:

- separation of tree and node parallelism: the degree of concurrency in both types of parallelism changes during the bottom-up traversal of the tree; fronts are relatively small at leaf nodes of the assembly tree and grow bigger towards the root node. On the contrary, tree parallelism provides a high level of concurrency at the bottom of the tree and only a little at the top part where the tree shrinks towards the root node. Since the node parallelism is delegated to an external multithreaded BLAS library, the number of threads dedicated to node parallelism and to tree parallelism has to be fixed before the execution of the factorization. Thus, a thread configuration that may be optimal for the bottom part of the tree will result in a poor parallelization of the top part and vice-versa.
- synchronizations: the assembly of a front is an atomic operation. This inevitably introduces synchronizations that limit the concurrency level in the multifrontal factorization.

3.2   A New, Fine-Grained Approach

The limitations of the classical approach discussed above can be overcome by employing a different parallelization technique based on fine granularity partitioning of operations combined with a data-flow model for the scheduling of tasks. This approach was already applied to dense matrix factorizations [4] and extended to the supernodal Cholesky factorization of sparse matrices [11].

In order to handle both tree and node parallelism in the same framework, a block-column partitioning of the fronts is applied and three elementary operations defined:

1. **panel**: this operation amounts to computing the QR factorization of a block-column;
2. **update**: updating a block-column with respect to a panel corresponds to applying to the block-column the Householder reflections resulting from the panel reduction;
3. **assemble**: assembles a block-column into the parent node (if it exists);

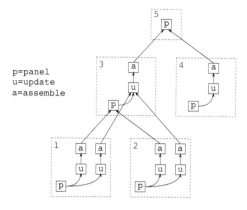

Fig. 3. The DAG associated to the problem in Figure 1

The multifrontal factorization of a sparse matrix can thus be defined as a sequence of tasks, each task corresponding to the execution of an elementary operation of the type described above on a block-column. The tasks are arranged in a Directed Acyclic Graph (DAG); the edges of the DAG define the dependencies among tasks and thus the order in which they have to be executed. These dependencies are defined according to the following rules:

- a block-column is fully assembled when all the corresponding portions of the contribution blocks from its children have been assembled into it. Once a block-column is fully assembled, any elementary operation can be performed on it (according to the other dependencies) even if the rest of the front is not yet assembled or if the factorization of its children is not completed;
- a panel factorization can be executed on a fully assembled block-column if the block-column is up-to-date with respect to all the previous panel factorizations in the same front;
- a fully assembled block-column can be updated with respect to panel i in its front if it is up-to-date with respect to all the panels $1, ..., i-1$ in the same front and if the panel factorization on block-column i has completed;
- a block-column can be assembled into the parent (if it exists) when it is up-to-date with respect to the last panel factorization to be performed on the front it belongs to.

Figure 3 shows the DAG associated with the problem in Figure 1 for the case where the block-columns have size one. The dashed boxes surround all the tasks that are related to a single front and the horizontal displacement of a task identifies the index, within the front, of the column on which the task is executed. In the figure and in the above discussion, the assembly of the matrix nonzero entries into the frontal matrices has been ignored for the sake of readability.

This DAG globally retains the structure of the assembly tree but expresses a higher degree of concurrency because tasks are defined on a block-column basis instead of a front basis. This allows us to handle both tree and node parallelism in a consistent way.

The execution of the tasks in the DAG is controlled by a data-flow model; a task is dynamically scheduled for execution as soon as all the input operands are available to it, i.e., when all the tasks on which it depends have finished. The scheduling of tasks can be guided by a set of rules that prioritize the execution of a task based on, for example,

- cache awareness: in order to maximize the reuse of data into cache memories, tasks may be assigned to threads based on a locality policy (see [11]);
- fan-out: the fan-out of a task in the DAG defines the number of other tasks that depend on it. Thus, tasks with a higher fan-out should acquire higher priority since they generate more concurrency. In the case of the QR method described above, panel factorizations are regarded as higher priority operations over the updates and assemblies.

4   Experimental Results

The method discussed in Section 3.2 was implemented in a software package referred to as qrm below. The code is written in Fortran95 and OpenMP is the technology chosen to implement the multithreading. Although there are many other technologies for multithreaded programming (e.g., pThreads, Intel TBB, Cilk or SMPSS), OpenMP offers the best portability since it is available on any relatively recent system. The current version of the code does not include cache-aware scheduling of tasks. The qrm code was compared to the SuiteSparseQR [7] (referred to as spqr) released by Tim Davis in 2009. For both packages, the COLAMD matrix permutation was applied in the analysis phase to reduce the fill-in and equivalent choices were made for other parameters related to matrix preprocessing (e.g., nodes amalgamation); as a result, the assembly trees produced by the two packages only present negligible differences. Both packages are based on the same variant of the multifrontal method (that includes the two optimization techniques discussed in Section 2) and, thus, the number of floating point operations done in the factorization and the number of entries in the resulting factors are comparable. The size of block-columns in qrm and the blocking size (in the classical LAPACK sense) in spqr were chosen to be the best for each matrix. The rank-revealing feature of spqr was disabled as it is not present in qrm.

The two packages were tested on a set of ten matrices with different characteristics from the UF Sparse Matrix Collection [5]; in this section, only results related to the matrices listed in Table 1 are presented as they are representative of the general behavior of the qrm and spqr codes measured on the whole test set. In the case of underdetermined systems, the transposed matrix is factorized, as it is commonly done to find the minimum-norm solution of a problem.

Experiments were run on two architectures whose features are listed in Table 2.

Table 1. Test matrices. nz(R), nz(H) and Gflops result from the `qrm` factorization

Mat. name	m	n	nz	nz(R)	nz(H)	Gflops
Rucci1	1977885	109900	7791168	184115313	1967908664	12340
ASIC_100ks	99190	99190	578890	110952162	53306532	729
ohne2	181343	181343	6869939	574051931	296067596	3600
mk11-b4	10395	17325	51975	21534913	42887317	396
route	20894	43019	206782	3267449	7998419	2.4

Table 2. Test architectures

Type	# of cores	freq.	mem. type	compilers	BLAS/LAPACK
Intel Xeon	8 (4-cores × 2-sockets)	2.8 GHz	UMA	Intel 11.1	Intel MKL 10.2
AMD Opteron	24 (6-cores × 4-sockets)	2.4 GHz	NUMA	Intel 11.1	Intel MKL 10.2

Figure 4 shows the speedup achieved by the `qrm` code for the factorization of the Rucci1 matrix on both test architectures compared to the `spqr` code; the curves plot the results in Tables 3 and 4 normalized to the sequential execution time.

On the Intel Xeon platform (Figure 4, left), a remarkable 6.9 speedup is achieved on eight cores which is extremely close to the value obtained by the LAPACK `dgeqrf` dense factorization routine; the `spqr` code only achieves a 3.88 speedup using eight cores on the Intel Xeon system.

On the AMD Opteron system (Figure 4, right), the `qrm` code still shows a good speedup when compared to `spqr` and `dgeqrf` although it must be noted that all of them exhibit some scalability limits; this is most likely caused by poor data locality due to the NUMA architecture of the memory subsystem. An ongoing research activity aims at investigating cache-aware task scheduling policies that may mitigate this problem.

Figure 5 shows the fraction of the `dgemm` matrix multiply routine performance that is achieved by the `qrm` and `spqr` factorizations.

Tables 3 and 4 show the factorization times for the test matrices on the two reference architectures. Analysis times are also reported for `qrm` in parentheses.

The number of threads participating in the factorization in the `spqr` code is given by the product of the number of threads that exploit the tree parallelism times the number of threads in the BLAS routines. As discussed in Section 3.1, this rigid partitioning of threads may result in suboptimal performance; choosing a total number of threads that is higher than the number of cores available on the system may yield a better compromise. This obviously does not provide any benefit to `qrm`. The last line in Tables 3 and 4 shows, for `spqr`, the factorization times for the best combination of tree and node parallelism; for example, for the ohne2 matrix, on the Intel Xeon system, the shortest factorization time is achieved by allocating five threads to the tree parallelism and three to the BLAS parallelism for a total of 15 threads.

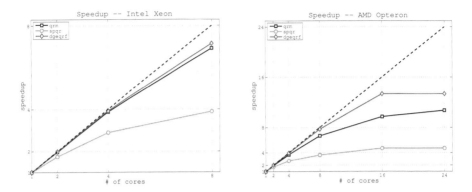

Fig. 4. Speedup for the `qrm` and `spqr` multifrontal factorization of the Rucci1 matrix compared to the LAPACK dense `dgeqrf` factorization routine. The dashed lines represent linear speedup.

The experimental results show that the proposed approach described in Section 3.2 achieves better scalability and better overall execution times on modern, multicore-based systems when compared to the classical parallelization strategy implemented in the `spqr` software. On the AMD Opteron architecture, the `qrm` code has consistently higher factorization times than `spqr` and a poor scaling for the *route* matrix: this is exclusively due to flaws in the implementation of the tasks scheduler and are not related to the proposed parallelization approach. The `qrm` tasks scheduler is currently undergoing a complete rewriting that aims at improving its efficiency by reducing the search space in the tasks DAG.

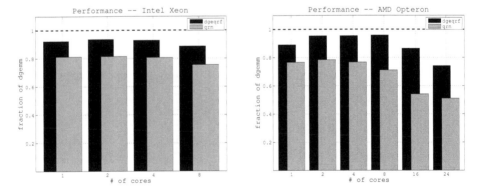

Fig. 5. Performance for the `qrm` multifrontal factorization of the Rucci1 matrix and the LAPACK `dgeqrf` dense matrix factorization routine compared to the BLAS `dgemm` dense matrix product routine

Table 3. Factorization times, in seconds, on the Intel Xeon system for qrm (*top*) and spqr (*bottom*). Analysis times are reported in parentheses for qrm.

				Intel Xeon		
	# th.	Rucci1	ASIC_100ks	ohne2	mk11-b4	route
qrm	1	1237.4 (3.0)	81.7 (0.5)	427.7 (6.6)	41.3 (0.1)	0.88 (0.1)
	2	629.5	41.9	218.3	21.2	0.54
	4	319.8	21.7	110.8	11.2	0.33
	8	179.4	12.4	60.8	6.6	0.21
	# th.	Rucci1	ASIC_100ks	ohne2	mk11-b4	route
spqr	1	1245.2	84.8	449.3	42.3	0.85
	2	714.9	50.3	271.3	24.4	0.53
	4	430.0	32.3	161.0	15.0	0.34
	8	320.7	25.0	111.9	10.8	0.31
	best	295.5	22.2	104.4	10.8	0.29

Table 4. Factorization times, in seconds, on the AMD Opteron system for qrm (*top*) and spqr (*bottom*). Analysis times are reported in parentheses for qrm.

				AMD Opteron		
	# th.	Rucci1	ASIC_100ks	ohne2	mk11-b4	route
qrm	1	1873.8 (2.5)	125.9 (0.3)	664.8 (4.1)	66.7 (0.1)	1.33 (0.1)
	2	969.0	64.7	338.8	34.8	0.76
	4	507.1	33.8	175.7	18.5	0.45
	8	281.7	18.3	92.2	11.4	0.31
	16	193.7	12.7	55.7	10.5	0.61
	24	175.4	12.2	46.0	9.9	0.97
	# th.	Rucci1	ASIC_100ks	ohne2	mk11-b4	route
spqr	1	2081.1	134.4	712.8	65.8	1.15
	2	1206.8	83.1	428.2	38.8	0.63
	4	773.2	54.2	279.4	25.4	0.37
	8	574.1	40.2	178.8	17.8	0.26
	16	443.4	31.1	138.0	17.0	0.21
	24	390.1	28.0	108.4	16.5	0.24
	best	379.5	26.5	107.1	16.5	0.21

Acknowledgments. I would like to thank the MUMPS team and, particularly, Chiara Puglisi and Patrick Amestoy for their precious help and support.

References

1. Amestoy, P.R., Duff, I.S., L'Excellent, J.-Y., Koster, J.: MUMPS: A General Purpose Distributed Memory Sparse Solver. In: Sørevik, T., Manne, F., Moe, R., Gebremedhin, A.H. (eds.) PARA 2000. LNCS, vol. 1947, pp. 121–131. Springer, Heidelberg (2001)

2. Amestoy, P.R., Duff, I.S., Puglisi, C.: Multifrontal QR factorization in a multi-processor environment. Int. Journal of Num. Linear Alg. and Appl. 3(4), 275–300 (1996)
3. Björck, Å.: Numerical methods for Least Squares Problems. SIAM, Philadelphia (1996)
4. Buttari, A., Langou, J., Kurzak, J., Dongarra, J.: A class of parallel tiled linear algebra algorithms for multicore architectures. Parallel Comput. 35(1), 38–53 (2009)
5. Davis, T.A.: University of Florida sparse matrix collection (2002),
   http://www.cise.ufl.edu/research/sparse/matrices
6. Davis, T.A.: Algorithm 832: UMFPACK V4.3 — an unsymmetric-pattern multifrontal method. ACM Trans. Math. Softw. 30(2), 196–199 (2004)
7. Davis, T.A.: Multifrontal multithreaded rank-revealing sparse QR factorization. Accepted for Publication on ACM Transactions on Mathematical Software (2009)
8. Duff, I.S., Reid, J.K.: The multifrontal solution of indefinite sparse symmetric linear systems. ACM Trans. Math. Softw. 9, 302–325 (1983)
9. George, A., Liu, J.W.H.: Householder reflections versus Givens rotations in sparse orthogonal decomposition. Linear Algebra and its Applications 88/89, 223–238 (1987)
10. George, J.A., Heath, M.T.: Solution of sparse linear least squares problems using Givens rotations. Linear Algebra and its Applications 34, 69–83 (1980)
11. Hogg, J., Reid, J.K., Scott, J.A.: A DAG-based sparse Cholesky solver for multicore architectures. Technical Report RAL-TR-2009-004, Rutherford Appleton Laboratory (2009)
12. Liu, J.W.H.: On general row merging schemes for sparse Givens transformations. SIAM J. Sci. Stat. Comput. 7, 1190–1211 (1986)
13. Liu, J.W.H.: The role of elimination trees in sparse factorization. SIAM Journal on Matrix Analysis and Applications 11, 134–172 (1990)
14. Matstoms, P.: Parallel sparse QR factorization on shared memory architectures. Technical Report LiTH-MAT-R-1993-18, Department of Mathematics (1993)
15. Rice, J.R.: PARVEC workshop on very large least squares problems and supercomputers. Technical Report CSD-TR 464, Purdue University, IN (1983)
16. Schreiber, R., Van Loan, C.: A storage-efficient WY representation for products of Householder transformations. SIAM J. Sci. Stat. Comput. 10, 52–57 (1989)
17. Silc, J., Robic, B., Ungerer, T.: Asynchrony in parallel computing: From dataflow to multithreading. Journal of Parallel and Distributed Computing Practices 1, 1–33 (1998)

Mixed Precision Iterative Refinement Methods for Linear Systems: Convergence Analysis Based on Krylov Subspace Methods

Hartwig Anzt, Vincent Heuveline, and Björn Rocker

Karlsruhe Institute of Technology (KIT)
Institute for Applied and Numerical Mathematics 4
Fritz-Erler-Str. 23, 76133 Karlsruhe, Germany
{hartwig.anzt,vincent.heuveline,bjoern.rocker}@kit.edu
http://numhpc.math.kit.edu/

Abstract. The convergence analysis of Krylov subspace solvers usually provides an estimation for the computational cost. Exact knowledge about the convergence theory of error correction methods using different floating point precision formats would enable to determine a priori whether the implementation of a mixed precision iterative refinement solver using a certain Krylov subspace method as error correction solver outperforms the plain solver in high precision. This paper reveals characteristics of mixed precision iterative refinement methods using Krylov subspace methods as inner solver.

Keywords: Mixed Precision Iterative Refinement, Linear Solvers, Krylov Subspace Methods, Convergence Analysis, GPGPU.

1 Introduction

In computational science, the acceleration of linear solvers is of high interest. Present-day coprocessor technologies like GPUs offer outstanding single precision performance. To exploit this computation power without sacrificing the accuracy of the result which is often needed in double precision, numerical algorithms have to be designed that use different precision formats. Especially the idea of using a lower precision than working precision within the error correction solver of an iterative refinement method has turned out to reduce the computational cost of the solving process for many linear problems without sacrificing the accuracy of the final result [2], [5], [9] and [10].

Although the free choice of the error correction solver type offers a large variety of iterative refinement methods, this work is focused on Krylov subspace methods, since they are used for many problems.

The combination of a given outer stopping criterion for the iterative refinement method and a chosen inner stopping criterion for the error correction solver has strong influence on the characteristics of the solver. A small quotient between

K. Jónasson (Ed.): PARA 2010, Part II, LNCS 7134, pp. 237–247, 2012.

outer and inner stopping criterion leads to a high number of inner iterations performed by the error correction solver and a low number of outer iterations performed by the iterative refinement method. A large quotient leads to a low number of inner iterations but a higher number of outer iterations, and therefore to a higher number of restarts of the inner solver. To optimize this trade-off, exact knowledge about the characteristics of both the solver and the linear system is necessary. Still, a theoretical analysis is difficult, since the convergence analysis of the iterative refinement solver is affected when using different precision formats within the method.

This paper presents results of numerical analysis concerning error correction methods based on Krylov subspace solvers. First the general mathematical background of iterative refinement methods is drafted, then the mixed precision approach is introduced and analyzed with respect to the theoretical convergence rate. A conclusion and prospects to future work complete the paper.

2 Mathematical Background

2.1 Iterative Refinement Methods

The motivation for the iterative refinement method can be obtained from Newton's method. Here f is a given function and x_i is the solution in the ith step:

$$x_{i+1} = x_i - (\nabla f(x_i))^{-1} f(x_i). \tag{1}$$

This method can be applied to the function $f(x) = b - Ax$ with $\nabla f(x) = A$, where $Ax = b$ is the linear system that should be solved.

By defining the residual $r_i := b - Ax_i$, one obtains

$$\begin{aligned} x_{i+1} &= x_i - (\nabla f(x_i))^{-1} f(x_i) \\ &= x_i + A^{-1}(b - Ax_i) \\ &= x_i + A^{-1} r_i. \end{aligned}$$

Denoting the solution update with $c_i := A^{-1} r_i$ and using an initial guess x_0 as starting value, an iterative algorithm can be defined, where any linear solver can be used as error correction solver.

Iterative Refinement Method

```
x₀:=rand();                    //initial guess as starting vector
r₀:=b-Ax₀;                     //compute initial residual
do {
    rᵢ:=b-Axᵢ;
    solve (Acᵢ=rᵢ);            //error correction equation
    xᵢ₊₁:=xᵢ + cᵢ;             //update solution
} while ( ‖ Axᵢ-b ‖ >ε ‖ r₀ ‖ );
```

(Iterative refinement algorithm in pseudocode)

In each iteration, the error correction solver searches for a c_i such that $Ac_i = r_i$ and the solution approximation is updated by $x_{i+1} = x_i + c_i$ until the outer residual stopping criterion with a given ε is fulfilled.

2.2   Error Correction Solver

Due to the fact that the error iterative refinement method makes no demands on the inner error correction solver, any backward stable linear solver can be chosen. Still, especially the Krylov subspace methods have turned out to be an adequate choice for many cases. These provide an approximation of the residual error iteratively in every computation loop, which can efficiently be used to control the stopping criterion of the error correction solver. The Krylov subspace methods used in our tests (see section 4) fulfill the demand of backward stability, [3] and [8].

2.3   Convergence Analysis of Iterative Refinement Methods

Based on the residual $r_i = b - Ax_i$ in the ith step, we can analyze the improvement associated with one iteration loop of the iterative refinement method.

Applying a solver to the error correction equation $Ac_i = r_i$ which generates a solution approximation with a relative residual error of at most $\varepsilon_{inner} \| r_i \|$, we get an error correction term c_i, fulfilling

$$r_i - Ac_i = d_i,$$

where d_i is the residual of the correction solver with the property

$$\| d_i \| \leq \varepsilon_{inner} \| r_i \| .$$

In the case of using a Krylov subspace method as inner solver, the threshold $\varepsilon_{inner} \| r_i \|$ can be chosen as residual stopping criterion ($\varepsilon \leq \varepsilon_{inner} < 1$).

Updating the solution $x_{i+1} = x_i + c_i$, we can obtain the new residual error term

$$
\begin{aligned}
\| r_{i+1} \| = {} & \| b - Ax_{i+1} \| \\
= {} & \| b - A(x_i + c_i) \| \\
= {} & \| \underbrace{b - Ax_i}_{=r_i} \underbrace{-Ac_i}_{=d_i - r_i} \| \\
= {} & \| d_i \| \leq \varepsilon_{inner} \| r_i \| .
\end{aligned}
$$

Hence, the accuracy improvements obtained by performing one iteration loop equal the accuracy of the residual stopping criterion of the error correction solver. Using this fact, we can prove by induction, that after i iteration loops, the residual r_i fulfills

$$\| r_i \| \leq \varepsilon_{inner}^i \| r_0 \| . \tag{2}$$

If we are interested in the number i of iterations that is necessary to get the residual error term r_i below a certain threshold

$$\| r_i \| \leq \varepsilon \| r_0 \| \tag{3}$$

we use the properties of the logarithm and estimate based on (2) and (3)

$$\varepsilon_{inner}^i \| r_0 \| \leq \varepsilon \| r_0 \|$$
$$\Leftrightarrow \quad \varepsilon_{inner}^i \leq \varepsilon$$
$$\Leftrightarrow \quad i \geq \frac{\log \varepsilon}{\log \varepsilon_{inner}}.$$

Since i has to be an integer, we use the Gaussian ceiling function and obtain

$$i = \left\lceil \frac{\log(\varepsilon)}{\log(\varepsilon_{inner})} \right\rceil \tag{4}$$

for the number of outer iterations that is necessary to guarantee an accuracy of $\| r_i \| \leq \varepsilon \| r_0 \|$.

3   Mixed Precision Iterative Refinement Solvers

3.1   Mixed Precision Approach

The underlying idea of mixed precision iterative refinement methods is to use different precision formats within the algorithm of the iterative refinement method, approximating the relative residual error and updating the solution approximation in high precision, but computing the error correction term in a lower precision format (see Fig. 1). This approach was also suggested by [2], [5], [9] and [10].

Using the mixed precision approach to the iterative refinement method, we have to be aware of the fact that the residual error bound of the error correction solver may not exceed the accuracy of the lower precision format.

Furthermore, each error correction produced by the inner solver in lower precision cannot exceed the data range of the lower precision format. This means that the smallest possible error correction is the smallest number ϵ_{low}, that can be represented in the lower precision. Thus, the accuracy of the final solution cannot exceed ϵ_{low} either. This can become a problem when working with very small numbers, because then the solution correction terms can not be denoted in low precision, but in most cases, the problem can be avoided by converting the original values to a higher order of magnitude. Instead of solving the error correction equation $Ac_i = r_i$, one applies the error correction solver to the system $Ac_i = 2^p r_i$ where p has to be chosen such that the solution update c_i can be represented in the used low precision format. In this case, the solution update in high precision becomes $x_{i+1} = x_i + 2^{-p} c_i$.

But there are also some more demands to the used low precision floating point format and the within used error correction solver with its respectively stopping criterion. While the low precision floating point format has to be chosen with respect to the condition number of the linear system such that the linear system is still solvable within this format, it has to be ensured that the used error correction solver in low precision converges for the given problem, and does not stagnate before the demanded accuracy of the solution update is achieved. At this point it should be mentioned, that the condition number of the low precision representation of the matrix A may differ from the condition number of the original system.

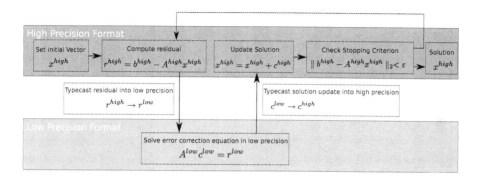

Fig. 1. Visualizing the mixed precision approach to an Iterative Refinement Solver

If the final accuracy does not exceed the smallest number that can be represented in the lower precision format, and if the condition number of the linear system is small enough such that the system is solvable in low precision and the used error correction solver converges and does not stagnate before the demanded accuracy is achieved, then the mixed precision iterative refinement method gives exactly the same solution approximation as if the solver was performed in the high precision format.

When comparing the algorithm of an iterative refinement solver using a certain Krylov subspace solver as error correction solver to the plain Krylov solver, we realize, that the iterative refinement method has more computations to execute due to the additional residual computation, solution updates and typecasts.

The goal is to analyze in which cases the mixed precision iterative refinement method outperforms the plain solver in high precision. Obviously this is the case if the additional operations (denoted with K) are overcompensated by the cheaper execution of the iterative error correction solver in low precision. Using an explicit residual computation the computational costs of K is in the magnitude of the matrix-vector multiplication. In case of an iterative update for the residual, the complexity is even lower.

3.2   Convergence Analysis of Mixed Precision Approaches

When discussing the convergence of the iterative refinement method in section 2.3, we derived a model for the number of outer iterations that are necessary to obtain a residual error below a certain residual threshold $\varepsilon \parallel r_0 \parallel$. Having a relative residual stopping criterion ε_{inner} of the Krylov subspace solver used as error correction solver, we need to perform, according to (4),

$$i = \left\lceil \frac{\log(\varepsilon)}{\log(\varepsilon_{inner})} \right\rceil$$

iterations to obtain an approximation x_i which fulfills

$$\parallel r_i \parallel = \parallel b - Ax_i \parallel \leq \varepsilon \parallel b - Ax_0 \parallel = \varepsilon \parallel r_0 \parallel .$$

It should be mentioned, that this stopping criterion can only be fulfilled, if the Krylov subspace solver converges in the respectively used floating point format.

If we use the iterative refinement technique in mixed precision, we have to modify this convergence analysis due to the floating point arithmetic. In fact, two phenomena may occur that require additional outer iterations.

1. Independently of the type of the inner error correction solver, the low precision format representations of the matrix A and the residual r_i contain representation errors due to the floating point arithmetic. These rounding errors imply that the error correction solver performs the solving process to a perturbed system $(A + \delta A)c_i = r_i + \delta r_i$. Due to this fact, the solution update c_i gives less improvement to the outer solution than expected. Hence, the convergence analysis of the iterative refinement method has to be modified when using different precision formats. To compensate the smaller improvements to the outer solution, we have to perform additional outer iterations.

2. When using a Krylov subspace method as inner correction solver, the residual is computed iteratively within the solving process. As floating point formats have limited accuracy, the iteratively computed residuals may differ from the explicit residuals due to rounding errors. This can lead to an early breakdown of the error correction solver. As in this case the improvement to the outer solution approximation is smaller than expected, the convergence analysis for iterative refinement methods using Krylov subspace solvers as error correction solvers has to be modified furthermore. It may happen, that additional outer iterations are necessary to compensate the early breakdowns of the error correction solver.

We denote the total number of additional outer iterations, induced by the rounding errors and the early breakdowns when using Krylov subspace methods for the inner solver, with g, and obtain

$$i_{total} \left\lceil \frac{\log \varepsilon}{\log \varepsilon_{inner}} \right\rceil + g \tag{5}$$

for the total number of outer iterations. It should be mentioned, that in fact g does not only depend on the type of the error correction solver, but also on the used floating point formats, the conversion and the properties of the linear problem including the matrix structure.

In order to be able to compare a mixed precision iterative refinement solver to a plain high precision solver, we derive a model serving as an upper bound for the computational cost. We denote the complexity of a Krylov subspace solver generating a solution approximation with the relative residual error $\tilde{\varepsilon}$ as $C_{solver}(\tilde{\varepsilon})$. We can obtain this complexity estimation from the convergence analysis of the Krylov subspace solvers [11]. Using this notation, the complexity $C_{mixed}(\varepsilon)$ of an iterative refinement method using a correction solver with relative residual error ε_{inner} can be displayed as

$$C_{mixed}(\varepsilon) = \left(\left\lceil \frac{\log(\varepsilon)}{\log(\varepsilon_{inner})} \right\rceil + g \right) \cdot (C_{solver}(\varepsilon_{inner}) \cdot s + K), \qquad (6)$$

where $s \leq 1$ denotes the speedup gained by performing computations in the low precision format (eventually parallel on the low precision device) instead of the high precision format. We denote the quotient between the mixed precision iterative refinement approach to a certain solver and the plain solver in high precision with $f_{solver} = \frac{C_{mixed}(\varepsilon)}{C_{solver}(\varepsilon)}$, and obtain

$$f_{solver} = \frac{\left(\left\lceil \frac{\log(\varepsilon)}{\log(\varepsilon_{inner})} \right\rceil + g \right) \cdot (C_{solver}(\varepsilon_{inner}) \cdot s + K)}{C_{solver}(\varepsilon).} \qquad (7)$$

Analyzing this fraction, we can state the following propositions:

1. If $f_{solver} < 1$, the mixed precision iterative refinement approach to a certain solver performs faster than the plain precision solver. This superiority of the mixed precision approach will particularly occur, if the speedup gained by performing the inner solver in a lower precision format (e.g. on a accelerator) overcompensates the additional computations, typecasts and the eventually needed transmissions in the mixed precision iterative refinement method.

2. The inverse $\frac{1}{f_{solver}}$ could be interpreted as *speedup factor* obtained by the implementation of the mixed precision refinement method with a certain error correction solver. Although this notation does not conform with the classical definition of the speedup concerning the quotient of a sequentially and a parallel executed algorithm, we can construe $\frac{1}{f_{solver}}$ as measure for the acceleration triggered by the use of the mixed precision approach (and the potentially hybrid system).

3. The iteration loops of Krylov subspace solvers are usually dominated by a matrix-vector multiplication, at least for large dimensions. Hence, using a Krylov subspace method as inner error correction solver, the factor f_{solver} is then for a constant condition number independent of the problem size. This can also be observed in numerical experiments (see section 4 and [1]).

Exact knowledge of all parameters would enable to determine a priori whether the mixed precision refinement method using a certain error correction solver outperforms the plain solver. The computational cost of a Krylov subspace solver depends on the dimension and the condition number of the linear system [11].

While the problem size can easily be determined, an approximation of the condition number of a certain linear system can be obtained by performing a certain number of iterations of the plain Krylov subspace solver, and analyzing the residual error improvement. Alternative method to obtain condition number estimations can for example be found in [12].

The only factor that poses problems is g, the number of additional outer iterations necessary to correct the rounding errors generated by the use of a lower precision format for the inner solver. As long as we do not have an estimation of g for a certain problem, we are not able to determine a priori, which solver performs faster.

To resolve this problem, an implementation of an intelligent solver suite could use the idea to determine a posterior an approximation of g, and then choose the optimal solver. To get an a posterior approximation of g, the solver executes the first iteration loop of the inner solver and then compares the improvement of the residual error with the expected improvement. Through the difference, an estimation for the number of additional outer iterations can be obtained, that then enables to determine the factor f_{solver} and choose the optimal version of the solver.

4   Numerical Experiments

In this section, we want to give a small set of experiments, showing three facts:

1. Depending on the condition number of the system, the plain solver or the mixed precision iterative refinement variant is superior.
2. The factor f_{solver} is for constant condition number independent of the dimension of the problem. This includes, that the mixed precision method works better for reasonably many problems.
3. The total number of outer iterations i_{total} (5) using limited precision usually differs only by a small value g from the theoretical value i (4). Hence approximating i_{total} by $i + 3$ is usually a reasonable estimation for the upper bound.

To show these results, we use a set of artificially created test-matrices with fixed condition number but increasing dimension.

To the linear system affiliated to these matrices, we apply a CG solver as well as a GMRES solver, and compare the performance to the respective mixed precision implementations. All solvers use the relative residual stopping criterion $\varepsilon = 10^{-10} \parallel r_0 \parallel_2$. Due to the iterative residual computation in the case of the plain solvers, the mixed precision iterative refinement variants usually iterate to a better approximation, since they compute the residual error explicitly, but as the difference is generally small, the solvers are comparable. For the mixed precision iterative refinement implementations, we use $\varepsilon_{inner} = 10^{-1}$. The GMRES algorithm, taken from [11], is equipped with a restart parameter of 10.

A more detailed description of the used test matrices, and a more extensive set of numerical experiments including physical applications, can be found in [1].

Table 1. Structure plots and properties of the artificial test-matrices

M1	M2	M3
$\begin{pmatrix} A & * & & & & * \\ * & A & & & & \\ & & \ddots & & & \\ & & & \ddots & & \\ & & & & \ddots & * \\ * & & & & * & A \end{pmatrix}$	$\begin{pmatrix} W & V & * & & & * \\ V & W & \ddots & & & \\ * & \ddots & \ddots & \ddots & & \\ & & \ddots & \ddots & \ddots & * \\ & & & \ddots & \ddots & V \\ * & & & & * & V & W \end{pmatrix}$	$\begin{pmatrix} H & -1 & & -1 & & 0 \\ -1 & H & \ddots & & 0 & \ddots \\ & \ddots & \ddots & & & -1 \\ -1 & & & \ddots & & \ddots \\ & 0 & & \ddots & & -1 \\ 0 & & -1 & & -1 & H \end{pmatrix}$
$A = 10 \cdot n$ $* = rand(0,1)$	$V = 10^3 \cdot n$ $W = 2 \cdot 10^3 \cdot n + n$ $* = rand(0,1)$	$H = 4 + 10^{-3}$
problem: artificial problem size: variable sparsity: $nnz = n^2$ cond. num.: $\kappa < 3$ storage format: MAS	problem: artificial problem size: variable sparsity: $nnz = n^2$ cond. num.: $\kappa \approx 8 \cdot 10^3$ storage format: MAS	problem: artificial problem size: variable sparsity: $nnz \approx 5n$ cond. num.: $\kappa \approx 8 \cdot 10^3$ storage format: CRS

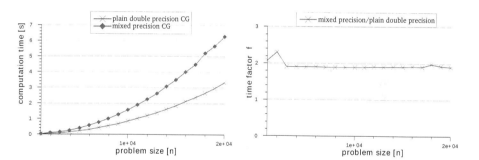

Fig. 2. Performance of CG/mixed CG applied to test case M1; $i_{total} = 5$

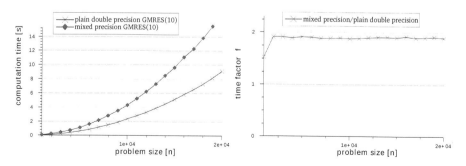

Fig. 3. Performance of GMRES/mixed GMRES applied to test case M1; $i_{total} = 3$

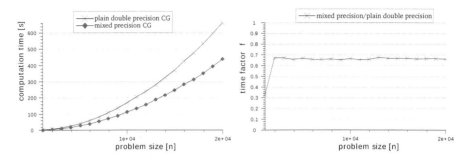

Fig. 4. Performance of CG/mixed CG applied to test case M2; $i_{total} = 13$

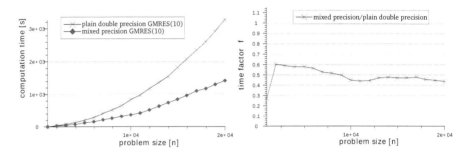

Fig. 5. Performance of GMRES/mixed GMRES applied to test case M2; $i_{total} = 12$

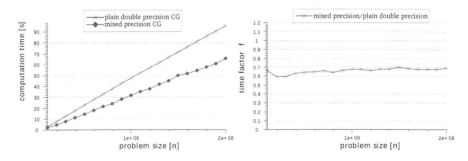

Fig. 6. Performance of CG/mixed CG applied to test case M3; $i_{total} = 10$

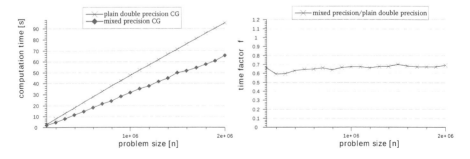

Fig. 7. Performance of GMRES/mixed GMRES applied to test case M3; $i_{total} = 12$

5 Conclusions and Future Work

This paper shows results of numerical analysis concerning the convergence theory of mixed precision iterative refinement methods. These results contribute to the possibility to control the usage of different precision formats within an error correction solver.

A problem still requiring a more satisfactory solution is to determine the exact dependency of the number of additional outer iterations on the characteristics of the linear system, the solver type, the inner and outer stopping criterion, and the used floating point precision formats. Further work in this field is necessary to enable an estimation depending on these parameters.

Technologies like FPGAs and application-specific designed processors offer a free choice of floating point formats. Controlling the usage of these precision formats within iterative refinement solvers is necessary for optimizing the performance.

References

1. Anzt, H., Rocker, B., Heuveline, V.: An Error Correction Solver for Linear Systems: Evaluation of Mixed Precision Implementations. EMCL Preprint Series, no. 2010-01, pages 11(2010)
2. Baboulin, M., Buttari, A., Dongarra, J., Langou, J., Langou, J., Luszcek, P., Kurzak, J., Tomov, S.: Accelerating Scientific Computations with Mixed Precision Algorithms. Computer Physics and Communications 180(12), 2526–2533 (2009)
3. Björck, Å., Elfving, T., Strakoš, Z.: Stability of Conjugate Gradient and Lanczos Methods for Linear Least Squares Problems. SIAM Journal on Matrix Analysis and Applications 19(3), 720–736 (1998)
4. Bai, Z., Demmel, J.W., Dongarra, J., Ruhe, A., van der Vorst, H.: Templates for the Solution of Algebraic Eigenvalue Problems. SIAM, Philadelphia (2000)
5. Buttari, A., Dongarra, J., Langou, J., Langou, J., Luszcek, P., Kurzak, J.: Mixed Precision Iterative Refinement Techniques for the Solution of Dense Linear Systems. IJHPCA 21(4), 457–466 (2007)
6. Demmel, J.W.: Applied Numerical Linear Algebra. SIAM, Philadelphia (1997)
7. Dongarra, J., Duff, I.S., Sorensen, D.C., van der Vorst, H.: Numerical Linear Algebra for High-Performance Computers. SIAM, Philadelphia (1998)
8. Drkošová, J., Greenbaum, A., Rozložník, M., Strakoš, Z.: Numerical Stability of the GMRES Method. BIT Numerical Mathematics 35(3), 309–330 (1995)
9. Göddeke, D., Strzodka, R., Turek, S.: Performance and accuracy of hardware-oriented native–, emulated– and mixed–precision solvers in FEM simulations. International Journal of Parallel, Emergent and Distributed Systems 22(4), 221–256 (2007)
10. Göddeke, D., Strzodka, R.: Performance and accuracy of hardware–oriented native–, emulated– and mixed–precision solvers in FEM simulations (Part 2: Double Precision GPUs). TU Dortmund, Fakultät für Mathematik, Techreport no. 370, pages 9 (2008)
11. Saad, Y.: Iterative Methods for Sparse Linear Systems. SIAM, Philadelphia (2003)
12. Higham, N.J.: Accuracy and Stability of Numerical Algorithms, 2nd edn. SIAM, Philadelphia (2002)

An Implementation of the Tile QR Factorization for a GPU and Multiple CPUs

Jakub Kurzak[1], Rajib Nath[1], Peng Du[1], and Jack Dongarra[1,2,3]

[1] University of Tennessee, Knoxville TN 37996, USA
[2] Oak Ridge National Laboratory, Oak Ridge, TN 37831, USA
[3] University of Manchester, Manchester, M13 9PL, UK
{kurzak,rnath1,du,dongarra}@eecs.utk.edu

Abstract. The tile QR factorization provides an efficient and scalable way for factoring a dense matrix in parallel on multicore processors. This article presents a way of efficiently implementing the algorithm on a system with a powerful GPU and many multicore CPUs.

1 Background

In recent years a tiled approach in applying Householder transformations has proven to be a superior method for computing the QR factorization of a dense matrix on multicore processors, including "standard" (x86 and alike) processors [7,8,10] and also the Cell Broadband Engine [9]. The basic elements contributing to the success of the algorithm are: processing the matrix by tiles of relatively small size, relying on laying out the matrix in memory by tiles, and scheduling operations in parallel in a dynamic, data-driven fashion.

2 Motivation

The efforts of implementing dense linear algebra on multicore and accelerators have been pursued in two different directions, one that emphasizes the efficient use of multicore processors [7,8,10], exemplified by the PLASMA project [3], and another that emphasizes the use of accelerators [15,16], exemplified by the MAGMA project [2]. While the former makes great usage of multicores, it is void of support for accelerators. While the the latter makes great usage of GPUs, it seriously underutilizes CPU resources.

The main problem of existing approaches to accelerating dense linear algebra using GPUs is that GPUs are used like monolithic devices, i.e., like another "core" in the system. The massive disproportion of computing power between the GPU and the standard cores creates problems in work scheduling and load balancing. As an alternative, the GPU can be treated as a set of cores, each of which can efficiently handle work at the same granularity as a standard CPU core.

K. Jónasson (Ed.): PARA 2010, Part II, LNCS 7134, pp. 248–257, 2012.

3   Implementation

All aspects of the tile QR factorization have been relatively well documented in recent literature [7,8,10,9]. Only a minimal description is presented here for the sake of further discussion. Figure 1a shows the basics of the algorithm and introduces the four sequential kernels relied upon. Kernel names are those used in the current release of the PLASMA library. Brief description of each kernel follows.

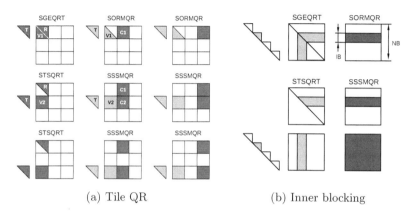

(a) Tile QR   (b) Inner blocking

Fig. 1. Tile QR with inner blocking (light gray - data being read, dark gray - data being modified)

SGEQRT: Performs standard (LAPACK-style) QR factorization within a tile. Replaces the upper triangle of the tile with the R factor. Conceptually, annihilates the entries in the lower triangle. Actually, replaces them with Householder reflectors used to zero these entries.

SORMQR: Applies the transformations of the SGEQRT kernel (the set of Householder reflectors) to a tile on the right of the diagonal tile. This kernel is used across all tiles in a row. (All invocations are independent, i.e., can be done in parallel.)

STSQRT: Performs (incremental) QR factorization of the matrix constructed by putting the diagonal tile, previously factored by SGEQRT, on top of another tile in the panel. Updates the previous R factor. Conceptually, annihilates the entries in the lower tile. Actually, replaces them with Householder reflectors used to zero the tile. (All invocations within one panel are dependent, i.e., have to be serialized.)

STSMQR: Applies the transformations of the STSQRT kernel (the set of Householder reflectors) to two tiles to the right of the panel. This kernel is used across all tiles in two rows. (All invocations are independent, i.e., can be done in parallel.)

One potential deficiency of the algorithm is the introduction of extra floating point operations not accounted for in the standard $4/3N^3$ formula. These operations come from accumulation of the Householder reflectors as reflected in the triangular T matrices in Figure 1a and amount to 25 % overhead if the T matrices are full triangles. The problem is remedied by internal blocking of the tile operations as shown in Figure 1b, which produces T matrices of triangular block-diagonal form and makes the overhead negligible.

The basic concept of the implementation presented here is laid out in Figure 2. It relies upon running the three complex kernels (SGEQRT, STSQRT, SORMQR) on CPUs and only offloading the performance critical SSSMQR kernel to the GPU. It is done in such a way that the Streaming Multiprocessor (SM) of the GPU is responsible for a similar amount of work as one CPU core. In one step of the factorization, the CPUs factorize one panel of the matrix (the SGEQRT and STSQRT kernels), update the top row of the *trailing submatrix* and also update a number of initial columns of the trailing submatrix (through a CPU implementation of the SSSMQR kernel). The GPU updates the trailing submatrix through a GPU implementation of the SSSMQR kernel (Figure 2a). As soon as some number of initial columns is updated, the CPUs can also initialize follow-up panel factorizations and updates, a concept known as *a lookahead* (Figure 2b). This way, when the GPU is finished with one update, the next panel is immediately ready for the following update, which keeps the GPU occupied all the time (avoiding GPU idle time). Also, at each step of the factorization, the GPU part shrinks by one column, and when the size of the trailing submatrix reaches the width of the lookahead, the work is continued by the CPUs only.

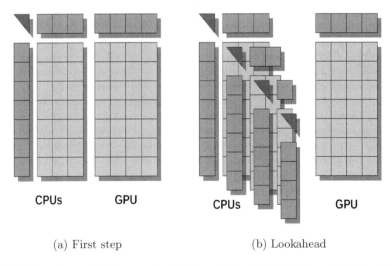

| (a) First step | (b) Lookahead |

Fig. 2. Splitting the work between the CPUs and the GPU. (Different shades of gray show different kernel operations.)

3.1   CPU Kernels

CPU implementations of all four kernels are taken directly from the publicly available *core BLAS* component of the PLASMA library. Ideally, core BLAS would be implemented as monolithic kernels optimized to the maximum for a given architecture. However, this amounts to a prohibitive coding effort, mainly due to the challenges of SIMD'zation for vector extensions ubiquitous in modern processors. Instead, these kernels are currently constructed from calls to BLAS and LAPACK, which is a suboptimal way of implementing them, but the only feasible one known to the authors. They are known to typically deliver about 75 % of the core's peak, while large matrix multiplication (GEMM) delivers up to 95 %.

3.2   GPU Kernel

The main building block of the SSSMQR kernel is matrix multiplication. The process of coding fast matrix multiplication for a GPU relies on a classic auto-tuning approach similar to the one utilized in the ATLAS library [17,1], where a code generator creates multiple variants of code and the best one is chosen through benchmarking. This is the approach taken by the MAGMA library and here the authors leverage this work by using MAGMA SGEMM (matrix multiply) kernels as building blocks for the SSSMQR kernel [11,12]. One shortcoming of this (initial) work is that the kernels were developed for the Nvidia G80 (Tesla) architecture and are used for the Nvidia GF100 (Fermi) architecture. All GPU development was done in Nvidia CUDA. OpenCL implementation was not explored.

The two required operations are $C = C - A^T \times B$ and $C = C - A \times B$. Figure 3 shows MAGMA implementations of these kernels. The first one is implemented as a 32×32 by $32 \times k$ matrix multiplication using a thread block of size 8×8 (Figure 3a). The second one is implemented as a 64×16 by $16 \times k$ matrix multiplication using a thread block of size 64×1 (Figure 3b).

Figure 4 shows the process of constructing the SSSMQR kernel. MAGMA SGEMM kernels allow for building an SSSMQR kernel for tile sizes $NB = 32, 64, 96, \ldots$ with inner blocking $IB = 32, 64, 96, \ldots$, such that IB divides NB (see the first paragraph of section 3 and Figure 1 for the explanation of inner blocking). It has been empirically tested that the combination (IB, NB) of (32, 256) provides the best performance on the GPU and is also a good combination for the CPUs.

The construction of the kernel can be explained in the following steps. Figure 4a shows the starting point. (This would be a CPU implementation of the kernel.) Operation 1 is a memory copy, which is trivial to implement in CUDA, and will not be further discussed. Same applies to operation 5, which is an AXPY operation, also trivial to implement. The first step is a vertical split of all operations (Figure 4b) to provide more parallelism. (What is being developed here is an operation for one thread block, and multiple thread blocks will run on a single Streaming Multiprocessor.) The next step is a conversion of the in-place

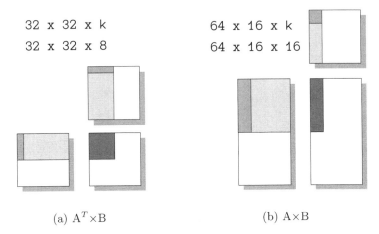

32 x 32 x k
32 x 32 x 8

64 x 16 x k
64 x 16 x 16

(a) $A^T \times B$ (b) $A \times B$

Fig. 3. GPU SGEMM kernels used as the building blocks for the SSSMQR kernel (medium gray - inner loop input data, light gray - outer loop input data, dark gray - in/out data)

triangular matrix multiplication (operation 3) to an out-of-place square matrix multiplication (Figure 4c). The last step is using MAGMA SGEMM kernels to implement operations 2, 3 and 4. The last step is done by incorporating the SGEMM kernels into the body of the SSSMQR kernel and a number of manual code adjustments such as reshaping pointer arithmetics and reshaping the thread block, a somewhat tedious process. A quicker alternative would be to rely on automatic function inlining. It turns out, however, that doing so results in a higher register usage, which leads to lower occupancy and lower overall performance. At the same time, forcing register usage with a compiler flag causes register spills to the memory and, again, lower performance.

Since tiles in a column have to be updated in a sequence, each thread block updates a stripe of the trailing submatrix of width $IB = 32$. This creates enough parallelism to keep the GPU busy for matrices of size 4000 and higher.

3.3   Scheduling

The next critical element of the implementation is dynamic scheduling of operations. Given the lookahead scheme presented in Figure 2b, keeping track of data dependencies and scheduling of operations manually would be close to impossible. Instead, the QUARK scheduler was used, the one used internally by the PLASMA library.

QUARK is a simple dynamic scheduler, very similar in design principles to projects like, e.g., Jade [14,5], StarSs [13] or StarPU [6,4]. The basic idea is the one of unrolling sequential code at runtime and scheduling tasks by resolving three basic data hazards: *Read After Write (RAW)*, *Write After Read (WAR)* and *Write After Write (WAW)*.

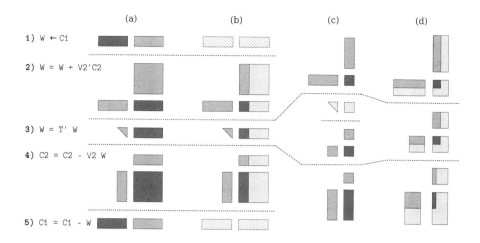

Fig. 4. Steps of the construction process of the SSSMQR GPU kernel (light gray - input data, dark gray - in/out data, hatching - simplifications)

The crucial concept here is the one of task aggregation. The GPU kernel is an aggregate of many CPU kernels, i.e., one invocation of the GPU kernel replaces many invocations of CPU kernels. In order to use the dynamic scheduler, the GPU kernel inherits all data dependencies of the CPU kernel it aggregates. This is done by a simple extension to the dynamic scheduler, where a task is initialized without any dependencies and dependencies are added to it in a loop nest. Figure 5a shows QUARK code for multicores only and Figure 5b shows QUARK code for multicores and a GPU (with lookahead).

Although scheduling is based on runtime resolution of data dependencies, it needs to be pointed out that GPU task placement is static. (It is predetermined which tasks are offloaded to the GPU, based on the lookahead.)

3.4 Communication

If the CPUs and the GPU were sharing a common memory system, the solution would be complete at this point. Since this is not yet the case, data has to be transferred between the CPUs' memory (the *host* memory) and the GPU memory (the *device* memory) through the slow PCIe bus. Despite the disparity between the computing power of a GPU and the communication power of the PCIe, a GPU can be used efficiently for dense linear algebra thanks to the *surface-to-volume effect* ($O(N^3)$ volume of computation and $O(N^2)$ volume of communication).

Here an approach is taken similar to the one of the MAGMA library. It can be referred to as *wavefront communication*, since at each step only a moving boundary region of the matrix is communicated. Initially, a copy of the entire matrix is made in the device memory. Then communication follows the scheme

```
for (k = 0; k < SIZE; k++)
{
  QUARK_Insert_Task(CORE_sgeqrt, ...

  for (m = k+1; m < SIZE; m++)
    QUARK_Insert_Task(CORE_stsqrt, ...

  for (n = k+1; n < SIZE; n++)
    QUARK_Insert_Task(CORE_sormqr, ...

  for (m = k+1; m < SIZE; m++)
    for (n = k+1; n < SIZE; n++)
      QUARK_Insert_Task(CORE_sssmqr, ...
}
```

```
for (k = 0; k < SIZE; k++)
{
  QUARK_Insert_Task(CORE_sgeqrt, ...

  for (m = k+1; m < SIZE; m++)
    QUARK_Insert_Task(CORE_stsqrt, ...

  for (n = k+1; n < SIZE; n++)
    QUARK_Insert_Task(CORE_sormqr, ...

  for (m = k+1; m < SIZE; m++)
    for (n = k+1; n < k+1+lookahead; n++)
      QUARK_Insert_Task(CORE_sssmqr, ...
  task = QUARK_Task_Init(cuda_sssmqr, ...
  for (m = k+1; m < SIZE; m++)
    for (n = k+1+lookahead; n < SIZE; n++)
    {
      QUARK_Task_Pack_Arg(task, &C1, INOUT);
      QUARK_Task_Pack_Arg(task, &C2, INOUT);
      QUARK_Task_Pack_Arg(task, &V2, INPUT);
      QUARK_Task_Pack_Arg(task, &T,  INPUT);
    }
  QUARK_Insert_Task_Packed(task);
}
```

(a) CPUs only   (b) CPUs + a GPU

Fig. 5. Simplified QUARK code

shown in Figure 6. Each GPU kernel invocation is preceded by bringing in to the device memory the panel, the column of T factors, and the top row associated with a given update (Figure 6a). Then, each GPU kernel execution is followed with sending back to the host memory the row brought in before the kernel execution, the first row and the first column of the update (Figure 6b). No additional communication is required when the factorization is completed. At that point the host memory contains the factorized matrix.

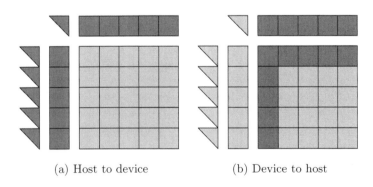

(a) Host to device   (b) Device to host

Fig. 6. *Wavefront* CPU-GPU communication. Dark gray tiles show the moving front of data transferred before each GPU kernel invocation (a) and after each invocation (b)

For simplicity, synchronous host-device communication was used. It did not introduce significant overhead, due to the wavefront pattern. Pinned memory allocation was used for maximum performance. A separate thread was used for GPU management. (One core was oversubscribed.)

4   Results and Discussion

Performance experiments were run using 4 sockets with 6-core AMD Opteron[TM] 8439 SE (Istanbul) processors clocked at 2.8 GHz and an Nvidia GTX 480 (Fermi architecture) graphics card clocked at 1.4 GHz. The core BLAS kernels relied on Intel MKL 11.1 for performance, which turned out to be faster than the AMD ACML library. GCC version 4.1.2 was used for compilation of the CPU code and CUDA SDK 3.1 for compilation of the the GPU code and GPU runtime. The system was running Linux kernel 2.6.32.3 x86_64.

The theoretical peak of the CPUs in single precision equals $2.8 \times 24 \times 8 = 537.6$ [Gflop/s] (clock $\times$ cores $\times$ floating point operations per clock per core). The theoretical peak of the GPU in single precision equals $1.4 \times 480 \times 2 = 1344$ [Gflop/s] (clock $\times$ CUDA "cores" $\times$ floating point operations per clock per core).

Figure 7a shows the performance results for CPUs-only runs, GPU-only runs and runs using both the 24 CPU cores and the GPU. GPU-only runs are basically CPU+GPU runs with *lookahead* = 1. This way the GPU is occupied most of the time, but the CPUs only perform the minimal part of the update to be able to factorize one consecutive panel, while the GPU performs the update so that the GPU does not stall waiting for the panel to be factorized. The CPU+GPU runs are runs with a deep level of lookahead, which keeps the CPUs occupied while the GPU performs the update. The optimal level of the lookahead was tuned manually and is reflected by the number on top of each performance point. Figure 7b shows the performance of each invocation of the GPU SSSMQR kernel throughout the largest factorization of a 19200×19200 (75×75 tiles) matrix with lookahead of 28 (since the number of stripes of $75 - 28 - 1 = 46$).

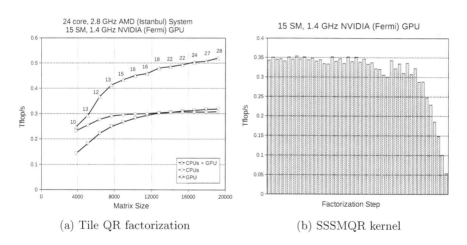

(a) Tile QR factorization   (b) SSSMQR kernel

Fig. 7. Performance results

Interestingly, for this setup, the CPU-only and GPU-only runs deliver very similar performance (slightly above 300 Gflop/s). One can clearly see the performance advantage of using both the CPUs and the GPU, delivering together the performance of 520 Gflop/s. Once again, the authors admit to using sub-optimal GPU kernels for the Fermi architecture. (The development of optimal Fermi kernels is underway.)

5   Conclusions

The results suggest that a system equipped with a high number of conventional cores and a GPU accelerator can be efficiently utilized for a classic dense linear algebra workload. The necessary components are a dynamic scheduler capable of task aggregation (accepting tasks with a very high number of dependencies) and a custom GPU kernel (not readily available in the CUBLAS library). Although a custom kernel is required, it can be built from blocks already available in a BLAS implementation in the CUDA language, such as the ones provided by MAGMA.

6   Future Work

The immediate objectives of the authors are to develop an optimized Fermi kernel for the SSSMQR operation (which should at least double the GPU performance) and generalize the work to multiple GPUs. One can observe that the latter can be accomplished by splitting the trailing submatrix vertically among multiple GPUs. In this case the wavefront communication will involve communication between each GPU and the CPUs and also communication between each pair of GPUs due to the shrinking size of the trailing submatrix and the necessity to shift the boundaries between the GPUs to balance the load.

References

1. ATLAS, http://math-atlas.sourceforge.net/
2. MAGMA, http://icl.cs.utk.edu/magma/
3. PLASMA, http://icl.cs.utk.edu/plasma/
4. StarPU, http://runtime.bordeaux.inria.fr/StarPU/
5. The Jade Parallel Programming Language, http://suif.stanford.edu/jade.html
6. Augonnet, C., Thibault, S., Namyst, R., Wacrenier, P.: StarPU: A unified plat-form for task scheduling on heterogeneous multicore architectures. Concurrency Computat. Pract. Exper. (2010) (to appear)
7. Buttari, A., Langou, J., Kurzak, J., Dongarra, J.J.: Parallel tiled QR factorization for multicore architectures. Concurrency Computat.: Pract. Exper. 20(13), 1573–1590 (2008)
8. Buttari, A., Langou, J., Kurzak, J., Dongarra, J.J.: A class of parallel tiled linear algebra algorithms for multicore architectures. Parallel Comput. Syst. Appl. 35, 38–53 (2009)
9. Kurzak, J., Dongarra, J.J.: QR factorization for the CELL processor. Scientific Programming, 1–12 (2008)

10. Kurzak, J., Ltaief, H., Dongarra, J.J., Badia, R.M.: Scheduling dense linear algebra operations on multicore processors. Concurrency Computat.: Pract. Exper. 21(1), 15–44 (2009)
11. Li, Y., Dongarra, J., Tomov, S.: A Note on Auto-Tuning GEMM for GPUs. In: Allen, G., Nabrzyski, J., Seidel, E., van Albada, G.D., Dongarra, J., Sloot, P.M.A. (eds.) ICCS 2009. LNCS, vol. 5544, pp. 884–892. Springer, Heidelberg (2009)
12. Nath, R., Tomov, S., Dongarra, J.: Accelerating GPU Kernels for Dense Linear Algebra. In: Palma, J.M.L.M., Daydé, M., Marques, O., Lopes, J.C. (eds.) VECPAR 2010. LNCS, vol. 6449, pp. 83–92. Springer, Heidelberg (2011)
13. Planas, J., Badia, R.M., Ayguad, E., Labarta, J.: Hierarchical task-based programming with StarSs. Int. J. High Perf. Comput. Applic. 23(3), 284–299 (2009)
14. Rinard, M.C., Lam, M.S.: The design, implementation, and evaluation of Jade. ACM Trans. Programming Lang. Syst. 20(3), 483–545 (1998)
15. Tomov, S., Dongarra, J., Baboulin, M.: Towards dense linear algebra for hybrid gpu accelerated manycore systems. Parellel Comput. Syst. Appl. 36(5-6), 232–240 (2010)
16. Tomov, S., Nath, R., Ltaief, H., Dongarra, J.: Dense linear algebra solvers for multicore with GPU accelerators. In: Proceedings of the 2010 IEEE International Parallel & Distributed Processing Symposium, IPDPS 2010, April 19-23, pp. 1–8. IEEE Computer Society, Atlanta (2010)
17. Whaley, R.C., Petitet, A., Dongarra, J.: Automated empirical optimizations of software and the ATLAS project. Parellel Comput. Syst. Appl. 27(1-2), 3–35 (2001)

Efficient Reduction from Block Hessenberg Form to Hessenberg Form Using Shared Memory

Lars Karlsson and Bo Kågström

Department of Computing Science and HPC2N,
Umeå University, Umeå, Sweden
{larsk,bokg}@cs.umu.se

Abstract. A new cache-efficient algorithm for reduction from block Hessenberg form to Hessenberg form is presented and evaluated. The algorithm targets parallel computers with shared memory. One level of look-ahead in combination with a dynamic load-balancing scheme significantly reduces the idle time and allows the use of coarse-grained tasks. The coarse tasks lead to high-performance computations on each processor/core. Speedups close to 13 over the sequential unblocked algorithm have been observed on a dual quad-core machine using one thread per core.

Keywords: Hessenberg reduction, block Hessenberg form, parallel algorithm, dynamic load-balancing, blocked algorithm, high performance.

1 Introduction

We say that a square $n \times n$ matrix A with zeroes below its r-th subdiagonal, i.e., $A(i, j) = 0$ if $i > j + r$, is in (upper) *Hessenberg form* if $r = 1$ and in (upper) *block Hessenberg form* if $r > 1$. The number of (possibly) nonzero subdiagonals is r.

The *Hessenberg decomposition* $A = QHQ^T$ of a square matrix $A \in \mathbb{R}^{n \times n}$ consists of an orthogonal matrix $Q \in \mathbb{R}^{n \times n}$ and a matrix $H \in \mathbb{R}^{n \times n}$ in upper Hessenberg form. The Hessenberg decomposition is a fundamental tool in numerical linear algebra and has many diverse applications. For example, a Schur decomposition is typically computed using the nonsymmetric QR algorithm with an initial reduction to Hessenberg form. Other applications include solving Sylvester-type matrix equations.

We focus in this paper on the special case where A is in block Hessenberg form. There are certainly applications where this case occurs naturally. However, our interest is primarily motivated by the fact that this case appears in the second stage of a two-stage approach for Hessenberg reduction of full matrices. The first stage is reduction to block Hessenberg form and efficient algorithms are described in, e.g., [5,9].

In the following, we describe and evaluate a new high-performance algorithm that computes a Hessenberg form $H \in \mathbb{R}^{n \times n}$ of a block Hessenberg matrix A with $1 < r \ll n$ nonzero subdiagonals.

K. Jónasson (Ed.): PARA 2010, Part II, LNCS 7134, pp. 258–268, 2012.

Algorithm 1. (Unblocked) Given a block Hessenberg matrix $A \in \mathbb{R}^{n \times n}$ with r nonzero subdiagonals, the following algorithm overwrites A with $H = Q^T A Q$ where $H \in \mathbb{R}^{n \times n}$ is in upper Hessenberg form and $Q \in \mathbb{R}^{n \times n}$ is orthogonal.

1: **for** $j = 1 : n - 2$
2:     $k_1 = 1 + \lfloor \frac{n-j-2}{r} \rfloor$
3:     **for** $k = 0 : k_1 - 1$
4:        $\alpha_1 = j + kr + 1; \quad \alpha_2 = \min\{\alpha_1 + r - 1, n\}$
5:        $\beta_1 = j + \max\{0, (k-1)r + 1\}; \quad \beta_2 = n$
6:        $\gamma_1 = 1; \quad \gamma_2 = \min\{j + (k+2)r, n\}$
7:        Reduce $A(\alpha_1 : \alpha_2, \beta_1)$ using a reflection Q_k^j
8:        $A(\alpha_1 : \alpha_2, \beta_1 : \beta_2) = (Q_k^j)^T A(\alpha_1 : \alpha_2, \beta_1 : \beta_2)$
9:        $A(\gamma_1 : \gamma_2, \alpha_1 : \alpha_2) = A(\gamma_1 : \gamma_2, \alpha_1 : \alpha_2) Q_k^j$
10:    **end for**
11: **end for**

2 Algorithms

Our blocked algorithm evolved from a known unblocked algorithm for symmetric band reduction [2] adapted to the nonsymmetric case. The band reduction technique [2] has also been applied to the reduction of a matrix pair in block Hessenberg-triangular form to Hessenberg-triangular form [3,1,4]. Since the understanding of the unblocked algorithm is crucial, we begin by describing it.

2.1 Unblocked Algorithm

Algorithm 1 reduces the columns of A from left to right [2,7,8]. Consider the first iteration of the outer loop, i.e., $j = 1$. In the first iteration of the inner loop, $k = 0$ and a Householder reflection, Q_0^1, of order r is constructed on line 7. Lines 8–9 apply a similarity transformation that reduces the first column and also introduces an $r \times r$ bulge of fill-in elements in the strictly lower triangular part of $A(r+2:2r+1, 2:r+1)$.

The next iteration of the inner loop, i.e., $k = 1$, constructs and applies the reflection Q_1^1 of order r which reduces the first column of the bulge. This introduces another bulge r steps further down the diagonal. The subsequent iterations of the inner loop reduce the first column of each newly created bulge.

The second iteration of the outer loop, i.e., $j = 2$, reduces the second column and the leftmost column of all bulges that appear. The new bulges align with the partially reduced bulges from the previous iteration and the bulges move one step down the diagonal.

Applying a Householder reflection of order r to a vector involves $4r$ flops. The flop count of Algorithm 1 is thus $2n^3$ plus lower order terms. During one iteration of the outer loop, each entry in the matrix is involved in zero to two reflections. Consequently, Algorithm 1 has a low arithmetic intensity (flops per byte transferred to/from memory) and its performance is ultimately bounded by the memory bandwidth.

Optionally, the reflections Q_k^j generated by Algorithm 1 can be accumulated into a orthogonal matrix Q such that $A = QHQ^T$ for a cost of $2n^3$ additional flops. An efficient accumulation algorithm is described in [2].

2.2   Blocked Algorithm

The key to increasing the arithmetic intensity of Algorithm 1, and thus creating a blocked algorithm, is to obtain the reflections from multiple consecutive *sweeps*[1] and then apply them in a different order. The reflections Q_k^j are generated by Algorithm 1 in the order of increasing j and increasing k for each j. A more efficient way to apply them, however, is in the order of decreasing k and increasing j for each k [2]. The reason for better efficiency is that the q reflections Q_k^j for q consecutive sweeps j and a fixed k touch only $r + q - 1$ unique entries in each row/column of the matrix. As a result, the arithmetic intensity can be increased almost by a factor of q. In [2], this reordering trick is used when accumulating Q. Below, we show that the same technique can be applied to the updating of A as well.

Overview of the Blocked Algorithm. The blocked algorithm consists of a sequence of iterations, each containing three steps. In the first step, all reflections from q consecutive sweeps are generated while only necessary updates are applied. In the second/third step, the remaining updates from the right-hand/left-hand side are applied. Due to dependencies, the three steps must be done in sequence. The first step is both time-consuming and sequential in nature, which leads to idle processors/cores. Therefore, we bisect the second and third steps and thus create a five-step iteration which supports one level of look-ahead.

The purpose of each step is explained below.

1. Generate the reflections from q consecutive sweeps (label: G).
2. Apply updates from the right-hand side to enable look-ahead (label: P_R).
3. Apply updates from the left-hand side to enable look-ahead (label: P_L).
4. Apply the remaining updates from the right-hand side (label: U_R).
5. Apply the remaining updates from the left-hand side (label: U_L).

Steps 4–5 are the most efficient in terms of memory traffic and parallelism. Steps 2–3 are less efficient but still worthwhile to parallelize. Step 1 is sequential in nature and cannot make efficient use of the cache hierarchy.

Figure 1 illustrates with an example which entries are touched by each step. Note that the G-, P_R-, and P_L-steps combined touch entries near the main diagonal (*medium and dark gray* in Figure 1). The thickness of the band depends on the number of subdiagonals, r, and the number of consecutive sweeps, q, but is independent of the matrix size, n. The algorithm as a whole therefore performs $O(n^3)$ flops cache-efficiently via the U_R- and U_L-steps, and $O(n^2)$ flops less efficiently via the G-, P_R-, and P_L-steps.

---

[1] A *sweep* corresponds to one iteration of the outer loop in Algorithm 1.

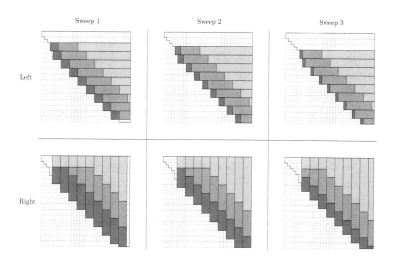

Fig. 1. Entries touched in each of the five steps: G (*dark gray*), P_R (*medium gray, bottom*), P_L (*medium gray, top*), U_R (*light gray, bottom*), and U_L (*light gray, top*). Transformations from the left-hand side (*top*) are shown separately from the transformations from the right-hand side (*bottom*). The $q = 3$ sweeps are shown separately from left to right.

Generate Reflections from Multiple Consecutive Sweeps.

The core of our algorithm is the G-step which is detailed in Algorithm 2.

This algorithm generates the reflections $Q_\star^j$ for q consecutive sweeps starting at column $j = j_1$ while updating only a few entries near the main diagonal (*dark gray* in Figure 1).

The structure of Algorithm 2 is as follows. The outer j-loop iterates over the q sweeps starting at column j_1. The inner k-loop iterates over the k_1 reflections contained in sweep j. First, the column that is about to be reduced is brought up-to-date by applying delayed updates from the left (if any) in the loop that starts on line 6. The actual reduction is performed on line 13. Finally, some of the updates from the right-hand side are applied on line 16.

Apply Updates from Consecutive Sweeps Efficiently.

Algorithm 3 implements the remaining steps P_R, P_L, U_R, and U_L. In this algorithm, the arithmetic intensity is increased by using the reordering trick described in [2].

To facilitate parallel execution, Algorithm 3 restricts the updates from the left-hand side to the column range $c_1 : c_2$ and the updates from the right-hand side to the row range $r_1 : r_2$. By partitioning the range $1 : n$ into p disjoint ranges, p threads can execute the same variant of Algorithm 3 in parallel without synchronization. Note that there is no accumulation of reflections in any of the steps and thus the blocked algorithm has the same flop count as the unblocked algorithm, i.e., $2n^3$ plus lower order terms.

Algorithm 2. (Generate) Given a block Hessenberg matrix $A \in \mathbb{R}^{n \times n}$ with r nonzero subdiagonals and where A is already in upper Hessenberg form in columns $1 : j_1 - 1$, the following algorithm generates $Q_\star^j$ for $j = j_1 : j_1 + q - 1$. The matrix A is partially overwritten by the similarity transformation implied by the q sweeps. The remaining updates can be applied using Algorithm 3.

```
 1: for j = j₁ : j₁ + q − 1
 2:     k₁ = 2 + ⌊(n−j−1)/r⌋
 3:     for k = 0 : k₁ − 1
 4:         α₁ = j + kr + 1;   α₂ = min{α₁ + r − 1, n}
 5:         β = j + max{0, (k − 1)r + 1}
 6:         for ĵ = j₁ : j − 1
 7:             α̂₁ = ĵ + kr + 1;   α̂₂ = min{α̂₁ + r − 1, n}
 8:             if α̂₂ − α̂₁ + 1 ≥ 2 then
 9:                 A(α̂₁ : α̂₂, β) = (Q_k^ĵ)ᵀ A(α̂₁ : α̂₂, β)
10:             end if
11:         end for
12:         if α₂ − α₁ + 1 ≥ 2 then
13:             Reduce A(α₁ : α₂, β) using a reflection Q_k^j
14:             A(α₁ : α₂, β) = (Q_k^j)ᵀ A(α₁ : α₂, β)
15:             γ₁ = j₁ + 1 + max{0, (k + j − j₁ − q + 2)r};   γ₂ = min{j + (k + 2)r, n}
16:             A(γ₁ : γ₂, α₁ : α₂) = A(γ₁ : γ₂, α₁ : α₂)Q_k^j
17:         end if
18:     end for
19: end for
```

3 Parallelization

To develop a shared-memory implementation of the new algorithm we have to decompose the five steps into independent tasks, map the tasks to threads, and synchronize the threads to obey the dependencies. Moreover, we must balance the workload to achieve high parallel efficiency.

3.1 Task Decomposition and Dependencies

Figure 2 illustrates all direct dependencies between the steps of four consecutive iterations. The steps within one iteration are layed out horizontally. Note in particular that the U_R-step is not dependent on the P_L-step and that the G-step of the next iteration can start as soon as the P_L-step of the current iteration completes, which enables one level of look-ahead.

The polygons in Figure 2 must execute sequentially, and we therefore implement the computation as a loop, which we call the *look-ahead loop*, in which each iteration corresponds to a polygon in Figure 2. The *prologue* and *epilogue* polygons correspond to the computations before and after the look-ahead loop, respectively. The steps PU_R and PU_L in the epilogue represent the union of P_R and U_R and the union of P_L and U_L, respectively.

Algorithm 3. (Update) Given a block Hessenberg matrix $A \in \mathbb{R}^{n \times n}$ with r nonzero subdiagonals, the following algorithm applies $Q_\star^j$ for $j = j_1 : j_1 + q - 1$ to both sides of A. This algorithm consists of four variants (P_R, P_L, U_R, and U_L). Together they complete the updates that remain after Algorithm 2. The ranges of rows and columns that are updated by the calling thread are $r_1 : r_2$ and $c_1 : c_2$, respectively.

1: $k_1 = 1 + \left\lfloor \frac{n - j_1 - 2}{r} \right\rfloor$
2: **for** $k = k_1 - 1 : -1 : 0$
3:     **for** $j = j_1 : j_1 + q - 1$
4:       $\alpha_1 = j + kr + 1;$ $\alpha_2 = \min\{\alpha_1 + r - 1, n\}$
5:       **if** $\alpha_2 - \alpha_1 + 1 \geq 2$ **then**
6:         **if** variant is P_R or U_R **then**
7:           **if** variant is P_R **then**
8:             $\gamma_1 = \max\{r_1, j_1 + (k + j - j_1 - 2q + 2)r + 1\}$
9:             $\gamma_2 = \min\{r_2, j_1 + \max\{0, (k + j - j_1 - q + 2)r\}\}$
10:          **else if** variant is U_R **then**
11:             $\gamma_1 = \max\{r_1, 1\}$
12:             $\gamma_2 = \min\{r_2, j_1 + (k + j - j_1 - 2q + 2)r\}$
13:          **end if**
14:          $A(\gamma_1 : \gamma_2, \alpha_1 : \alpha_2) = A(\gamma_1 : \gamma_2, \alpha_1 : \alpha_2) Q_k^j$
15:         **else if** variant is P_L or U_L **then**
16:           **if** variant is P_L **then**
17:             $\beta_1 = \max\{c_1, j_1 + q, j_1 + q + (k - 1)r + 1\}$
18:             $\beta_2 = \min\{c_2, j_1 + q + (k + q - 1)r\}$
19:           **else if** variant is U_L **then**
20:             $\beta_1 = \max\{c_1, j_1 + q + (k + q - 1)r + 1\}$
21:             $\beta_2 = \min\{c_2, n\}$
22:          **end if**
23:          $A(\alpha_1 : \alpha_2, \beta_1 : \beta_2) = (Q_k^j)^T A(\alpha_1 : \alpha_2, \beta_1 : \beta_2)$
24:         **end if**
25:       **end if**
26:     **end for**
27: **end for**

3.2 Parallel Execution

Following the dependencies in Figure 2, we see that the P_R-step must be completed before either the P_L-step or the U_R-step starts, which implies a barrier-style synchronization. Next, we could potentially do both P_L and U_R concurrently. Heuristically, however, we would like to start the G-step as soon as possible. Therefore, we compute the P_L-step before starting the steps G and U_R in parallel. Note that the G-step must not start until P_L has completed, but U_R can start at any time. Thus, the synchronization implied at this point is weaker than a barrier. Proceeding to the U_L-step, we see that both U_R and P_L must be completed, which again implies a barrier-style synchronization. At the end of the iteration there is a third and final barrier since all steps must complete before the next iteration starts.

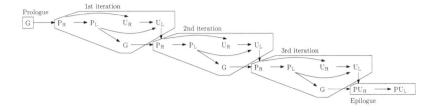

Fig. 2. Dependence graph for a problem consisting of four iterations (separated vertically). The steps of one iteration are layed out horizontally. Polygons enclose the steps within the prologue, epilogue, and look-ahead iterations, respectively.

The large number of barrier-style synchronization points, namely three per iteration, makes it difficult to use dynamic scheduling with fine-grained tasks. The reasons are that (i) fine-grained tasks cannot be executed as efficiently and cause more overhead than large-grained tasks, and (ii) the barriers introduce overhead due to idle processors/cores. We therefore use a model-driven dynamic load-balancing scheme with coarse-grained tasks.

Task Mapping. We have chosen to execute the G-step on a single thread since it is difficult to parallelize. The remaining steps are parallelized by partitioning the rows and columns as described in Algorithm 3.

The main problem is how to map the tasks to the threads to minimize the idle time caused by the synchronization points. Furthermore, thread p_0 should participate in (i) both U_R and U_L, (ii) only U_L, or (iii) none of them. The choice depends on the duration of the G-step. In Figure 3(a), the G-step is the limiting factor and p_0 should not take part in any update. In Figure 3(b), the G-step finishes half-way into the U_L-step and a (small) piece of the U_L-step should therefore be mapped to thread p_0. In Figure 3(c), the G-step finishes before the U_R-step and now a (small) piece of U_R as well as a (large) piece of U_L should be mapped to p_0.

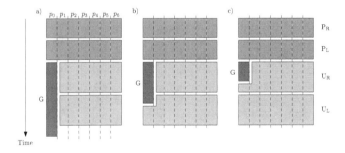

Fig. 3. Three scenarios for parallel execution of one look-ahead iteration

Dynamic Load Balancing. Given p threads and a step (P_R, P_L, U_R, or U_L), we want to find a row/column partitioning corresponding to p tasks mapped to p threads such that if thread p_i starts executing its task at time a_i, then all threads finish their respective task at the same time.

Let $f(x)$ be a function defined at the integer values $x = 1:n$ such that $f(x)$ is the number of flops needed to update the entries on row/column x by the step under consideration. Hence, the total number of flops is $W = \sum_{x=1}^{n} f(x)$.

Suppose that p_i executes its next task with an average speed of s_i flops/s. Then, p_i completes its task at time $t_i = a_i + w_i/s_i$, where w_i is the task's flop count. The optimal execution time, $t_{\min}$, is obtained by solving $\int_0^x s(t)\, dt = W$ for x, where $s(t) = \sum_{a_i < t} s_i$. The optimal task flop counts are computed from $w_i = \max\{0, s_i(t_{\min} - a_i)\}$. Finally, a partitioning of the range $1:n$ is constructed by solving $\sum_{x=1}^{x_i} f(x) = \sum_{j=0}^{i-1} w_j$ for x_i where $i = 1:p-1$. Using the convention that $x_0 = 1$ and $x_p = n + 1$, the partition mapped to p_i is $x_i : x_{i+1} - 1$.

The accuracy of the scheme above depends on how well we can guess the speeds s_i. We adaptively model the speeds measuring the actual execution time $\hat{t}_i$ and using the estimate $s_i = w_i/\hat{t}_i$ for the next iteration.

Synchronization. The many dependencies between steps (see Figure 2) simplify the synchronization of threads. There are three barriers and one weaker form of synchronization per iteration. A single mutex guards all the synchronization variables, which consist of four task counters (one for each step except G) and three condition variables (one for each of P_R, P_L, and U_L). The counters count the number of remaining tasks and the condition variables are signalled when the corresponding counter reaches zero, which guarantees that the step is complete.

4   Computational Experiments

Experiments were run on Akka at HPC2N. Akka is a cluster with dual Intel Xeon L5420 nodes (4 cores per socket) with a double precision theoretical peak performance of 80 Gflops/s (10 Gflops/s per core and 8 cores in total).

We compare four different implementations to demonstrate the impact of various aspects of our implementation. The first is an implementation of Algorithm 1 which we labeled Unblocked. The second, labeled Basic, is a variant of our blocked algorithm without both look-ahead and adaptive modeling of the speeds. The third, labeled Basic + Adapt, includes adaptive modeling. The fourth implementation, labeled Look-ahead + Adapt, includes both look-ahead and adaptive modeling.

The unblocked implementation is sequential while the others are parallel. All parallel executions use one thread per core (8 threads in total). We performed each experiment twice and selected the shortest execution time.

The performance, which we calculate as $2n^3/t$ where t is the execution time, is illustrated in Figure 4. As expected, the unblocked implementation is quite slow since it is sequential and causes a lot of (redundant) memory traffic. The

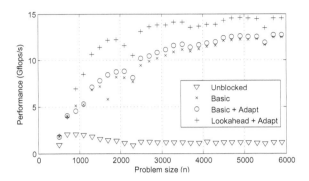

Fig. 4. Performance ($r = 12$, $q = 16$)

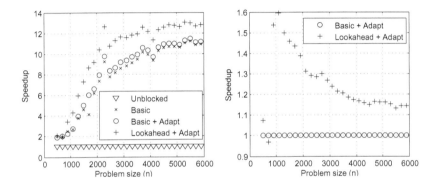

Fig. 5. Left: Speedup over Unblocked ($r = 12$, $q = 16$). Right: Speedup over Basic+Adapt ($r = 12$, $q = 16$).

most advanced implementation, Look-ahead + Adapt, comes out on top and peaks close to 15 Gflops/s (19% of the theoretical peak). The corresponding speedups over Unblocked are shown in Figure 5. The look-ahead technique adds a significant performance boost (15–50%) as shown in Figure 5.

The idle time gets substantially reduced by the addition of adaptation and the look-ahead implementation incurs relatively small amounts of idle time. Figure 6 illustrates this by showing the measured idle time per iteration for a problem with $n = 3000$, $r = 12$, and $q = 16$. The improvements obtained when going from Basic to Basic + Adapt are obvious. However, it is not possible to isolate the effect of the adaptive modeling in the look-ahead implementation since the adaptation is an integral part of the look-ahead approach (see Figure 3).

The partial trace in Figure 7 demonstrates that the load-balancing scheme works in practice. Note that all threads arrive almost simultaneously at all synchronization points (see Section 3.2). Moreover, the size of the tasks mapped to p_0 adapts to the time it takes to complete the G-step.

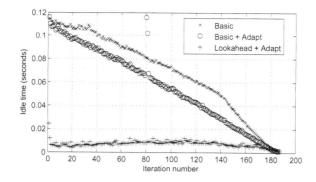

Fig. 6. Idle time per iteration ($n = 3000$, $r = 12$, $q = 16$)

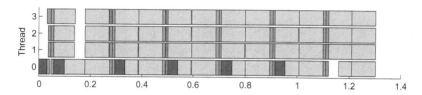

Fig. 7. Trace of iterations 1–6 (out of 313) from a run on Akka using four threads ($n = 5000$, $r = 12$, $q = 16$). Dark gray corresponds to the G-steps, medium gray corresponds to the P_R- and P_L-steps, and light gray corresponds to the U_R- and U_L-steps.

The load-balancing scheme estimates the per-thread speed of each kernel. Initially, no reliable estimates are available and this causes the idle time visible just prior to the 0.2s mark in Figure 7. The difficulty is that the predicted time for the first U_R-step (guessed speed) cannot be related to the predicted time for the second G-step (reliably adapted speed) and hence thread p_0 is assigned a disproportionally large chunk of the U_R-step. Note, however, that the adaptation model solves this problem already after the first iteration.

5   Conclusion

We have presented a new blocked high-performance shared-memory implementation of a Householder-based algorithm for reduction from block Hessenberg form to Hessenberg form. The implementation delays updates and applies them cache-efficiently in parallel. One level of look-ahead in conjunction with an adaptive coarse-grained load-balancing scheme significantly improves the performance. A performance of 15 Gflops/s (19% of the theoretical peak) has been observed on a dual quad-core machine. This corresponds to a speedup close to 13 over a sequential unblocked algorithm.

5.1   Related Work

A different approach to increase the arithmetic intensity of the reduction from block Hessenberg form to Hessenberg form is presented in [6]. A sliding computational window is employed to obtain a set of reflections while at the same time delaying most of the updates. The off-diagonal blocks are then updated efficiently. However, the focus in [6] is on algorithms which are theoretically I/O-efficient and it is also primarily concerned with efficiency in the asymptotic sense. Specifically, the proposed blocking appears to be impractical for matrices that fit entirely in main memory since the computational window would have to be relatively large in order to obtain even modest improvements to the arithmetic intensity.

Acknowledgements. We would like to thank Daniel Kressner for his support and motivating discussions. This research was conducted using the resources of High Performance Computing Center North (HPC2N) and was supported by the Swedish Research Council under grant VR7062571, by the Swedish Foundation for Strategic Research under grant A3 02:128, and the EU Mål 2 project UMIT.

In addition, support has been provided by eSSENCE, a collaborative e-Science programme funded by the Swedish Research Council within the framework of the strategic research areas designated by the Swedish Government.

References

1. Adlerborn, B., Dackland, K., Kågström, B.: Parallel Two-Stage Reduction of a Regular Matrix Pair to Hessenberg-Triangular Form. In: Sørvik, T., et al. (eds.) PARA 2000. LNCS, vol. 1947, pp. 92–102. Springer, Heidelberg (2001)
2. Bischof, C.H., Lang, B., Sun, X.: A framework for symmetric band reduction. ACM Trans. Math. Software 26(4), 581–601 (2000)
3. Dackland, K., Kågström, B.: Blocked algorithms and software for reduction of a regular matrix pair to generalized Schur form. ACM Trans. Math. Software 25(4), 425–454 (1999)
4. Kågström, B., Kressner, D., Quintana-Orti, E., Quintana-Orti, G.: Blocked algorithms for the reduction to Hessenberg-triangular form revisited. BIT Numerical Mathematics 48(1), 563–584 (2008)
5. Ltaief, H., Kurzak, J., Dongarra, J.J., Badia, R.M.: Scheduling two-sided transformations using tile algorithms on multicore architectures. Scientific Programming 18(1), 35–50 (2010)
6. Mohanty, S.: I/O Efficient Algorithms for Matrix Computations. Ph.D. thesis, Indian Institute of Technology Guwahati (2009)
7. Murata, K., Horikoshi, K.: A new method for the tridiagonalization of the symmetric band matrix. Information Processing in Japan 15, 108–112 (1975)
8. Rutishauser, H.: On Jacobi rotation patterns. In: Proc. Sympos. Appl. Math., vol. XV, pp. 219–239. Amer. Math. Soc., Providence (1963)
9. Tomov, S., Dongarra, J.J.: Accelerating the reduction to upper Hessenberg form through hybrid GPU-based computing. Tech. Rep. UT-CS-09-642, University of Tennessee Computer Science (May 2009); also as LAPACK Working Note 219

Cache-Oblivious Algorithms
and Matrix Formats for Computations
on Interval Matrices

Rafał Dabrowski and Bartłomiej Jacek Kubica

Institute of Control and Computation Engineering,
Warsaw University of Technology, Poland
{R.Dabrowski,bkubica}@elka.pw.edu.pl

Abstract. The paper considers the use of cache-oblivious algorithms and matrix formats for computations on interval matrices. We show how the efficient use of cache is of less importance in interval computations than in traditional floating-point ones. For interval matrices there are more important factors, like the number of rounding modes switches or the number of times we have to check if an interval contains zero or not. Yet the use of cache still plays some role.

Keywords: cache-oblivious matrix formats, matrix multiplication, interval computations, interval matrices.

1 Introduction

In recent years cache oblivious matrix formats have been intensively studied (see e.g. [8], [5], [3], [9], [10]) and resulted in several useful approaches, like the Rectangular Full Packed Format (RFP), Recursive Packed Format (RPF), Z-order format, etc. The most basic idea is – crudely speaking – to support recursive algorithms that split a matrix into submatrices subsequently [7], [5].

Very little effort on the other hand was put in studying such formats for interval matrices; the only known exception is the Master's thesis of the first author [4]. For example, the popular C-XSC interval library [1] uses only the simplest representation, storing all elements of a matrix row-by-row. On the other hand, the Profil/BIAS library [16] provides Basic Interval Arithmetic Operations (analogously to BLAS) for matrices and they do not make use of cache-oblivious formats. In this paper, we investigate the use of recursive algorithms together with cache-oblivious matrix formats in order to improve the performance and caching efficiency in interval arithmetic.

It might seem the formats and algorithms that are useful for floating-point matrices should be useful for matrices containing elements from any other domain. The aim of this paper is to investigate the usefulness of some of the common matrix formats and a few new ones for interval matrix multiplication.

K. Jónasson (Ed.): PARA 2010, Part II, LNCS 7134, pp. 269–279, 2012.

2   Interval Computations

Interval computations (see, e.g., [11], [12], [14] and other textbooks) find applications in several branches of engineering, applied mathematics, control theory, economical sciences, physics, etc.

The most basic tool is interval arithmetic, i.e. arithmetic operations (and other basic operations) defined on intervals instead of numbers. Such operations should fulfill the following condition.

Assume we have two intervals: $\mathbf{a} = [\underline{a}, \overline{a}]$, $\mathbf{b} = [\underline{b}, \overline{b}]$, where $\underline{a}, \overline{a}, \underline{b}, \overline{b} \in \mathbb{R}$. Assume, we have an arithmetic operator $\odot \in \{+, -, \cdot, /\}$.

Then:

$$a \in \mathbf{a}, \ b \in \mathbf{b} \quad \Rightarrow \quad a \odot b \in \mathbf{a} \odot \mathbf{b} . \tag{1}$$

The arithmetic operations, satisfying (1) may be formulated as follows:

$$
\begin{aligned}
[\underline{a}, \overline{a}] + [\underline{b}, \overline{b}] &= [\underline{a} + \underline{b}, \overline{a} + \overline{b}] , \\
[\underline{a}, \overline{a}] - [\underline{b}, \overline{b}] &= [\underline{a} - \overline{b}, \overline{a} - \underline{b}] , \\
[\underline{a}, \overline{a}] \cdot [\underline{b}, \overline{b}] &= [\min(\underline{a}\underline{b}, \underline{a}\overline{b}, \overline{a}\underline{b}, \overline{a}\overline{b}), \max(\underline{a}\underline{b}, \underline{a}\overline{b}, \overline{a}\underline{b}, \overline{a}\overline{b})] , \\
[\underline{a}, \overline{a}] / [\underline{b}, \overline{b}] &= [\underline{a}, \overline{a}] \cdot [1 / \overline{b}, 1 / \underline{b}] , \qquad 0 \notin [\underline{b}, \overline{b}] .
\end{aligned}
\tag{2}
$$

To make sure we enclose the actual result of the operation, the lower bound of the interval is rounded downwards and the upper bound upwards.

3   Specific Features of Interval Arithmetic

The arithmetic defined by formulae (2) has several interesting and counterintuitive features. In contrast to real numbers, intervals do not form a ring nor even a group (neither addition nor multiplication of intervals are invertible). In particular, $[a, b] - [a, b]$ is not necessarily equal to zero, but:

$$[a, b] - [a, b] = [a - b, b - a] , \tag{3}$$

which always *contains* zero, but *is equal to* zero only when $a = b$. In general: when an interval expression contains the same term a few times, it is likely to be overestimated.

Similarly, the multiplication is not distributive with respect to addition (it is subdistributive only, i.e. $\mathbf{a} \cdot (\mathbf{b} + \mathbf{c}) \subseteq \mathbf{ab} + \mathbf{ac}$). Obviously, for both variants the true result will be enclosed, but the overestimation may be different.

We can say that issues with accuracy of floating-point operations are raised to a new level for interval computations. For the floating-point, the numerical inaccuracies are usually very small (unless the problem is ill-conditioned) and for interval computations the diameters of initial intervals may be large already. This makes some linear transformations very significant, e.g., the Strassen algorithm applied to interval matrices computes very overestimated results (as it will be discussed in Section 7).

Please note also that the multiplication as defined by (2) is a particularly costly operation – the operations necessary to multiply two intervals are at least 8 floating-point multiplications (4 in the downwards rounding mode and 4 in the upwards one), a min() and a max() operation and some switching of the rounding mode. According to [16] in the Profil/BIAS library a different procedure is used to multiply intervals. It can be described by the following pseudocode:

```
if a ≥ 0 then
      if b ≥ 0 then [c, c̄] = [ab, āb̄];
      if b̄ ≤ 0 then [c, c̄] = [āb, ab̄];
      else [c, c̄] = [āb, āb̄];
if ā ≤ 0 then
      if b ≥ 0 then [c, c̄] = [ab̄, āb];
      if b̄ ≤ 0 then [c, c̄] = [āb̄, ab];
      else [c, c̄] = [ab̄, ab];
else
      if b ≥ 0 then [c, c̄] = [ab̄, āb̄];
      if b̄ ≤ 0 then [c, c̄] = [āb, ab];
      else
            if ab̄ < āb then c = ab̄;
            else c = āb;
            if ab > āb̄ then c̄ = ab;
            else c̄ = āb̄;
```

This procedure reduces the required number of floating-point multiplications. Note that for this algorithm the number of floating-point operations depends heavily on operands values; the most expensive case is when they both contain zero. This implies that when considering the efficiency of interval operations there can be more important factors than proper caching – the number of floating-point operations (comparisons, multiplications, etc.) can be higher or lower.

Finally, as mentioned in Section 2, interval arithmetic requires switching the rounding modes of the processor. This operation – rarely used for floating-point computations – is also time consuming.

Opposite Trick. To reduce the number of rounding mode switches a special representation of an interval was proposed by some authors (see, e.g., [6]): storing an interval $[a, b]$ as a record $\langle -a, b \rangle$. Thanks to the new representation, all operations can be done with the upwards rounding mode (rounding $-a$ upwards is equivalent to rounding a downwards). It is called an "opposite trick".

4   Interval Linear Algebra

Linear algebra operations are commonly used in interval algorithms (see [11], [12], [14]). One important is solving systems of linear equations $Ax = b$, where A is an interval matrix and b is an interval vector. Such systems arise from linearization of nonlinear equation systems (e.g., the interval Newton operator in global optimization) and in problems with uncertain parameters.

Methods of solving such systems (often quite different from methods for non-interval parameters) are out of the scope of this paper (see, e.g., [12], [14]), but a tool they commonly use is – as for non-interval methods – multiplication of two matrices.

Multiplication of two matrices (of numbers, intervals or yet other objects) requires multiplications and additions of their elements (and some intermediate terms, possibly). As it was discussed in Section 2, multiplication of interval quantities is time consuming, which makes matrix multiplication even more costly. This makes several acceleration or rationalization tools very welcome or even necessary. Indeed, available interval packages do not multiply matrices naively, but try to execute this operation in a more efficient way.

4.1   Profil/BIAS

This package decomposes the process of multiplication by an interval matrix to several multiplications by its elements. This approach – completely improper for numbers – is a quite good solution for intervals. It allows to check the sign of interval bounds only once – as in the pseudocode at the end of Section 2 – and multiply the whole matrix without checking it again.

4.2   C-XSC

Until recently, the XSC package offered only naive matrix multiplication. Moreover, a long accumulator was used to store partial results of the product – to execute the calculations with highest possible accuracy. Consequently, matrix multiplication was very precise, but very slow, also.

From version 2.3.0 the package started offering the DotK algorithm [17], which uses the error-free transformation to compute dot products of vectors exactly. The algorithm is quite efficient when implemented properly (at least on some architectures) and probably it will be included in a future IEEE standard for interval arithmetic [18].

Rump Algorithm. A yet more recent addition (from version 2.4.0) to XSC libraries is the possibility of using the Rump algorithm [19] for matrix multiplication. This algorithm requires a matrix in the midpoint-radius representation, i.e. each interval is represented by a midpoint-radius pair, instead of a pair of end points. This representation allows reducing the problem of interval matrix multiplication to four floating-point matrix multiplications with upwards rounding only.

By this virtue, we can set the proper rounding mode and use the highly optimized BLAS routines to multiply the matrices. Obviously, matrices have to be transformed to the midpoint-radius representation and then back again.

The possibility of using BLAS routines comes however at a yet much higher price: the result of the operation in the midpoint-radius representation is highly overestimated – up to the factor of 1.5. Therefore, the use of the Rump algorithm with BLAS routines is a difficult design decision – it is neither the only option nor the default one. In the authors' opinion, the choice of DotK algorithm is superior and this should be even more true for future architectures as there might be hardware support for the exact dot product operation (see, e.g., [13], [18]).

Recently, Nguyen [15] proposed an alternative to the Rump algorithm that requires 9 floating-point matrix multiplications, but the overestimation factor is bounded by 1.18.

Algorithm variants, described in this paragraph will not be considered in the remainder of the paper. Their comparison may be an interesting subject of further research.

4.3   New Possibilities

Both packages offer some optimized procedures to multiply interval matrices. However both of them are far from being optimal as none of them considers making efficient use of the cache. Following [4] we are going to adapt cache-oblivious matrix formats to interval matrices.

As it was already mentioned, these formats and algorithms will not be identical to the ones used for floating-point matrices as they have to take checking sign and switching the rounding mode into account. In addition, for interval matrices we may have a trade-off: the algorithm can compute the result more efficiently, but less accurately or vice-versa [4].

5   Considered Formats

This section presents selected possibilities of applying cache-oblivious formats to interval matrices. More considerations and extensive results can be found in [4].

The following traditional matrix formats will be investigated for the interval content: Z-order, RBR and RPF (Fig. 1).

Also, we consider some ideas specific for interval computations:

- the "opposite trick", mentioned in Section 3, i.e. storing an interval $[a, b]$ as a record $\langle -a, b \rangle$ [6],
- separate matrices of lower and upper endpoints,
- interval Strassen algorithm (the Winograd version) [2].

Several tests have been executed for different combinations of algorithm features.

For recursive formats, the used algorithms are recursive, too, obviously. The recursion is terminated for sufficiently small blocks, which is 64 for the Z-order format and 100 – for other formats. According to [4], these block sizes were close to optimal. Investigation of other block sizes and exploration of the space of algorithms' other parameters may be subject to further research.

Fig. 1. Matrices in Z-order Format, Row-wise Block-Recursive Format (RBR) and Recursive Packed Format (RPF)

6 The Configuration Used for Experiments

All computational experiments were performed on a machine with Intel Core 2 Duo processor (2 GHz) with 3 GB of memory and the following caches:

- two L1 caches, 32 KB each,
- an L2 cache of 4096 KB.

The line length was 64 bytes in both cases. The computer worked under supervision of a Linux operating system – openSUSE 11.0. All algorithms were implemented in C++, using the Profil/BIAS library for interval arithmetic operations. Changing the rounding modes was done in a way known from BIAS routines [4], i.e. using `asm volatile ("fldcw _BiasCwRoundDown")` and `asm volatile ("fldcw _BiasCwRoundUp")` assembler calls, passing proper coprocessor control words.

7 Selected Experiments and Results

First, we present the comparison of variants using the "opposite trick" or not in the multiplication algorithm used in the Profil/BIAS package (see Subsection 4.1). The comparison is presented in Figure 2. The notation is as follows:

- Profil/BIAS 1 – the traditional interval representation with matrices containing zero in all elements,
- Profil/BIAS 1OT – the "opposite trick" interval representation with matrices containing zero in all elements,
- Profil/BIAS 2 – the traditional interval representation with purely positive matrices,
- Profil/BIAS 2OT – the "opposite trick" interval representation with purely positive matrices.

Clearly, use of the opposite trick results in significant speedup.

But for what multiplication algorithms (and what matrix formats) should it be used to optimize the performance? Let us compare the traditional Profil/BIAS algorithm with the Z-order.

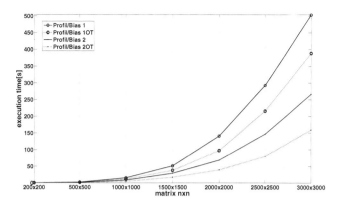

Fig. 2. Comparison of Profil/BIAS algorithms using traditional and "opposite trick" interval representations

Here we use the following notation:

- Profil/BIAS 1OT – the row-major order and matrices containing zero in all elements (non-recursive algorithm),
- Z-order 1OT – the Z-order format and matrices containing zero in all elements (recursive algorithm),
- Profil/BIAS 2 – the row-major order and purely positive matrices (non-recursive algorithm),
- Z-order 2OT – the Z-order format and purely positive matrices (recursive algorithm).

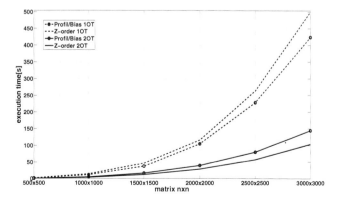

Fig. 3. Comparison of Profil/BIAS algorithms using "opposite trick" interval representation on row-major order and Z-order format

According to Figure 3, matrix multiplication was slightly slower with Z-order than with Profil/BIAS for matrices containing 0 and slightly faster for purely positive matrices.

For the RPF format two variants of the multiplication procedure have been investigated:

– RPF-1 – recursion is used only for triangular submatrices,
– RPF-2 – recursion is used for both triangular and rectangular submatrices.

The comparison with each other and with alternative tools is presented in Figure 4.

The results in Figure 4 show that the Profil/BIAS routine and the RPF-1 format are much slower than the RPF-2 and the RBR formats. This shows that recursion is crucial for the efficiency of this operation, but specific reasons of this phenomenon are uncertain.

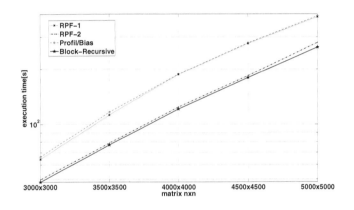

Fig. 4. Comparison of algorithms using RPF, Block-Recursive Format and Profil/BIAS method with "opposite trick"

Yet another representation was considered – two distinct matrices of lower and upper bounds stored separately. All matrices and interval matrices are stored in RBR format. Matrix multiplications are applied in a blocked recursive manner. Figure 5 presents results for the following variants of the separation strategy (for matrices containing zeros):

– Standard – using standard interval format for A, B and C without separation,
– Sep-1 – using distinct matrices of upper and lower bounds for matrices A, B and C; lower and upper bounds for matrix C are computed at the same time,
– Sep-2 – using distinct matrices of upper and lower bounds for C only; A and B are stored in standard interval format without separation; lower bounds on all elements of C are computed, then all upper bounds,
– Sep-3 – using distinct matrices of upper and lower bounds for A, B and C; lower bounds on all elements of C are computed, then all upper bounds.

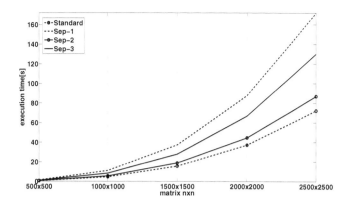

Fig. 5. Results for separate matrices of lower and upper bounds

Figure 5 clearly shows that versions using distinct matrices of upper and lower bounds were slow – at least for matrices containing zeros. Apparently, such a separation requires too many reads from the memory.

Please note that computing lower and upper bounds at the same time (Sep-1) was slower than computing all lower bounds and then all upper bounds (Sep-2 and Sep-3), so minimizing the number of the rounding mode switches is important. Yet the idea of separating matrices in the memory occurred to be inefficient; the version when only matrix C is stored in this way (Sep-2) is more efficient than the version storing all three matrices as pairs of boundary matrices (Sep-3). And the version using no separation at all (Standard) is the most efficient. Apparently, the caching penalty is more important than the speedup related to smaller number of rounding mode switches.

The last version to be tested is the interval Winograd (Strassen) algorithm [2]. Figure 6 compares it with the standard recursive algorithm using rectangular block-recursive data format. Clearly, the Strassen algorithm is less efficient. According to [4], it is less accurate, also. Even for floating-point arithmetic, the Strassen algorithm happens to be numerically inaccurate; features of interval arithmetic (see Section 3) result in highly overestimated (and thus poor) results for this method.

It occurs that the efficient use of cache is significant only when other problems are solved:

- In particular when the "opposite trick" is used.
- For purely positive (or purely negative) matrices caching is more efficient than for the ones that contain zeros.

The interval Strassen algorithm results in overestimated results (dependency problem) and was not efficient either (Figure 6).

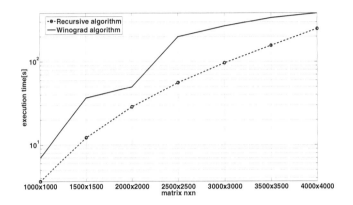

Fig. 6. Results for the interval Strassen algorithm

8   Conclusions

Efficiently and accurately computing the multiplication of interval matrices is a more difficult problem than this operation for floating-point matrices. Commonly used interval libraries try to optimize this operation. The procedures of BIAS – thanks to being somewhat optimized – allow doing this operation far quicker than by naive calculations (see discussion in Section 4), but they do not take caching effects into account.

It turns out that, as caching is not as important for interval matrix multiplication as it is for floating-point matrices, it can still improve the performance of computations. The improvement varied between a few and about 30% (e.g. Figure 4).

Unfortunately, the main hopes of the authors that were to use these formats in interval global optimization algorithms failed. The importance of proper caching becomes noticeable only for matrix sizes far larger than the ones resulting from global optimization problems that are tractable for present algorithms and computers. Nevertheless, for solving large linear systems with interval parameters the proposed formats are going to be very useful.

Comparing the results of the presented algorithms with the methods of Rump or Nguyen and the ones based on error-free transformations might be an interesting subject of future research.

Acknowledgments. The research has been supported by the Polish Ministry of Science and Higher Education under grant N N514 416934.

The author is very grateful to Fred Gustavson, Jerzy Wasniewski and Bo Kågström for support and valuable comments. Also, thanks to anonymous referees for helping to improve the quality of the paper.

References

1. C-XSC interval library, http://www.xsc.de
2. Chatterjee, S., Lebeck, A.R., Patnala, P.K., Thottethodi, M.: Recursive array layouts and fast parallel matrix multiplication. In: Proceedings of Eleventh Annual ACM Symposium on Parallel Algorithms and Architectures, pp. 222–231 (1999)
3. Andersen, B.S., Wasniewski, J., Gustavson, F.G.: A recursive formulation of Cholesky factorization of a matrix in packed storage. ACM Trans. Math. Soft. 27, 214–244 (2001)
4. Dabrowski, R.J.: Cache-oblivious representations of interval matrices. Master's thesis, Warsaw University of Technology (2009) (in polish)
5. Elmroth, E., Gustavson, F., Jonsson, I., Kågström, B.: Recursive blocked algorithms and hybrid data structures for dense matrix library software. SIAM Review 46(1), 3–45 (2004)
6. Goualard, F.: Fast and correct SIMD algorithms for interval arithmetic. In: PARA 2010 Proceedings. Accepted for publication in LNCS (2010)
7. Gustavson, F.G.: Recursion leads to automatic variable blocking for dense linear-algebra algorithms. IBM Journal of Res. Develop. 41(6), 737–755 (1997)
8. Gustavson, F., Henriksson, A., Jonsson, I., Kågström, B., Ling, P.: Recursive Blocked Data Formats and BLAS's for Dense Linear Algebra Algorithms. In: Kågström, B., Elmroth, E., Waśniewski, J., Dongarra, J. (eds.) PARA 1998. LNCS, vol. 1541, pp. 195–206. Springer, Heidelberg (1998)
9. Gustavson, F.G., Waśniewski, J.: Rectangular Full Packed Format for LAPACK Algorithms Timings on Several Computers. In: Kågström, B., Elmroth, E., Dongarra, J., Waśniewski, J. (eds.) PARA 2006. LNCS, vol. 4699, pp. 570–579. Springer, Heidelberg (2007)
10. Gustavson, F.G., Wasniewski, J., Dongarra, J.J., Langou, J.: Rectangular full packed format for Cholesky's algorithm: factorization, solution, and inversion. ACM Trans. Math. Soft. 37(2), article no. 18 (2010)
11. Hansen, E., Walster, W.: Global Optimization Using Interval Analysis. Marcel Dekker, New York (2004)
12. Kearfott, R.B.: Rigorous Global Search: Continuous Problems. Kluwer, Dordrecht (1996)
13. Kulisch, U.: Very fast and exact accumulation of products. Computing 91, 397–405 (2011)
14. Neumaier, A.: Interval Methods for Systems of Equations. Cambridge University Press, Cambridge (1990)
15. Nguyen, H.D.: Efficient Implementation of Interval Matrix Multiplication. In: Jónasson, K. (ed.) PARA 2010, Part II. LNCS, vol. 7134, pp. 179–188. Springer, Heidelberg (2012)
16. Knüppel, O.: Profil/BIAS v 2.0. Technical report, Technische Universität Hamburg-Harburg (1999)
17. Ogita, T., Rump, S.M., Oishi, S.: Accurate sum and dot product. SIAM Journal on Scientific Computing 26, 1955–1988 (2005)
18. Pryce, J. D. (ed.): P1788, IEEE Standard for Interval Arithmetic, http://grouper.ieee.org/groups/1788/email/pdfOWdtH2mOd9.pdf
19. Rump, S.M.: Fast and parallel interval arithmetic. BIT 39(3), 534–544 (1999)

Parallel Solution of Narrow Banded Diagonally Dominant Linear Systems

Carl Christian Kjelgaard Mikkelsen and Bo Kågström

Department of Computing Science and HPC2N
Umeå University, Sweden
{spock,bokg}@cs.umu.se

Abstract. ScaLAPACK contains a pair of routines for solving systems which are narrow banded and diagonally dominant by rows. Mathematically, the algorithm is block cyclic reduction. The ScaLAPACK implementation can be improved using incomplete, rather than complete block cyclic reduction. If the matrix is strictly dominant by rows, then the truncation error can be bounded directly in terms of the dominance factor and the size of the partitions. Our analysis includes new results applicable in our ongoing work of developing an efficient parallel solver.

Keywords: Narrow banded, diagonally dominant linear systems, block cyclic reduction, parallel algorithms, ScaLAPACK.

1 Introduction

Let $A = [a_{ij}]$ be a real n by n matrix and consider the solution of the linear system

$$Ax = f,$$

where $f \in \mathbb{R}^n$. The matrix A is banded with bandwidth $2k + 1$ if

$$|i - j| > k \Rightarrow a_{ij} = 0.$$

The matrix A is diagonally dominant by rows if $|a_{ii}| \geq \sum_{j \neq i} |a_{ij}|$ for all i. If the inequality is sharp, then A is strictly diagonally dominant by rows. If A is diagonally dominant by rows and nonsingular, then $a_{ii} \neq 0$ and the dominance factor ϵ is defined by

$$\epsilon = \max_i \left\{ \frac{1}{|a_{ii}|} \sum_{j \neq i} |a_{ij}| \right\} \in [0, 1].$$

Narrow banded linear systems which are strictly diagonally dominant can be found throughout the physical sciences. In particular, the solution of parabolic PDEs using compact finite difference methods is a rich source of examples.

In general, we cannot assume $\epsilon \ll 1$. However, in this paper we argue that if ϵ is not too close to 1, then incomplete cyclic reduction becomes a viable

K. Jónasson (Ed.): PARA 2010, Part II, LNCS 7134, pp. 280–290, 2012.

alternative to the ScaLAPACK algorithm [2], which is a special case of block cyclic reduction.

We begin by stating a few results on matrices which are strictly diagonally dominant by rows in Section 2. We review the cyclic reduction algorithm by R. W. Hockney and G. H. Golub [4] and provide an elementary extension of a result by Heller [3] in Section 3. We state and prove our main results in Section 4.

This paper builds on the analysis of the truncated SPIKE algorithm by Mikkelsen and Manguoglu [5] and it requires a good understanding of the routines PDDBTRF/PDDBTRS [2,1] from ScaLAPACK, as well as the work of Heller [3]. The truncated SPIKE algorithm (introduced by Polizzi and Sameh [6,7]) also applies to systems which are banded and strictly diagonally dominant by rows.

2   Basic Properties

The following results (proved in [5]) are central to our analysis.

Lemma 1. *Let A be an m by m matrix which is strictly diagonally dominant by rows with dominance factor ϵ. Let $A = LU$ be the LU factorization of A. Then U is strictly diagonally dominant by rows and ϵ_U, the dominance factor of U, satisfies $\epsilon_U \leq \epsilon$.*

Lemma 2. *Let A be an m by m matrix and let B be an m by n matrix such that $[A, B]$ is strictly diagonally dominant by rows with dominance factor ϵ. Then $C = [I, A^{-1}B]$ is strictly diagonally dominant by rows and ϵ_C, the dominance factor of C, satisfies $\epsilon_C \leq \epsilon$.*

Lemma 3. *Let A be a banded matrix with bandwidth $2k + 1$ which is strictly diagonally dominant by rows with dominance factor ϵ, and let A be partitioned in the block tridiagonal form*

$$A = \begin{bmatrix} D_1 & F_1 & & \\ E_2 & \ddots & \ddots & \\ & \ddots & \ddots & F_{m-1} \\ & & E_m & D_m \end{bmatrix} \tag{1}$$

with block size $\mu = qk$ for an integer $q > 0$. Moreover, let $\begin{bmatrix} U_i & V_i \end{bmatrix}$ be the solution of the linear system

$$D_i \begin{bmatrix} U_i & V_i \end{bmatrix} = \begin{bmatrix} E_i & F_i \end{bmatrix},$$

where E_1, U_1, F_m, and V_m are undefined and should be treated as zero, and let

$$U_i = \left(U_{i,1}^T, U_{i,2}^T, \ldots, U_{i,q}^T\right)^T, \quad V_i = \left(V_{i,1}^T, V_{i,2}^T, \ldots, V_{i,q}^T\right)^T$$

be a partitioning of U_i and V_i into blocks each consisting of k rows. Then

$$\|U_{i,j}\|_\infty \leq \epsilon^j, \quad \|V_{i,j}\|_\infty \leq \epsilon^{q-(j-1)}, \quad j = 1, 2, \ldots, q.$$

3   Block Cyclic Reduction

Mathematically, the algorithm used by PDDBTRF/PDDBTRS is a special case of block cyclic reduction [4] which we briefly review below. In addition, we present an extension of a relevant result by D. Heller on incomplete block cyclic reduction [3].

Let A be an m by m block tridiagonal matrix in the form (1) which is also strictly diagonally dominant by rows, and let D be the matrix given by

$$D = \mathrm{diag}(D_1, D_2, \ldots, D_m)$$

and consider the auxiliary matrix B defined by

$$B = D^{-1}(A - D) = \begin{bmatrix} 0 & D_1^{-1}F_1 & & & \\ D_2^{-1}E_2 & 0 & D_2^{-1}F_2 & & \\ & D_3^{-1}E_3 & \ddots & & \ddots & \\ & & \ddots & & \ddots & D_{m-1}^{-1}F_{m-1} \\ & & & D_m^{-1}E_m & 0 \end{bmatrix}.$$

The norm of the matrix B measures the significance of the off diagonal blocks of A. Specifically, let $f \in \mathbb{R}^n$ and let x and y be the solutions of the linear systems

$$Ax = f, \quad Dy = f.$$

Then,

$$x - y = x - D^{-1}Ax = (I - D^{-1}A)x = D^{-1}(D - A)x = -Bx,$$

which for all $x \neq 0$ implies that

$$\frac{\|x - y\|_\infty}{\|x\|_\infty} \leq \|B\|_\infty.$$

Therefore, if $\|B\|_\infty$ is sufficiently small, then y is a good approximation of x. The linear systems $D_i y_i = f_i$, $i = 1, 2, \ldots, m$ can be solved concurrently on different processors without any communication. Therefore, the block diagonal linear system $Dy = f$ is even more suitable for parallel computing than the original linear system $Ax = f$.

We illustrate block cyclic reduction in the case of $m = 7$. Let P denote the matrix which represents the usual odd-even permutation σ of the blocks, i.e.

$$\sigma = (1, 3, 5, 7, 2, 4, 6),$$

and define A' by

$$A' := PAP^T = \left[\begin{array}{cccc|ccc} D_1 & & & & F_1 & & \\ & D_3 & & & E_3 & F_3 & \\ & & D_5 & & & E_5 & F_5 \\ & & & D_7 & & & E_7 \\ \hline E_2 & F_2 & & & D_2 & & \\ & E_4 & F_4 & & & D_4 & \\ & & E_6 & F_6 & & & D_6 \end{array}\right] = \begin{bmatrix} A'_{11} & A'_{12} \\ \hline A'_{21} & A'_{22} \end{bmatrix}.$$

The Schur complement of A' is the block tridiagonal matrix $A^{(1)}$ given by

$$A^{(1)} := A'_{11} - A'_{21} A'^{-1}_{22} A'_{12} = \begin{bmatrix} D_1^{(1)} & F_1^{(1)} & \\ E_2^{(1)} & D_2^{(1)} & F_2^{(1)} \\ & E_3^{(1)} & D_3^{(1)} \end{bmatrix},$$

where

$$D_i^{(1)} = D_{2i} - E_{2i} D_{2i-1}^{-1} F_{2i-1} - F_{2i} D_{2i+1}^{-1} E_{2i+1}$$

and

$$E_i^{(1)} = -E_{2i} D_{2i-1}^{-1} E_{2i-1}, \quad F_i^{(1)} = -F_{2i} D_{2i+1}^{-1} F_{2i+1}.$$

Heller [3] showed that if A is strictly diagonally dominant by rows, then block cyclic reduction is well defined and the new auxiliary matrix

$$B^{(1)} = D^{(1)^{-1}} (A^{(1)} - D^{(1)}),$$

with $D^{(1)} = \operatorname{diag}(D_1^{(1)}, D_2^{(1)}, D_3^{(1)})$, satisfies

$$\|B^{(1)}\|_\infty \leq \|B^2\|_\infty \leq \|B\|_\infty^2.$$

In addition, Heller [3] showed that if A is strictly diagonally dominant by rows, then the initial matrix B satisfies $\|B\|_\infty < 1$ and the significance of the off diagonal blocks decays quadratically to zero.

We have found that it is possible to explicitly incorporate the dominance factor into the analysis. For the sake of notational simplicity we define U_i and V_i as the solution of the linear system

$$D_i \begin{bmatrix} U_i & V_i \end{bmatrix} = \begin{bmatrix} E_i & F_i \end{bmatrix},$$

where E_1, U_1, F_m, and V_m are undefined and should be treated as zero. It follows that

$$\|B^{(1)}\|_\infty = \max \|Z_i\|_\infty,$$

where

$$Z_i = (I - U_{2i} V_{2i-1} - V_{2i} U_{2i+1})^{-1} \begin{bmatrix} U_{2i} U_{2i-1}, & V_{2i} V_{2i+1} \end{bmatrix}.$$

Therefore,

$$Z_i = \begin{bmatrix} U_{2i} & V_{2i} \end{bmatrix} \begin{bmatrix} V_{2i-1} \\ U_{2i+1} \end{bmatrix} Z_i + \begin{bmatrix} U_{2i} & V_{2i} \end{bmatrix} \begin{bmatrix} U_{2i-1} & 0 \\ 0 & V_{2i+1} \end{bmatrix}$$

$$= \begin{bmatrix} U_{2i} & V_{2i} \end{bmatrix} \begin{bmatrix} V_{2i-1} & U_{2i-1} & 0 \\ U_{2i+1} & 0 & V_{2i+1} \end{bmatrix} \begin{bmatrix} Z_i \\ I \end{bmatrix}.$$

The right hand side can be estimated using Lemma 2. We have

$$\|Z_i\|_\infty \leq \epsilon^2 \max\{\|Z_i\|_\infty, 1\}. \tag{2}$$

However, if we assume $\|Z_i\|_\infty \geq 1$, then (2) reduces to

$$\|Z_i\|_\infty \leq \epsilon^2 \|Z_i\|_\infty$$

which forces the contradiction $\|Z_i\|_\infty = 0$, simply because $\epsilon < 1$. Therefore, $\|Z_i\|_\infty < 1$, which inserted into (2) yields

$$\|Z_i\|_\infty \leq \epsilon^2.$$

It follows, that

$$\|B^{(1)}\|_\infty \leq \epsilon^2. \tag{3}$$

This estimate is tight and equality is achieved for matrices of the form

$$A = \begin{bmatrix} I_k & \epsilon I_k & & & \\ & \ddots & \ddots & & \\ & & \ddots & \epsilon I_k \\ & & & I_k \end{bmatrix}.$$

4   Preliminary Analysis of the ScaLAPACK Routine PDDBTRF

The ScaLAPACK routine PDDBTRF can be used to obtain a factorization of a narrow banded matrix A which is diagonally dominant by rows, [1].

Mathematically, the algorithm is block cyclic reduction applied to a special partitioning of the matrix, which is designed to exploit the banded structure. Specifically, the odd numbered blocks are very large, say, of dimension $\mu = qk$, where $q \gg 1$ is a large positive integer, while the even numbered blocks have dimension k.

The large odd numbered diagonal blocks can be factored in parallel without any communication. It is the construction and factorization of the Schur complement $A^{(1)}$ which represents the parallel bottleneck. Obviously, the factorization can be accelerated, whenever the off diagonal blocks can be ignored. Now, while we do inherit the estimate

$$\|B^{(1)}\|_\infty \leq \epsilon^2$$

from the previous analysis, this estimate does not take the banded structure into account. We have the following theorem.

Theorem 1. *Let A be a tridiagonal matrix which is strictly diagonally dominant by rows. Then the significance of the off diagonal blocks of the initial Schur complement is bounded by*

$$\|B^{(1)}\|_\infty \leq \epsilon^{1+q}.$$

Proof. We must show that Z_i given by

$$Z_i = (I - U_{2i}V_{2i-1} - V_{2i}U_{2i+1})^{-1} [U_{2i}U_{2i-1},\ V_{2i}V_{2i+1}]$$

satisfies

$$\|Z_i\|_\infty \leq \epsilon^{1+q}.$$

We solve this optimization problem by partitioning it into two subproblems, which can be solved by induction.

We begin by making the following very general estimate

$$\|Z_i\|_\infty \leq \frac{\|U_{2i}\|_\infty\|U_{2i-1}^{(b)}\|_\infty + \|V_{2i}\|_\infty\|V_{2i+1}^{(t)}\|_\infty}{1 - \|U_{2i}\|_\infty\|V_{2i-1}^{(b)}\|_\infty - \|V_{2i}\|_\infty\|U_{2i+1}^{(t)}\|_\infty},$$

where the notation $U^{(t)}$ ($U^{(b)}$) is used to identify the matrix consisting of the top (bottom) k rows of the matrix U. This estimate is easy to verify, but it relies on the zero structure of the matrices U_{2i} and V_{2i}. Now, let

$$\alpha = \|U_{2i-1}^{(b)}\|_\infty, \quad \beta = \|V_{2i-1}^{(b)}\|_\infty, \quad \gamma = \|U_{2i+1}^{(t)}\|_\infty, \quad \delta = \|V_{2i+1}^{(t)}\|_\infty,$$

and define an auxiliary function

$$g(x,y) = \frac{\alpha x + \delta y}{1 - \beta x - \gamma y},$$

where the appropriate domain will be determined shortly. If

$$x = \|U_{2i}\|_\infty, \qquad y = \|V_{2i}\|_\infty$$

then by design

$$\|Z_i\|_\infty \leq g(x,y).$$

In general, Lemma 1 implies that

$$\|[U_{2i}, V_{2i}]\|_\infty \leq \epsilon$$

but in the current case of $k = 1$, this follows directly from the definition of strict diagonal dominance. Regardless, we see that the natural domain for g is the closure of the set Ω given by

$$\Omega = \{(x,y) \in \mathbb{R}^2 : 0 < x \land 0 < y \land x + y < \epsilon\}.$$

It suffices to show that $g(x,y) \leq \epsilon^q$ for all $(x,y) \in \bar{\Omega}$. It is clear, that g is well defined and $g \in C^\infty(\bar{\Omega})$, simply because $\beta \leq \epsilon$, $\gamma \leq \epsilon$ and $\epsilon < 1$, so that we never divide by zero.

Now, does g assume its maximum within Ω? We seek out any stationary points. We have

$$\frac{\partial g}{\partial x}(x,y) = \frac{\alpha - (\alpha\gamma - \beta\delta)y}{(1 - \beta x - \gamma y)^2}, \quad \text{and} \quad \frac{\partial g}{\partial y}(x,y) = \frac{\delta + (\alpha\gamma - \beta\delta)x}{(1 - \beta x - \gamma y)^2}.$$

Therefore, there are now two distinct scenarios, namely

$$\alpha\gamma - \beta\delta = 0 \quad \text{or} \quad \alpha\gamma - \beta\delta \neq 0.$$

If $\alpha\gamma - \beta\delta = 0$, then there are no stationary points, unless $\alpha = \delta = 0$, in which case $g \equiv 0$ and there is nothing to prove. If $\alpha\gamma - \beta\delta \neq 0$, then

$$(x_0, y_0) = \left(\frac{\alpha}{\alpha\gamma - \beta\delta}, -\frac{\delta}{\alpha\gamma - \beta\delta} \right)$$

is the only candidate, but $(x_0, y_0) \notin \Omega$, simply because

$$x_0 y_0 = -\frac{\alpha\delta}{(\alpha\gamma - \beta\delta)^2} \leq 0$$

is not strictly positive. In both cases, we conclude that the global maximum for g is assumed on the boundary of Ω.

The boundary of Ω consists of three line segments. We examine them one at a time. We begin by defining

$$g_1(y) = g(0, y) = \frac{\delta y}{1 - \gamma y}, \quad y \in [0, \epsilon].$$

Then

$$g_1'(y) = \frac{\delta(1 - \gamma y) - \delta y(-\gamma)}{(1 - \gamma y)^2} = \frac{\delta}{(1 - \gamma y)^2} \geq 0$$

and we conclude that

$$g_1(y) \leq g_1(\epsilon) = g(0, \epsilon) = \frac{\delta\epsilon}{1 - \gamma\epsilon}.$$

Similarly, we define

$$g_2(x) = g(x, 0) = \frac{\alpha x}{1 - \beta x}, \quad x \in [0, \epsilon].$$

Then

$$g_2'(x) = \frac{\alpha(1 - \beta x) - \alpha x(-\beta)}{(1 - \beta x)^2} = \frac{\alpha}{(1 - \beta x)^2} \geq 0$$

which allows us to conclude that

$$g_2(x) \leq g_2(\epsilon) = \frac{\epsilon\alpha}{1 - \epsilon\beta}.$$

Finally, we let $s \in [0, \epsilon]$ and define

$$g_3(s) = g(s, \epsilon - s) = \frac{\alpha s + \delta(\epsilon - s)}{1 - \beta s - \gamma(\epsilon - s)} = \frac{(\alpha - \delta)s + \delta\epsilon}{1 - \gamma\epsilon - (\beta - \gamma)s}.$$

Then

$$g_3'(s) = \frac{(\alpha - \delta)(1 - \gamma\epsilon) + \delta\epsilon(\beta - \gamma)}{(1 - \gamma\epsilon - (\beta - \gamma)s)^2}.$$

Therefore, g_3 is either a constant or strictly monotone. In either case

$$g_3(s) \leq \max\{g_3(0), g_3(\epsilon)\}.$$

We can now conclude that for all $(x, y) \in \bar{\Omega}$: $g(x, y) \leq \max\{g(\epsilon, 0), g(0, \epsilon)\}$. We will only show that

$$g(\epsilon, 0) = \frac{\alpha\epsilon}{1 - \beta\epsilon} = \frac{\epsilon\|U_{2i-1}^{(b)}\|_\infty}{1 - \epsilon\|V_{2i-1}^{(b)}\|_\infty} \leq \epsilon^{1+q}$$

simply because the other case is similar. The proof is by induction on q, i.e. the size of the odd numbered partitions. Let

$$\begin{bmatrix} c_1 & a_1 & b_1 & & \\ & c_2 & a_2 & b_2 & \\ & & \ddots & \ddots & \ddots \\ & & & c_q & a_q & b_q \end{bmatrix}$$

be a representation of the $(2i - 1)$th block row of the original matrix A. Using Gaussian elimination without pivoting we obtain the matrix

$$\begin{bmatrix} c_1' & a_1' & b_1' & & \\ c_2' & & a_2' & b_2 & \\ c_3' & & & a_3' & b_3 & \\ \vdots & & & & \ddots & \ddots \\ c_q' & & & & & a_q' & b_q \end{bmatrix}.$$

Now, let V be the set given by

$$V = \left\{ j \in \{1, 2, \ldots, q\} \ : \ \frac{\epsilon|c_j'|/|a_j'|}{1 - \epsilon|b_j|/|a_j'|} \leq \epsilon^{1+j}, \quad j = 1, 2, \ldots, q \right\}.$$

We claim that $V = \{1, 2, \ldots, q\}$. We begin by showing that $1 \in V$. Let

$$x = |c_1'|/|a_1'|, \quad y = |b_1|/|a_1'|.$$

Then $(x, y) \in \bar{\Omega}$ and it is straightforward to show that

$$\frac{\epsilon x}{1 - \epsilon y} \leq \epsilon^2.$$

Now, suppose that $j \in V$ for some $j < q$. Does $j + 1 \in V$? We have

$$a_{j+1}' = a_{j+1} - c_{j+1}\frac{b_j}{a_j'}, \quad c_{j+1}' = -c_{j+1}\frac{c_j'}{a_j'},$$

which implies

$$\frac{\epsilon|c_{j+1}'|/|a_{j+1}'|}{1 - \epsilon|b_{j+1}|/|a_{j+1}'|} = \frac{\epsilon|c_{j+1}'|}{|a_{j+1}'| - \epsilon|b_{j+1}|} = \frac{\epsilon|c_{j+1}|(|c_j'|/|a_j'|)}{|a_{j+1} - c_{j+1}(b_j/a_j')| - \epsilon|b_{j+1}|}$$

$$\leq \frac{\epsilon|c_{j+1}|(|c_j'|/|a_j'|)}{|a_{j+1}| - |c_{j+1}||b_j|/|a_j'| - \epsilon|b_{j+1}|}$$

$$= \frac{\epsilon(|c_{j+1}|/|a_{j+1}|)(|c_j'|/|a_j'|)}{1 - (|c_{j+1}|/|a_{j+1}|)(|b_j|/|a_j'|) - \epsilon(|b_{j+1}|/|a_{j+1}|)}.$$

We simplify the notation by introducing

$$\nu = |c'_j|/|a'_j|, \quad \mu = |b_j|/|a'_j|, \quad x = |c_{j+1}|/|a_{j+1}|, \quad y = |b_{j+1}|/|a_{j+1}|,$$

and defining

$$h(x,y) = \frac{\epsilon x \nu}{1 - x\mu - \epsilon y}.$$

By the strict diagonal dominance of A, we have $(x,y) \in \bar{\Omega}$ and it is easy to see that $h(x,y) \le h(\epsilon,0)$. Therefore

$$\frac{\epsilon x \nu}{1 - x\mu - \epsilon y} \le \frac{\epsilon^2 \nu}{1 - \epsilon\mu} = \epsilon \left(\frac{\epsilon \nu}{1 - \epsilon\mu} \right) \le \epsilon \cdot \epsilon^{1+j} = \epsilon^{1+(j+1)}$$

and $j + 1 \in V$. By the well ordering principle, $V = \{1, 2, \ldots, q\}$.

In view of Theorem 1 we make the following conjecture.

Conjecture 1. The auxiliary matrix corresponding to the initial Schur complement generated by the ScaLAPACK routine PDDBTRF satisfies

$$\|B^{(1)}\|_\infty \le \epsilon^{1+q},$$

where $\mu = qk$ is the size of the odd number partitions and $\epsilon < 1$ is the dominance factor.

This is one possible generalization of the case $q = 1$ to the case $q > 1$ and it does reduce to Theorem 1 in the case of $k = 1$. The proof of Conjecture 1 for the general case is ongoing work. So far, we have derived the following results.

Theorem 2. *If A is strictly diagonally dominant by rows and banded with bandwidth $(2k + 1)$ then*

$$\|B^{(1)}\|_\infty \le \frac{\epsilon^{1+q}}{1 - \epsilon^2}.$$

Proof. By definition

$$Z_i = [U_{2i}, \, V_{2i}] \left\{ \begin{bmatrix} V_{2i-1} \\ U_{2i+1} \end{bmatrix} Z_i + \begin{bmatrix} \tilde{U}_{2i-1} & 0 \\ 0 & \tilde{V}_{2i+1} \end{bmatrix} \right\}.$$

Therefore

$$\|Z_i\|_\infty \le \epsilon\{\epsilon\|Z_i\|_\infty + \epsilon^q\} = \epsilon^2\|Z_i\|_\infty + \epsilon^{1+q}$$

and the proof follows immediately from the fact that $\epsilon < 1$.

It is the elimination of the singularity at $\epsilon = 1$ which is proving difficult in the case of $k > 1$. Specifically, it is the decomposition into two separate subproblems which is difficult to achieve for $k > 1$.

In the case of matrices which are both banded and triangular Conjecture 1 is trivially true.

Theorem 3. *If A is a strictly upper (lower) triangular banded matrix with dominance factor ϵ and upper (lower) bandwidth k, then*

$$\|B^{(1)}\|_\infty \le \epsilon^{1+q},$$

where $\mu = qk$ is the size of the odd numbered partitions.

In general, we have the following theorem.

Theorem 4. *Dropping the off diagonal blocks in the initial Schur complement is equivalent to replacing the original matrix A with a perturbed matrix $A + \Delta A$ for which*

$$\|\Delta A\|_\infty \le \epsilon^{1+q}\|A\|_\infty.$$

Proof. By definition

$$\left[E_i^{(1)}, \ F_i^{(1)}\right] = -\left[E_{2i}D_{2i-1}^{-1}E_{2i-1}, \ F_{2i}D_{2i+1}^{-1}F_{2i+1}\right] = -\left[E_{2i}U_{2i-1}, \ F_{2i}V_{2i+1}\right].$$

However, the zero structure of E_{2i} and F_{2i} implies that

$$[E_{2i}U_{2i-1}, \ F_{2i}V_{2i+1}] = \left[E_{2i}\tilde{U}_{2i-1}, \ F_{2i}\tilde{V}_{2i+1}\right],$$

where

$$\tilde{U}_{2i-1} = \begin{bmatrix} 0 \\ U_{2i-1}^{(b)} \end{bmatrix}, \quad \tilde{V}_{2i+1} = \begin{bmatrix} V_{2i+1}^{(t)} \\ 0 \end{bmatrix}$$

isolate the bottom k rows of U_{2i-1} and the top k rows of V_{2i+1}. By Lemma 3

$$\|U_{2i-1}^{(b)}\|_\infty \le \epsilon^q, \quad \|V_{2i+1}^{(t)}\|_\infty \le \epsilon^q.$$

It follows that

$$\left[E_i^{(1)}, \ F_i^{(1)}\right] = -D_{2i}\left[U_{2i}, \ V_{2i}\right]\begin{bmatrix} \tilde{U}_{2i-1} & 0 \\ 0 & \tilde{V}_{2i+1} \end{bmatrix}$$

which implies

$$\left\|\left[E_i^{(1)}, \ F_i^{(1)}\right]\right\|_\infty \le \epsilon^{1+q}\|D_{2i}\|_\infty \le \epsilon^{1+q}\|A\|_\infty$$

and the proof is complete.

Conjecture 1 is interesting in precisely those cases where the relative backward error bound given in Theorem 4 is small, but not small enough, to satisfy the demands of the user. If the conjecture is correct, then the sequence of Schur complements generated by the ScaLAPACK algorithm will satisfy

$$\|B^{(k)}\|_\infty \le (\epsilon^{1+q})^{2^{k-1}}, \quad k = 1, 2, \ldots$$

Therefore, if ϵ is not too close to 1, then a few steps of cyclic reduction will permit us to drop the off diagonal blocks, thus facilitating a parallel solve. In addition, if we increase the number of processors by a factor of 2, then we must replace q with $q' \approx q/2$, but the accuracy can be maintained by executing a single extra step of cyclic reduction.

5   Future Work

We have shown that incomplete cyclic reduction is applicable to tridiagonal $(k = 1)$ linear systems which are diagonally dominant by rows and we identified the worst case behavior. Ongoing work includes extending our analysis to the general case of $k > 1$. In addition, we are developing a parallel implementation of incomplete, rather than complete cyclic reduction for narrow banded systems. The factorization phase will feature an explicit calculation of the auxiliary matrices, in order to determine the minimal number of reduction steps necessary to achieve a given accuracy.

Acknowledgments. The authors would like to thank Lars Karlsson as well as Meiyue Shao from the Umeå Research Group in Parallel and Scientific Computing for their input during several related discussions. The work is supported by eSSENCE, a collaborative e-Science programme funded by the Swedish Research Council within the framework of the strategic research areas designated by the Swedish Government. In addition, support has been provided by the Swedish Foundation for Strategic Research under the frame program A3 02:128 and the EU Mål 2 project UMIT.

References

1. Arbenz, P., Cleary, A., Dongarra, J., Hegland, M.: A Comparison of Parallel Solvers for Diagonally Dominant and General Narrow-Banded Linear Systems II. In: Amestoy, P., Berger, P., Daydé, M., Duff, I., Frayssé, V., Giraud, L., Ruiz, D. (eds.) Euro-Par 1999. LNCS, vol. 1685, pp. 1078–1087. Springer, Heidelberg (1999)
2. Blackford, L.S., Choi, J., Cleary, A., D'Azevedo, E., Demmel, J., Dhillon, I., Dongarra, J., Hammarling, S., Henry, G., Petitet, A., Stanley, K., Walker, D., Whaley, R.C.: ScaLAPACK User's Guide. SIAM, USA (1997)
3. Heller, D.: Some aspects of the cyclic reduction algorithm for block tridiagonal linear systems. SIAM J. Numer. Anal. 13(4), 484–496 (1976)
4. Hockney, R.W.: A fast direct solution of Poisson's equation using Fourier analysis. J. Assoc. Comput. Mach. 12(1), 95–113 (1965)
5. Mikkelsen, C.C.K., Manguoglu, M.: Analysis of the truncated SPIKE algorithm. SIAM J. Matrix Anal. Appl. 30(4), 1500–1519 (2008)
6. Polizzi, E., Sameh, A.: A parallel hybrid banded system solver: The SPIKE algorithm. Parallel Comput. 32(2), 177–194 (2006)
7. Polizzi, E., Sameh, A.: SPIKE: A parallel environment for solving banded linear systems. Comput. Fluids 36(1), 113–120 (2007)

An Approach for Semiautomatic Locality Optimizations Using OpenMP

Jens Breitbart

Research Group Programming Languages / Methodologies
Universität Kassel
Kassel, Germany
jbreitbart@uni-kassel.de

Abstract. The processing power of multicore CPUs increases at a high rate, whereas memory bandwidth is falling behind. Almost all modern processors use multiple cache levels to overcome the penalty of slow main memory; however cache efficiency is directly bound to data locality. This paper studies a possible way to incorporate data locality exposure into the syntax of the parallel programming system OpenMP. We study data locality optimizations on two applications: matrix multiplication and Gauß-Seidel stencil. We show that only small changes to OpenMP are required to expose data locality so a compiler can transform the code. Our notion of tiled loops allows developers to easily describe data locality even at scenarios with non-trivial data dependencies. Furthermore, we describe two optimization techniques. One explicitly uses a form of local memory to prevent conflict cache misses, whereas the second one modifies the wavefront parallel programming pattern with dynamically sized blocks to increase the number of parallel tasks. As an additional contribution we explore the benefit of using multiple levels of tiling.

1 Introduction

In the last years multicore CPUs became the standard processing platform for both end user systems and clusters. However, whereas the available processing power in CPUs continues to grow at a rapid rate, the DRAM memory bandwidth is increasing rather slowly. CPUs use multiple levels of fast on-chip cache memory to overcome the penalty of the slow main memory. Caches, however, are only useful if the program exposes data locality, so reoccurring accesses to the data may be fetched from cache instead of main memory.

Parallel programming without considering data locality provides far from optimal performance even on systems based on Intel's Nehalem architecture, which provides a highly improved memory subsystem compared to all previous Intel architectures. In this paper we use two well-known applications – a matrix-matrix multiplication and a Gauß-Seidel stencil – to demonstrate tiling and its effect. We utilize two levels of tiling to increase performance in scenarios for which single-level tiling is not already close to the maximum memory/cache bandwidth. Explicit tiling is a technique to overcome conflict cache misses. Explicit

K. Jónasson (Ed.): PARA 2010, Part II, LNCS 7134, pp. 291–301, 2012.

Table 1. Memory bandwidth of the test system measured with the STREAM benchmark

Test	copy	scale	add	triad
Bandwidth (MB/s)	18463	19345	17949	18356

tiling requires duplicating data, but still offers an overall increase in performance. Furthermore we provide an example to demonstrate how the shown locality optimizations can be integrated into e. g. the OpenMP syntax with only small changes. The approach however is not limited to OpenMP and could be implemented into other pragma-based systems or as a standalone system, as well. As a result of our OpenMP enhancements, developers must only hint the compiler at what kind of changes are required to achieve the best performance. We expect this to ease locality optimizations. At the time of this writing, the changes have not been implemented in a compiler, but the required transformations are described in detail.

The paper is organized as follows. First, Sect. 2 describes the hardware used for our experiments. Our applications, and the used locality optimizations and results are shown in Sect. 3 for matrix multiplication and in Sect. 4 for the Gauß-Seidel stencil, respectively. Section 5 describes the changes to OpenMP to allow compilers to automatically change code the way we have described in Sects. 3 and 4. The paper finishes with an overview of related work and conclusions, in Sects. 6 and 7, respectively.

2   Hardware Setup

All experiments shown were conducted at an Intel Core i7 920, which is a CPU based on the Nehalem architecture. It is a quad core CPU running at 2.67 GHz using a three-ary cache hierarchy. The level 3 cache can store up to 8 MB of data and is shared by all cores of the CPU. In contrast every core has its own level 2 (256 KB) and level 1 cache (32 KB for data). The caches are divided into cache lines of 64 bytes, so in case a requested data element is not in the cache, the system fetches the cache line storing it from main memory. Both level 1 and level 2 cache are 8-way associative. The level 3 cache is 16-way associative. The Nehalem is Intel's first x86 architecture featuring an on-chip memory controller to lower memory access latency. Table 1 shows memory bandwidth measured with the STREAM benchmark [8] using 4 threads. We used three memory channels supplied with DDR3-1333 memory modules. Each CPU core provides additional hardware support to run up to two threads efficiently (SMT), so the CPU supports up to 8 threads effectively.

3   Matrix Multiplication

We describe our first set of locality optimizations, for which OpenMP pragmas are shown in Sect. 5. Their performance is shown by implementing them in a trivial matrix-matrix multiplication calculating $A * B = C$ with A, B, and C being square matrices of size n. Our trivial implementation uses three nested loops, which are ordered i-k-j with the i- and j-loop looping over C and the k-loop being used to access a column/row of A/B. This loop order allows for easy automatic vectorization and good data locality, for details see e.g. [11]. The parallelization of the matrix multiplication is obvious, as each matrix element can be calculated independently.

The trivial matrix multiplication algorithm is well-known to have low memory locality, since successive accesses to elements of A, B are too far away to keep data in cache [11]. A well-known optimization to improve data locality and cache reuse in such scenarios is tiling [2], also known as loop blocking. Tiling divides the loop iteration space into tiles and transforms the loop to iterate over the tiles. Thereby a loop is split into two loops: an outer loop which loops over tiles and an inner loop which loops over the elements of a tile. We applied loop tiling to our matrix multiplication and refer to the tiles as A_{sub}, B_{sub} and C_{sub}, respectively.

Tiling improves the performance of our matrix multiplication in almost all cases. Measurements (not depicted here) show that there are hardly any L2 cache misses, but the performance suffers from regular L1 cache misses. The phenomenon is measured at any block size, even when choosing a block size that fits into the L1 data cache and its impact is at maximum, when the matrix size is a power of two. Therefore, we expect these misses to be conflict misses, as array sizes of a power of two are worst case scenarios for current CPU caches [4]. Conflict misses arise due to the associativity of the cache, meaning that there are multiple data elements in A_{sub} and B_{sub} that are mapped to the same position in the cache and thereby replace each other even though there is enough space in the cache available.

To resolve this problem we explicitly copy the data used during the calculation into a newly allocated array – in our example we allocate an array of size $blocksize * blocksize$ to store both A_{sub} and B_{sub} and copy the submatrix from A/B to the newly allocated array. In the calculation we only access the newly allocated arrays. We refer to this optimization as explicit blocking. By using explicit block, we no longer suffer from conflict cache misses during the calculations.

To improve the performance further, we added a second level of tiling. This is a well known technique, which we refer to as multilevel tiling during this work. We combined small explicitly tiled blocks that fit into L1 cache with larger tiles that help to increase data locality in L2 and L3 cache. Our final implementation thereby uses large blocks that are split into smaller subblocks.

Figure 1 shows the performance of our code with the matrix data being stored in a one dimensional array, compiled with the Intel Compiler version 11.1. The performance was measured using matrices of size 8192^2 at both single and double precision. We compare our results with two libraries: Intel's Math Kernel Library[1] (MKL) and TifaMMy [3]. Both libraries use hand-tuned code and

---

[1] http://software.intel.com/en-us/intel-mkl/

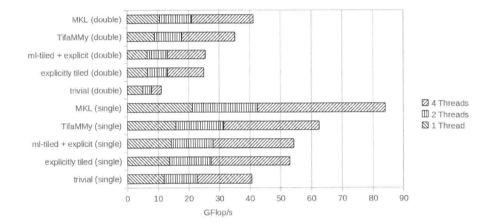

Fig. 1. Matrix multiplication performance (matrix size: 8192^2)

thereby obviously outperform our implementation, which purely relies on the Intel compiler to generate optimized code. TifaMMy has not been optimized for Nehalem-based CPUs and therefore falls behind Intel's MKL. Our trivial version consists only of three nested loops, however the compiler automatically applies tiling to the loop.

The code changes to achieve the performance increase are rather trivial, but yet require writing multiple lines of code. For example applying explicit tiling requires developers to create two new arrays and to copy data. Even though the implementation is simple, it is still a source of bugs. Furthermore, the compiler did not automatically detect that these transformations should be applied and OpenMP does not offer us a way to express these. Some compilers offer ways to specify loop tiling, however using compiler specific options obviously is not-compiler compatible and even interfere with the OpenMP parallelization in an unpredictable way.

4   Gauß-Seidel Stencil

The calculations of the Gauß-Seidel stencil are being applied on a two-dimensional data matrix V with the borders having fixed values. The non-border element with the coordinates (x, y) in the kth iteration is calculated by

$$
V_{x,y}^k = \frac{V_{x-1,y}^k + V_{x+1,y}^{k-1} + V_{x,y-1}^{k-1} + V_{x,y+1}^k}{4}.
$$

This calculation is repeated for a fixed number of steps or until it converges. Considering the low arithmetic intensity of the calculations, it should come at no surprise that the runtime is limited by memory bandwidth and not by the available processing power. We consider a bandwidth-limited application an

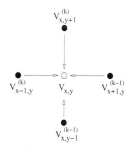

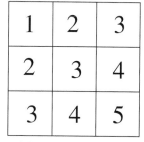

(a) Gauß-Seidel data dependencies

(b) Wavefront pattern

Fig. 2. Gauß-Seidel

interesting scenario, as the gap between available processing power and memory bandwidth is increasing and more scenarios will become bandwidth-limited in the future.

We measure the performance of the Gauß-Seidel stencil by stencils computed per second (Ste/s). We again applied the locality optimizations described in the last section, to our sequential implementation and as a result we increased the performance from 283 MSte/s to 415 MSte/s.

Figure 2(a) shows a visualization of the data dependencies of the Gauß-Seidel stencil. It is important to notice that the upper and the left values are from the current iteration, whereas the right and bottom value are from the previous iterations. The parallelization requires using the wavefront pattern [9]. In a wavefront, the data matrix is divided into multiple diagonals as shown in Fig. 2(b). The elements in one diagonal can be calculated in parallel. Tiling can again be applied to increase data locality by creating tiles of V.

We implemented both a strict and relaxed version of the wavefront pattern. The strict version directly follows the wavefront pattern and only calculates the diagonals in parallel, whereas the relaxed version breaks up the diagonals. In the relaxed version we split up the matrix in x-dimension in multiple columns and assign these columns to threads with a round robin scheduling and use one counter per column, which indicates how much of that column has already been updated in the current iteration. These counters are shared by all threads and are used to identify how *deep* the current thread can calculate the current column, before it has to wait on the thread calculating the column left from it. We have implemented two different ways to prevent the race conditions at accessing the shared counters. The first version uses OpenMP lock variables to guard the counters. See Listing 1 for the source code. The second version uses an atomic function provided by the host system to update the counters. We cannot use OpenMP atomic operations as OpenMP does not allow threads to read a variable that has been updated by another thread without synchronizing. When using the host system atomic operation to update the counter, the read must only be joined by a flush/fence. Tiling to x- and y-dimensions is applied in

---

Algorithm 1. Manually blocked Gauß-Seidel

```
1  int nb_blocks = size/block_size;
2  int *counter = new int[nb_blocks+1];
3  counter[0] = size -1;
4  //initialize all other counters with 0
5  omp_lock_t *locks = new omp_lock_t[nb_blocks+1];
6  //call omp_init_lock for all locks
7  #pragma omp parallel for
8  for (int x=1; x<size-1; x+=block_size) {
9      int y = 1;
10     const int x_block = x/block_size;
11     while (y<size-1) {
12        omp_set_lock (&locks[x_block]);
13        const int lcounter = counter[x_block];
14        omp_unset_lock (&locks[x_block]);
15        for (; y<lcounter; y+=block_size) {
16           for (int xx=x; xx<x+block_size; ++xx)
17              for (int yy=y; yy<y+block_size; ++yy)
18                 V[yy][xx] = (V[yy][xx-1] + V[yy][xx+1] +
                                 V[yy-1][xx] + V[yy+1][xx])/4;
19           omp_set_lock (&locks[x_block+1]);
20           counter[x_block+1] += block_size;
21           omp_unset_lock (&locks[x_block+1]);
22        }
23     }
24  }
```

---

both versions. Tiling to the y-dimension reduces the number of times the shared counters are updated, so larger tiles decrease the number of lock operations, however also reduces the chance of having the rows y, $y-1$ and $y+1$ in the cache.

To achieve better performance, we applied multi-level tiling. We continue to use large tiles to reduce the number of lock operations and subdivide the large tiles into smaller tiles to expose the best data locality. However multilevel tiling increases the overhead and decreases performance when running with 8 threads.

As a final optimization we moved from fixed to dynamic tile sizes for the large tiles. Our dynamic implementation chooses smaller tiles at the beginning and the end of the data matrix and larger ones in the middle. Dynamic tile sizes allow more work to be done in parallel, since e.g. the calculation of the second column can only be started if the first tile of the first column is calculated. However, this optimization imposes additional overhead, while on our test system the performance loss due to data dependencies is rather small. As a result, the performance is almost identical to the one using static tile sizes. We expect dynamic tile sizes to be useful on systems with more cores and additional memory bandwidth to satisfy all cores.

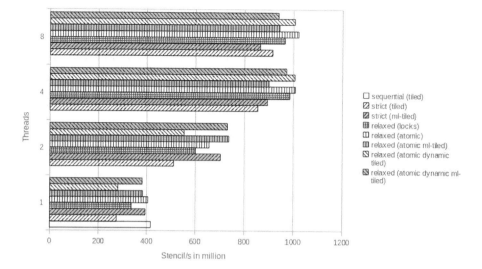

Fig. 3. Gauß-Seidel performance

Figure 3 shows the achieved bandwidth with the different versions for calculating a matrix of size 16386^2 with 100 iterations. We use double precision for our calculations. Our fully optimized implementation sustains a performance of over 1000 MSte/s. In the best case, the Gauß-Seidel stencil requires both one element read and one written to/from main memory per stencil. With 16 bytes transfered between the CPU and main memory our implementation uses a bandwidth of over 16 GB/s, which is about 83% of the performance measured with the STREAM benchmark (Tab. 1). The best performance is achieved when using 8 threads with single level tiling. The additional overhead of multilevel tiling reduces performance with 4 or 8 threads, however the increased data locality helps to outperform single level tiling with less threads. We can see that using locks instead of atomic operations decrease our performance by about 6% when using 4 cores. With one thread the parallel atomic version performs almost identical to the sequential version.

5   Locality-Aware OpenMP Syntax

In the last two sections, we have described our experiences with locality optimizations in two scenarios. The first scenario was matrix-matrix multiplication, which only consists of a perfect loop nest and a trivial parallelization. The loop transformations could have automatically been applied by a compiler, however our compilers did not. Our second scenario also benefits from tiling by both increased data locality and decreased number of lock operations. The parallelization of the second scenario was more complex and the code is not a perfect loop nest. We expect that it is hardly possible for a compiler to identify all required

---

Algorithm 2. Matrix multiplication with improved OpenMP

```
1 #pragma omp parallel for schedule(blocked, 64)
2 for (int i=0; i<size; ++i)
3   #pragma omp for schedule(blocked, 64)
4   for (int k=0; k<size; ++k)
5     #pragma omp for schedule(blocked, 64)
6     for (int j=0; j<size; ++j)
7       #pragma omp block
8       C[i][j] += A[i][k] * B[k][j];
```

---

optimizations, however we show in the rest of this section that the OpenMP syntax can be enhanced to let developers tell the compiler which optimizations should automatically be applied. Furthermore we sketch how compilers can do the required transformations and discuss the impact on other parts of OpenMP.

We start with the matrix-matrix multiplication as an easy example to describe our advanced OpenMP syntax. Algorithm 2 shows the new syntax for the matrix multiplication. We introduce both a new scheduling variant called `blocked` and the ability to nest the `#pragma omp for` pragma without using nested parallelism. The OpenMP `for` pragma tells the compiler that the loop iterations can be carried out in any order, which is true for all loops of the example, and the new `blocked` schedule tells the compiler to apply tiling to this loop.

Compilers tile the loops annotated with `schedule(blocked)`. The size of the tiles can be specified as a second parameter of the scheduling-clause or may be automatically determined by a compiler. Having the compiler determine the tile size may not result in the optimal result, but in a reasonable good outcome. Multilevel tiling may be specified by adding not one tile size, but multiple tile sizes to the schedule clause. Explicit tiling can be specified by an additional pragma parameter to identify the variables to which it should be applied. The outer loop, which is generated when tiling is applied to the original loop, remains at the position in the code where the original loop was. The inner loop, in contrast, gets moved directly in front of what we call the *instruction block*. The instruction block should contain only the code that must be executed in every loop iteration. It is identified by `#pragma omp block` and there may only be one instruction block in a tiled loop. In Alg. 2 only line 8 is the instruction block and all newly created loops will be moved in front of this line. When multiple loops are defined as `schedule(blocked)`, the loop order in front of the block is identical to the one of the original loops. It is expected that users make sure that the modified code is still correct, that is e. g. only annotate loops that be tiled. In perfect loop nests where the innermost loop body is the instruction block – as it is the case for the matrix multiplication – this is given if all loop iterations can be carried out in any order. If the loop body consists of more than the instruction block, every code outside the instruction block will only be executed once per tile. We discuss this behavior based on our Gauß-Seidel code next.

Algorithm 3. Gauß-Seidel with improved OpenMP

```
 1 int *counter;
 2 omp_lock_t *locks;
 3 #pragma omp parallel for schedule(blocked)
 4 for (int x=0; x<size; ++x) {
 5   #pragma omp single
 6   {
 7     counter = new int [omp_num_blocks()+1];
 8     counter[0] = size;
 9     //initialize all other counters with 0
10     locks = new omp_lock_t [omp_num_blocks()+1];
11     //call omp_init_lock for all locks
12   }
13   int y = 0;
14   int x_block = omp_block_num();
15   while (y<size) {
16     omp_set_lock (&locks[x_block]);
17     int lcounter = counter[x_block];
18     omp_unset_lock (&locks[x_block]);
19     #pragma omp for schedule(blocked)
20     for (; y<lcounter; ++y) {
21       #pragma omp block
22       V[y][x] = (V[y][x-1] + V[y][x+1] + V[y-1][x] +
                    V[y+1][x])/4;
23       omp_set_lock (&locks[x_block+1]);
24       counter[x_block+1] += omp_block_size();
25       omp_unset_lock (&locks[x_block+1]);
26     }
27   }
28 }
```

The extensions introduced up till now do not allow user to specify dependencies between tiles, as they are used in the Gauß-Seidel example. Listing 3 shows the code with new library functions that overcome this limitation. `omp_num_blocks()` returns the number of tiles a tiled loop is split into, `omp_block_num()` returns the number of the tile currently calculated by the calling thread and `omp_block_size()` returns the size of tile. The functions are always bound to the tiled loop they are directly part of, meaning in our example `omp_block_size()` (Alg. 3, line 24) is bound to the second for-loop. The code to be generated based on Alg. 3 can be found in Alg. 1.

These functions allow users to specify dependencies between tiles. For example in the Gauß-Seidel example one counter and lock per tile in the x-dimension is allocated. In the y-dimension the counter is only updated once per tile, since the update of the code is not part of the instruction-block. To achieve this, first the two created inner loops must be moved in front of the instruction block and the instruction block must be modified so that it no longer uses the old loop indexes but the indexes of the newly created loops. Furthermore the library functions must be created, that is for example `omp_block_num()` must returns the index of the outer tiling loop and `omp_block_size()` the size of the tiles.

The newly suggested `blocked` loop scheduling and the existing `static` scheduling both offer a form of loop tiling. It would have been possible to reuse the `static` schedule for our modified tiling approach – e. g. the existing behavior is always used when there is no instruction block present. However since, in contrast to the existing scheduling variants, `blocked` may influence the correctness of the program, we decided to not reuse the existing name. The extensions play well with all data sharing clauses, however the concept will not ease of tiling in SPMD style OpenMP programs using the `threadprivate` directive for data privatization, since one thread will only execute a subset of tiles and not the whole loop iteration space. We do not expect the extensions to interfere with the existing synchronization concept. Adding the new functions to the runtime system should be rather trivial, as they mostly must only return a value being made available by the tiling.

It is left for future work to analyze the usability of the new extensions for upcoming many-core architectures. However we expect that an user controlled tiling mechanism will be needed for all tile-based many-core architectures, as for example Intel's Single-chip Cloud Computer (SCC) or GPUs. In tile based many-core system it may for example be possible to have a set of closely coupled cores working on a single tile. The new extensions do not tackle the problem of NUMA like remote memory, but concentrate on a way to easily improve cache usage in loop based code. Further research is necessary to identify ways to support NUMA remote memory.

6   Related Work

A similar extension for OpenMP has been suggested by Gan et al. [7], however it offers a subset of the functionality we present. Especially they focus only on perfect loop nests and do not offer direct access to the blocks nor allow using tiles beyond data locality. Compilers like IBMs XLC/C++ and SGI MIPSpro C/C++ offer directive-based support for loop tiling. Extensions for the ZPL [6] and SAC [10] language provide tiling, even though again not with direct access to the tiles. High Performance Fortran (HPF) offers tiling as part of the language. Loop tiling in general has been worked on for several years and discussed in several aspects, e. g. when a compiler can automatically incorporate loop tiling [1]. Optimizations on stencil computation have been analyzed again for several years, a recent study has e. g. been done by Datta et al. [5].

7   Conclusion / Future Work

We demonstrated tiling on two scenarios with the result of increased performance. The main performance increase resulted from the increased data locality of tiling, however tiling also reduced the number of lock and atomic operations. We furthermore demonstrated that using the optimization technique of processor local storage, which is well-known in the GPGPU realm, is beneficial on current CPUs as well. As an addition, we experimented with dynamic tile size in the wavefront pattern to increase the amount of work that can be done in parallel. As a major contribution we showed that these optimizations techniques could be added to OpenMP with only small changes.

Making the notion of tiles available in OpenMP will not only enable developers to specify data locality and thereby increase performance on current CPUs, but lays out the foundation for future work to effectively deploy OpenMP on hardware which natively requires blocks, e. g. GPUs. Further work is required to check if the suggested extensions may result in ambiguous situations, in scenarios different from ones shown in this paper.

References

1. Ahmed, N., Mateev, N., Pingali, K.: Tiling imperfectly-nested loop nests. In: Supercomputing 2000: Proceedings of the 2000 ACM/IEEE Conference on Supercomputing (CDROM), page 31. IEEE Computer Society, Washington, DC (2000)
2. Bacon, D.F., Graham, S.L., Sharp, O.J.: Compiler transformations for high-performance computing. ACM Comput. Surv. 26(4), 345–420 (1994)
3. Bader, M., Franz, R., Günther, S., Heinecke, A.: Hardware-Oriented Implementation of Cache Oblivious Matrix Operations Based on Space-Filling Curves. In: Wyrzykowski, R., Dongarra, J., Karczewski, K., Wasniewski, J. (eds.) PPAM 2007. LNCS, vol. 4967, pp. 628–638. Springer, Heidelberg (2008)
4. Culler, D., Singh, J., Gupta, A.: Parallel Computer Architecture: A Hardware/Software Approach, 1st edn. The Morgan Kaufmann Series in Computer Architecture and Design. Morgan Kaufmann (1998)
5. Datta, K., Kamil, S., Williams, S., Oliker, L., Shalf, J., Yelick, K.: Optimization and performance modeling of stencil computations on modern microprocessors. SIAM Review 51(1), 129–159 (2009)
6. Deitz, S.J., Chamberlain, B.L., Snyder, L.: High-level language support for user-defined reductions. J. Supercomput. 23(1), 23–37 (2002)
7. Gan, G., Wang, X., Manzano, J., Gao, G.R.: Tile Reduction: The First Step Towards Tile Aware Parallelization in OpenMP. In: Müller, M.S., de Supinski, B.R., Chapman, B.M. (eds.) IWOMP 2009. LNCS, vol. 5568, pp. 140–153. Springer, Heidelberg (2009)
8. McCalpin, J.D.: Memory bandwidth and machine balance in current high performance computers. In: IEEE Computer Society Technical Committee on Computer Architecture (TCCA) Newsletter, pp. 19–25 (December 1995)
9. Pfister, G.F.: In search of clusters, 2nd edn. Prentice-Hall, Inc., Upper Saddle River (1998)
10. Scholz, S.-B.: On defining application-specific high-level array operations by means of shape-invariant programming facilities. In: APL 1998: Proceedings of the APL 1998 Conference on Array Processing Language, pp. 32–38. ACM, New York (1998)
11. Wolfe, M.J.: High Performance Compilers for Parallel Computing. Addison-Wesley Longman Publishing Co., Inc., Boston (1995)

Memory-Efficient Sierpinski-Order Traversals on Dynamically Adaptive, Recursively Structured Triangular Grids

Michael Bader[1], Kaveh Rahnema[1], and Csaba Vigh[2]

[1] Institute of Parallel and Distributed Systems, Universität Stuttgart, Germany
{Michael.Bader,Kaveh.Rahnema}@ipvs.uni-stuttgart.de
[2] Department of Informatics, Technische Universität München, Germany
vigh@in.tum.de

Abstract. Adaptive mesh refinement and iterative traversals of unknowns on such adaptive grids are fundamental building blocks for PDE solvers. We discuss a respective integrated approach for grid refinement and processing of unknowns that is based on recursively structured triangular grids and space-filling element orders. In earlier work, the approach was demonstrated to be highly memory- and cache-efficient. In this paper, we analyse the cache efficiency of the traversal algorithms using the I/O model. Further, we discuss how the nested recursive traversal algorithms can be efficiently implemented. For that purpose, we compare the memory throughput of respective implementations with simple stream benchmarks, and study the dependence of memory throughput and floating point performance from the computational load per element.

Keywords: adaptive mesh refinement, cache oblivious algorithms, space-filling curves, memory-bound performance, partial differential equations.

1 Mesh-Based PDE Solvers as Memory-Bound Applications

Partial differential equations (PDE) are ubiquitous as modelling tools in numerical simulation. A large family of PDE solvers, such as Finite Difference, Finite Volume and Finite Element methods, as well as variants, such as discontinuous Galerkin methods, is based on computational meshes to discretise the spatial domain. *Grid traversals*, i.e., updating all cells or unknowns of such a mesh, are thus essential building blocks for PDE solvers: grid traversals occur in explicit time-stepping methods to solve time-dependent PDEs, in iterative solvers for systems of equations obtained from discretisation, and in refinement and coarsening of the mesh, itself – including evaluation of error estimators, interpolation operators, and maintaining certain balancing or conformity criteria of the mesh.

The computational load of such traversals, i.e., the ratio between the number of operations and the accessed data is often small – usually a constant effort per grid cell. Respective implementations are therefore typically *memory-bound*:

K. Jónasson (Ed.): PARA 2010, Part II, LNCS 7134, pp. 302–312, 2012.

their performance is limited by the latency and bandwidth of the memory hardware and its ability to feed the data into the arithmetics units of the CPU. To speed up memory-bound problems, we therefore have to improve how algorithms use and access memory:

- Reducing the *memory footprint* of an implementation, and thus the bandwidth requirements, can already speed up a memory-bound application.
- Modern memory hardware is hierarchical and parallel, with multiple cache levels that are shared between CPU cores and with non-uniform memory access (NUMA) on compute nodes with multiple CPUs. Maximal performance is only achieved, if cache and memory are optimally exploited.
- CPUs and memory systems are optimised for rather simple memory access patterns: pipelining, for example, requires stream-based memory access. In contrast, random memory access (to unstructured data) is heavily penalised.

In earlier work [3,2], we introduced an approach that builds on bisection-based, fully adaptive triangular grids represented via refinement trees and addresses the described memory issues via a space-filling curve approach. The recursively structured grid is traversed along a Sierpinski curve, which imposes an inherently local, sequential order on the grid cells and corresponding unknowns. Element-oriented traversals (in Sierpinski order) of the corresponding refinement tree require only stacks and streams as basic data structures, and thus avoid random access to memory entirely. The stream-based approach also allows to retain the locality properties of data structure and traversal algorithms despite dynamic refinement and coarsening of the grid.

In Section 2, we will recapitulate the Sierpinski-curve-based approach for adaptive triangular grids. An analogous approach for quadrilateral, octree-type grids has been developed by Mehl et al. [11,6]. In Section 3, we present an analysis of the cache misses using the I/O model, in order to quantify the previously observed excellent cache hit rates for the algorithm. As our approach is inherently recursive, special care has to be taken to avoid implementation overheads. We will introduce loop-based techniques to implement our traversals in Section 4, and present a study that demonstrates that a performance close to the available memory bandwidth is achieved even for traversals on adaptive grids.

2 Sierpinski-Order Traversals on Triangular Grids

Our grid generation approach is based on recursive bisection of triangles, which can be represented via a corresponding refinement tree. Respective approaches were presented by Mitchell, for example [12]. Figure 1 illustrates such an adaptive grid together with its refinement tree. Note that to encode the refinement status of the grid (and the tree), a single bit per node of the refinement tree is sufficient. These refinement bits can be stored as a bitstream in depth-first order.

Left and right children in the refinement tree are defined such that a depth-first traversal of the refinement tree will lead to a grid traversal in Sierpinski order.

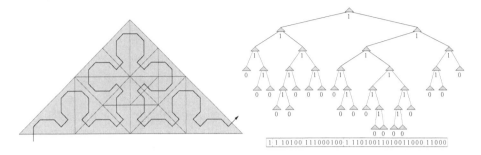

Fig. 1. Adaptive triangular grid and corresponding refinement tree: the tree structure is coded as a bit stream that carries the refinement status of each tree node in depth-first-traversal order

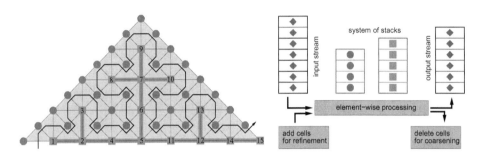

Fig. 2. Stack property of the Sierpinski-curve traversal (left image, note the last-in-first-out access to the vertices 5–10 during a cell-oriented traversal) and resulting stack-and-stream-based processing scheme (right image). The stream-based approach also allows to introduce and delete cells for refinement and coarsening of the grid.

In PDE solvers, cell-local unknowns will be stored and accessed in Sierpinski order, such that we obtain an optimal, stream-based memory access. However, we still need to provide efficient access to unknowns located on cell vertices and edges. Such unknowns are accessed by several grid cells throughout the traversal, so we need to store intermediate values. Figure 2 illustrates that these multiple accesses can be organised via stacks. In each grid cell, unknowns that are accessed for the first time are obtained from an input stream, while unknowns that have already been accessed by neighbouring cells will be retrieved from one of two *colour stacks*. When leaving a cell during the traversal, unknowns are put onto the output stream or onto one of the stacks, depending on whether the respective unknown will be accessed again during the traversal. The resulting stack&stream-based processing scheme is illustrated in Figure 2. For detailed traversal algorithms, including the rules which unknowns to place on which stack or stream, see [3] (for classical Finite-Element computations using cell-, egde-, and vertex-located unknowns) and [2] (for edge-based unknowns).

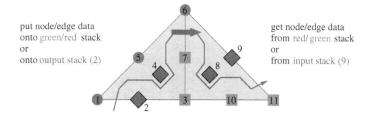

Fig. 3. Stack and stream access in two subsequent grid cells. We assume that the element adjacent to edge 5 will still be visited during the traversal.

3   Cache Misses of the Traversals in the I/O-Model

To estimate the number of cache misses during a Sierpinski traversal, we use the classical *I/O model*, as defined by Aggarwal and Vitter[1]. There, we assume an arbitrarily large external (main) memory. The single CPU, however, can only work on a smaller working memory (cache), which consists of M words organised as M/B cache lines of B words each. In this model, we count the number of read and write transfers between external memory and cache, assuming an intelligent strategy to use the cache lines. A similar approach would be to assume an ideal cache that can foresee which cache lines are no longer required (note that a least-recently-used strategy will perform as an ideal cache for stack data structures).

Figure 3 provides a schematic view on the stream- and stack-based access to unknowns when moving from one cell to the next during a traversal. The cell-located unknowns 4 are written back to the output stream. Similar, the unknowns 8 are read from the input stream. The unknowns on 2 are finished with processing, and are written to the output stream. In contrast, unknowns on 1, 5, and 6 will be used again, and are thus written to the green stack. Similar, unknowns on 3 and 7 are written to the red stack. The right cell obtains unknowns on 6 from the green stack, and unknowns from 3, 7, 10, and 11 from the red stack. The unknowns on 9 are accessed for the first time, and are thus read from input stream. For our later study on cache misses, we denote the following table that specifies the (re-)use of cache locations for all involved unknowns:

in left cell	1	2	3	4	5	6	7
in right cell	11	10	3	8	9	6	7

3.1   Cache Misses on a Uniformly Refined Grid

We now assume that the two cells in Figure 3 represent uniformly refined blocks of a certain depth, which is chosen such that all unknowns on the boundary of such a block fit into the cache. Our strategy is then to treat all interior unknowns in a stream fashion: reading them from an input stream and writing them to an output stream should require only a single cache line. The major part of the cache will be used to hold the unknowns on the block boundaries.

For simpler presentation, we will disregard unknowns on vertices, and assume that we do two refinement steps at once from each refinement level to the next. Hence, a uniformly refined block of depth k (in short: *depth-k-block*) will consist of 4^k elements, and each of the three block boundaries will have 2^k adjacent elements. A uniform grid of depth n will consist of 4^{n-k} such depth-k-blocks. We further stick to the case of one unknown (requiring one word of memory) per element edge. To make sure that all unknowns on the boundary just fit into cache, we therefore have to chose the depth k of our blocks such that $3 \cdot 2^k \leq M$, but $3 \cdot 2^{k+1} > M$. Assuming that we have at least 4 cache lines (i.e., $M/B \geq 4$), we can derive the following scheme to use the cache lines for a depth-k-block:

- The unknowns on 7 shall stay in cache and will not cause any loads or stores.
- The unknowns on 2 will be written to the output stream (located in main memory) and are replaced by the unknowns on 10.
- The unknowns on 5 are written to the green stack, but cannot be held in cache, as the respective lines are required by the unknowns on 9.
- Interior unknowns follow a stream access, so we get by with a single cache line (4 is successively replaced by 8).

The last statement deserves some further attention, because during the recursive traversal of the depth-k-block, many of the interior unknowns (all unknowns located on a cell edge) will be stored on a colour stack. However, as the unknowns on 2 are successively written to the output stream and the unknowns on 10 will not be read, yet, in the left triangle, we have free cache lines available. The same argument holds for the unknowns on 5 (which are written to the green stack) and on 9. This additional cache capacity is sufficient to hold interior unknowns, if the green and red stack, during a traversal of a grid block, together never contain more unknowns than are located on the boundary. This property is quite easily proved by induction. As a result, we can count the loads and stores to and from main memory during the traversal of each depth-k-block:

- Writing back the unknowns on 5 to the green stack (to a part located in main memory) requires $2^k/B$ stores (as B unknowns fit into a cache line). Similar, reading the unknowns on 10 from the red stack requires $2^k/B$ loads.
- Writing 2 to the output stream and reading the unknowns on 9 will require $2^k/B$ loads and stores, as well. However, these are *compulsory* reads and stores that cannot be avoided: in any traversal, we have to read and write each unknown at least once.
- Similar, all loads and stores of interior unknowns, 4 and 8, (located on element edges in the interior of the depth-k-blocks) are compulsory.

We thus require $2^k/B$ non-compulsory loads and stores for each of the $4^{(n-k)}$ depth-k blocks. As we chose $2^k \leq M/3$, the total number of non-compulsory loads and stores is at most $4^{(n-k)} \cdot \frac{M}{3B}$. Compared to the number of elements $N = 4^{(n-k)} \cdot 4^k$, we have at most $4^{(n-k)} \cdot 4^k \cdot \frac{M}{3B} \cdot 4^{-k} = N \cdot \frac{M}{3B}/(2^k)^2$ non-compulsory loads and stores. As we chose k as large as possible in order to exploit the entire cache, we required $3 \cdot 2^{k+1} > M$, which implies $2^k > \frac{1}{6}M$. Altogether, the number of non-compulsory cache misses is at most $12\frac{N}{MB} \in \mathcal{O}\left(\frac{N}{MB}\right)$.

In addition, we require exactly one compulsory load and store for each un-known, which leads to a total number of compulsory loads and stores of $\alpha N/B$, where αN is the number of unknowns in all N elements. Assuming 1 unknown per edge plus 1 interior unknown in each element, we obtain approximately $2N$ edge-located unknowns plus N element-local unknowns, such that $\alpha \approx 3$.

If we consider only vertex-located unknowns in our scheme, we have only $2^k - 1$ unknowns on the edge of each block. In return, we need to consider the unknowns on 1, 3, 6, and 11, such that we basically obtain the same result. For unknowns on both vertices and edges (even multiple unknowns per edge, as in higher-order methods), we will obtain estimates of $\mathcal{O}\left(\frac{N}{MB}\right)$ non-compulsory loads and stores. Hence, in total, we require $\alpha N/B + \mathcal{O}\left(\frac{N}{MB}\right)$ loads and stores.

3.2   Extension to Non-uniform Grids

So far, we required a uniformly refined grid to determine the loads and stores for the Sierpinski traversal. However, the upper bound of $\mathcal{O}\left(\frac{N}{MB}\right)$ will hold as long as the ratio of boundary unknowns vs. number of cells in a cell block stays asymptotically the same. Consider a partitioning of a non-uniform grid of N cells into $\frac{N}{K}$ partitions, each consisting of K contiguous cells in Sierpinski order. We demand that each such K-block has at most $c\sqrt{K}$ element edges and that all boundary unknowns fit into cache. Hence, K is chosen such that $c\sqrt{K} < M$. Then, the number of non-compulsory cache misses per K-Block is definitely less than $c\sqrt{K}/B$ (on average $\frac{c}{3}\sqrt{K}/B$, as only one of the edges will cause cache misses), such that we obtain $\frac{N}{K} \cdot c\sqrt{K}/B = c\frac{N}{B\sqrt{K}}$ cache misses for an entire grid traversal. As K was chosen relative to the cache size M, $c\sqrt{K} < M$, we again obtain $\mathcal{O}\left(\frac{N}{MB}\right)$ non-compulsory loads and stores.

Hence, the upper bound of $\mathcal{O}\left(\frac{N}{MB}\right)$ cache misses holds as long as a block of K cells has only $\mathcal{O}(\sqrt{K})$ cell edges. It is rather simple, however, to construct a counter-example: take a triangular block, where only the (geometrically) left-most cell is bisected in each level – the resulting block then consists of $k+1$ cells and $k+3$ edges, and we obtain $\mathcal{O}(N/B)$ non-compulsory cache misses.

3.3   Discussion and Related Work

As long as the ratio of $K : \sqrt{K}$ between interior elements and edges of a K-block holds, the Sierpinski traversal will exploit caches of any size M – including multi-ple layers of caches – and lead to $\alpha N/B + \mathcal{O}\left(\frac{N}{MB}\right)$ cache misses (compulsory + non-compulsory). This $K : \sqrt{K}$ ratio also reflects the Hölder continuity of the Sier-pinski curve, and is known as an argument for the asymptotically optimal quality of parallel partitions induced by space-filling curves [14] (where it is also known that this property does not hold for degenerate adaptive meshes).

There is not too much previous work on cache misses caused by traver-sal algorithms on adaptive grids. For the case of structured, uniformly refined grids, respective bounds are straightforward for simple *vertex-oriented*, row-wise traversals. There, in each vertex, we need to access neighbouring, vertex-located unknowns of three successive rows. If the unknowns of one row do not fit into cache, we cause non-compulsory misses for 2 of the 3 rows plus compulsory misses for the third row, which leads to $3\alpha\left(\frac{N}{B}\right)$ loads and stores. Intelligent blocking of the grid, together with combining multiple traversals, leads to cache-oblivious algorithms [7] that produce $\mathcal{O}(N/\sqrt{M})$ cache misses in the I/O model. These algorithms are not available for adaptive grids, however.

If element orders exist, where any two successive triangular elements share a common edge, then *element-oriented* traversals along this order will only cause compulsory misses for the unknowns on the common edges. Only accesses to the unknowns on the third vertex will cause non-compulsory misses, which leads to $\alpha N/B + \mathcal{O}\left(\frac{N}{B}\right)$ cache misses. Note that the cache size M is not exploited at all, and that the same bound holds for our Sierpinski traversals for "degenerate" adaptive grids. In computer graphics, such element orders are known as *triangle strips*, and are also used for unstructured meshes [9,5] (however, triangle strips cannot be constructed for any mesh). Several groups discussed the construction of triangle strips on adaptive grids by using Sierpinski curves [13,10,8].

For the context of rendering unstructured triangle meshes, Bar-Yehuda and Gotsman [4] presented an algorithm that requires only compulsory misses, provided the cache size is $\mathcal{O}(\sqrt{N})$. Their approach is based on efficient algorithms to find balanced partitions of size $\mathcal{O}(K)$, with a separator of length $\mathcal{O}(\sqrt{K})$. For limited cache size M, their traversal algorithm, similar to our Sierpinski traversals, leads to $\mathcal{O}\left(\frac{N}{M}\right)$ non-compulsory misses. However, generating these cache-efficient rendering sequences requires $\mathcal{O}(N^2)$ work, and the resulting algorithm does not consider cache lines (i.e., assumes a cache line size of $B = 1$).

4   Implementation and Performance Results

The performance of recursive implementations of our Sierpinski traversals was already examined for multigrid solvers [3] and for PDE solvers with explicit time-stepping [2]. We observed excellent cache hit rates ($> 99\,\%$ level-1 hits), however, the results also revealed that quite some overhead is caused by the recursive implementation and by the data movement to and from the stacks. This is partly caused by the effort to execute recursive calls and if-statements (to check for leaf-level or end of a stack frame, e.g.), but also because all non-leaf cells are visited during a refinement-tree traversal. Visiting interior nodes, in a binary tree, requires twice as many recursive calls as we have elements (leaves of the tree); hence, our goal was to particularly reduce this overhead.

The loop-based implementation exploits that we can infer the cell size and orientation from the entry and exit points of the Sierpinski curve. For example, if we leave a cell via its hypotenuse and enter via a leg, the entered cell has to be coarser by 1 level. As the entry and exit points determine the stack access rules, this classification is available from our refinement tree. For loop-based traversals, we store the classification in an array using an initialisation traversal. We thus only traverse the leaves of the refinement tree, which, together with saving overheads for recursive calls, accounts for a substantial performance improvement. As an initialisation overhead, we have to rebuild the structures to enable loop-based traversals in each global refinement and coarsening step. However, as we see in the following section, the additional execution time is affordable.

4.1   Traversals for Grid Refinement and Coarsening

As a first test, we measured the execution time for an "empty" grid traversal without any numerical unknowns. In these traversals, only the required data movements are executed (as they would be required for traversals that ensure conforming refinement and coarsening of an adaptive grid), but no actual computations are performed. For a uniformly refined grid of approximately 2 million (2^{21}) grid cells, such a traversal requires 0.13 seconds for the recursive implementation. The loop-based traversal requires 0.054 seconds, which is more than twice as fast. This test, as well as all the following tests, were executed on a single core of an Intel Core2 Duo processor (T7700, 2.4 GHz, 4MB L2 cache).

The following table lists the execution times for loop-based traversals for grid-refinement, starting with a uniform grid (with 2^{21} cells) and an a-priori adaptive grid with roughly the same number of grid cells:

grid	number of cells	empty traversal	synchron. traversals	cells added	grid rebuild	total time (incl. interp.)
uniform, 2^{21}	2,097,152	0.053 s	0.09 s – 0.11 s	1,049,600	0.66 s	1.42 s
a-priori ref.	2,121,520	0.053 s	0.09 s – 0.11 s	1,055,080	0.72 s	1,78 s

In particular, we wanted to compare the time required for synchronisation traversals, which check the refinement status of each edge and ensure conforming grid refinement, and we measured the total time spent for grid refinement, including set up of the mesh and interpolation of the unknowns (here: trivial interpolation for piecewise constant basis functions). As expected, the performance for uniform vs. adaptive grids does not vary significantly (the a-priori-refined grid requires more synchronisation traversals, which explains the increased total time for refinement). Column "grid rebuild" lists the execution time to rebuild all grid data structures after refinement and coarsening, including the set-up of the data structures for loop-based traversals ($\approx$15–20 % of the rebuild time). Hence, switching to loop-based traversals will already pay off, if only 3–5 traversals are performed between two refinement steps (for comparison: a forth-order Runge-Kutta scheme would require 4 traversals).

We also measured the execution time of refinement traversals for grids that were refined along a circle line with growing radius (mimicking an inflating shock front, e.g.) – our implementation takes $\approx$ 0.05–0.06 s per million grid cells for the synchronisation traversals, and $\approx$ 0.6 s per million grid cells for the entire refinement and coarsening step including interpolation of unknowns.

4.2   Traversals with Floating Point Operations

To determine the achievable performance of the recursive and of the loop implementation of traversals for PDE solvers, we studied a set-up that artificially

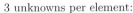

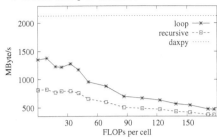
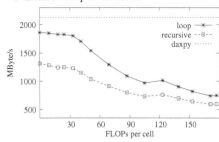

Fig. 4. Memory throughput for loop-based and recursive Sierpinski traversals with a varying amount of floating-point operations per element

increases the computational load per element, and measured the effect on memory throughput (in MB/s) and floating point performance (in MFlop/s). Both implementations include all stack operations to synchronise refinement data and compute numerical data on common edges. In addition, they perform a predefined number of floating point operations that mimic the execution of local element operators in each cell. Increasing the number of the artificial operations per cell should approach a situation where the performance is purely computation-bound, and the MFlop/s ratio is only limited by the performance of the CPU core. For very few operations per element, our traversal will rather be memory-bound, and reach a throughput that is hopefully close to the available bandwidth.

From our analysis of cache misses, we can expect that our traversals can at best be as efficient as a simple stream operation that causes the same number of compulsory misses. We chose the BLAS `daxpy` subroutine, i.e. a simple vector update $y = y + \beta x$, as benchmark, because it will cause $\alpha N/B$ compulsory misses (here $\alpha = 2$), similar to our traversals, and because it is known to achieve a memory performance close to the available bandwidth.

Figure 4 plots the memory throughput of loop-based vs. recursive traversals for increasing number of artificial operations per element. For 9 unknowns per cell (corresponding to the use of piecewise linear basis functions for a system of PDE with three components), we observe that up to 30–40 floating point operations per cell are executed "for free" – here, performance is memory-bound and execution with fewer operations per cell does not give any speedup. The best memory throughput achieved for the loop-based traversal is quite close to the achievable throughput, as determined by the `daxpy` operation. The recursion-based implementation behaves similarly, but reaches a substantially lower maximal throughput. The plot for 3 unknowns per cell reveals lower throughput and slightly erratic behaviour, which demonstrates that a low memory footprint and a low computational load per cell make it harder to achieve good performance.

Figure 5 shows that the MFlop/s performance increases "automatically" for increasing work load per element, but apparently reaches a "limit performance" of only about 1 GFlop/s, even for rather large computational load.

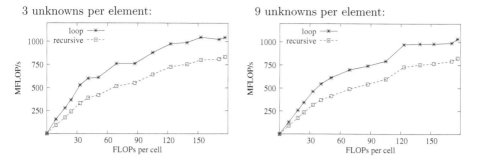

Fig. 5. MFlop/s for the loop-based and the recursive Sierpinski traversals with a varying amount of floating-point operations per element

4.3   Traversals for the Shallow Water Equation

Finally, we tested the performance of the loop-based traversal within our desired target application, a dynamically adaptive solution of the shallow water equations using the discontinuous Galerkin method for discretisation (see [2] for details). With piecewise constant basis functions, we require three unknowns (water height plus two velocity components) per grid cell. The inner kernels to compute the flux terms and update the unknowns in each time step require 90 floating point operations per element. With that setting, a simple forward Euler timestep requires 0.23 s, which is equivalent to a memory throughput of 778 MB/s, and to executing 833 MFlop/s. In comparison, solving only the transport equation for the wave height requires only 30 floating point operations per element, which reduces the execution time to 0.20 s on the same grid – equivalent to a memory throughput of 895 MB/s and 320 MFlop/s. For the shallow-water problem, both memory throughput and MFlop/s performance are well within the performance range of the idealised traversal with artificial flops. For the simpler transport problem, we are a bit below these values, which might be due to slightly worse optimisation of the required operations (additional if-statements to implement boundary conditions, less effective use of vectorisation capabilities of the CPUs, etc.). The study in section 4.2 tells us that this performance is only half of the theoretically available performance in terms of throughput. However, we also see that improved results can be expected for higher order discretisations with larger data volume and higher computational load per element.

5   Conclusion

Our goal is to provide a grid generation and processing scheme for fully adaptive PDE solvers on dynamically refined grids with minimal memory requirements. Our scheme is based on space-filling orders on the grid cells and uses stack and stream data structures only. The interplay of locality properties induced by the space-filling curve and inherently cache-friendly data structures leads to a cache-oblivious algorithm with only $\mathcal{O}(N/MB)$ non-compulsory cache misses.

Via optimised, loop-based traversals of these normally recursive traversals, we can also achieve high computational performance in practice. We have provided a performance analysis that demonstrates that the memory throughput of our traversals comes close to the available memory bandwidth (and thus close to the optimal achievable performance) for low computational load per element. For problems such as simple transport equations, we achieved a memory throughput of up to 75 % of the throughput measured for simple vector operations. For problems with larger computational load, such as solving the shallow water equations, we achieved more than 800 MFlop/s for piecewise constant basis functions, and can expect more than 1 GFlop/s for the piecewise linear case, which would be an exceptional performance for a matrix-free PDE solver on fully adaptive grids.

References

1. Aggarwal, A., Vitter, J.: The input/output complexity of sorting and related problems. Commun. ACM 31, 1116–1127 (1988)
2. Bader, M., Böck, C., Schwaiger, J., Vigh, C.: Dynamically adaptive simulations with minimal memory requirement – solving the shallow water equations using Sierpinksi curves. SIAM J. Sci. Comput. 32(1), 212–228 (2010)
3. Bader, M., Schraufstetter, S., Vigh, C., Behrens, J.: Memory efficient adaptive mesh generation and implementation of multigrid algorithms using Sierpinski curves. Int. J. Comput. Sci. Eng. 4(1), 12–21 (2008)
4. Bar-Yehuda, R., Gotsman, C.: Time/space tradeoffs for polygon mesh rendering. ACM Transactions on Graphics 15(2), 141–152 (1996)
5. Bogomjakov, A., Gotsman, C.: Universal rendering sequences for transparent vertex caching of progressive meshes. Comput. Graphics Forum 21(2), 137–148 (2002)
6. Bungartz, H., Mehl, M., Neckel, T., Weinzierl, T.: The PDE framework Peano applied to fluid dynamics. Comput. Mech. 46(1) (2010)
7. Frigo, M., Strumpen, V.: Cache oblivious stencil computations. In: ICS 2005: Proc. of the 19th Annual International Conference on Supercomputing, pp. 361–366. ACM (2005)
8. Gerstner, T.: Multiresolution Compression and Visualization of Global Topographic Data. GeoInformatica 7(1), 7–32 (2003)
9. Hoppe, H.: Optimization of mesh locality for transparent vertex caching. In: SIGGRAPH 1999: Proc. of the 26th Annual Conf. on Computer Graphics and Interactive Techniques, pp. 269–276 (1999)
10. Lindstrom, P., Pascucci, V.: Terrain simplification simplified: A general framework for view-dependent out-of-core visualization. Technical Report UCRL-JC-147847 (2002)
11. Mehl, M., Weinzierl, T., Zenger, C.: A cache-oblivious self-adaptive full multigrid method. Numer. Lin. Alg. Appl. 13(2-3), 275–291 (2006)
12. Mitchell, W.: A refinement-tree based partitioning method for dynamic load balancing with adaptively refined grids. J. Parallel Distrib. Comput. 67(4), 417–429 (2007)
13. Pajarola, R.: Large scale terrain visualization using the restricted quadtree triangulation. In: VIS 1998: Proc. of the Conf. on Visualization 1998, pp. 19–26. IEEE Computer Society Press (1998)
14. Zumbusch, G.: On the quality of space-filling curve induced partitions. Z. Angew. Math. Mech. 81(suppl. 10), 25–28 (2001)

Fast Wavelet Transform Utilizing a Multicore-Aware Framework

Markus Stürmer, Harald Köstler, and Ulrich Rüde

System Simulation Group, University of Erlangen-Nuremberg, Germany
markus.stuermer@informatik.uni-erlangen.de
http://www10.informatik.uni-erlangen.de

Abstract. The move to multicore processors creates new demands on software development in order to profit from the improved capabilities. Most important, algorithm and code must be parallelized wherever possible, but also the growing memory wall must be considered. Additionally, high computational performance can only be reached if architecture-specific features are made use of. To address this complexity, we developed a C++ framework that simplifies the development of performance-optimized, parallel, memory-efficient, stencil-based codes on standard multicore processors and the heterogeneous Cell processor developed jointly by Sony, Toshiba, and IBM. We illustrate the implementation and optimization of the Fast Wavelet Transform and its inverse for Haar wavelets within our hybrid framework, using OpenMP, and using the Open Compute Language, and analyze performance results for different platforms.

Keywords: cache blocking, parallelization, CBEA, OpenCL, OpenMP.

1 Introduction

Different approaches are viable to write fast, parallel code, each with different support for or restrictions to algorithm, optimization techniques and target platform, and different requirements on development time and experience. As today even mobile computers are able to execute several instruction streams concurrently in hardware, sequential software can only utilize a fraction of any modern machine.

A drawback of conservative multicore approaches is the growing gap between computational power and available memory bandwidth per socket. Many algorithms tend to be heavily memory bound on such systems after having been parallelized, which is especially true for many stencil-based scientific computing or image processing codes operating on large data sets. To some extent memory layout optimizations and blocking techniques can increase spatial and temporal locality of memory references, at the cost of increased development time and code complexity.

The Cell Broadband Engine Architecture[9] takes a heterogeneous approach, providing two diversified types of cores. PowerPC Processor Elements (PPEs)

K. Jónasson (Ed.): PARA 2010, Part II, LNCS 7134, pp. 313–323, 2012.

are general purpose cores based on IBM's Power architecture, meant to control the program flow and run the operating system. The Synergistic Processor Elements (SPEs) are specialized cores optimized for computation and throughput. Each SPE uses a small and fast local storage (LS) of 256 KiB instead of caches, and the instruction code must therefore explicitly exchange data between a remote address space (main memory) with its own using asynchronous copy operations. By this specialization, Cell processors provide impressive peak performance and memory throughput with low power consumption, but require tailored and optimized software.

GPUs take the multicore approach one step further and typically offer tens or hundreds of specialized compute units operating on dedicated memory. Therefore they reach outstanding compute and memory performance and are more and more used for compute-intensive applications, often called general purpose programming on graphics processing units (GPGPU). They are best suitable for massively-data parallel algorithms, inadequate problems, that e. g. require a high degree of synchronization or provide only limited parallelism, are left to the host CPU.

Currently the only way to write code that runs on all three hardware platforms is to use the Open Compute Language (OpenCL)[1]. OpenCL provides a runtime library running on a general-purpose core (host) that offers facilities to manage memory objects, to compile kernels written in the OpenCL kernel language —a derivative of C99—, and to issue many instances arranged on a logical grid to compute cores, which may be CPU cores, SPEs or compute units on GPUs. Although providing concepts representing features of certain platforms, like SIMD operations on many general purpose CPUs and SPEs, local memories on GPUs and SPEs, and asynchronous copy operations on the latter, it does not allow low-level access to underlying hardware.

Besides OpenCL, vendors of GPUs also offer proprietary environments for GPGPU. NVIDIA, e. g., provides the possibility to write single-source programs that execute kernels written in a subset of C and C++ on their Compute Unified Device Architecture (C for CUDA).

The OpenMP language extension[2] allows for simple parallelization and makes no restrictions on low-level optimizations, but the coarse control over synchronization and coherency limits its applicability. Unfortunately, it cannot be used directly on the CBEA except on the general purpose PPE.[3]

Manual parallelization using thread and synchronization primitives, e. g. using pthreads, leaves all freedom to the programmer, but is tedious and results in software for either common multicore or for Cell processors. On the CBEA, separate code for PPE and SPE is required: SPE code must issue copy operations between main memory and the respective LS, while the PPE code must employ libraries to load binary code onto SPEs and control them.

---

[1] See http://www.khronos.org/opencl/

[2] See http://openmp.org/

[3] IBM's XL C/C++ for Multicore Acceleration compiler can to a limited extend delegate work to SPEs in parallel OpenMP constructs.

Our framework written in C++ is designed to relief the programmer from cumbersome manual parallelization and its pecularities on the CBEA while leaving all freedom in writing low-level optimized kernels. It allows to write cache-efficient, parallel codes mainly for stencil-based algorithms. Generic C++ code can be compiled on common multicore as well as on Cell processors, but highly optimized versions can be made for compute-intensive parts.

While ours leaves as much control to the programmer as possible —also requiring him or her to provide optimized versions where appropriate— other frameworks try to automatically tune high-level descriptions. As an example, [13] map OpenMP directives to CUDA, [4] presents an auto-tuning software to generate code for standard CPUs, CBEA, and CUDA.

Limited to the CBEA, and therefore not considered here, two more solutions are offered: Cell Superscalar[4] allows to write annotated single-source programs whose compute-intensive tasks are executed on SPEs. IBM offers the Accelerated Library Framework (ALF) which orchestrates the execution of so-called tasks on SPEs to manipulate memory objects.

The paper is organized as follows: In section 2 we briefly describe the concepts of our multicore-aware hybrid framework that allows to develop code running on standard multicore CPUs and on the CBEA. Section 3 summarizes the Wavelet algorithm we have chosen as example application, and in section 4 we present implementation details and performance results for different architectures before concluding the paper in section 5.

2   Hybrid Framework

This section presents a novel framework design that simplifies the process of writing performance-optimized, parallel codes for stencil-based algorithms on regular grids for multicore and shared memory multiprocessor machines, including the CBEA. It is not a black box that tries to create fast code, but a tool that lets the experienced programmer focus on the design of data structures and kernels while program flow and synchronization are taken care of in a configurable way.

Algorithms that are based on stencils or show comparable access patterns appear in many areas, from filtering in image processing to iterative solvers in large-scale simulations, but the majority of them have similar well-understood properties, so that a framework approach is reasonable and appropriate for them. A former version of the framework not supporting the CBEA is described in [14].

Conception. The framework was designed to

1. enable writing parallel programs that can run on multicore as well as Cell processors, without or with only minor modifications.
2. support cache blocking techniques, but without narrowing applicability to certain types of stencils or data dependencies.
3. allow for kernels that employ machine-specific optimizations.

---

[4] Developed by the Barcelona Supercomputing Center (CellSs), http://www.bsc.es/

The peculiar design of SPEs make it necessary to impose certain restrictions on hybrid code. As SPE code works on local copies, this must become the default case, even for homogeneous CPUs. Furthermore, it is necessary to separate the control code from the compute kernels, as it must be possible to compile the latter separately for the CBEA. Consequently, the framework must provide means to exchange information between control part and kernels, and provide kernels the ability to learn of and access data structures created by the control part.

Control and kernel components are implemented as C++ classes that make use of the framework by inheritance and through utility classes and functions. As data structures, only two-dimensional arrays of arbitrary type are supported currently, as finding the best approach for extension to 3D requires further investigation.

When executing, the control component first configures the thread setup, especially how many threads should be spawned, and set up required data structures through the framework. The actual work is done by letting the framework guide one or multiple so-called *sweeps*, which will be detailed below, of the working threads through a rectangular logical grid. As no synchronization or ordering of memory accesses is guaranteed during a sweep, data can either be read by multiple threads, or accessed by a single thread if it is to be modified. This is basically what pthreads and OpenMP enforce when beginning and end of a sweep are the only synchronization points, and what OpenCL assumes during a kernel execution.

Parallel Blocking Technique. Cache blocking techniques change the order of operations in a way that increases cache locality to reduce main memory accesses. Temporal blocking techniques carry this further by performing multiple operations on data that is already available in the cache hierarchy. Such methods can also be applied to codes executed on SPEs despite their lack of cache, which is why we refer more general to *blocking techniques*.

Different approaches to cache blocking can, e. g., be found in [12,11,17,2,3] and have also been examined thoroughly in the DiME[5] project[1,10,16].

The framework splits the rectangular domain into multiple *stripes* and a thread is then responsible to compute all results within one stripe at a time in an efficient way. In practice, whole stripes cannot be expected to fit into LS or caches. Therefore, a stripe is divided horizontally into equally wide tiles to implement a spatial or temporal blocking technique. For the considered Haar wavelets, no overlapping data dependencies exist like in many stencil-based algorithms, for temporal blocking in more complex cases we point to [14].

Sweeps. If the control component starts a number of sweeps, the framework wakes the worker threads, which have been waiting or suspended by the kernel base class. On SPEs, the kernel base class will also have to copy up-to-date management information or input parameters to Local Storage – on general purpose cores this information is directly accessible by the kernels. The worker

---

[5] http://www10.informatik.uni-erlangen.de/Research/Projects/DiME/

threads cooperatively distribute stripes and make use of virtual function calls to run the actual kernel implementation. Buffer structures, each being able to hold data for computing a tile, are managed in a ring-like manner.

Kernel classes need to implement members to fill buffer structures with data from persistent storage in main memory, perform computation, and store computed data in buffers back. Data movement functions get a description of the respective tile, a reference to the buffer structure to use, and an utility object for reading and writing data. They are required on the CBEA, since SPEs need to manage dynamic data structures and associate DMA transfers with a tag. The actual computation has only access to a configurable number of tiles and their descriptions.

Without local memory, this approach can only work effectively if the whole ring buffer remains in the cache hierarchy most of the time, because otherwise main memory access would increase. Fortunately, the relevant cache lines are touched regularly, and on most architectures cache pollution can be avoided if special load and store instruction are used for copying. But even without that support by the ISA, the approach works surprisingly well in most cases due to increasing size and associativity of caches. Padding can help for toxic cases.

The horizontal, tile-wise traversal is not only important to prevent redundant computations. For cache-based architectures it can be expected that data of neighboring tiles has already been allocated either due to hardware prefetchers or if lines of a tile do not fully coincide with cache line boundaries. On the CBEA, latencies of the asynchronous DMA transfers can only be hidden if they are queued long before their results are required, so the next tile to be computed must be known before the preceding one has been finished.

Clearly, additional time is spent for pure data movement on general purpose cores. But various optimizations can often more than compensate for this: Intermediate values, e. g. coefficients, can be managed in the ring buffer and will not account to main memory traffic. Optimized copy routines can exploit special features of the hardware more easily, like non-temporal moves available on the Intel 64 architecture, which prevent unnecessary allocation of cache lines and reduce cache coherency traffic.

3 Application: Fast Haar Wavelet Transform

Wavelets have a variety of applications, like analysis, denoising or compression in image and signal processing, and a lot of research is done to accelerate them on different architectures, e. g. [15,5,7,6]. Here, the simple Haar wavelet[8] transform and its inverse are used to illustrate and evaluate implementation and optimization of hybrid codes with OpenMP, in OpenCL, and within our hybrid framework.

The fast wavelet transform and its inverse are simple algorithms, but their optimization is challenging in that it is mainly memory bound and allows only for limited saving of memory bandwidth using blocking techniques. For simplicity, we restrict ourselves to input with sizes of $2^n \times 2^n$ and perform horizontal and vertical transforms in alternation. The algorithm can be seen as a multi-level algorithm, as it applied recursively to smaller and smaller regions of the input, and inverses increasingly larger regions for the reconstruction.

4   Performance Results

The following section describes the implementations within the different environments and presents performance results on appropriate architectures. Two scenarios are considered: In scenario A, all data and computations are done in single floating point precision. In scenario B, the input of the transform is 8 bit and the inverse 16 bit integral data, while all computations and intermediate data remain in single precision.

For measurements, the following test platforms have been chosen:

i7:   A workstation housing an Intel Core i7 940 processor running at 2.93 GHz, with four cores sharing 8 MiB of third level cache, exposing eight logical cores using SMT. Three channels of DDR3-1066 memory provide a theoretical peak bandwidth of 25.5 GB/s. The machine is running openSUSE 11.2 and programs were compiled using the GCC compiler suite in version 4.5.1.

Cell: A Sony PlayStation 3, equipped with an STI Cell/B.E. running at 3.2 GHz, having access to two channels of Rambus XDR memory with a theoretical peak bandwidth of 25.6 GB/s. The machine is running Fixstar's Yellow Dog Linux 6.2, code was compiled with the GCC compilers in version 4.1.1 distributed with IBM's Cell SDK 3.1. In this configuration, only six SPEs can be used.

Fermi: An NVIDIA Tesla C2050 GPGPU, providing 448 CUDA compute units operating on GDDR5 memory with a theoretical peak bandwidth of 144 GB/s. However, data needs to be transfered between host CPU and accelerator through a 16x PCIe. NVIDIA's OpenCL 1.0 implementation shipping with CUDA SDK 3.1 is used, host code is compiled with the GNU compiler suite in version 4.4.1.

4.1   OpenMP Implementation

Transform and inverse have been implemented in plain C99 first before adding OpenMP pragmas. A horizontal and vertical transform have been combined, so that the innermost loop body of a kernel reads from four positions and writes four results. Persistent memory arrays are used to hold temporary data to allow for non-destructive and parallel processing . The conversions required for scenario B are performed on-the-fly. Two SIMD-vectorized variants for Intel 64 have been written using compiler intrinsics, one using usual memory operations, the other memory operations optimized for streaming. Especially so-called non-temporal stores can lead to a higher achievable memory bandwidth by removing the need of allocating lines in cache that are only written to.

Table 1. Time for wavelet transform and reconstruction on the $i7$ test platform in μs in the OpenMP implementation. Cases in which four threads performed better than eight are marked with an *. NT indicates the usage of non-temporal memory operations.

| | scenario A | | | scenario B | | |
size	scalar	SIMD	SIMD NT	scalar	SIMD	SIMD NT
512	236	*194*	377	276	*156*	333
1024	1605	1567	*1474*	1123	858	1259
2048	7450	7533	*5763*	5875	5836	*4591**
4096	33804	32057	*23609**	25364	24449	*19044**

Table 1 shows the performance achievable on the $i7$ test platform. SIMD-vectorization alone is only beneficial for problems that fully fit into the last cache level. If this is not the case any more —for scenario A starting with a size of 1024^2, for scenario B of 2048^2— only minor differences to the scalar kernels can be observed. Streaming memory operations can increase the performance for such large cases by about 30%, but can also about half the performance for small problems. None of the optimized versions is able to surpass the simple scalar version for all sizes, and an optimal implementation would need to carefully choose non-temporal and caching memory operations.

Tests for the OpenMP-parallel scalar version have also been conducted on the PPU of the *Cell* platform, but the heterogeneous Cell design intends high memory bandwidth and fast computation for the SPEs only. As it performed consistently more than an order of magnitude slower than $i7$, no results are shown here.

4.2   OpenCL

The kernel for OpenCL basically corresponds to the inner loop body of the scalar implementation, with the major difference that OpenCL natively only supports one-dimensional arrays and the index for each access must be computed explicitly. Additionally, the OpenCL kernel must check if it executes within the bounds of the data to be processed. For the *Fermi* platform and its Single Instruction Multiple Threads (SIMT) design, this simple kernel proved to be the fastest. Similar to the OpenMP implementation, additional kernels are required for scenario B that additionally convert from and to integral data.

Table 2. Time for wavelet transform and reconstruction on the *Fermi* test platform in μs in the OpenCL implementation, either only counting the computation or considering also the data transfer of input and reconstruction

| | scenario A | | scenario B | |
size	computation	w/ transfer	computation	w/ transfer
512	430	1918	423	1270
1024	611	3371	574	2244
2048	1185	10186	1033	5130

Performance results for OpenCL on *Fermi* are shown in table 2. Obviously, data transfer takes at least two third of the time. One would expect the runtime to scale about with the problem size, but the superlinear scaling when only computation is considered indicates that overhead for the kernels operating only on few data, and therefore cannot exploit the whole GPGPU, and the runtime overhead is predominant.

The very same kernel successfully executes on $i7^6$ and $Cell^7$, table 3 compares performance for a certain problem size. For $i7$, one would expect performance similar to the scalar implementation for OpenMP, but is about 40% slower even if looking only at the computation. On *Cell*, the OpenCL compiler, not surprisingly, fails in transmogrifying scalar operations on global data into SIMD operations on copies performed using large DMA transfers.

Table 3. Time for wavelet transform and reconstruction in μs for 2048^2, scenario A, using OpenCL on different platforms

	Fermi	i7	Cell
computation	1185	10382	200863
w/ transfer/mapping	10186	14444	435505

4.3   Hybrid Framework

The implementation within the hybrid framework requires to write the control part that provides the required configuration and parameters and instructs the necessary sweeps, and the functionality to copy data into and out of the buffer structures and to do the actual computation. For the transform, only a rectangular region from the input array must be copied from main memory. Computation yields three temporary buffers containing wavelet coefficients that will be copied into the output array, and a forth that will need to be transformed again and needs to be stored in a temporary array. Temporal blocking can be facilitated by immediately performing the next transform(s) recursively and preventing storage of temporary data in main memory. Obviously, this approach can be applied inverse in the reconstruction phase. For $i7$ as well as *Cell*, performing three transform operations in a single sweep were fastest.

Similar to the OpenMP implementation, SIMD-optimized versions of the scalar kernels were written. In contrast to the OpenMP implementation, conversions for scenario B are not performed on the fly, but employ temporary buffer structures. As copying between buffers and persistent storage is done by optimized routines of the framework, non-temporal stores can be combined with unoptimized scalar kernels. To be able to chose the intensity of temporal blocking and the size of tiles at runtime, it was necessary to implement custom dynamic memory management based on memory pools. For the cache-based $i7$, it ensures

---

[6] Using AMD's Stream SDK 2.2 on the CPU.

[7] Using IBM's OpenCL 0.2.

Table 4. Time for wavelet transform and reconstruction on the $i7$ test platform in μs in the hybrid framework implementation. Cases, in which four threads performed better than eight are marked with an asterisk.

	scenario A				scenario B			
size	scalar	scalar NT	SIMD	SIMD NT	scalar	scalar NT	SIMD	SIMD NT
512	412	521	309	421*	521	591	296	399
1024	1727	1758	1463	1374	1782	2012	946	1222
2048	7031	6419	6117	4930	7388	7360	4624	4147
4096	27849	25589	24651	20304	29197	29354	18208	16214

Table 5. Time for wavelet transform and reconstruction on the *Cell* test platform in μs in the hybrid framework implementation

	scenario A		scenario B	
size	scalar	SIMD	scalar	SIMD
512	889	338	1710	301
1024	3268	1209	6552	1025
2048	12269	4299	25142	3580

that memory in cache is reused again. On *Cell*, SPEs allocate memory on their LS anyway, but the default dynamic memory management implementation is very slow.

Table 4 shows the performance for the framework implementation on the $i7$ platform. Similar to the OpenMP implementation, usage of streaming memory operations can severely reduce performance for small problems. Performance on the *Cell* platform is shown in table 5. As a PlayStation 3 provides only 256 MiB of main memory, data of size 4096^2 could not be tested. Please note that on the same user code is used, except for the tailored SIMD kernels. The difference between scalar and manually SIMD-vectorized code, however, is much larger on *Cell*: the compiler fails in SIMD-vectorizing the scalar description automatically and therefore must create intricate code to imitate scalar operation with the SIMD-centralized ISA of SPEs. This also explains why scenario B performs worse although it requires less memory transfer: The generated code for the scalar description is compute bound.

4.4 Comparison

Table 6 gives an overview of the shortest runtimes measured. Except for the smallest data size evaluated, the computation on *Fermi* using OpenCL takes by far least time, but the transfer through the PCIe makes it an uneconomical approach unless further operations on that data profit from GPGPU. The hybrid framework is able to save memory transfer, but introduces additional overhead for copying between persistent storage and temporary buffers. As a consequence,

Table 6. Overview of time for wavelet transform and reconstruction. The best results were taken on the respective platforms for the hybrid framework (HWF), OpenMP and OpenCL, for the latter the time for only the computation is given in parentheses.

scenario	size	HFW i7	OpenMP i7	HFW Cell	OpenCL Fermi
A	512	309^T	194^T	338	1918 (430)
A	1024	1374^{NT}	1474^{NT}	1209	3371 (611)
A	2048	4930^{NT}	5763^{NT}	4299	10186 (1185)
A	4096	20304^{NT}	23609^{NT}	–	–
B	512	296^T	156^T	301	1270 (423)
B	1024	946^T	858^T	1025	2244 (574)
B	2048	4147^{NT}	4591^{NT}	3580	5130 (1033)
B	4096	16214^{NT}	19044^{NT}	–	–

it is outperformed by OpenMP for small problems and is only beneficial for large ones. Surprisingly, the two years older *Cell* beats *i7* except for small problems that fit in cache, confirming its suitability for many image processing algorithms.

5   Conclusions

We outlined performance-efficient implementations of the fast wavelet transform for Haar wavelets using OpenMP, OpenCL and our hybrid framework, and presented performance results on three platforms – involving a general-purpose Intel Core i7 processor, an STI Cell/B.E., and an NVIDIA Fermi GPGPU.

Not surprisingly, OpenMP was the most simple solution to use, and produced very good results after optimization of the compute routines, especially SIMD-vectorization and the usage of streaming operations for larger problems. However, OpenMP is limited to general-purpose CPUs.

OpenCL enforces a separation between control and compute code. It became obvious that the OpenCL kernel language is not meant to write code that performs great on any platform. Instead it provides non-orthogonal constructs that match particular features of a certain platform which might be slow to imitate on others. Especially with the SIMT design of NVIDIA's GPUs, straight-forward scalar code performs great for simple algorithms. A drawback of OpenCL is the cumbersome management of contexts, command queues, memory, program and kernel objects, and substantial overhead current implementations involve.

Our hybrid framework performed well on *i7* and could outperform OpenMP on it by exploiting temporal blocking. The control code was simpler to write than for OpenCL, but the compute code required a more complicated management of temporary arrays. SIMD-vectorization was equally simple as for OpenMP. The main contribution of our framework is the possibility to write code working equally well on common multicore CPUs as well as for the CBEA.

References

1. Abschlussbericht des Projekts Ru 422/7-5 (DiME-2). Lehrstuhl für Informatik 10 (Systemsimulation), Friedrich-Alexander-Universität Erlangen-Nürnberg (2008)
2. Christen, M., Schenk, O., Neufeld, E., Messmer, P., Burkhart, H.: Parallel data-locality aware stencil computations on modern micro-architectures. In: Proceedings of the 2009 IEEE International Symposium on Parallel & Distributed Processing, pp. 1–10. IEEE Computer Society (2009)
3. Datta, K., Kamil, S., Williams, S., Oliker, L., Shalf, J., Yelick, K.: Optimization and performance modeling of stencil computations on modern microprocessors. SIAM Review 51(1), 129–159 (2009)
4. Datta, K., Murphy, M., Volkov, V., Williams, S., Carter, J., Oliker, L., Patterson, D., Shalf, J., Yelick, K.: Stencil computation optimization and auto-tuning on state-of-the-art multicore architectures. In: International Conference for High Performance Computing, Networking, Storage and Analysis, SC 2008, pp. 1–12 (2009)
5. Franco, J., Bernabé, G., Fernández, J., Acacio, M.: A Parallel Implementation of the 2D Wavelet Transform Using CUDA. In: Parallel, Distributed and Network-Based Processing, pp. 111–118 (2009)
6. Franco, J., Bernabé, G., Fernández, J., Ujaldón, M.: Parallel 3D fast wavelet transform on manycore GPUs and multicore CPUs. Procedia Computer Science 1(1), 1095–1104 (2010)
7. Garcia, A., Shen, H.: GPU-based 3D wavelet reconstruction with tileboarding. The Visual Computer 21(8), 755–763 (2005)
8. Haar, A.: Zur Theorie der orthogonalen Funktionensysteme. Mathematische Annalen 69, 331–371 (1910)
9. International Business Machines Corporation, Sony Computer Entertainment Incorporated, Toshiba Corporation: Cell Broadband Engine Architecture 1.02 (2007)
10. Kowarschik, M.: Data Locality Optimizations for Iterative Numerical Algorithms and Cellular Automata on Hierarchical Memory Architectures (2004)
11. McKinley, K.S., Carr, S., Tseng, C.W.: Improving data locality with loop transformations. ACM Trans. Program. Lang. Syst. 18(4), 424–453 (1996)
12. Mohiyuddin, M., Hoemmen, M., Demmel, J., Yelick, K.: Minimizing communication in sparse matrix solvers. In: SC 2009: Proceedings of the Conference on High Performance Computing Networking, Storage and Analysis. pp. 1–12. ACM, New York (2009)
13. Ohshima, S., Hirasawa, S., Honda, H.: OMPCUDA: OpenMP Execution Framework for CUDA Based on Omni OpenMP Compiler. In: Beyond Loop Level Parallelism in OpenMP: Accelerators, Tasking and More, pp. 161–173 (2010)
14. Stürmer, M., Rüde, U.: A framework that supports in writing performance-optimized stencil-based codes. Tech. Rep. 10-5, Lehrstuhl für Informatik 10 (Systemsimulation), Friedrich-Alexander-Universität Erlangen-Nürnberg (2010)
15. Tenllado, C., Setoain, J., Prieto, M., et al.: Parallel implementation of the 2d discrete wavelet transform on graphics processing units: Filter bank versus lifting. IEEE Transactions on Parallel and Distributed Systems 19(3), 299–310 (2008)
16. Weiß, C.: Data Locality Optimizations for Multigrid Methods on Structured Grids. Ph.D. thesis, Lehrstuhlr für Rechnertechnik und Rechnerorganisation, Institut für Informatik, Technische Universität München, Munich, Germany (2001)
17. Wellein, G., Hager, G., Zeiser, T., Wittmann, M., Fehske, H.: Efficient temporal blocking for stencil computations by multicore-aware wavefront parallelization. In: Proceedings of the 2009 33rd Annual IEEE International Computer Software and Applications Conference, vol. 01, pp. 579–586. IEEE Computer Society (2009)

Direct Sparse Factorization
of Blocked Saddle Point Matrices

Claude Lacoursière[1,*], Mattias Linde[2], and Olof Sabelström[3]

[1] HPC2N/UMIT, UmeåUniversity, Sweden
claude@hpc2n.umu.se
[2] Algoryx Simulation AB, Umeå, Sweden
[3] Department of Computing Science, UmeåUniversity, Sweden

Abstract. We present a parallel algorithm for the direct factorization of sparse saddle-point matrices of moderate size coming from real-time multibody dynamics simulations. We used the specific structure of these problems both for *a priori* construction of supernodes and to avoid all dynamic permutations during factorization. For the latter, we present a technique we call "leaf swapping" which performs permutations of the supernodes in the elimination tree without any reference to numerical values. The results compare favorably with currently available high performance codes on our problem sets because of the high overhead necessary to process very large problems on increasingly complex supercomputers.

1   Introduction

We consider the direct factorization of symmetric indefinite sparse matrices for the solution of linear systems of the form

$$Hz = b, \text{ where } H = \begin{bmatrix} M & G^T \\ G & -T \end{bmatrix}, \tag{1}$$

where the vectors $z^T = [x^T \quad y^T]$ and $b^T = [c^T \quad d^T]$, have compatible dimensions with the block matrices in H. Matrix M is symmetric, positive definite and assumed to be well conditioned. Matrix T is symmetric positive semi-definite with very small entries in comparison to those in M. In particular, the diagonal elements T_{ii} are considered to be logical zeros. The rectangular matrix G is such that $[G \quad -T]$ has full row rank. The resulting factorization yields

$$H = PLDL^T P^T, \tag{2}$$

where L is unit lower triangular, D is diagonal, and P is a permutation matrix. The latter is computed *a priori* to reduce fill-ins in the L factor and to avoid pivots involving the diagonal elements T_{ii} which are assumed to be very small. We seek to exclude dynamic permutations based on the numerical values of individual entries during factorization. This is usually difficult to achieve for the

---

* Corresponding author.

K. Jónasson (Ed.): PARA 2010, Part II, LNCS 7134, pp. 324–335, 2012.

symmetric indefinite matrices. We also seek to preserve the natural block structure inherent to certain classical mechanical problems to construct supernodes. This increases performance since the smaller number of supernodes speeds up the analysis stage, and since operations on supernodes involve BLAS3 kernels as opposed to scalar operations on the variables themselves. The detailed structure of matrix H is given below in Sec. 4.

The problem with factorization of symmetric indefinite matrices is easily illustrated with a 2×2 matrix and its two symmetric permutations, yielding the factorizations

$$\begin{bmatrix} m & g \\ g & -t \end{bmatrix} = \begin{bmatrix} 1 & 0 \\ g/m & 1 \end{bmatrix} \begin{bmatrix} m & 0 \\ 0 & -t - g^2/m \end{bmatrix} \begin{bmatrix} 1 & g/m \\ 0 & 1 \end{bmatrix} \tag{3}$$

and

$$\begin{bmatrix} -t & g \\ g & m \end{bmatrix} = \begin{bmatrix} 1 & 0 \\ -g/t & 1 \end{bmatrix} \begin{bmatrix} -t & 0 \\ 0 & m + g^2/t \end{bmatrix} \begin{bmatrix} 1 & -g/t \\ 0 & 1 \end{bmatrix}. \tag{4}$$

Clearly, the second ordering is not numerically stable as $t \to 0$, and the matrix itself is badly conditioned if $|g| \approx 0$. Diagonal block matrices of the form in Eqn. (3) with $g \neq 0$ are called 2×2 pivots. The Bunch-Kaufman (BK) procedure [6] is commonly used to locate the pivots when a scalar one is unsuitably small.

Our application context is the numerical integration of mechanical systems made of discrete elements subject to kinematic constraints, namely, multibody systems dynamics. This requires the solutions of saddle point problems of the form given in Eqn. (1), since as a consequence of the least action principle, the discretized trajectories are in fact solutions of constrained optimization problems [11].

Our multibody dynamics simulations are at the heart of interactive applications such as virtual environment-based training systems. These have proved effective for educating heavy machinery operators and surgeons among others. Due to physiological and psychological factors, these simulations must deliver visual updates at 60Hz, giving at most sixteen milliseconds available time for computations. There is constant demand for more realism and more functionality and this translates to a demand for faster numerical codes. Given the stagnation of clock speeds, both on CPUs and data buses, parallelism is the only way to achieve this.

Our main end user applications requires direct solutions of the linear systems in double precision because of the mass ratios and stiffness involved. Unfortunately, existing libraries designed for high performance computing computing (HPC) incur too much overhead to be usable and this is why we developed our own.

The performance of direct sparse factorization codes depends strongly on good reorderings for the minimization of fill-ins, the introduction of blocking to maximize utilization of BLAS3 kernels, the reduction or elimination of the need for dynamic pivoting, and the optimization of memory alignment. For parallel codes, it is also necessary to reorganize the variables in evenly sized, weakly coupled groups to minimize communication and balance the work load. This reorganization can compete with fill-in reduction and decrease speed up. The main tradeoff is between the time taken for symbolic analysis vs that taken for numerical factorization.

In what follow, we briefly summarize the state of the art in Sec. 2, explain the principles behind our design in Sec. 3 and provide the specifics applicable to multibody problems in Sec. 4, which also includes our main results regarding numerically safe static pivoting which we call "leaf swapping". Performance results and a perspective on future work are presented in Sec. 5 and Sec. 6.

2   State of the Art

There are two main parallel codes which can process saddle point matrices, namely, MUMPS [1] and PARDISO [12,13]. The both provide *a priori* stability preserving reordering for saddle point matrices, eliminating the need for most dynamic pivoting [5,13]. These strategies are based on the numerical values of the matrix elements and rely on the solution of maximum matching problems. Both MUMPS and PARDISO are HPC codes designed to handle very large problems and even offer out-of-core functionality. They are also designed for general problems and make no assumption on structure. For maximum flexibility, they use standard input matrix formats such as triplet, row, or column packed formats.

Recent performance analysis [7] indicate that MUMPS is a good representative choice for our performance tests, especially since it provides extensive support for saddle-point matrices.

The analysis phase of direct sparse factorization codes consists of extracting the graph structure of the matrix from the input format, computing a sparsity preserving reordering based on that, discovering supernodes by matching sparsity patterns between columns, then computing the elimination tree, and then computing a schedule for balancing the work load. The size of the supernodes and the exact data layout of the data actually used during factorization is outside of the user's control. In addition, the assumption made in MUMPS is that the frontal matrices are dense and large which is good since this increases the BLAS3 fraction. This is not true for the problems we consider however since the overhead of BLAS3 library calls for the relatively small frontal matrices is prohibitive.

After experimenting with MUMPS we concluded that our problems which involve a few thousand variables at the present time, are simply too small to benefit from parallelism and other features of that code because of the excessive overhead necessary for general purpose software. Given the current trends in HPC software, this can only get worse and for good reasons. Factorization is relatively expensive and thus for very large problems, any amount of analysis is easily amortized. The converse of this is that the HPC codes are likely to become less usable for smaller problems.

3   Our Strategy: The Sabre Solver

Multibody systems, lumped, discrete, and finite element models share a common structure, namely, bodies or elements interacting via forces or connected with kinematic constraints. This corresponds to a bipartite graph. Each body, element or interaction carries a number of variables which defines a natural block structure. The computation of forces and the time integration requires traversing this bipartite graph. For strongly coupled systems, it is also necessary to solve systems of linear equations whose matrices have essentially the same graph. The difference being that the graph of the matrix has one node for each variable, but the graph of the system has one node for each body, element, or interaction.

Having direct access to this bipartite graph makes it possible to construct the elimination tree of the matrix and perform the fill-reducing analysis considering only a fraction of the variables. It is then possible to allocate memory with optimal layout before numerical values are computed and thus avoid all data copying. We also use symbolic analysis only to avoid bad numerical pivots prior to factorization using our "leaf swapping" technique described below in Sec. 4.

In terms of data layout, our matrices are composed of small, dense, rectangular blocks in row major format. These blocks are contiguous, i.e., the leading dimension is the largest dimension of the block except for a small amount of padding necessary for keeping good alignment. We use our own hand-coded BLAS3 inline kernels to perform matrix-matrix multiplication and symmetric rank-k updates (GEMM and SYRK) for small dense matrices of various sizes using SSE3 instructions. The blocks themselves are then stored in column packed formats. This is a simple two level hierarchical data layout.

Our factorization is based on common multifrontal techniques [8] and uses threads since we target multi-CPU multi-core systems. We also used both Approximate Minimum Degree (AMD[2]) and the well-known nested dissection code METIS [9] as the basis for reordering.

Considering that our matrices have moderate size, we separate the tasks as purely parallel and purely sequential ones, the latter assigned to the master thread and performed last. The parallel tasks factor independent subtrees in the elimination tree. To select the subtrees and assign them to the different threads, we first traverse the elimination tree from the leaves upward and compute an estimate of the work required by each node, which includes the sum of that required by its children. The negative of this estimate is used as a key in a heap [4]. Then, starting at the top of the heap, we delete the root and insert its children, and repeat this procedure until the number of nodes in the heap is sufficiently large given the number of available threads. These nodes, which are subtrees in the elimination tree and entirely independent of each other, are then assigned to the threads in a greedy manner. Once the master thread has completed its set of parallel tasks, it continues with the serial one. This serial task traverses the elimination tree from the root down to the fully factored subtrees, and processes the nodes that were deleted from the heap during the load balancing computations on the way back up to the root. This is the only task that involves synchronization. The factorization of larger matrices would require further refinement and parallelization of the serial task.

We coined our software SABRE after Zorro's main weapon which is light and effective.

4   Specialization to Multibody Systems

The physical systems considered consist of a number of bodies with generalized coordinates q and generalized velocities v subject to kinematic constraints of the form $g(q) = 0$. The numerical integration of the equations of motion requires the solution of systems of linear equations of the form given in Eqn. (1), in which M is the mass matrix of the system, and $G = \partial g/\partial q$ is the Jacobian matrix of the constraints. Matrix T is a diagonal perturbation with very small non-negative elements and so any T_{ii} is considered as an unsafe pivot. It is for these elements that we require the 2×2 pivots. The solution of Eqn. (1) yields the discrete acceleration $\dot{v}$ as well as the vector of Lagrange multipliers λ which produce the constraint force $G^T\lambda$.

Given n bodies, the mass matrix has the form

$$M = \mathrm{diag}(M^{(1)}, M^{(2)}, \dots, M^{(n)}) \tag{5}$$

where the blocks $M^{(i)}$ are the $k^{(i)} \times k^{(i)}, i = 1, 2, \dots, n$ mass matrices of the individual bodies, each with $k^{(i)}$ degrees of freedom.

The constraints $g^{(i)}, i = \{1, 2, \dots, m\}$ come in blocks of $l^{(i)}$ equations, each acting on $r^{(i)}$ bodies. The Jacobian matrix then has the block form

$$G^{(i)} = \left[\dots G^{(i)}_{b_{j_1}} \dots G^{(i)}_{b_{j_2}} \dots G^{(i)}_{b_{j_s}} \dots \right], \tag{6}$$

with $s = r^{(i)}$, and where the ellipsis represent sequences of zero blocks. Each block matrix $G^{(i)}_b$ is rectangular of size $l^{(i)} \times k^{(b)}$. The perturbation matrix Matrix T has the block diagonal form

$$T = \mathrm{diag}(T^{(1)}, T^{(2)}, \dots, T^{(m)}), \tag{7}$$

with one block for each constraint.

For rigid multibody systems for instance, a hinge joint brings a block of five rows containing two nonzero blocks of six columns each in G. This corresponds to the number of individual constraint equations imposed by the geometry of the hinge joint in this case, and the number of degrees of freedom of each of the two rigid bodies connected by the constraint. The asymptotic fill ratio of matrix H is $O(1/(n \cdot f + m \cdot p))$, where f is the average number of degrees of freedom of the bodies, and p is the average number of bodies interacting via a constraint or a force. For short range forces, p is small.

The computation of a fill reducing reordering of matrix H is a graph theoretical problem[2,9], which can be addressed directly by considering the connectivity of the physical system, instead of the numerical matrix itself. But since fill-reducing reordering makes no consideration for numerical values, numerically unstable permutations can be generated.

This is illustrated in Fig. 1 for the case of a slider-crank mechanism. The same figure contains the elimination tree obtained after an AMD reordering which has precisely the form we want to avoid. The degree of a node here is the number of connections to other nodes. Indeed, the constraints at the end of the mechanisms have degree 1 as opposed to degree 2 for all other nodes in the graph, and so, one of them becomes a leaf. The resulting reordered matrix H in Eqn. (8) cannot be factored safely without permutation since the first block is very small, i.e., $T_{11} \approx 0$ and thus an unsafe pivot. The labels used in Eqn. (8) apply to blocks as they would have been assigned in the block form showed in Eqn. (1) once each block is subdivided in the manner described, i.e., individual bodies and constraints each introduce a block. This means also that G_{31} and G_{32} are the constraint blocks from constraint 3 which apply to body 1 and 2 respectively. Reordering the leaf node C_1 with its first "body" ancestor as shown in Fig. 1 yields matrix $\tilde{H}$ in Eqn. (8) with only one more fill-in at position $(3, 2)$, marked with $\star$, but is numerically stable, provided M_{22} is large enough, and that is our assumption.

$$
H = \begin{bmatrix}
-T_{11} & & & & \\
G_{12}^T & M_{22} & & & \\
& G_{32} & -T_{33} & & \\
& & G_{34}^T & M_{44} & \\
& & & G_{54} & -T_{55}
\end{bmatrix}, \quad
\tilde{H} = \begin{bmatrix}
M_{22} & & & & \\
G_{12} & -T_{11} & & & \\
G_{32} & \star & -T_{33} & & \\
& & G_{34}^T & M_{44} & \\
& & & G_{54} & -T_{55}
\end{bmatrix}. \tag{8}
$$

We call this procedure "leaf swapping". In general, a T leaf is permuted upward until its child is an M node, and this procedure is repeated until there are no T leaves. This reordering is very quickly performed and involves no data motion and no indirect addressing, just simple permutation operations on a tree. A larger example of leaf swapping is shown in Fig. 2.

After leaf swapping, the blocks on the main diagonal have the form

$$
H_{ii} = \begin{bmatrix} \bar{M}^{(i)} & \bar{G}^{(i)T} \\ \bar{G}^{(i)} & -\bar{T}^{(i)} \end{bmatrix} \text{ or } H_{ii} = \bar{M}^{(i)}, \tag{9}
$$

and these can be factored without without fear of division by zero as shown in Eqns. (3) and (4). There might still be stability issues in the blocks H_{ii} though, and this can be addressed by using any of the stable algorithms that include pivoting [6], or by introducing perturbations dynamically as is done in PARDISO for instance [12,13], and the errors thus introduced can then be removed using iterative refinement if needed. The computational cost of using complete pivoting here would be small since the blocks H_{ii} have moderate size and are dense.

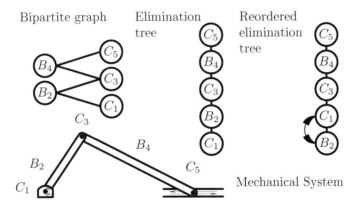

Fig. 1. The slider crank mechanism schematics across the bottom, the corresponding bipartite graph above, and the elimination tree produced by AMD to the right

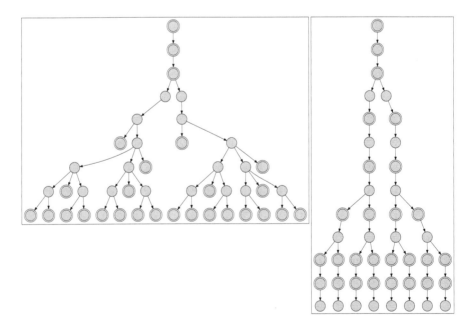

Fig. 2. Leaf swapping on a matrix. The disks with a white band are the unsafe T nodes, the others are safe M nodes. The reordering makes the tree taller, which translates to both less parallelism and more fill-ins.

Table 1. Timing data in milliseconds for selected matrices of size 2867, 2278 and 2328, respectively. The prefix labels A, F, and T stand for analysis, factor, and total time, respectively. Labels S and M stand for SABRE and MUMPS, respectively, and the suffix 1 and 2 denote AMD and METIS reordering, respectively. MUMPS uses a reordering based on METIS followed by stability preserving permutations.

#Threads	Lines on deck				Tractor with rocks				Fluid matrix			
	1	2	3	4	1	2	3	4	1	2	3	4
A-S1	2.07	2.40	2.41	2.41	0.29	0.44	0.45	0.46	3.43	4.94	4.95	5.00
A-S2	3.44	3.84	3.84	3.85	1.29	1.45	1.46	1.46	9.76	12.52	12.52	12.48
A-M	12.73	13.27	13.37	13.47	14.64	15.16	15.38	15.40	38.13	39.39	39.51	41.19
F-S1	1.88	0.98	0.81	0.72	0.96	0.61	0.41	0.40	12.92	6.65	7.20	8.07
F-S2	2.07	1.08	0.76	0.76	1.09	0.60	0.43	0.39	39.30	20.49	18.30	19.33
F-M	5.51	5.13	4.70	5.13	4.76	6.28	6.46	7.05	18.73	22.58	17.39	13.83
T-S1	3.95	3.38	3.22	3.13	1.25	1.05	0.86	0.86	16.35	11.59	12.15	13.07
T-S2	5.51	4.92	4.60	4.61	2.38	2.05	1.89	1.85	49.06	33.01	30.82	31.81
T-M	18.24	18.40	18.07	18.60	19.40	21.44	21.84	22.45	56.86	61.97	56.90	55.02

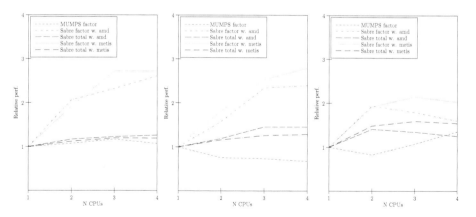

Fig. 3. Scalability analysis. The representative problems are the "line on deck" with 2867 variables on the left, the "tractor with rocks" with 2278 variables in the center, and the "fluid" with 2328 variables.

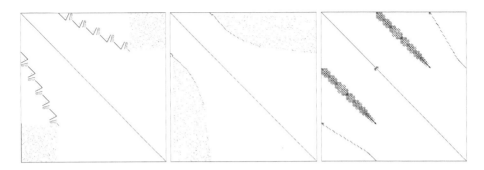

Fig. 4. Matrices investigated here in their natural, unpermuted form. From left to right, these are the "line on deck", "tractor with rocks", and "fluid" problems, with 2867, 2278 and 2328 variables, respectively.

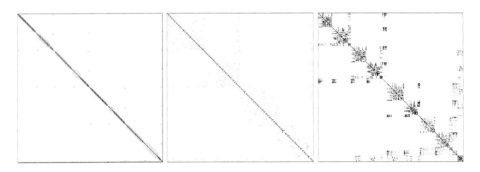

Fig. 5. Sparsity pattern after factorization and reordering with METIS followed by leaf swaping of matrices shown in Fig. 4

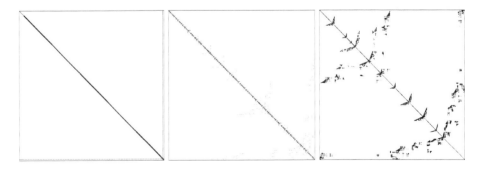

Fig. 6. Sparsity pattern after factorization and reordering with AMD followed by leaf swaping of matrices shown in Fig. 4

5   Experimental Results

We coded our libraries in C++ and used pthreads on a Linux desktop. We used an Intel® Core™3.07GHz i7-950 processor which has four cores. Our specific model has L1 I and D caches of 32KB, L2 cache of 256KB, and L3 cache of 8192KB. The main memory was 6GB accessed by a 1066MHz bus. The OS was Linux, kernel 2.6.32 (Ubuntu 10.04). We used GNU gcc for our code and gfortran for MUMPS and linked against OPEN MPI.

We chose three representative matrices extracted from three different simulations. The first relates to marine cables and anchors in contacts with ships and oil platforms [14]. The second is a wheel loader which shovels small stones. The third example is from a fluid simulation based on smoothed particle hydrodynamics, but including kinematic constraints to preserve volume and thus produce incompressibility [3]. All three problems have different connectivity structure. Common to the three simulations is a numerical integration method designed for constrained systems [10]. The typical block sizes in these problems are 6×6 for the first two, but there is a significant fraction of 1×3 blocks for the last. For such node sizes, it makes no sense to use standard BLAS routines but specialized kernels are very effective nevertheless.

The matrices in default ordering are shown in Fig. 4. The result of fill-reducing and leaf swapping reordering with METIS and AMD, respectively, are shown in Fig. 5 and Fig. 6, respectively. Nested dissection algorithms such as METIS are usually better at producing balanced elimination trees than AMD, though they usually produce more fill-ins. Unfortunately, the leaf-reordering procedure breaks this balance and introduces even more fill-ins. The reordering is followed by memory allocation, load balancing computation, and then factorization. Timing results of the different phase of the factorization process are provided in Table 1 which validates our claims. We obtain good scaling for the factorization itself as shown in Fig. 3, but due to the cost of the analysis and preparation phase, the overall result is sublinear. Yet there is almost no speedup for the factorization phase of MUMPS. This is not surprising as the user's manual clearly indicates that parallelization is beneficial only for problems which take at least a few seconds to solve with the sequential version. A revealing aspect is the amount of time required by MUMPS to perform the analysis, which seems to increase when there are more threads available. When this is considered, our solution is up to 20 times faster in some cases, and still more than 13% faster in the worst case. More to the point, MUMPS timings are never within the real-time requirement of 16.67ms, as seen from Table 1.

6   Conclusion

Our results validate the effort invested in constructing this new specialized code and demonstrate that the current trend in the development of parallel sparse factorization software leaves a gap in functionality. The increasing overhead required to solve very large problems efficiently significantly degrades the performance on smaller ones.

The techniques presented here are usable for other physical problems. Our assumptions are only that the matrices are symmetric and have explicit block structure defined by the physical models. Our leaf swapping technique is applicable to any symmetric indefinite matrix as long as nodes can be explicitly identified as M or T nodes, referring to the sub-matrices in Eqn. (1).

The analysis phase in our code is still a significant fraction of the total cost which indicates where to focus our future work. We are currently considering fusing the computation of fill-in reducing and leaf swapping permutations together with the partitioning for load balancing, since these are all related graph problems. There are also aspects of the connectivity structure common to several problems in mechanics which could be exploited in this analysis.

We are also investigating recursive blocked data structures to store the sparse columns to improve data locality, as well as possible GPGPU implementations.

Acknowledgments. This research was supported by High Performance Computing Center North (HPC2N), Swedish Foundation for Strategic Research grant (A3 02:128), EU Mål 2 Structural Funds (UMIT-project), VINNOVA/ProcessIT Innovations, and by Algoryx Simulation AB. Critical comments from our reviewers helped improve our manuscript significantly.

References

1. Amestoy, P.R., Duff, I.S., L'Excellent, J.Y.: Multifrontal parallel distributed symmetric and unsymmetric solvers. Computer Methods in Applied Mechanics and Engineering 184(2–4), 501–520 (2000)
2. Amestoy, P.R., Enseeiht-Irit, Davis, T.A., Duff, I.S.: Algorithm 837: AMD, an approximate minimum degree ordering algorithm. ACM Trans. Math. Softw. 30(3), 381–388 (2004)
3. Bodin, K., Lacoursière, C., Servin, M.: Constraint fluids. IEEE Trans. on Visualization and Computer Graphics (2010) (to appear)
4. Cormen, T.H., Leiserson, C.E., Rivest, R.L.: Introduction to Algorithms. The MIT Electrical Engineering and Computer Science Series. McGraw-Hill, New York (1990)
5. Duff, I.S., Pralet, S.: Towards stable mixed pivoting strategies for the sequential and parallel solution of sparse symmetric indefinite systems. SIAM J. on Mat. Anal. and Appl. 29(3), 1007–1024 (2007)
6. Golub, G.H., Van Loan, C.F.: Matrix Computations, 3rd edn. Johns Hopkins Studies in the Mathematical Sciences. Johns Hopkins Press, Baltimore (1996)
7. Gould, N.I.M., Scott, J.A., Hu, Y.: A numerical evaluation of sparse direct solvers for the solution of large sparse symmetric linear systems of equations. ACM Trans. Math. Softw. 33(2), 10 (2007)
8. Gupta, A., Karypis, G., Kumar, V.: Highly scalable parallel algorithms for sparse matrix factorization. IEEE Trans. on Par. and Dist. Syst. 8(5), 502–520 (1997)
9. Karypis, G., Kumar, V.: A fast and high quality multilevel scheme for partitioning irregular graph. SIAM J. Sci. Comp. 20(1), 359–392 (1998)

10. Lacoursière, C.: Regularized, stabilized, variational methods for multibodies. In: Bunus, D.F.P., Führer, C. (eds.) The 48th Scandinavian Conference on Simulation and Modeling (SIMS 2007). Linköping Electronic Conference Proceedings, pp. 40–48. Linköping University Electronic Press, Linköping (2007)
11. Marsden, J.E., West, M.: Discrete mechanics and variational integrators. Acta Numer. 10, 357–514 (2001)
12. Schenk, O., Gaertner, K.: On fast factorization pivoting methods for sparse symmetric indefinite systems. Elec. Trans. Numer. Anal. 23, 158–179 (2006)
13. Schenk, O., Waechter, A., Hagemann, M.: Matching-based preprocessing algorithms to the solution of saddle-point problems in large-scale nonconvex interior-point optimization. Computational Optimization and Applications 36(2–3), 321–341 (2007)
14. Servin, M., Lacoursière, C., Nordfelth, F., Bodin, K.: Hybrid, multiresolution wires with massless frictional contacts. IEEE Transactions on Visualization and Computer Graphics (2010) (in press)

Multi-Target Vectorization
with MTPS C++ Generic Library

Wilfried Kirschenmann[1,3], Laurent Plagne[1], and Stéphane Vialle[2,3]

[1] SINETICS Department, EDF R&D, France
{wilfried.kirschenmann,laurent.plagne}@edf.fr
[2] SUPELEC - UMI 2958, France
stephane.vialle@supelec.fr
[3] AlGorille INRIA Project Team, France

Abstract. This article introduces a C++ template library dedicated at vectorizing algorithms for different target architectures: Multi-Target Parallel Skeleton (MTPS). Skeletons describing the data structures and algorithms are provided and allow MTPS to generate a code with optimized memory access patterns for the choosen architecture. MTPS currently supports x86-64 multicore CPUs and CUDA enabled GPUs. On these architectures, performances close to hardware limits are observed.

Keywords: GPU, SSE, Vectorization, C++ Template Metaprogramming, Performances.

1 Introduction

In many scientific applications, computation time is a strong constraint. Optimizing these applications for the rapidly changing computer hardware is a very expensive and time consuming task. Emerging hybrid architectures tend to make this process even more complex.

The classical way to ease this optimization process is to build applications on top of High Performance Computing (HPC) libraries that are available for a large variety of hardware architectures. Such scientific applications, whose computing time is mostly consumed within such HPC library subroutines, then automatically exhibit optimal performances for various hardware architectures.

However, most classical HPC libraries implement fixed APIs like BLAS which may make them too rigid to match the needs of all client applications. In particular, classical APIs are limited to manipulate rather simple data structures like dense linear algebra matrices. As a more complex issue, general sparse matrices cannot be represented with a unified data structure and various formats are proposed by more specialized libraries. In the extreme case, structured sparse matrices cannot be efficiently captured by any of the classical library data structures. Several neutron transport codes developed at EDF R&D rely on such complex matrices that another kind of library is required.

Following the model of the C++ Standard Template Library (STL), template based *generic libraries* such as Blitz++ [13] provide more flexible APIs and

K. Jónasson (Ed.): PARA 2010, Part II, LNCS 7134, pp. 336–346, 2012.
© Springer-Verlag Berlin Heidelberg 2012

extend the scope of library-based design for scientific applications. Such generic libraries allow to define Domain Specific Embedded Languages (DSELs) [2].

Legolas++ (*GLASS* in [11]), a basis for several HPC codes at EDF, is a C++ DSEL dedicated to structured sparse linear algebra. In order to meet EDF's industrial quality standards, a *multi-target* version of Legolas++, currently under development, will provide a unified interface for the different target architectures available at EDF, including clusters of heterogeneous nodes (i.e., with both multi-core CPUs and GPUs). This article presents MTPS (Multi-Target Parallel Skeletons), a C++ generic library dedicated to *multi-target* vectorization that is used to write the *multi-target* version of Legolas++. Only developments concerning a single heterogeneous node are presented here.

The next section presents the principles of Legolas++ and Section 3 introduces MTPS. Its optimization strategies and the achieved performances are discussed in Section 4. Finally, conclusions are drawn in Section 5.

2   Towards a Multi-Target Linear Algebra Library

Legolas++ is a C++ DSEL developed at EDF R&D to build structured sparse linear algebra solvers. Legolas++ provides building bricks to describe structured sparse matrix patterns and the associated vectors and algorithms. This library separates the actual implementation of the Linear Algebra (LA) computationnal kernels from the physics issues.

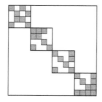   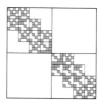

Fig. 1. Block Matrix Patterns. A block diagonal matrix pattern is represented on the left while a block diagonal matrix with tridiagonal blocks is represented on the right.

Block decomposition is a common linear algebra operation that allows to describe the sparsity pattern of a given matrix from one or several basic sparsity patterns. For example Fig. 1(left) represents a matrix with a block diagonal sparsity pattern that can help to identify the optimal algorithm for dealing with this matrix. If a matrix block can itself be block-decomposed, the matrix is said to be a multi-level block matrix. Fig. 1(right) represents such a multi-level block matrix which is diagonal with tridiagonal blocks. Legolas++ is a C++ library developed at EDF R&D for structured sparse linear algebra problems. The central issue in this domain is to describe efficiently multi-level block matrices as combinations of basic sparsity patterns. Legolas++ allows to access the block elements of a block matrix in the same manner as if it was a simple matrix. For example A(i,j) returns a reference to the (i,j) matrix element which can be either a scalar if A is a matrix, or a block if A is a block matrix. In the latter case, this block can be seen as a simple sub-matrix and provides the same interface. This means that A(i,j)(k,l) returns a reference to the (k,l) scalar element of the (i,j) block.

Such a block matrix naturally operates with 2D vectors. For example let us consider the following matrix-vector product $Y = Y + A * X$ where A is a block matrix and X and Y are vectors. Legolas++ allows to implement this product as:

```
1 for (int i=0 ; i < A.nrows() ; i++)
2     for (int j=0 ; j < A.ncols() ; j++)
3         Y[i]+=A(i,j)*X[j];
```

In this case, each elementary operation Y[i]+=A(i,j)*X[j] corresponds to a simple matrix-vector sub-product and the C++ compiler statically transforms the previous handwritten algorithm into the following generated block algorithm:

```
1 for (int i=0 ; i < A.nrows() ; i++)
2     for (int j=0 ; j < A.ncols() ; j++)
3         for (int k=0 ; k < A(i,j).nrows() ; k++)
4             for (int l=0 ; l < A(i,j).ncols() ; l++)
5                 Y[i][k]+=A(i,j)(k,l)*X[j][l];
```

The previous algorithm shows that the vectors X and Y, corresponding to the block matrix A, are two-dimensional. In order to simplify the Legolas++ vocabulary, one describes a block matrix like A as a 2-level Legolas++ matrix that operates on 2D Legolas++ vectors. The main objective of the Legolas++ library is to provide tools for the users to explicitly define their n-level matrices and corresponding nD vectors. For instance, Fig. 2 shows the sparsity pattern of a 5-level Legolas++ matrix block that belongs to the 7-level matrix of our neutron transport code [11,6,8].

The explicit GPU parallelization of a neutron transport code resulted in speed-ups around 30 over the sequential Legolas++ CPU implementation [6,8]. To generalize this gain of performances to other Legolas++ based applications, a parallel and multi-target version of Legolas++ is to be developed. As the parallel CPU and GPU versions exhibit strong similarities, Legolas++ developments for

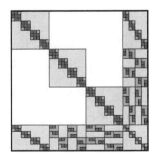

Fig. 2. Sparsity Pattern of a 5-level Legolas++ block matrix

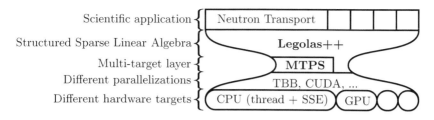

Fig. 3. Our hourglass software architecture to achieve a multi-target Legolas++: a minimal MTPS library adapts the code for different hardware architectures

the different targets are factorized into an intermediate layer between Legolas++ and the different hardware architectures, namely MTPS (see Fig. 3). Note that examples provided in the following of this paper correspond only to MTPS code as the multi-target version of Legolas++ is currently under development.

3 Introduction to MTPS

3.1 Related Work

Many libraries parallelize for different architectures from a single source code. A complete bibliography is beyond the scope of this article; only some examples based on C++ meta-programming techniques are introduced.

Some libraries, like TrilinosNode [1], Quaff [4] or Intel TBB [12], require their users to explicitly express the parallelism within the application by using *parallel skeletons*. This expression of available parallelism can be encapsulated into specialized and implicitly parallel STL-like *containers* and *algorithms*, as in Thrust[1] and Honei [3].

Our goal is to provide implicit parallelism within Legolas++ *containers* and *algorithms*. To ease the writing of its *containers* and *algorithms*, Legolas++ relies on MTPS which follows a *parallel skeletons* based approach. Then MTPS optimizes the code for the different architectures.

As this article presents MTPS, only code using MTPS is shown. However Legolas++ will hide MTPS details in its *containers* and *algorithms* so its user do not need to be aware of MTPS.

3.2 *Collections* and *Vectorizable Algorithms*

This section introduces the notions of *collection* and *vectorizable algorithm* on which MTPS relies.

In C++, a *Plain Old Data* (POD) is a data structure that is represented only as passive collections of field values, without using encapsulation or other object-oriented features. POD non-static data members can only be integral types or

---

[1] Thrust: http://code.google.com/p/thrust/

PODs. As a POD have neither constructor nor desctructor, it can be copied or moved in memory [5]. This particularity allow MTPS to copy a POD from one memory space to another (e.g., GPU memory space). In the following code snippet, MyPOD is a POD with three float data members:

```
1 struct MyPOD{ float a,b,c; };
```

Let a *collection* be a data structure containing different instances of the same POD and f be a *pure* function (i.e., f has no side effects). An algorithm applying f to all elements of a *collection* is said to be *vectorizable*. To parallelize such algorithms, MTPS provides two parallel skeletons optimized for different target architectures: *map* and *fold* which correspond to a *parallel for loop* and to a *parallel reduction* respectively.

An algorithm applied to a set of data is *vectorizable* if and only if this set of data is considered as a *collection* and if the algorithm can be decomposed as a *pure* function applied to each element of the *collection*. We say that an algorithm is *vectorizable* in reference to a given *collection*. For instance, an algorithm operating on each row of a matrix is *vectorizable* only if the matrix is considered as a *collection* of rows. The same algorithm is not *vectorizable* if the matrix is considered as a *collection* of columns: the matrix must be transformed (i.e., transposed). Two algorithms *vectorizable* in reference to the same *collection* are said to be in the same *vectorial context*. On the contrary, if two consecutive algorithms are not *vectorizable* in reference to the same *collection*, a *context switch* (the matrix transposition in our example) is required. In a distributed memory system, *context switches* correspond to communications.

3.3   Linear Algebra *Hello World* of MTPS: saxpy

This section presents how to use MTPS to implement the saxpy operation and to execute it efficiently on different target architectures. The saxpy operation is part of the BLAS interface and its C implementation is:

```
1 float *X, *Y, a;
2 for(int i=0; i<N; ++i) Y[i]+=a*X[i];
```

First, the iteration-dependent data are gathered in a POD XYData whose members correspond to X[i] and Y[i]. The types of the two members (float) are passed as template arguments to MTPS::POD and their names (x and y) are given in the Fields enum:

```
1 struct XYData: public MTPS::POD<float,float>{
2     enum Fields{x, y};
3 };
```

Second, a *collection* of XYData elements, xyCol, is built using MTPS containers. Optimized containers are provided as member of the class corresponding to the target architecture. Two levels of parallelism are available on CPUs: thread parallelism and SIMD parallelism. The choice for each level is made by passing two arguments to the CPU template class. Thread can be one of

`MTPS::Sequential`, `MTPS::OMP` (openMP) or `MTPS::TBB` (Intel TBB). `SIMD` can be one of `MTPS::Scalar` (no SIMD instruction generated) or `MTPS::SSE` (SIMD instruction generated using SSE intrinsics). On CUDA-enabled GPUs, only the SIMD parallelism is provided.

```
1 //typedef MTPS::GPU::CUDA Target;          // To use the GPU
2
3 //typedef MTPS::Sequential Thread;         // Single threaded
4 //typedef MTPS::TBB Thread;                // TBB parallelism
5 typedef MTPS::OMP Thread;                  // OpenMP ↵
       parallelism
6 //typedef MTPS::Scalar SIMD;               // Disable SIMD ↵
       units
7 typedef MTPS::SSE SIMD;                    // Enable SSE units
8 typedef MTPS::CPU<Thread, SIMD> Target;   // To use the CPU
9
10 Target::collection<XYData> xyCol(N);
```

Third, the function that is to be applied to all elements of the collection must be written as a functor class: `AxpyOp`. The coefficient `a` is common for all elements of the collection and is stored as a member of the `AxpyOp` functor class:

```
1 struct AxpyOp{
2     float a_;
3     template <template <class> class View>
4          INLINE void operator()(View<XYData> xy) const {
5         typedef View<XYData> XYV;
6         int x = XYV::x;
7         xy(XYV::y)+=a_*xy(x);
8     }
9 };
```

As `XYData` elements may not be stored identically on different target architecture, `AxpyOp::operator()` does not take an `XYData` as argument. A `View` is provided instead. `XYData` members can be accessed with the `operator()` of the `View` which takes an `int` as argument. This `int` identify the data member that is to be accessed; either `X[i]` or `Y[i]` in our exemple. Elements of the `Fields` enum can be used either to initialize an `int` (line 7) or directly (line 8). The declaration of `AxpyOp::operator()` must be preceeded by the `INLINE` macro which defines target-dependent keywords (e.g. `__device__` for CUDA).

Finally, the functor is passed to the `map` and `fold` parallel skeletons provided by the *collection* container:

```
1 AxpyOp axpyOp; axpyOp.a_ =...;
2 xyCol.map(axpyOp);
3 ...
4 DotOp dotOp;
5 float dot = xyCol.fold(dotOp);
```

Although more verbose and harder to use than the approaches presented in Section 3.1, this formalism allow MTPS to be the only library at our knowledge that optimizes the data layout for different architectures as Section 4 will show.

4   Optimization of Performances

For each architecture, the specific optimizations required to enable good performances will be presented. The implementation of a more complex example will then be discussed.

4.1   Multi-target Performance Optimizations

Parallelizing a *vectorizable algorithm* is straightforward. However, achieving good performances on different hardware architectures is not: modifications of the *collection* data structure may be required. Indeed, achieving efficient usage of memory bandwidth on a given hardware architecture requires specific access patterns [7]. Fig. 4 shows a block-diagonal matrix of 8 TriDiagonal Symmetric Matrix blocks (TDSM) of size 4 (left). Assuming that this matrix is considered as a *collection* of TDSM blocks in reference to an algorithm, Fig. 4 shows how to store it on three different architectures to optimize the access pattern (right):

- on CPU (top), maximizing data locality is required to avoid cache misses. As the TDSM blocks are independent, data locality only matters inside a TDSM block. Hence, the best performances are achieved when each TDSM block is stored in a contiguous chunk of memory;
- on GPU (bottom), memory accesses have to be *coalesced* to achieve good performances (see the CUDA programming guide [10]). This implies that the accesses made by two threads i and $i + 1$ must correspond to two elements at index j and $j + 1$. As the same function is applied in a SIMD fashion to the different TDSM blocks, all elements A are accessed at the same time and they have to be stored in a contiguous chunk of memory;
- using the SSE units (middle) requires to pack the data into *vectors* containing 4 independent elements that have to be accessed together. Although a GPU ordering would fill this need, this would break the data locality and imply CPU cache misses. Finally, an intermediate storage between the two previous is optimal.

Performances achieved thanks to this optimization will be shown in Section 4.3. This optimization is made in MTPS *collection* container. To construct a *collection*, MTPS user must define both the size per POD-element of each field (4 for the diagonal field on Fig. 4) and the number of POD-elements. Using this information, MTPS optimizes the storage for each target architecture.

As a *context* corresponds to a storage pattern, a *context* switch imply a data reordering. For instance, switching a *collection* of matrix rows to a *collection* of matrix columns modifies the effective storage (i.e. the matrix is transposed). MTPS provides some switch skeletons.

4.2   Implementation of a Linear System Resolution

The example presented in this section corresponds to a basic operation that represents the major part of the execution time of a neutron transport code [6,8].

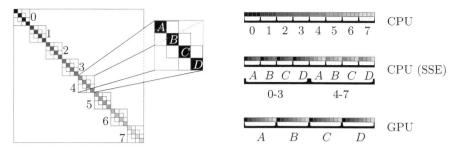

Fig. 4. The storage of the diagonal is adapted by MTPS for the target architecture

Let $\mathbf{A}$ be a block-diagonal matrix with TDSM blocks. The $\mathbf{AX} = \mathbf{B}$ linear system can be seen as a *collection* of smaller block systems $ax = b$ that can be solved independently. To solve one $ax = b$ system, the matrix a is factorized in-place with a LDL^T decomposition and a forward and backward substitution is then applied on x. Only the code for the factorization is shown here.

First, let us introduce TData which represents a TDSM block. TData elements are stored in two vectors corresponding to the diagonal and the lower diagonal:

```
1 struct TData: public MTPS::POD<float,float>{
2     enum Fields{diag, low};
3     typedef typename MTPS::POD<float,float>::Shape Shape;
4     static Shape createShape(int size){
5         Shape out;
6         out[diag] = size;                          out[low] ↩
            = size-1;
7     }
8 };
```

The Shape type of line 5 contains the effective sizes of the two fields. All elements of a *collection* have the same shape.

Second, to build a *collection* of TData, one has to provide both the number of TData elements and their shape. With these elements, the storage pattern of tCol can be optimized according to the target architecture (see Fig. 4):

```
1 TData::Shape s=TData::createShape(size);
2 Target::collection<TData> tCol(N, s);
```

Third, the TLDLtOp functor class that factorizes the matrix a in-place using a LDL^T decomposition has to be provided:

```
1 struct TLDLtOp{
2     template <template <class> class View>
3         INLINE void operator()(View<TData> a) const{
4         typedef View<TData> TV;
5         int low = TV::low, diag = TV::diag;
6         typename TV::template Type<low>::Type low_i_1;
7         int size = a.shape()[diag];
```

```
8          for (int i = 1 ; i < size ; i++){
9              a(low, i−1)=a(low, i−1)/a(diag, i−1);
10             a(diag, i)−=a(diag, i−1)*a(low, i−1)*a(low, ↩
                   i−1);
11         }
12     }
13 };
14 TLDLtOp op;
15 tCol.map(op);
```

The elements of a field are accessed by passing their index as the second argument of the view `operator()`. If no index is provided as in the line 8 of the `AxpyOp` example of Section 3.3, the first element is returned. Line 7 shows how the type of the field elements can be retrieved.

4.3   Performances

Table 1 shows the performances obtained to compute the solution of the $\mathbf{AX} = \mathbf{B}$ system from Section 4.2 with $\mathbf{A}$ having 10^5 blocks of size 100×100. The matrix and vector are directly constructed on the target architecture and do not require further reordering to fit the target architecture. Speed-ups take the sequential scalar CPU version as reference. CPU tests are run on a machine with two 2.933 GHz Intel X5670 hexa-core processors. GPU tests are run on a Nvidia Quadro C2050 card. Both architectures were launched in 2010. On CPU, icpc 11.1 and g++ 4.5 provide the same performances. On GPU, nvcc 3.2 has been used. Computation performances are given in GFlops. Data throughput is given in GB/s and takes into account the effective data transfers to and from the memory. Consequently, an element remaining in the cache memory between two loads is considered to have been loaded only once.

For each architecture, the achieved performances are compared to the expected performances corresponding to the best observed performances on the given architecture. Expected computational power are measured with large BLAS matrix-matrix multiplications (`sgemm`): 11.2 GFlops with one CPU core and 126.5 GFlops with the 12 CPU cores using Intel MKL. The MKL uses the SSE units. As these units can execute 4 single precision floating point operations, we define expected computational performances without the SSE units as ¼ of the SSE performances (i.e., 2.8 GFlops and 31.6 GFlops respectively). On GPU, the expected computational power measured is 435 GFlops. On CPU, the expected memory throughputs are measured with an extended version of the stream benchmark [9]. This version adds a new subroutine containing 9 memory accesses (instead of 3 for the `Triad` routine) and shows 12.4 GB/s for single threaded execution and 35.0 GB/s for the parallel execution using openMP. On GPU, the expected memory throughput is measured with the CUDA SDK bandwidth benchmark on GPU: 86.3 GB/s.

To provide comparable results in spite of the hardware differences, the specifications of the hardware have been taken into account. For the computational power, the difference relies in the number of cycles to evaluate a floating point division: on GPU, 1 cycle is required but 15 are required on CPU. Thus, on CPU, the division is considered as 15 floating point operations. For the memory throughput, only accesses to the main memory are counted. In other words, accesses to a piece of data that have already been loaded in cache are considered as free. On CPU, 3 accesses are saved whereas no accesses are saved on GPU. Evaluating the computational power and the memory throughputs this way allow us to make fair comparison to the expected performances.

The performances of a code on a given architecture are limited either by the computational power or by the memory throughput. Bold figures in Table 1 correspond to the limiting factor for the corresponding target architecture. On CPU, when no or few parallelism is used, the performances are limited by the computational power: the performances achieved are between 75% and 93% of the expected computational performances. When both the threading parallelism and the SIMD parallelism are enabled, the performances are limited by the memory throughput: almost 100% of the best observed throughput is obtained. On GPU, all the available parallelism is used and the performances are thus limited by the memory throughput: 95% of the best observed throughput is reached.

Table 1. Performances of MTPS for the TDSM example. Computation are carried out in single precision floating point.

Thread	SIMD	Time (ms)	Speed Up	Computational Power			Data Throughput		
				GFlops	Expected	%Exp.	GB/s	Expected	%Exp.
sequential	scalar	131.9	1.0	2.5	2.8	**88**	1.7	12.4	14
	SSE	37.3	3.5	8.7	11.2	**78**	6.0	12.4	48
intel TBB	scalar	12.1	11.	27.0	31.6	**85**	18.5	35.0	53
	SSE	6.6	20.	49.4	126.5	39	33.9	35.0	**97**
openMP	scalar	11.1	12.	29.4	31.6	**93**	20.1	35.0	57
	SSE	6.5	20.	50.1	126.5	40	34.4	35.0	**98**
CUDA C		4.1	32.	15.9	435.	4	81.8	86.3	**95**

The limitation of the performances by the memory throughput shows the importance of optimizing the memory accesses. Finally, Table 1 shows that by abstracting the memory access pattern to the target architecture the the performances of a given code can near the hardware limits on different architectures.

5   Conclusions and Perspectives

We have presented MTPS, a C++ generic library simplifying the parallelization and the optimization of *vectorizable algorithms* for different architectures. Although MTPS semantics and syntax remain complex, the end user should not be aware of this complexity: MTPS is designed to be generated, especiallly with the C++ template metaprogramming approach. Finaly, an algorithm written once with MTPS can be compiled to be executed on the SSE units of multicore

CPUs or on CUDA-enabled GPUs and obtain performances close to hardware limits: more than 95% of peak performances were observed.

For further developments of MTPS, the design of an new version of Legolas++ on top of MTPS will allow to validate the set of skeletons provided by MTPS, especially concerning the *context switches*. The implementation of a neutron transport solver [6,8] with this version of Legolas++ will automaticaly provide a multi-target version of this solver. Efforts will be made to keep the portability of the performances currently available with MTPS.

Acknowledgement. Authors want to thank Region Lorraine and ANRT for supporting this research.

References

1. Baker, C.G., Carter Edwards, H., Heroux, M.A., Williams, A.B.: A light-weight API for Portable Multicore Programming. In: PDP 2010: Proceedings of The 18th Euromicro International Conference on Parallel, Distributed and Network-Based Computing. IEEE Computer Society, Washington, DC (2010)
2. Czarnecki, K., O'Donnell, J.T., Striegnitz, J., Taha, W.: DSL Implementation in MetaOCaml, Template Haskell, and C++. In: Lengauer, C., Batory, D., Blum, A., Vetta, A. (eds.) Domain-Specific Program Generation. LNCS, vol. 3016, pp. 51–72. Springer, Heidelberg (2004)
3. Dyk, D.V., Geveler, M., Mallach, S., Ribbrock, D., Göddeke, D., Gutwenger, C.: HONEI: A collection of libraries for numerical computations targeting multiple processor architectures. Computer Physics Communications 180(12), 2534–2543 (2009)
4. Falcou, J., Sérot, J., Chateau, T., Lapresté, J.T.: Quaff: efficient C++ design for parallel skeletons. Parallel Computing 32(7-8), 604–615 (2006)
5. ISO: ISO/IEC 14882:2003: Programming languages — C++. International Organization for Standardization, Geneva, Switzerland (2003), (§3.9)
6. Kirschenmann, W., Plagne, L., Ploix, S., Ponçot, A., Vialle, S.: Massively Parallel Solving of 3D Simplified P_N Equations on Graphic Processing Units. In: Proceedings of Mathematics, Computational Methods & Reactor Physics (May 2009)
7. Kirschenmann, W., Plagne, L., Vialle, S.: Multi-target C++ implementation of parallel skeletons. In: POOSC 2009: Proceedings of the 8th Workshop on Parallel/High-Performance Object-Oriented Scientific Computing. ACM, New York (2009)
8. Kirschenmann, W., Plagne, L., Vialle, S.: Parallel sp_n on multi-core cpus and many-core gpus. Transport Theory and Statistical Physics 39(2), 255–281 (2010)
9. McCalpin, J.D.: Memory Bandwidth and Machine Balance in Current High Performance Computers. In: IEEE Computer Society Technical Committee on Computer Architecture (TCCA) Newsletter, pp. 19–25 (December 1995)
10. NVIDIA: NVIDIA CUDA C Programming Guide 3.1 (2010)
11. Plagne, L., Ponçot, A.: Generic Programming for Deterministic Neutron Transport Codes. In: Proceedings of Mathematics and Computation, Supercomputing, Reactor Physics and Nuclear and Biological Applications, Palais des Papes, Avignon, France (September 2005)
12. Reinders, J.: Intel threading building blocks. O'Reilly & Associates, Inc., Sebastopol (2007)
13. Veldhuizen, T.L.: Arrays in Blitz++. In: Caromel, D., Oldehoeft, R.R., Tholburn, M. (eds.) ISCOPE 1998. LNCS, vol. 1505, pp. 223–230. Springer, Heidelberg (1998)

Analysis of Gravitational Wave Signals on Heterogeneous Architectures

Maciej Cytowski

Interdisciplinary Centre for Mathematical and Computational Modelling,
University of Warsaw
m.cytowski@icm.edu.pl

Abstract. Heterogeneous architectures and programming techniques will be very important in the development of exascale HPC applications. Adapting heterogeneous programming techniques to scientific programming is not always straightforward. Here we present an in-depth analysis of an astrophysical application used for performing an all-sky coherent search for periodic signals of gravitational waves in narrowband detector data. The application was first ported to the PowerXCell8i architecture and then on the basis of achieved performance it was again redesigned and programmed in a heterogeneous model. Moreover presented heterogeneous techniques could be easily adopted for other scientific computational problems involving FFT computations.

Keywords: hybrid computing, parallel computations, gravitational waves.

1   Introduction

Nowadays using specialized hardware architectures or accelerators for specific computational problems is very common. For large scientific codes it usually means that special programming techniques have to be applied to offload some of the computationally intensive parts of the application on given hardware. Such techniques are usually called heterogeneous computing.

In this work we present an in-depth analysis of an astrophysical application used for performing an all-sky coherent search for periodic signals of gravitational waves in narrowband detector data. The application was ported to a prototype hybrid platform based on the IBM PowerXCell8i architecture. The resulting implementation can be compiled and used as a standalone x86-64 application, standalone Cell application or hybrid x86-64/Cell application. The IBM Cell processor was designed to bridge the gap between general purpose processors and specialized computer architectures like GPUs. The architecture was already extensively described i.e. in [1], [2] and [3]. Supercomputer architectures like Roadrunner [4] or Nautilus [9] utilize the IBM PowerXCell8i processor as an accelerator for calculations running on x86-64 cores. One of the programming techniques available for such heterogeneous architectures is the IBM DaCS library which has proven to be useful in several scientific applications developed

K. Jónasson (Ed.): PARA 2010, Part II, LNCS 7134, pp. 347–356, 2012.

for the Roadrunner and Nautilus supercomputers. The described application is one of the codes implemented with the use of DaCS for execution on Nautilus.

The application itself and its reference setup is briefly discussed in Chapter 2. A detailed description of the programming techniques used for implementation of both the PowerXCell8i and hybrid versions of the code is presented in Chapter 3. In this chapter we also show how the custom designed and implemented data conversion mechanism improves the overall performance of the hybrid application. In the last chapter we discuss the results and formulate the conclusions and perspective of code development on described computer architectures.

2   Compute Problem

2.1   Scientific Background

A gravitational wave is a physical phenomenon which arises from Einstein's theory of general relativity and is defined as a fluctuation in the curvature of spacetime which propagates as a wave, traveling outward from the source. Gravitational waves are radiated by objects whose motion involves acceleration. This class of objects include binary star systems composed of white dwarfs, neutron stars, or black holes. Compared to standard methods used for observing the universe, like visible light or radio telescope observations, gravitational waves have two important unique properties. First of all gravitational waves can be emitted by a binary system of uncharged black holes without presence of any type of matter nearby. Secondly gravitational waves can pass through any intervening matter without being scattered significantly. Both of these features allow researchers to explore astronomical phenomena which have never before been observed by humans. The observations of gravitational waves are usually done by ground-based interferometers. Big research projects like LIGO [5] or VIRGO [6] usually involve operationally running interferometers and produce a significant amount of observational data. This data is subject to further analysis by computer codes developed by research groups involved in those projects. The main computational tasks to be performed on those observational data sets are usually described by algorithms for searching of periodicity or quasiperiodicity.

2.2   Compute Algorithm

In this work we present an in-depth analysis of an astronomical application used for performing an all-sky coherent search for periodic signals of gravitational waves in narrowband data from a detector. The search is based on the maximum likelihood statistics called the F-statistics as proposed by P.Jaranowski, A. Królak and B.F.Schultz [8]. The computer code developed by the Polgraw group [7] was used in an operational manner for analysis of observational data from the NAUTILUS [9] and VIRGO [6] detectors. The algorithm implemented in the code was designed for gravitational wave signals generated by rotating neutron stars. The mathematical description of the algorithm was given in previous works of the code authors ([8],[10],[11],[12]). Here we will just shortly describe the most important parts of the code. The simplified flowchart of the code is presented in Fig. 1.

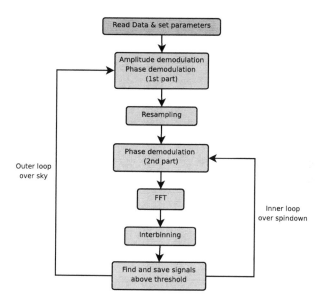

Fig. 1. Flowchart of the application

The code begins with reading the observational data sets and setting appropriate parameters for the searching algorithm. After that an outer loop across the sky begins. The very first step in this loop is an ampiltude and phase demodulation. The following step, the so-called resampling of the signal, is performed in two stages: a Fourier interpolation and a spline interpolation. The most computationally expensive part of the whole code is a loop over spindown which can have a length between 0 and 1000 depending on the signal currently analyzed. This loop consists of four main steps:

1. **Phase demodulation (2nd part)** - double precision sin/cos computations, double precision complex multiplications
2. **FFT computations** - double precision 1 dimensional complex forward transforms of size $N = 524288$
3. **Interbinning step** - interpolation algorithm, double precision complex subtractions and divisions, double precision real square root computations
4. **Finding and saving signals** - nonlinear optimization for finding a maximum, saving the resulted signals with values below threshold

3   Methodology

The very first attempt to accelerate the execution of the described application was to port some of its functional parts to the Synergistic Processing Units (SPUs) of the PowerXCell8i processor. In order to select appropriate parts of the application the programmer usually has to perform the following tasks:

- compile, run and measure the performance of the application on a general purpose architecture (i.e. x86-64)
- compile, run and measure the performance of the application on the PPU of the PowerXCell8i processor,
- identify the most computationally intensive parts of the application,
- check the suitability of the selected parts for execution on SPUs.

Unfortunately the usual result of the first two tasks listed above is that the application's performance is much higher on the single core of the x86-64 architecture than on the PPU, which is related to the fact that the Power Processing Unit of the Cell processor was not designed and optimized for computations. This was also valid for the described application. Executions on the PPU were approximately 3 times slower than on a corresponding single core of the x86-64 chip. This observation is of crucial importance to the overall performance of the application on the Cell processor even if some of its parts were already optimized for executions on SPUs. One of the ways to overcome this issue is to use a heterogeneous programming model where only the well optimized parts of the application are executed on the Cell processor whereas the application itself is running on an x86-64 core. In this chapter we present the performance results of the application ported to the Cell processor. Since not all of the parts/algorithms are well suited for execution on the SPUs we decided to use a heterogeneous environment to increase the overall performance. This is briefly described in the second section of this chapter.

3.1   SPU Implementation

We have identified 3 functional parts of the application that were especially well suited for execution on the SPU architecture: the 2nd part of phase demodulation, FFT computations and the interbinning step. However to achieve a certain level of granularity for computations on the SPUs we needed to redesign the whole program. We decided to make use of the available RAM memory (8GB for each IBM QS22 Cell blade) and perform all 3 functional parts in seperate loops over spindown. The new resulting flowchart of the program is presented in Fig. 2.

This small change turned out to be very important for the final performance of the application on the Cell chip. Here we will describe the effort we have made to optimize the code for this architecture in detail.

We have implemented a parallel version of the phase demodulation and interbinning step on the SPUs with the use of the libspe2 library [17]. We have used a double buffering scheme for DMA transfers between Local Store of the

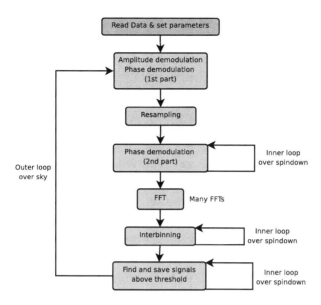

Fig. 2. Flowchart of the redesigned application

SPUs and main memory. We have then compared the resulting performance with the use of 8 SPUs (one PowerXCell8i chip).

The FFT step was initially implemented with the use of the Fastest Fourier Transform in the West (FFTW) library [13]. We have decided to use the same library in our implementation since FFTW was already ported to the Cell processor by a group of programmers at IBM Austin Research Laboratories.

A performance comparison between a single core AMD Opteron 2216 processor and a single PowerXCell8i chip for the described functional parts is presented in Fig. 3. It should be also mentioned that all computations involved in the algorithm are based on double precision arithmetic. Therefore the maximum performance rate that could be achieved on one PowerXCell8i chip is 102.4 GFlops. The single core of the corresponding AMD Opteron 2216 processor has a maximum performance rate of 9.6 GFlops.

We were able to speed up few parts of the application with the use of multiple SPUs. However not all steps of the implemented algorithm could be ported and optimized on the Cell architecture. One example is the so-called "Finding signals" step based on the maximum likelihood statistics. Moreover as we mentioned before computational performance of the PPU core is very poor, thus usually the fragments of the code that are not accelerated on SPUs slow down the overall performance. In particular the "Finding signals" step takes in average 2.67 sec. on the Cell and 0.5 sec. on a single core of the x86-64 architecture chip. The final result we obtained was approximately **3.24 speedup** of the 8 SPU version compared to a single core x86-64 version.

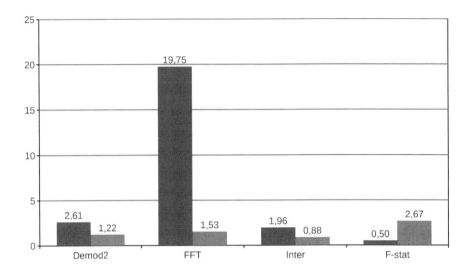

Fig. 3. Performance of the described functional parts of the algorithm on the Cell architecture (blue - single core x86-64, red - PowerXCell8i)

3.2   Hybrid Implementation

The observation that some of the parts of the application have much better performance on the x86-64 architecture encouraged us to prepare a heterogeneous version of the code where we use the PowerXCell8i processor as an accelerator to compute only the functional parts of the application that were optimized for execution on multiple SPUs. For implementing such a scheme we have chosen to use a hybrid library developed by IBM [14].

The Data Communication and Synchronization (DaCS) [14] library and runtime was designed to support the development of applications for heterogeneous systems based on the PowerXCell8i and x86-64 architectures. The DaCS API provides an architecturally neutral layer for application developers. It serves as resource and process manager for applications that use different computing devices. With the use of specific DaCS functions we can execute different remote processes and initiate data transfers or synchronization between them.

One of the main concepts of DaCS is a hierarchical topology which enables application developers to choose between a variety of hybrid configurations. First of all it can be used for programming applications for the Cell processor by exploiting its specific hybrid design. In such a model developers use DaCS to create and execute processes on the PPU and SPUs and to initiate data transfers or synchronization between those processes. It should be stated that developers can choose between a few other programming concepts for Cell processor and that the DaCS model is for sure not the most productive and efficient one. However the DaCS library is much more interesting as a tool for creating hybrid applications that use two different processor architectures, in this case: AMD Opteron and

PowerXCell8i. In such a model DaCS can support the execution, data transfers, synchronization and error handling of processes on three different architectural levels: x86-64 cores, PPUs and SPUs. Additionally the programmer can decide to use DaCS with any available Cell programming model on the level of PPU process. PPU process can execute SPU kernels implemented within optimized libraries or created originally by developers with the use of programming models like libspe2 [17], Cell SuperScalar [18] or OpenMP [19].

The DaCS library has a much wider impact on high performance computing since it was designed to support highly parallel applications where the MPI library is used between heterogeneous nodes and the DaCS model is used within those nodes. It is presented schematically on Figure 4. Such programming model was used for application development on the Roadrunner and Nautilus supercomputers ([15],[16]).

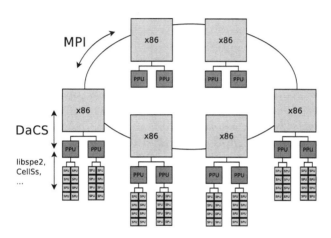

Fig. 4. Scheme of multi-level DaCS programming model for heterogeneous architectures

The resulting heterogeneous scheme of the application is presented in Fig. 5. The application is executed on the x86-64 architecture. The initialization of the DaCS library is performed in the very beginning of the code together with the allocation of specific memory regions reserved for synchronized data transfers between hybrid processes. At this time the corresponding Cell process is executed on the PPU via the DaCS library and hangs its execution waiting for proper signals. The application parts performed on the Cell processor are only those that presented good performance and were optimized for execution on SPUs (the demodulation step, FFTs and the interbinning step). It should be stated here that such an implementation introduces memory transfers between both processes. The size of such data transfers is reaching 1 GB per each outer loop step and thus the performance rate of the interconnect is of crucial importance for the overall performance of the application.

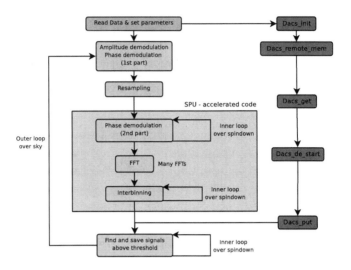

Fig. 5. Flowchart of the Hybrid DaCS version of the application

We have used two heterogeneous systems for testing purposes. Both of them were composed of one IBM LS21 blade and two IBM QS22 blades and they differ in the type of interconnect used for data transfers between those blades. The basic development system installed at ICM uses Gigabit Ethernet. The other system is a node of IBM's triblade cluster (RoadRunner-like prototype system) located at Rochester, USA and uses PCI for data transfers. Like we have assumed the performance of data transfers used for moving data from x86-64 to the Cell architecture turned out to be very important for the overall speedup of the application. On Gigabit Ethernet the maximum speedup was approx. **1.5**. First measurements made on PCI showed that the performance gain from the hybrid approach is rather small reaching a maximum speedup of **3.56** compared to the previously mentioned **3.24** result in the non-hybrid Cell version of the application. Thus we decided to take a closer look at the performance of the DaCS data transfers on PCI. Those data transfers always involve DMA operations followed by byte swapping applied to the binary data being sent (different endianness of processing devices). In the DaCS library you can decide to turn the byte swapping functionality on and off. We have measured that the byte swapping operation limits the performance of the data transfers over PCI to approx. 280 MB/s. Transfers that don't involve this step can reach performance of more than 1100 MB/s. Following this observation we have decided to implement an optimized version of byte swapping on the Cell processor itself which resulted in much better data transfer performance reaching up to 900 MB/s for big data sizes. Such a high performance level was achieved thanks to a number of Cell programming techniques, mainly: parallel processing on SPUs and SIMD vector processing. Our custom implementation of the byte swapping step is universal and could be used as library by other applications. The following Fig. 6 shows the performance gain of data transfers for different data sizes.

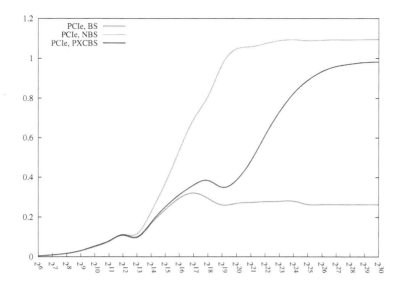

Fig. 6. DaCS data transfer performance with optimized byte swapping library (blue) compared to non-byte swapping (green) and DaCS byte swapping (red) versions. Data size in bytes is depicted on the x-axis. Performance measured in GB/s is depicted on the y-axis.

The application described in this work achieved a speedup of **4.5** with the use of our optimized byte swapping technique and this performance level is our final result. Moreover in our opinion it is the highest performance level that could be achieved on the corresponding computer architectures.

4   Results and Conclusions

We have successfully implemented a heterogeneous version of the presented application on RoadRunner-like prototype systems. The application described in this work achieved the highest speedup of **4.5** with the use of a custom optimized byte swapping technique and IBM DaCS programming model. Methods used for the implementation can also be adopted and used in many different scientific applications, especially those involving large FFT computations. The main disadvantage in programming applications for heterogeneous systems is usually related to the necessity of creating few programs dedicated for different computational devices. Therefore our future work on the presented topic will be addressing the development of a heterogeneous parser library for Fourier computations. Such a tool could be used with minor code modifications within scientific codes that make use of the FFTW library. All the important heterogeneous programming issues like data transfers and byte swapping can be hidden behind the library interface. The overall performance of such a heterogeneous tool will be based on the DaCS model and most importantly on heterogeneous techniques developed and used in this work.

Acknowledgements. We would like to thank dr Peter Hofstee (IBM Systems and Technology Group, Austin), Minassie Tewoldebrhan (IBM Rochester) and Maciej Remiszewski (IBM Central and Eastern Europe) for making the IBM triblades available remotely for testing. We have also used the computing resources of the Interdisciplinary Centre for Mathematical and Computational Modelling (ICM), University of Warsaw within research grant G33-19.

References

1. Gschwind, M., IBM, Hofstee, H.P., Flachs, B., Hopkins, M., Watanabe, Y., Yamazaki, T.: Synergistic Processing in Cell's Multicore Architecture. IEEE Micro 26(22), 10–24 (2006)
2. Kahle, J.A., Day, M.N., Hofstee, H.P., Johns, C.R., Maeurer, T.R., Shippy, D.: Introduction to the Cell Multiprocessor. IBM Journal of Research and Development 49(4), 589–604 (2005)
3. Kistler, M., Perrone, M., Petrini, F.: Cell Multiprocessor Communication Network: Build for Speed. IEEE Micro 26(3), 10–23 (2006)
4. Roadrunner supercomputer site, http://www.lanl.gov/roadrunner/
5. Ligo project, http://www.ligo.caltech.edu/
6. Acernese, F., et al.: Virgo status. Class. Quantum Grav. 25, 184001 (2008)
7. Polgraw group, http://www.astro.uni.torun.pl/~kb/allsky/virgo/polgraw.htm
8. Jaranowski, P., Królak, A., Schultz, B.F.: Data analysis of gravitational-wave signals from spinning neutron stars: The signal and its detection. Phys. Rev. D 58, 063001 (1998)
9. Astone, P., et al.: All-sky search of NAUTILUS data. Class. Quantum Grav. 25, 184012 (2008)
10. Jaranowski, P., Królak, A.: Data analysis of gravitational-wave signals from spinning neutron stars. II. Accuracy of estimation of parameters. Phys. Rev. D 59, 063003 (1999)
11. Jaranowski, P., Królak, A.: Data analysis of gravitational-wave signals from spinning neutron stars. III. Detection statistics and computational requirements. Phys. Rev. D 61, 062001 (2000)
12. Astone, P., et al.: Data analysis of gravitational-wave signals from spinning neutron stars. IV. An all-sky search. Phys. Rev. D 65, 042003 (2002)
13. Frigo, M., Johnson, S.G.: The Design and Implementation of FFTW3. Proceedings of the IEEE 93(2), 216–231 (2005)
14. Data Communication and Synchronization for Cell BE Programmer's Guide and API Reference IBM SC33-8408-01, Publication Number: v3.1
15. Barker, K.J., Davis, K., Hoisie, A., Kerbyson, D.J., Lang, M., Pakin, S., Sancho, J.C.: Entering the Petaflop Era: The Architecture and Performance of Roadrunner. In: Proceedings of the 2008 ACM/IEEE Conference on Supercomputing (2008)
16. Nautilus supercomputer site, http://cell.icm.edu.pl/index.php/nautilus
17. International Business Machines Corporation. Programming tutorial. Technical document SC33-8410-00. Software Development Kit for Multicore Acceleration Version 3.1
18. BSC. Cell Superscalar (CellSs) User's Manual, Version 2.1 (May 2008)
19. OpenMP Application Program Interface, Version 3.0 (May 2008)

Towards Efficient Execution of Erasure Codes on Multicore Architectures

Roman Wyrzykowski, Lukasz Kuczynski, and Marcin Wozniak

Institute of Computer and Information Sciences,
Czestochowa University of Technology,
Dabrowskiego 73, 42-201 Czestochowa, Poland
{roman,lkucz,marcell}@icis.pcz.pl

Abstract. Erasure codes can improve the availability of distributed storage in comparison with replication systems. In this paper, we focus on investigating how to map systematically the Reed-Solomon and Cauchy Reed-Solomon erasure codes onto the Cell/B.E. and GPU multicore architecture. A method for the systematic mapping of computation kernels of encoding/decoding algorithms onto the Cell/B.E. architecture is proposed. This method takes into account properties of the architecture on all three levels of its parallel processing hierarchy. The performance results are shown to be very promising. The possibility of using GPUs is studied as well, based on the Cauchy version of Reed-Solomon codes.

Keywords: Erasure codes, Reed-Solomon codes, Cauchy Reed-Solomon codes, multicore architectures, Cell/B.E., GPU.

1   Introduction

There is a rapid increase in sensitive data, such as biomedical records or financial data. Protecting such data while in transit as well as while at rest is crucial [6]. An example are distributed data storage systems in grids [18], that have different security concerns than traditional file systems. Rather than being concentrated in one place, data are now spread across multiple hosts. Failure of a single host or an adversary taking control of a host could lead to loss of sensitive data, and compromise the whole system. Consequently, suitable techniques, e.g. cryptographic algorithms and data replication, should be applied to fulfill such key requirements as confidentiality, integrity, and availability [18,19].

A classical concept of building fault-tolerant systems consists of replicating data on several servers. Erasure codes can improve the availability of distributed storage by splitting up the data into n blocks, encoding them redundantly using m blocks, and distributing the blocks over various servers [2]. As was shown in [15], the use of erasure codes reduces "mean time of failures by many orders of magnitude compared to replication systems with similar storage and bandwidth requirements".

There are many ways of generating erasure codes. A standard way is the use of the Reed-Solomon (or RS) codes [10]. The main disadvantage of this approach

K. Jónasson (Ed.): PARA 2010, Part II, LNCS 7134, pp. 357–367, 2012.

is a large computational cost because all operations, including multiplications, are performed over the Galois field $GF(2^w)$ arithmetic, which is not supported by modern microprocessors, where $2^w \geq n + m$. In this context, an interesting alternative are the Digital Fountain Codes, or more generally, Low-Density Parity-Check (LDPC) codes [7]. Their implementation can be reduced to a series of bitwise XOR operations. However, this potential advantage of LDPC codes is not always realized in practice [13], when relatively small values of n are often used. In particular, it was shown that for the encoding ratio $r = n/(n+m) = 1/2$, the performance of RS codes is not worse than that of LDPC codes if $n \leq 50$. This relationship depends on the ratio between the performance of a network, and performance of processing units used for encoding/decoding. For a constant network performance, increasing the performance of processing units gives advantage to the RS codes.

The last conclusion is especially important nowadays when multicore architectures begin to emerge in every area of computing [16]. Furthermore, an important step in the direction of improving the performance of RS codes has been done recently, when a Cauchy version of these codes was proposed [14]. In particular, this new class of codes (CRS codes, for short) does not require performing any multiplication using the Galois field arithmetic; a series of bitwise XOR operations is executed instead.

In this work, we focus on investigating how to systematically map the RS and CRS erasure codes onto the Cell/B.E. architecture [1]. This innovative heterogeneous multicore chip is significantly different from conventional multiprocessor or multicore architectures. The Cell/B.E. integrates nine processor elements (cores) of two types: the Power processor element (PPE) is optimized for control tasks, while the eight synergistic processor elements (SPEs) provide an execution environment optimized for data-intensive processing. Each SPE supports vector processing on 128-bit words, implemented in parallel by two pipelines. Each SPE includes 128 vector registers, as well as a a private local store for fast instruction and data access. The EIB bus connects all the cores with a high-performance communication subsystem. Also, the Cell/B.E. offers an advanced, hardware-based security architecture [19]. The impressive computational power of Cell/B.E., coupled with its security features, make it a suitable platform to implement algorithms aimed at improving data confidentiality, integrity, and availability [18,19].

In the last part of this paper, we study the possibility of using another, very promising type of multicore architectures which are GPUs (Graphics Processing Units) [5,17]. Basic features of GPUs include utilization of a large number of relatively simple processing units which operate in the SIMD fashion, as well as hardware supported, advanced multithreading. For example, Nvidia Tesla C1060 is equipped with 240 cores, delivering the peak performance of 0.93 TFLOPS. A tremendous step towards a wider acceptation of GPUs in general-purpose computations was the development of software environments which made it possible to program GPUs in high-level languages. The new software developments, such as Nvidia CUDA [8] and OpenCL [9], allow programmers to implement algorithms on existing and future GPUs much easier.

2   Reed-Solomon Codes and Linear Algebra Algorithms

More precisely, an erasure code works in the following way. A file F of size $|F|$ is partitioned into n blocks (stripes) of size B words each, where $B = |F|/n$. Each block is stored on one of n data devices $D_0, D_1, ..., D_{n-1}$. Additionally, there are m checksum devices $C_0, C_1, ..., C_{m-1}$. Their contents are derived from contents of data devices, using a special encoding algorithm. This algorithm has to allow for restoring the original file from any n (or a bit more) of $n + m$ storage devices $D_0, D_1, ..., D_{n-1}, C_0, C_1, ..., C_{m-1}$, even if m of these devices failed, in the worst case.

The application of the RS erasure codes includes [10,11] two stages: (i) encoding, and (ii) decoding. At the encoding stage, an input data vector $\mathbf{d}_n = [d_0, d_1, ..., d_{n-1}]^T$, containing n words, each of size w bits, is multiplied by a special matrix

$$\mathbf{F}_{(n+m) \times n} = \begin{bmatrix} \mathbf{I}_{n \times n} \\ \mathbf{F}^*_{m \times n} \end{bmatrix}. \tag{1}$$

Its first n rows correspond to the identity matrix, while the whole matrix is derived as a result of transforming an $(n + m) \times n$ Vandermonde matrix, with elements defined over the Galois field $GF(2^w)$.

The result of the encoding is an $(n + m)$ column vector

$$\mathbf{e}_{n+m} = \mathbf{F}_{(n+m) \times n} \times \mathbf{d}_n = \begin{bmatrix} \mathbf{d}_n \\ \mathbf{c}_m \end{bmatrix}, \tag{2}$$

where:

$$\mathbf{c}_m = \mathbf{F}^*_{m \times n} \times \mathbf{d}_n. \tag{3}$$

Therefore, the encoding stage can be reduced to performing many times the matrix-vector multiplication (3), where all operations are carried out over $GF(2^w)$.

At the decoding stage, the following expression is used to reconstruct failed data from non-failed data and checksum devices:

$$\mathbf{d}_n = \phi^{-1}_{n \times n} \times \mathbf{e}^*_n, \tag{4}$$

where the inverse matrix $\phi^{-1}_{n \times n}$ is computed from those rows of the matrix $\mathbf{F}_{(n+m) \times n}$ that correspond to non-failed data and checksum devices.

3   Mapping Reed-Solomon Erasure Codes and Their Cauchy Version onto Cell/B.E. Architecture

3.1   Mapping Reed-Solomon Codes

In our investigation, we focus on mapping the following expression:

$$\mathbf{C}_{m \times B} = \mathbf{F}^*_{m \times n} \mathbf{D}_{n \times B}, \tag{5}$$

which is obtained from Eqn. (3) taking into consideration the necessity to process not a single vector $\mathbf{d}_n$, but B such vectors. An expression of the same kind is used at the decoding stage. Moreover, in this work we neglect the influence of computing the inverse matrix $\phi_{n \times n}^{-1}$ on the performance of the whole algorithm. The cost of this operation can be neglected for relatively small values of m, which are of our primary interest in the case of distributed data storage in grids [18].

In this work, we propose a method for the systematic mapping of Eqn. (5) onto the Cell/B.E. architecture. This method takes into account properties of the architecture on all three levels of its parallel processing hierarchy, namely:

1. eight SPE cores running independently, and communicating via the EIB bus;
2. vector (SIMD) processing of 16 bytes in each SPE core;
3. executing instructions by two pipelines (*odd* and *even*) in parallel.

For this aim, Eqn. (5) is decomposed into a set of matrix-matrix multiplications:

$$\mathbf{C}_{m \times 16} = \mathbf{F}^*_{m \times n} \mathbf{D}_{n \times 16} . \tag{6}$$

To compute each of these multiplications within a corresponding SPE core using its SIMD parallel capabilities, the following vectorization algorithm is proposed:

$$
\begin{aligned}
&\textbf{for} \quad i = 0, 1, \ldots, m-1 \quad \textbf{do} \quad \{ \\
&\qquad \mathbf{c}_i = [0, 0, \ldots, 0] \\
&\qquad \textbf{for} \quad j = 0, 1, \ldots, n-1 \quad \textbf{do} \\
&\qquad\qquad \mathbf{c}_i := \mathbf{c}_i \oplus \mathbf{f}^*_{i,j} \odot \mathbf{d}_j \tag{7}
\end{aligned}
$$

$\}$

where:

- vector $\mathbf{f}^*_{i,j}$ is obtained by copying element (byte) $f^*_{i,j}$ of matrix $\mathbf{F}^*_{n \times mj}$ onto all 16 elements (bytes) corresponding to a vector register of SPE;
- operation $\odot$ is the element-by-element multiplication of two vectors, implemented over $GF(2^8)$;
- $\oplus$ denotes the bitwise XOR operation.

Furthermore, to execute this algorithm efficiently on a SPE core, the multiplication operation of the form $c = f * d$ is implemented using table lookups [11], based on the following formula:

$$c = gflog(gflog(f) + gflog(d)) . \tag{8}$$

Here $gflog$ and $gfilog$ denote respectively logarithms and antilogarithms, defined over $GF(2^w)$. Their values are stored in two tables, whose length does not exceed 256 bytes. Following our previous work [19], the efficient implementation of table lookups required by Eqn. (8) is based on utilization of *shufb* permutation instruction, which performs 16 simultaneous byte table lookups in a 32-entry table. Larger tables are addressed using a binary-tree process on a series of 32-entry table lookups, when successive bits of the table indices are used as a selector (using *selb* instruction) to choose the correct sub-table value.

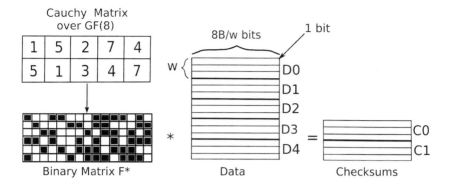

Fig. 1. An example of encoding using the Cauchy binary matrix (based on [14])

3.2   Mapping Cauchy Reed-Solomon Codes

In case of the Cauchy version of RS codes, the matrix $\mathbf{F}^*_{m \times n}$ is transformed into a $wm \times wn$ binary matrix. An example of such a Cauchy matrix is shown in Fig. 1, for $GF(2^3)$, $n = 5$, $m = 2$. As a result, any multiplication over the Galois field is reduced to a series of bitwise XOR operations. For example, the following expression is used to compute the checksum $\mathbf{c}_{1,2}$ from Fig. 1:

$$\mathbf{c}_{1,2} = \mathbf{d}_{0,0} \oplus \mathbf{d}_{1,2} \oplus \mathbf{d}_{2,1} \oplus \mathbf{d}_{2,2} \oplus \mathbf{d}_{3,0} \oplus \mathbf{d}_{3,2} \oplus \mathbf{d}_{4,0} \oplus \mathbf{d}_{4,1} \,, \tag{9}$$

where $\mathbf{d}_{j,l}$ denotes the l-th package of the j-th data device, $l = 0, 1, \ldots, w - 1$.

The mapping method proposed in Subsection 3.1 can be applied in this case as well, providing that properties of binary Cauchy matrices are taken into account. In particular, the vectorization algorithm takes the following form:

$$
\begin{aligned}
&\textbf{for}\ \ i = 0, 1, \ldots, m - 1\ \ \textbf{do}\ \ \{ \\
&\quad \textbf{for}\ \ k = 0, 1, \ldots, w - 1\ \ \textbf{do}\ \ \{ \\
&\quad\quad \mathbf{c}_{i,k} = [0, 0, \ldots, 0] \\
&\quad\quad \textbf{for}\ \ j = 0, 1, \ldots, n - 1\ \ \textbf{do}\ \ \{ \\
&\quad\quad\quad \textbf{for}\ \ l = 0, 1, \ldots, w - 1\ \ \textbf{do} \\
&\quad\quad\quad\quad \mathbf{c}_{i,k} := \mathbf{c}_{i,k} \oplus f^*_{i,k,j,l} * \mathbf{d}_{j,l} \\
&\quad\quad \} \\
&\quad \} \\
&\}
\end{aligned}
\tag{10}
$$

where:

- coefficients $f^*_{i,k,j,l}$ are equal to 1 or 0;
- depending on $f^*_{i,k,j,l}$, the innermost loop operation reduces to a XOR operation with either the vector $\mathbf{d}_{j,l}$, or $[0, 0, \ldots, 0]$.

One of the most important conclusions from this algorithm is the necessity to consider a sparse format of representing the Cauchy matrix, besides the dense format. By introducing some additional overhead, the use of sparse format [16] allows us to avoid operations with nonzero coefficients $f^*_{i,k,j,l}$.

4   Performance Results on Cell/B.E. Processor

4.1   Using Reed-Solomon Codes

In Table 1, we present the performance results achieved for three different implementations of the encoding procedure (7). The pair of values $n, m = 4$ was applied as one of the most promising options to be used for our distributed data storage system in the ClusteriX grid [18]. This table shows the number L_C of clock cycles necessary to process by a single SPE core either one ($L_B = 1$) or ten ($L_B = 10$) data packages each of size $n \times 16$ bytes. The variants correspond to different optimization of the program code performed manually.

Based on Table 1, we can estimate the maximum bandwidth for encoding data on all 8 SPEs as:

$$b_8^{RS} = (8 \times 3.2 \times L_B \times n \times 16)/L_C = 9.58 \ [GB/s] . \tag{11}$$

Such a high value of the bandwidth b_8^{RS} means that in real circumstances it is no longer a constraint for performance of the whole system. For example, in the above-mentioned ClusteriX grid this performance is constrained by the bandwidth of 2×10 Gb/s available in the wide-area network PIONIER, which is used to connect local clusters.

4.2   Using Cauchy Reed-Solomon Codes

Using the open-source Jerasure library [12] for $m, n = 4$, $w = 3$, we generate the Cauchy matrix $\mathbf{F}^*_{wm \times wn} = \mathbf{F}^*_{12 \times 12}$, which contains 88 nonzero elements among all 144 elements. For the Cauchy version of RS codes, Eqn. (11) takes the following form:

$$b_8^{CRS} = (8 \times 3.2 \times L_B \times n \times w \times 16)/L_C \ [GB/s] . \tag{12}$$

Table 1. Performance results (number L_C of clock cycles) for different variants of implementing Reed-Solomon encoding

Compiler option	variant 1		variant 2		variant 3	
	$L_B = 1$	$L_B = 10$	$L_B = 1$	$L_B = 10$	$L_B = 1$	$L_B = 10$
O1	223	2118	201	2078	214	2211
O2	215	1990	198	1710	215	1770
O3	215	1990	198	1710	215	1770

Then by substituting here the values of L_C achieved experimentally for the dense and sparse formats of representing the Cauchy matrix, we obtain the following estimations, respectively:

$$b_{8,D}^{CRSC} = 13.65\,GB/s\,;$$

$$b_{8,S}^{CRSC} = 62.2\,GB/s\,.$$

These results confirm that the sparse format allows for achieving a much higher performance than the dense one. However, the current implementation for the sparse format is not flexible. A further investigation is required in order to combine the flexibility of a program code with high performance.

Keeping in mind the achieved value b_8^{RS} of bandwidth for the classic RS codes, we can also conclude that for the experimental setting considered in this Section, it does not make sense to use the Cauchy version of RS codes instead of the classic one. Also, in practice the estimated value of $b_{8,S}^{CRSC}$ is constrained by the maximum bandwidth of access to the main memory of Cell/B.E. processor, which is equal to 25.6 GB/s. However, the rationale for utilization of the CRS erasure codes could have a place when considering other multicore architectures than the Cell/B.E. processor, or other distributed storage systems characterized by different values of parameters m and n than those studied in this Section.

5   Implementing CRS Codes on Nvidia Tesla C1060 GPU

The CUDA programming environment [8] makes it possible to develop parallel applications for both the Windows and Linux operating systems, giving access to a well-designed programming interface in the C language. On a single GPU, it is possible to run several CUDA and graphics applications concurrently. However, the utilization of GPUs in an everyday practice is still limited. The main reason is the necessity of adapting implemented applications and algorithms to a target architecture, in order to match its internal characteristics. This paper deals with the problem of how to perform such an adaptation efficiently for the encoding stage in the Cauchy version of Reed-Solomon codes.

The CUDA software architecture includes two modules dedicated respectively to a general-purpose CPU, and a graphic processor. This allows for utilization of GPU as an application accelerator, when a part of the application is executed on a standard processor, while another part is assigned to GPU, as a so-called "kernel". The allocation of GPU memory, data transfers, and kernel execution are initialized by the CPU. Each data item used in the GPU needs to be copied from the main memory to the GPU memory; each of such transfers is a source of latency which affects the resulting performance negatively [4]. These performance overheads can be reduced in CUDA using the stream processing mechanism. It allows for overlapping kernel computations with data transfers between the main memory and the GPU memory using the asynchronous CUDA API, which immediately returns from CUDA calls before their completion

Another key feature of modern GPUs is their hierarchical memory organization, which includes several levels with different volume and access time. First of all, GPUs are equipped with a *Global Memory* accessible by all threads (read and write). However, access to this relatively large memory is rather expensive. Other types of GPU memory, accessible for all the threads running on a graphics card, are *Constant Memory* and *Texture Memory*. Their access time is shorter, but threads are allowed only to read from these memories. Threads within a particular CUDA block share a fast *Shared Memory*, which is used for communication and synchronization among threads across a block. Finally, after being initialized, each thread obtains access to a pool of *registers*.

5.1   Mapping CRS Codes onto GPU Architecture

The issue of how to implement erasure codes on GPUs using the Reed-Solomon approach was investigated in [3,4]. The necessity to perform expensive multiplications over the Galois Field $GF(2^w)$ limits the performance achieved for such an approach. Therefore, we decided to investigate the possibility of using the Cauchy version of Reed-Solomon codes for GPUs. For this aim, we have implemented a modified version of the encoding algorithm (10), which is shown below:

$$
\begin{aligned}
&\textbf{for} \quad j = 0, 1, \ldots, n-1 \quad \textbf{do} \quad \{ \\
&\quad \textbf{for} \quad l = 0, 1, \ldots, w-1 \quad \textbf{do} \quad \{ \\
&\qquad \rule{3cm}{0.4pt} \quad GPU\ kernel \quad \rule{1cm}{0.4pt} \\
&\qquad \textbf{for} \quad i = 0, 1, \ldots, m-1 \quad \textbf{do} \\
&\qquad\quad \textbf{for} \quad k = 0, 1, \ldots, w-1 \quad \textbf{do} \\
&\qquad\qquad \mathbf{c}_{i,k} := \mathbf{c}_{i,k} \oplus f^*_{i,k,j,l} * \mathbf{d}_{j,l} \qquad\qquad (13) \\
&\quad \} \\
&\}
\end{aligned}
$$

where $\mathbf{d}_{j,l}$ and $\mathbf{c}_{i,k}$ are data and checksums vectors, respectively; each of them consists of $L_E = |F|/(4*n*w)$ elements of *int* type. The total number L_T of created threads should be greater or equal to L_E. For example, when encoding a file of size 192 MB with $m, n = 4$, $w = 3$, this constraint gives $L_T = 192 * 1024 * 1024/(4 * 4 * 3) = 201326592/48 = 4194304$ threads. Assuming the maximum number of threads within a single block (512 threads), we should create 8192 blocks of threads.

The proposed modification of the encoding algorithm allows us to utilize the stream processing mechanism, when transfer of a certain data stream (vector $\mathbf{d}_{j,l}$) is performed in a particular step. After copying the vector $\mathbf{d}_{j,l}$ to the GPU memory, the GPU kernel is invoked. Each of the GPU threads created in this way is responsible for the execution of XOR operations for a single element of the vector $\mathbf{d}_{j,l}$, and corresponding elements of checksum vectors $\mathbf{c}_{i,k}$. The resulting distribution of computation among threads, as well as the organization of data in the GPU memory, allows us to optimize access to the available *Global Memory*, since consequent threads access data in contiguous areas of memory. Moreover, each thread fetches the vector $\mathbf{d}_{j,l}$ only once, utilizing it $m * w$ times for computations.

The Cauchy matrix $\mathbf{F}^*$ is small and constant; it is located in the *Texture Memory* in order to speed up fetching elements of the matrix by GPU threads. In this

work, both the dense and sparse formats of representing the Cauchy binary matrix
were implemented. In particular, to represent the sparsity structure of the Cauchy
matrix, the standard Compressed Sparse Row (CSR) format [16,17] was used.

5.2   Performance Results on Nvidia Tesla C1060

The performance experiments were carried out for the platform containing Tesla
C1060 CPU, and AMD PhenomII X4 3.12GHz CPU, with CUDA 2.2 as a software environment. In this platform, GPU and CPU are coupled through the PCIe
x16 bus (version 2.0), which provides the maximum bandwidth of 8 GB/s. The
experimental results are presented in Tables 2 and 3, for two sets of parameters:
(i) $n, m = 4$, $w = 3$, and (ii) $n = 8$, $m = 4$, $w = 4$, respectively.

These tables show the real bandwidth of data encoding on a CPU accelerated
by graphics processor (GPU + CPU bandwidth). When measuring this bandwidth, we take into account the following phases: (i) memory allocation and data
copying from the main memory to the GPU memory, and (ii) encoding on the
GPU. The phase of transferring results back to the main memory is not considered because this phase can be overlapped with transfer of data from CPU
to GPU, for the next file. Also, we measure the bandwidth achieved when the
general-purpose CPU is used solely (last column), as well as the performance
achieved by the GPU kernel (without any interaction with the CPU).

In general, the advantage of using a GPU as an accelerator against a solely
CPU-based implementation is reduced by the overhead caused by data transfers
between the CPU and the GPU. The results of the experiments confirm that this
overhead is compensated by using the GPU's parallel processing capabilities
even for relatively short files, with size of several hundred kilobytes. For files
containing several megabytes, the accelerated environment processes data more

Table 2. Bandwidth achieved for GPUs and CPUs when encoding files with different
size ($n, m = 4$, $w = 3$)

File size [MB]	Number of CUDA blocks	dense format		sparse format		CPU bandwidth [MB/s]
		GPU kernel bandwidth [MB/s]	GPU+CPU bandwidth [MB/s]	GPU kernel bandwidth [MB/s]	GPU+CPU bandwidth [MB/s]	
0.05	2	74	57	79	64	241
0.09	4	155	126	158	128	117
0.19	8	293	238	314	249	47
0.38	16	561	407	563	409	48
0.75	32	847	609	827	598	49
1.5	64	1139	899	1141	902	48
3	128	1466	1251	1456	1248	48
6	256	1504	1375	1507	1377	45
12	512	2047	1911	2038	1906	44
24	1024	2097	2021	2080	2002	47
96	4096	2136	2110	2127	2103	44
384	16384	2144	2132	2142	2131	47

Table 3. Bandwidth achieved for GPUs and CPUs when encoding files with different size ($n = 8$, $m = 4$, $w = 4$)

File size [MB]	Number of CUDA blocks	dense format		sparse format		CPU bandwidth [MB/s]
		GPU kernel bandwidth [MB/s]	GPU+CPU bandwidth [MB/s]	GPU kernel bandwidth [MB/s]	GPU+CPU bandwidth [MB/s]	
0.06	2	42	38	39	36	42
0.13	4	83	75	81	74	36
0.25	8	166	149	161	145	18
0.5	16	286	249	292	249	16
1	32	437	379	441	381	6
2	64	580	525	580	527	6
4	128	740	689	740	690	5
8	256	754	725	755	727	5
16	512	1107	1075	1106	1075	5
32	1024	1124	1106	1125	1107	5
128	4096	1165	1158	1164	1158	5
512	16384	1168	1162	1168	1164	5

than ten times faster than CPU, for $n, m = 4$. When encoding large files, up to several hundred megabytes, it becomes possible to achieve more than 2.1 GB of bandwidth. For $n = 8$, $m = 4$, the advantage achieved by the accelerated platform against the general-purpose CPU is even larger.

Another conclusion is a similar efficiency of using the dense and the sparse formats of representing the Cauchy matrix $\mathbf{F}^*$. The use of sparse representation reduces the amount of computations. However, this reduction is balanced with the additional overhead related to indirect addressing of elements of the checksum vectors $\mathbf{c}_{i,k}$. This effect is not surprising since the degree of sparsity of Cauchy matrices is relatively low. For the matrices $\mathbf{F}^*$ corresponding to Tables 2 and 3, less than 50% of all the elements of these matrices (41.9% and 48.2%, respectively) are zeros.

6   Conclusions

Erasure codes can radically improve the availability of distributed storage in comparison with replication systems. In order to realize this thesis, efficient implementations of the most compute-intensive parts of the underlying algorithms should be developed. The investigation carried out in this work confirms the advantage of using modern multicore architectures for the efficient implementation of the classic Reed-Solomon erasure codes, as well as their Cauchy modification.

References

1. Chen, T., Raghavan, R., Dale, J.N., Iwata, E.: Cell Broadband Engine Architecture and its first implementation: A performance view. IBM Journal of Research and Development 51(5), 559–572 (2007)

2. Collins, R., Plank, J.: Assessing the Performance of Erasure Codes in the Wide-Area. In: Proc. 2005 Int. Conf. Dependable Systems and Networks, DSN 2005, pp. 182–187. IEEE Computer Society (2005)
3. Curry, M.L., Skjellum, A., Ward, H.L., Brightwell, R.: Arbitrary Dimension Reed-Solomon Coding and Decoding for Extended RAID on GPUs. In: Proc. 3rd Petascale Data Storage Workshop, PDSW 2008 (2008)
4. Curry, M.L., Skjellum, A., Ward, H.L., Brightwell, R.: Accelerating Reed-Solomon coding in RAID systems with GPUs. In: IPDPS 2008, pp. 1–6. IEEE Press (2008), http://www.bibsonomy.org/bibtex/2e7f39d74179b4d96fea4d89df77c5d6b/dblp
5. Fatahalian, K., Houston, M.: GPUs: A Closer Look. Comm. ACM 51, 50–57 (2008)
6. Kher, V., Kim, Y.: Securing Distributed Storage: Challenges, Techniques, and Systems. In: ACM Workshop on Storage Security and Survivability, pp. 9–25 (2005)
7. MacKay, D.J.C.: Fountain Codes. IEE Proc. – Communications 152(6), 1062–1068 (2005)
8. Nickolls, J., Buck, I., Garland, M., Skadron, K.: Scalable Parallel Programming with CUDA. Queue 6(2), 40–53 (2008)
9. OpenCL - The open standard for parallel programming of heterogeneous systems, http://www.khronos.org/opencl
10. Plank, J.: A tutorial on Reed-Solomon coding for fault-tolerance in RAID-like systems. Software – Practice & Experience 27(9), 995–1012 (1997)
11. Plank, J., Ding, Y.: Note: Correction to the 1997 tutorial on Reed-Solomon coding. Software – Practice & Experience 35(2), 189–194 (2005)
12. Plank, J., Simmerman, S., Schuman, C.: Jerasure: A Library in C/C++ Facilitating Erasure Coding for Storage Applications, https://www.cs.utk.edu/~plank/plank/papers/CS-08-627.pdf
13. Plank, J., Thomason, M.: A Practical Analysis of Low-Density Parity-Check Erasure Codes for Wide-Area Storage Applications. In: Proc. 2004 Int. Conf. Dependable Systems and Networks, pp. 115–124. IEEE Comp. Soc. (2004)
14. Plank, J., Xu, L.: Optimizing Cauchy Reed-Solomon codes for fault-tolerant network storage applications. In: NCA 2006: 5th IEEE Int. Symp. Network Computing Applications, pp. 173–180 (2006)
15. Weatherspoon, H., Kubiatowicz, J.D.: Erasure Coding Vs. Replication: A Quantitative Comparison. In: Druschel, P., Kaashoek, M.F., Rowstron, A. (eds.) IPTPS 2002. LNCS, vol. 2429, pp. 328–338. Springer, Heidelberg (2002)
16. Williams, S., Oliker, L., Vuduc, R., Shalf, J., Yelick, K., Demmel, J.: Optimization of sparse matrix-vector multiplication on emerging multicore platforms. Parallel Computing 35, 178–194 (2009)
17. Wozniak, M., Olas, T., Wyrzykowski, R.: Parallel Implementation of Conjugate Gradient Method on Graphics Processors. In: Wyrzykowski, R., Dongarra, J., Karczewski, K., Wasniewski, J. (eds.) PPAM 2009. LNCS, vol. 6067, pp. 125–135. Springer, Heidelberg (2010)
18. Wyrzykowski, R., Kuczynski, L.: Towards Secure Data Management System for Grid Environment Based on the Cell Broadband Engine. In: Wyrzykowski, R., Dongarra, J., Karczewski, K., Wasniewski, J. (eds.) PPAM 2007. LNCS, vol. 4967, pp. 825–834. Springer, Heidelberg (2008)
19. Wyrzykowski, R. Kuczynski, L., Rojek, K.: Mapping AES Cryptography and Whirlpool Hashing onto Cell/B.E. Architecture. In: Proc. PARA 2008 (2010) (accepted for publication)

Communication-Efficient Algorithms for Numerical Quantum Dynamics

Magnus Gustafsson, Katharina Kormann, and Sverker Holmgren

Division of Scientific Computing, Uppsala University,
Box 337, SE-75105 Uppsala, Sweden
{magnus.gustafsson,katharina.kormann,sverker.holmgren}@it.uu.se

Abstract. The time-dependent Schrödinger equation (TDSE) describes the quantum dynamical nature of molecular processes. However, numerical simulations of this linear, high-dimensional partial differential equation (PDE) rapidly become computationally very demanding and massive-scale parallel computing is needed to tackle many interesting problems. We present recent improvements to our MPI and OpenMP parallelized code framework HAParaNDA for solving high-dimensional PDE problems like the TDSE. By using communication-efficient high-order finite difference methods and Lanczos time propagators, we are able to accurately and efficiently solve TDSE problems in up to five dimensions on medium-sized clusters. We report numerical experiments which show that the solver scales well up to at least 4096 computing cores, also on computer systems with commodity communication networks.

Keywords: Lanczos algorithm, high-order finite difference, parallel scalability.

1 Introduction

The time-dependent Schrödinger equation (TDSE) provides a description of dynamic processes at the quantum-mechanical level, e.g., "particles" (nuclei, electrons) in atoms and molecules. The number of spatial dimensions d in the TDSE is in principal given by $d = 3n$, where n denotes the number of particles involved. This makes the numerical solution challenging since the number of degrees of freedom for a grid-based numerical method will grow exponentially with the number of particles. Therefore, the demands on memory as well as computing power rapidly becomes huge, and it is necessary to employ massive parallelism.

The dimensionality of TDSE problems can be somewhat reduced by ignoring global degrees of freedom of the system (translation, rotation). Also further modeling can be introduced where some dimensions are removed or treated in another way. An important model of this type is the Born–Oppenheimer approximation where the nuclear and electronic degrees of freedom in a molecular system are separated, leaving the nuclei to evolve on potential surfaces determined by the electrons.

K. Jónasson (Ed.): PARA 2010, Part II, LNCS 7134, pp. 368–378, 2012.

The d-dimensional TDSE reads

$$i\hbar\frac{\partial}{\partial t}\psi(x,t) = \underbrace{\left(-\sum_{i=1}^{d}\frac{\hbar^2}{2m_i}\frac{\partial^2}{\partial x_i^2} + V(x,t)\right)}_{:=\hat{H}}\psi(x,t), \quad \psi(x,0)=\psi_0, \quad x\in\mathbb{R}^d,$$

where the Hamiltonian $\hat{H}$ consists of a second-derivative part describing the kinetic energy of the system and the potential $V(x,t)$ modeling interactions within the system as well as with its surroundings (for instance, with electromagnetic radiation). The reduced Planck constant is denoted by $\hbar$, the particle mass corresponding to dimension i by m_i, and the wave function by ψ. The TDSE is posed on the whole of $\mathbb{R}^d$ and for a numerical simulation, we have to restrict our attention to a finite domain. In order to cover boundary effects correctly, absorbing boundary condtions might be needed for unbounded systems (e.g., scattering and dissociation problems). Modeling of such boundary conditions and the solution of the corresponding TDSE can be complicated since physical properties, like conservation of total probability, do not hold anymore. For bounded systems, i.e., when the total probability is conserved, it is often preferable to pose the problem on a sufficiently large d-orthope such that the wave function can be considered to vanish at the boundaries. The actual boundary conditions are then unimportant for the accuracy of the model (but can be important for the stability of the numerical method). A standard choice for bounded problems is to use periodic boundary conditions, and the current version of our code framework HAParaNDA [1] is focused on bounded systems with periodic boundary closure. However, our plan is to extend the implementation with support for unbounded problems and other boundary conditions in HAParanNDA and we have already developed some prototype code for doing this.

Our discretization is based on the standard method-of-lines approach where we first introduce a spatial discretization and then solve the corresponding large-scale ordinary differential equation problem in time. For a TDSE problem with a time-independent Hamiltonian, the solution of this semi-discrete problem is given by

$$u(t) = e^{-\frac{i}{\hbar}H\cdot t}u(0),$$

where the Hamiltonian matrix H is the spatially discretized Hamiltonian operator and u the discrete wave function. For a time-dependent Hamiltonian, the exponential form can still be used on small time intervals but the Hamiltonian has to be time-averaged using the Magnus expansion (see [2] for details). This means that deriving numerical methods for time propagation of the TDSE (and other linear PDE problems) in effect boils down to finding efficient and accurate schemes for computing the matrix exponential of a large matrix.

Even though the parallelization results reported in this article mainly focus on a test problem with a time-independent Hamiltonian, our code framework is mainly aimed at handling problems with explicit time dependence. For this class of problems, small time steps are necessary for accuracy and the matrix exponential can be efficiently computed by the short-iterative Lanczos propagator (cf. [3]). The Lanczos algorithm is a Krylov subspace method and each iteration involves both multiplying the Hamiltonian matrix with a wave function vector and performing inner products of such vectors. In a parallel implementation, these operations will involve communication and make the parallelization challenging on large-scale computers.

In quantum dynamics, the standard approach for discretizing the spatial derivatives and computing the action of the Hamiltonian matrix is to use a (pseudo)spectral discretization and the Fast Fourier Transform or explicit multiplication with a dense differentiation matrix in each spatial direction. However, such schemes involve heavy global communication in the parallel implementation and scalability on massive-scale computers will not be achievable. Instead, we have chosen to compute the spatial derivatives using a qth order finite difference (FD) method, where q an even number. The numerical dispersion error introduced by the FD scheme can be controlled by using the formulas presented in [4]. The FD computations require nearest-neighbor communication only, which can be implemented in a scalable way as described in Sec. 2. Compared to the multiplication with the Hamiltonian matrix, the arithmetic work and the amount of communicated data for the inner product computations in the Lanczos scheme is small. However, since these computations do involve global reductions, they may potentially affect scalability for massive-scale systems. In Sec. 3, we investigate different versions of the Lanczos algorithm where the number of global communication points is reduced. Performance tests are reported in Sec. 4, and some conclusions are presented in Sec. 5.

2   Parallel Matrix-Vector Product

In HAParaNDA, the d-dimensional structured spatial grid is divided into equally sized blocks, which are distributed among the compute nodes (a node consists of one or more CPUs, where each CPU typically has several cores). In the current implementation, the spatial grid is equidistant and each node hosts a single MPI process handling a single grid block, but future versions will include spatial adaptivity and a more general way of distributing the MPI processes and grid blocks over the compute nodes.

Along the borders of a block, the finite difference stencils depend on function values in neighboring blocks which are located in remote nodes. To retrieve these data values to the local memory of each node as needed, we use a standard scheme exploiting explicit message passing and local buffers referred to as ghost regions. Since we only consider tensor-product stencils, each grid block will have to exchange data with $2d$ other blocks that are logical neighbors.

In order to optimize the computation of the matrix-vector product, we use non-blocking communication and attempt to overlap all inter-block communication delays with computations. First, each block issues non-blocking send and receive operations in all directions. While waiting for the remote values to arrive, each block performs all the calculations that are possible without having access to any ghost region values. Thus, along the block border, the FD stencil is only partially applied at first. Finally, the missing FD computations are filled in as the ghost region blocks are received, regardless of the order in which they arrive for maximum latency hiding.

An important feature of our code framework is that the explicit partitioning of the grid into different MPI processes is performed at the level of nodes in the computer system. Within each block, further parallelization is achieved by using OpenMP threads which map to the cores within a node. The threads share memory and no explicit replication of data is needed, which results in a significant reduction of the memory overhead compared to an MPI-only implementation since the amount of memory that has to be allocated for the ghost region buffers is reduced. Since the grid blocks are of fixed size and since all gridpoints require the same amount of arithmetic operations, we split the work within a block statically and equally among the available threads.

For efficient computation of the stencil operator in the matrix-vector product, we must take care to maximize reuse of the data that we bring into the caches. Here, we use cache-tiling which has proven to be a useful approach to do blocking. Tiling makes use of the hardware prefetchers that are prevalent in modern architectures and allows us to use larger block sizes than would be feasible in simpler blocking techniques [5]. Furthermore, we do explicit inlining of all computations in the kernels and, as far as possible, we have merged loops to minimize the number of passes through data and combined parallel regions to minimize the overhead of forking and joining threads.

3 Parallel Lanczos Propagators

The idea of the Lanczos propagator is to project the solution $w = \exp\left(-\frac{i}{\hbar} H \Delta t\right) u$ onto the Krylov subspace $\mathcal{K}_p(H, u)$ (spanned by $\{u, Hu, \ldots, H^{p-1}u\}$). This is an iterative process that successively builds an orthonormal basis, $V_p = [v_1, \ldots, v_p]$, of $\mathcal{K}_p(H, u)$ and a tridiagonal projection matrix, L_p, of the matrix H,

$$
L_p = \begin{pmatrix} \alpha_1 & \beta_1 & & & \\ \beta_1 & \alpha_2 & \beta_2 & & \\ & \ddots & \ddots & \ddots & \\ & & \beta_{p-2} & \alpha_{p-1} & \beta_{p-1} \\ & & & \beta_{p-1} & \alpha_p \end{pmatrix} = V_p^T H V_p \in \mathbb{R}^{p \times p}.
$$

The vector w is then approximated by

$$
w \approx \|u\| V_p \exp\left(-\frac{i}{\hbar} L_p \Delta t\right) e_1,
$$

with $e_1 = (1, 0, \ldots, 0) \in \mathbb{R}^p$. The steps to compute (L_p, V_p) are summarized in Algorithm 1.

Algorithm 1. THE LANCZOS ALGORITHM

1: $v_1 = u \ / \ \|u\|_2$, $v_0 = 0$, $\beta_0 = 0$
2: **for** $j = 1, 2, \ldots, p$ **do**
3:     $r = Hv_j - \beta_{j-1}v_{j-1}$
4:     $\alpha_j = (r, v_j)$
5:     $r = r - \alpha_j v_j$
6:     $\beta_j = \|r\|_2$
7:     $v_{j+1} = r \ / \ \beta_j$
8: **end for**

If the grid functions v and r are distributed among several nodes, the matrix-vector product requires communication as described in the previous section. Furthermore, the global communication required to perform the inner products in Algorithm 1 will become increasingly expensive as the number of nodes grows. Thus, in a parallel environment, scalability will eventually be hampered by the fact that two inner products have to be computed at different locations in each iteration. When aiming at scalability for massively parallel systems, it might therefore be better to rearrange the algorithm as described by Kim and Chronopoulos [6], see Algorithm 2. At the expense of one extra vector operation, the number of synchronization points is then reduced to one per iteration. We refer to this scheme as *FS-Lanczos*.

Algorithm 2. THE FEW SYNCHRONIZATION (FS) LANCZOS ALGORITHM

1: $v_0 = u \ / \ \|u\|_2$, $q_0 = 0$
2: **for** $j = 0, 1, 2, \ldots, p$ **do**
3:     $r = Hv_j$
4:     $u = (r, v_j)$
5:     $a = (v_j, v_j)$
6:     $\beta_j = \sqrt{a}$
7:     $\alpha_{j+1} = u/a$
8:     $q_{j+1} = v_j/\beta_j$
9:     $v_{j+1} = r/\beta_j - \beta_j q_j - \alpha_{j+1}q_{j+1}$
10: **end for**

Algorithm 2 still requires p synchronization points per time step. In the s-step Lanczos method [6], presented in Algorithm 3, blocks of s consecutive steps are executed with only one synchronization step required for each block. We have transferred these ideas to the Lanczos propagator method.

3.1   Error Control and Stability

When using Lanczos procedures, the number of iterations p has to be chosen with care to avoid loss of orthogonality of the basis vectors and possibly also numerical instabilities. Therefore, it is common to choose p adaptively and to make

---

Algorithm 3. THE s-STEP LANCZOS ALGORITHM

1: $v_1^1 = u \,/\, \|u\|_2$
2: $\bar{V}_1 = [\, v_1^1, H v_1^1, \ldots, H^{s-1} v_1^1 \,]$
3: Compute $2s$ inner products $(H^j v_1^1, v_1^1)$, $j = 0, \ldots, 2s - 1$
4: **for** $k = 1, \ldots, \lceil m/s \rceil$ **do**
5:     Form $W_k = \bar{V}_k^T \bar{V}_k$; solve $W_{k-1} \gamma_{k-1}^i = c_{k-1}^i$ and $W_k \alpha_k^i = d_k^i$, $i = 1, \ldots, s$
6:     $v_{k+1}^1 = H v_k^1 \; - \bar{V}_{k-1} \gamma_{k-1}^s - \bar{V}_k \alpha_k^s$
7:     Compute $H v_{k+1}^1, H^2 v_{k+1}^1, \ldots, H^s v_{k+1}^1$
8:     Compute $2s$ inner products $(H^j v_k^1, v_k^1)$, $j = 0, \ldots, 2s - 1$
9:     Solve $W_k t_k^j = b_k^j$, $j = 2, \ldots, s$
10:    $v_{k+1}^j = H^{j-1} v_{k+1}^1 - \bar{V}_k t_k^j$, $j = 2, \ldots, s$
11: **end for**

---

sure the iterations are stopped once the Lanczos residual R_p is small enough. Residual-based stopping criteria have been extensively studied for the standard Lanczos method (cf. Saad [7] and Hochbruck, Lubich & Selhofer [8]), and the same adaptivity criterion can be applied for the FS-Lanczos scheme. For the Schrödinger equation, the residual after p iterations reads

$$R_p = \Delta t \left(\exp\left(-\frac{\mathrm{i}}{\hbar} \Delta t L_p \right) \right)_{p,1} \beta_p \|u\|.$$

We can compute the residual R_p after completion of the pth iteration. However, in the FS-Lanczos scheme β_p is not known until after the $(p+1)$th matrix-vector product has been computed. Of course, β_p could be computed before — but it is this rearrangement that avoids one of the synchronization points. In order to avoid computing an extra matrix-vector product, we combine both versions of the algorithm to get an adaptive FS-Lanczos scheme: We start with the FS-Lanczos and then switch to the standard Lanczos as soon as the residual comes close to the tolerance. Since we are repeatedly using the Lanczos algorithm in consecutive time steps, we can use the number of iterations needed in the previous time step m as a guess for the number of iterations needed in the present step. We then do $m - 2$ iterations with the FS-Lanczos before switching to the standard one. We refer to this version as *hybrid Lanczos* in the following.

For the s-step variant introducing adaptivity gets more complicated. After M outer loops with the s-step algorithm, it holds that

$$H \cdot V_M = V_M \bar{T}_M + v_{M+1}^1 e_{sM},$$

where $V_M = [\bar{V}_1, \ldots, \bar{V}_M]$, $\bar{T}_M$ is a block matrix built from the α_k^i and γ_{k-1}^i in line 5 of Algorithm 3, and $e_{sM} = (0, \ldots, 0, 1) \in \mathbb{R}^{sM}$. Based on this identity, one can show, following the argumentation for the standard case, that the remainder can be computed by

$$\Delta t \left(\exp\left(-\frac{\mathrm{i}}{\hbar} \Delta t \bar{T}_M \right) \right)_{p,1} \|v_{M+1}^1\| \|u\|.$$

Hence, checking the residual in the s-step algorithm requires additional computational work as well as synchronization: We have to execute lines 5 and 6 for loop index $M + 1$. Furthermore, we have to synchronize computations after line 6 in order to decide whether or not we have to continue the iteration.

3.2   Scalability of Various Lanczos Variants

In order to better understand the scalability of the different Lanczos variants, we compare the number of floating point operations required to take blocks of s Lanczos steps. The results are summarized in Table 1 (cf. [6]).

Lanczos and FS-Lanczos are very similar in terms of computational complexity, with the only difference that FS-Lanczos requires an extra vector update in each iteration compared to Lanczos. We expect that this extra amount of computation will at least be outweighed by doing half the amount of synchronization.

Clearly, the most dominant arithmetic operation is the matrix-vector multiplication, and its complexity will grow with the number of dimensions in the TDSE as well as the order of the finite difference operator. Since the s-step algorithm requires an extra matrix-vector multiplication in each iteration block, we expect it to be slower than the other two algorithms when synchronization latency is low. As we distribute the computations to more and more processors (due to larger problems), synchronization delays will grow, and at some point we expect the more communication efficient method to be faster.

Table 1. Comparison of the complexity of the different Lanczos variants (one block for s-step)

	FLOPs*	Lanczos	FS	s-step
Inner products	$4^\dagger$	$2s$	$2s$	$2s$
Vector updates	4	$5s$	$6s$	$2s(s + 1)$
Mat-vec multiplies	$4d(q + 1)^\ddagger$	s	s	$s + 1$
All-to-all reductions		$2s$	s	1

*Required number of FLOPs per gridpoint.

$\dagger$In the Lanczos algorithm, the complex inner product is always pure real.

$\ddagger$Computations are reordered for optimal cache efficiency, not minimal FLOPs.

4   Performance Tests

In order to demonstrate the performance of a massively parallel simulation of the TDSE based on a FD-Lanczos discretization, we have conducted several simulations on two different clusters which are referred to as Kalkyl and Akka. The hardware configurations are summarized in Table 2. We have used the same combination of compilers and libraries on both machines; Intel icc (v11.1), OpenMPI (v1.4.2), and Intel MKL (v10.1.1 on Kalkyl, v10.2.5 on Akka).

Table 2. Overview of the clusters that have been used for the computations

	Kalkyl	Akka
No. Nodes	348	672
No. CPUs/node:	2	2
DRAM/node:	24 GB	16 GB
Interconnect:	InfiniBand:	InfiniBand:
	20 Gbps, 4:1 oversubscribed	10 Gbps, full bisectional bw
CPU:	Intel Xeon E5520 (Nehalem)	Intel Xeon L5420 (Harpertown)
No. Cores:	4	4
Clock rate:	2.26 GHz	2.50 GHz
L1-cache:	32kB/32kB (I/D)	32kB/32kB (I/D)
L2-cache:	256 KB	2×6 MB
L3-cache:	8MB	–

For the experiments, we use a model for a harmonic oscillator with

$$V(x) = \frac{1}{2} \sum_{i=1}^{d} m_i \omega_i^2 x_i^2, \qquad \omega \in \mathbb{R}.$$

This problem suits our scaling experiments well since it is bounded and can be formulated in an arbitrary number of dimensions. Furthermore, the analytical solution is known so we can verify the correctness of the solution. In quantum physics, harmonic oscillators are common since they approximate the behavior of more complicated potentials close to a stable equilibrium [9, Chapt. 2.3].

Fig. 1 shows weak scaling results for the two clusters for 3D and 4D problems. The curves for all four experiments show a similar behavior. The slopes of the curves are very moderate, i.e., we get a good parallel speed-up for all the methods. Overall one can state that the timings for the FS-Lanczos are slightly superior. The timings for the standard Lanczos version are worse than the other implementations, especially on Akka, indicating that the effect of synchronization is slightly more pronounced on Akka than on Kalkyl, where the computationally more intense s-step method performs the worst. On Akka, we also note that the gain from reducing the amount of synchronization is less in 4D than in 3D. This seems reasonable since the higher the dimension, the more points in the FD stencil, which increases the complexity of the matrix-vector product and its proportion of the total cost. We expect the same effect to show up also when the order of the FD stencil is changed. Indeed, we did perform some experiments with a second order stencil where the impact of communication reduction on computing times was considerably higher than with the 8th order stencil.

The wall clock timings are quite similar for Lanczos and s-step Lanczos. Since the computational complexity of the s-step variant is $1/s$ times higher, the rearrangements of the algorithm appear to have an effect. However, the timings cannot be explained by the latencies for the global reduction/synchronization operations. Typically, for small amounts of data, the node-to-node synchronization latency of InfiniBand is about 10 μs [10], and [11] reports 60 μs latency for global reduction operations. Since each time step in our scaling experiments takes a few seconds, the synchronization latencies should not be noticeable. More important are cache effects, since vector operations are merged into dense matrix operations. Cache effects due to loop merging explain the fact that the FS-Lanczos method performs slightly better than the standard version even on a single compute node where no inter-node communication is involved.

In the next set of results, we adaptively select the size of the Krylov subspace. Due to the problems with estimating the error for the s-step version (as discussed in Sec. 3.1) together with the fact that its scaling does not seem to improve the one observed with the FS version, we do not include the s-step Lanczos in these tests. Instead, we compare the standard Lanczos and the hybrid version in Fig. 2. For most of the experiments, the hybrid method slightly outperforms the standard version. It is also obvious that reducing the synchronization again pays off more in the 3D example than is the case for 4D. Note that Fig. 2 does not show the results of a simple scaling experiment: When discretizing more accurately in space, more Lanczos steps are needed to retain accuracy. Thus, the slopes of the curves are due to the fact that more matrix-vector products are performed and not due to inefficient parallelization.

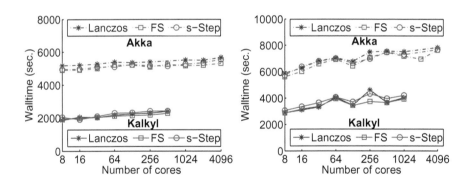

Fig. 1. Weak scaling of a model problem with a fixed workload per node. FD stencil of order 8, 1000 timesteps with 6 Lanczos iterations (fixed) per step. Left: 3D, 240^3 gridpoints per node. Right: 4D, 60^4 gridpoints per node.

In order to demonstrate that our implementation is indeed capable of solving more complex problems with several coupled potential surfaces, time-dependent potentials and five spatial dimensions, we finally present results for a TDSE describing two coupled 5D harmonic oscillators computed on Akka. In each spatial coordinate, we discretize with 90 points and distribute the wave function over 256 computing nodes. It took 17h 25min to perform 1517 time steps with 9162 matrix-vector multiplications. We have chosen the time step size as well as the Krylov space adaptively following the theory developed in [12]. Both parameters

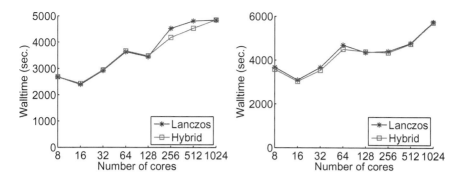

Fig. 2. Comparison of Lanczos and Hybrid Lanczos. FD stencil of order 8, 1000 timesteps, the number of Lanczos iterations adaptively chosen. Left: 3D, 240^3 gridpoints per node. Right: 4D, 60^4 gridpoints per node.

are shown in Fig. 3(a). The time step is chosen based on the Magnus expansion error only. The number of Krylov iterations needed to meet the same tolerance is moderate (between 6 and 8), which means that the time step sizes are of a magnitude where the Krylov method is efficient and not too memory intense. Fig. 3(b) shows the estimated errors due to truncation in the Magnus expansion and the Krylov space. Both errors are below the local tolerance (1e-8) at any time.

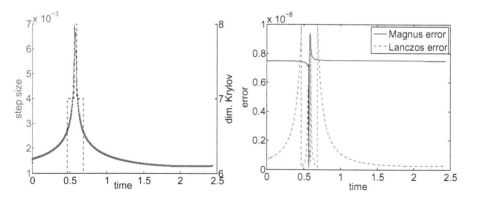

Fig. 3. Time-dependent test example in 5D. Left: step sizes (solid) and number of Krylov vectors (dashed). Right: estimated error due to truncation in Magnus expansion (solid) and due to the Lanczos algorithm (dashed).

5   Conclusions

We have presented the HAParaNDA package providing a hybrid MPI–OpenMP implementation to solve the time-dependent Schrödinger equation based on finite differences in space and Lanczos propagation in space. Our performance experiments show that our code scales very nicely for massively parallel applications. On clusters with InfiniBand interconnect, we find the matrix-vector product to be the by far most time-consuming part and that a reduction of the number of synchronizations in the Lanczos algorithm only pays off as long as no extra matrix-vector product is involved. We also solve a quantum system in 5D that includes an oscillatory time-dependent potential. For this problem, we have implemented adaptive time-stepping.

Acknowledgments. The computations were performed on resources provided by SNIC through Uppsala Multidisciplinary Center for Advanced Computational Science (UPPMAX) and High Performance Computing Center North (HPC2N). The authors acknowledge discussions with Hans O. Karlsson, Quantum Chemistry, Uppsala University.

References

1. Gustafsson, M., Holmgren, S.: An implementation framework for solving high-dimensional PDEs on massively parallel computers. In: Pro. of ENUMATH 2009. Springer, Heidelberg (2010)
2. Kormann, K., Holmgren, S., Karlsson, H.: Accurate time propagation for the Schrödinger equation with an explicitly time-dependent Hamiltonian. J. Chem. Phys. 128, 184101 (2008)
3. Leforestier, C., Bisseling, R.H., Cerjan, C., Feit, M.D., Friesner, R., Guldberg, A., Hammerich, A., Jolicard, G., Karrlein, W., Meyer, H.D., Lipkin, N., Roncero, O., Kosloff, R.: A Comparison of Different Propagation Schemes for the Time Dependent Schrödinger Equation. J. Comput. Phys. 94, 59–80 (1991)
4. Holmgren, S., Peterson, C., Karlsson, H.O.: Time-marching methods for the time-dependent Schrödinger equation. In: Proc. of the International Conference on Computational and Mathematical Methods in Science and Engineering 2004, pp. 53–56. Uppsala University (2004)
5. Datta, K., Kamil, S., Williams, S., Oliker, L., Shalf, J., Yelick, K.: Optimization and Performance Modeling of Stencil Computations on Modern Microprocessors. SIAM Rev. 51, 129–159 (2009)
6. Kim, S., Chronopoulos, A.: A class of Lanczos-like algorithms implemented on parallel computers. Parallel Comput. 17 (1991)
7. Saad, Y.: Analysis of some Krylov subspace approximations to the matrix exponential operator. SIAM J. Numer. Anal. 29, 209–228 (1992)
8. Hochbruck, M., Lubich, C., Selhofer, H.: Exponential integrators for large systems of differential equations. SIAM J. Sci. Compt. 19, 1552–1574 (1998)
9. Griffiths, D.J.: Introduction to Quantum Mechanics, 2nd edn. Pearson Education, Upper Saddle River (2005)
10. Latency Results from Pallas MPI Benchmarks,
    http://vmi.ncsa.uiuc.edu/performance/pmb_lt.php
11. Mamidala, A.R., Liu, J., Panda, D.K.: Efficient Barrier and Allreduce on Infini-Band clusters using multicast and adaptive algorithms. In: 2004 IEEE International Conference on Cluster Computing, pp. 135–144. IEEE, New York (2004)
12. Kormann, K., Holmgren, S., Karlsson, H.O.: Global Error Control of the Time-Propagation for the Schrödinger Equation with a Time-Dependent Hamiltonian. Technical Report 2009-021, Uppsala University (2009)

Efficiently Implementing Monte Carlo Electrostatics Simulations on Multicore Accelerators

Marcus Holm and Sverker Holmgren

Department of Information Technology, Uppsala University
{marcus.holm,sverker}@it.uu.se

Abstract. The field of high-performance computing is highly dependent on increasingly complex computer architectures. Parallel computing has been the norm for decades, but hardware architectures like the Cell Broadband Engine (Cell/BE) and General Purpose GPUs (GPGPUs) introduce additional complexities and are difficult to program efficiently even for well-suited problems. Efficiency is taken to include both maximizing the performance of the software and minimizing the programming effort required. With the goal of exposing the challenges facing a domain scientist using these types of hardware, in this paper we discuss the implementation of a Monte Carlo simulation of a system of charged particles on the Cell/BE and for GPUs. We focus on Coulomb interactions because their long-range nature prohibits using cut-offs to reduce the number of calculations, making simulations very expensive. The goal was to encapsulate the computationally expensive component of the program in a way so as to be useful to domain researchers with legacy codes. Generality and flexibility were therefore just as important as performance. Using the GPU and Cell/BE library requires only small changes in the simulation codes we've seen and yields programs that run at or near the theoretical peak performance of the hardware.

Keywords: Monte Carlo, GPU, Cell, electrostatics.

1 Introduction

The goals of this paper are twofold. First, we intend to characterize the challenges that a domain scientist faces when programming different heterogeneous multi-core architectures. This is done by examining a Monte Carlo code for molecular electrostatics simulations. Second, we briefly describe an efficient library we've developed for physical chemists interested in these simulations. In this introduction, we'll briefly cover the relevant concepts in chemistry and describe the hardware.

1.1 Monte Carlo Simulations in Physical Chemistry

In the field of physical chemistry, certain problems have characteristics that make simulations difficult or expensive. Whether the problem lies with the size of the

K. Jónasson (Ed.): PARA 2010, Part II, LNCS 7134, pp. 379–388, 2012.

system, the convergence rates of the available algorithms, or the sensitivity to errors, they need a lot of computational resources and it is therefore of continual interest to implement the latest numerical techniques on the latest hardware. One such problem is the simulation of large electrostatic molecular systems. These suffer from the curse of dimensionality and are therefore commonly solved using Monte Carlo (MC) methods. These problems are of scientific value because the behavior of many important systems in biology and industry is dominated by electrostatic forces, for example polyelectrolytes like DNA [5].

1.2   Multicore Accelerators

We chose two specific architectures to represent two directions that heterogeneous multicore architecture design has taken. They promise very good performance per Watt and per dollar, but are so different from ordinary processors that traditional programming techniques are ineffective. The Cell Broadband Engine (Cell/BE or Cell) was jointly developed by Sony, Toshiba and IBM, and combines an "ordinary" CPU with eight SIMD coprocessors on a single chip. The GPU is a highly parallel coprocessor with a large memory bank of its own, and is connected much less closely to the CPU. In this section, we discuss some key features of these architectures that must be taken into account when designing high-performance applications.

Graphics Processing Units. The GPU is a true many-core architecture, containing hundreds of computational units called stream processors grouped into a number of multiprocessors. Each multiprocessor has one instruction unit, a private memory bank shared between the stream processors, and a bus connecting to the device's main memory. High hardware utilization is achieved by spawning many hundreds of threads to run in parallel. These threads are partitioned into groups called warps, which run in lock-step fashion on a single stream processor and share the processor's private memory.

Stream processors have a relatively high-bandwidth connection to the device memory, capable of multiple concurrent accesses. However, the ratio of floating point rate (flops) to memory bandwidth is still quite low, such that data reuse (e.g. by using the warp shared memory) is an important factor in performance. Bandwidth between device memory and host memory is very much smaller. Efficient use of memory is therefore critical to performance. [4]

Cell Broadband Engine. The Cell B/E is often called a "supercomputer on a chip". It features an ordinary but stripped-down CPU called the Power Processing Element (PPE) and eight SIMD coprocessors called Synergistic Processing Elements (SPEs), connected via a high-bandwidth bus called the Element Interchange Bus (EIB). Standard usage is to use the PPE for control processing and to offload compute-intensive tasks to the SPEs. Parallelism is achieved mainly by using the SPEs in parallel and leveraging SIMD instructions. Since each SPE works on data stored in its own small private memory bank, the local store (LS), data must be sent to and from the SPEs. Each SPE has a Memory Flow

Controller (MFC), a processor that handles data transfers while the arithmetic unit performs calculations. This allows computation and communication to be effectively overlapped. [1]

There are many similarities between the Cell and GPU. On a high level, both achieve high flop rates by trading large automatic caches for parallel arithmetic units, requiring the user to manage the flow of data. For both architectures, applications must be arithmetically intense so communication time can be hidden by computation. Blocking schemes are usually necessary to achieve data reuse and minimize communication, often yielding very similar-looking program designs.

In actual practice, however, programming the Cell differs from GPU programming in a number of ways. First, the number of threads that can run concurrently on the Cell is much smaller, and threads must be scheduled by hand. On the GPU, communication time is automatically hidden by the scheduler, while the Cell requires the user to use e.g. double-buffering. While the Cell has actual SIMD registers, the GPU multiprocessor has ordinary registers but multiple processors per instruction unit in a SIMT design. Cell SPEs can communicate and synchronize with each other, while GPU multiprocessors have little or no such functionality. These differences can make for very different-looking program codes.

2   Model and Implementation

While the library we implemented can be made to work with any application that involves electrostatic interactions between point charges, the application we chose as a platform to test with is a Monte Carlo simulation of a charged polymer (polyelectrolyte). The polyelectrolyte is described as a simple bead-and-stick model with N monomers (see Fig. 1). Each monomer is taken to be a hard, uniformly charged ball that is connected to its neighbors by a bond of constant length. The main program loop effects a change in the conformation of this string of monomers, calculates the change in energy, and accepts or rejects the change according to a probability function. In this way, the conformational space of the polyelectrolyte is explored and various properties can be extracted.

The main functionality of the library is to take a vector of particles and calculate the electrostatic potential, U, of the system, given as

$$U = \sum_{i=1}^{N} \sum_{j=i}^{N} \frac{q_i q_j}{r_{ij}} \tag{1}$$

where q_i is the charge of particle i, r_{ij} is the distance between particles i and j, and N is the number of particles. A particle is represented by a position in 3-space and a charge. The basic algorithm to be implemented is simply a nested loop that calculates all the pairwise interactions.

Fig. 1. Example polyelectrolyte conformations

2.1   Library Design

The goal in creating this library was to amortize the programming effort involved in writing Cell and GPU code over as many applications as possible. One of the main considerations was therefore to reformulate the computationally expensive routines of the polyelectrolyte code into a more general form suitable for a wider variety of molecular electrostatics codes. Program 1 gives the pseudocode for the polyelectrolyte code. The functions `pivot_move` and `calculate_delta_U` are written specifically for polyelectrolyte simulations and aren't usable in other cases. These functions change the system conformation and calculate the change in electrostatic energy, respectively. While these can readily be implemented on the Cell and GPU, the utility of such an effort is much less than a more general implementation.

Program 1 Polyelectrolyte MC program

```
current_sys = init()
pivot = 1

while(! done)
    pivot = (pivot+1)%N
    proposed = pivot_move(current_sys, pivot)
    delta_U = calculate_delta_U(current_sys,
        proposed, pivot)

    if evaluate(delta_U)
        current_sys = proposed
end
```

A general program for electrostatics simulations is shown in Program 2. The `pivot_move` function is replaced by a more general `move` function, and the `calculate_delta_U` function is replaced by a function that calculates an electrostatic potential for the entire system. This formulation is quite general and with a varying but small amount of work we've been able to rewrite three different simulation codes to fit this program. Therefore, we believe that an acceleration library that targets this generalized program can be of widespread utility in Monte Carlo molecular electrostatics simulations.

Program 2 Electrostatics MC program

```
current_system = init()
current_U = calculate_U( current_system )

while ( ! done )
    proposed = move( current_system )
    proposed_U = calculate_U( proposed )
    if evaluate( current_U, proposed_U )
        current_U = proposed_U
        current_system = proposed
end
```

GPU Implementation. The implementation was based on the nbody example from Nvidia's SDK, described in Chapter 31 of GPU Gems 3 [7] and is written in CUDA. Figure 2 illustrates the way the calculation is divided and work is performed. One thread per particle is spawned and run concurrently. The array of threads is split into blocks of equal size. Each of these blocks uses shared memory to reduce the number of accesses to global memory. In a block, each thread loads the data of a particle into shared memory, and then each thread computes the interaction of its own particle with the particles in shared memory. When the block runs out of data in shared memory, a synchronization point is reached and a new set of coordinates are loaded into shared memory. The program loops over the particles until all are consumed, summing the results. Unfortunately, this computes each pairwise interaction twice, which could be avoided by not calculating tiles that have no "new" results. Doing twice the necessary work is obviously inefficient, but restructuring the calculations to avoid recomputation is not a trivial task and is left as a future optimization.

Determining the right block size to use is essential for achieving good performance. Unfortunately, this depends a little on the program and a lot on the specific hardware. The optimal block size corresponds to maximum hardware utilization and varies from graphics card to graphics card because it involves balancing the need for a large number of blocks to hide communication latency against the need to minimize the number of synchronization points.

Cell/BE Implementation. The Cell/BE implementation first partitions the vector of particles among the available SPEs. Each SPE is responsible for calculating the interactions of its own particles with all the particles, analogous to a block in the CUDA implementation. Since the local store of each SPE is of limited size, the SPE divides its particles into smaller sub-blocks which will fit in memory. For each sub-block, the SPE retrieves a series of blocks of the whole particle array, summing the results of each set of interactions. A double-buffering scheme is used to hide communication. This is illustrated in Fig. 3.

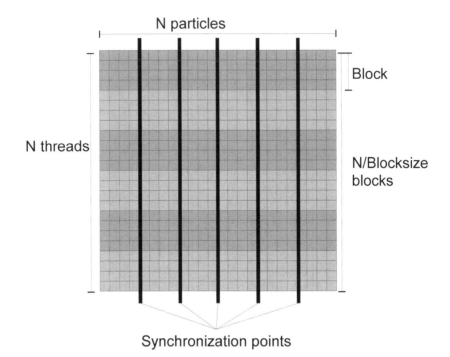

Fig. 2. GPU program design

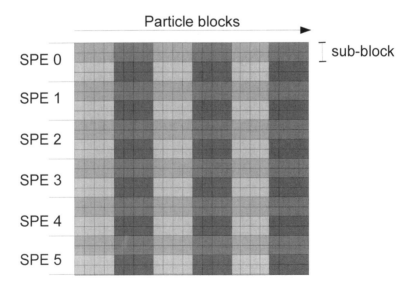

Fig. 3. Cell/BE program design

The PPE's role is to execute the lightweight portion of the program, providing the SPEs with the address of their chunk of the array of particles. When the SPEs are done, the PPE sums the partial results from each. Work is distributed and results are returned using mailbox messages, but SPEs get particle coordinates with DMA transfers.

Having to write a double-buffering scheme, keeping track of memory alignment when transferring parts of an array, and properly using the vector registers can be very challenging. These elements are in themselves not novelties in high-performance computing, but there is a marked difference in that these techniques are now required for a code to work rather than optimizations introduced into a working code. We'll discuss the implications of this difference in programming style below.

3   Results

The results of our work can be divided into two parts. First, we show that the implementations are efficient and yield high performance. Second, and of at least equal importance, we try to characterize the effort required to use the library and we discuss its limitations in a research environment.

To measure the performance of the codes in as general a way as possible, we chose as a metric the number of particle-particle interactions processed per second. This rate is effectively the limiting factor in any n-body simulation and can be directly translated to an estimated simulation runtime. However, this metric does not capture issues of numerical accuracy, which may influence the real runtime of a simulation.

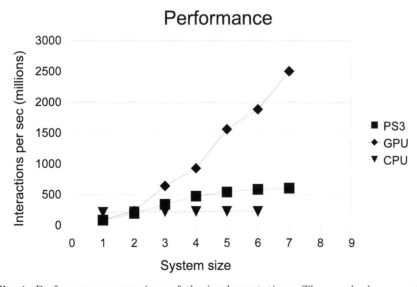

Fig. 4. Performance comparison of the implementations. The graph shows particle-particle interactions per second vs. number of particles.

Table 1. Comparing actual performance to theoretical peak performance. Units are interactions per second. The peak measured performance for the GPU (in parentheses) was limited by the maximum problem size that our program could handle, as opposed to any asymptotic limit.

System	PS3	GPU	CPU
Peak	600	28000	440
Measured	600	(8487)	222
Efficiency	100%	(30%)	50%

The test system for the Cell/BE was a PlayStation 3 (PS3). Other Cell platforms exist with more resources, but the PS3 has received some attention because its extremely low price makes for a low barrier to entry. The CPU and GPU runs were on a machine with an Intel i7 920 CPU with four cores at 2.67GHz and a Nvidia GeForce GTX260. We use a well-optimized single-threaded CPU code to compare with to avoid artifacts arising from different parallelization methods. Existing parallel MC codes for x86 processors typically scale well on multiple cores and on clusters. [6]

The scaling of raw performance to problem size is shown in Fig. 4. Unsurprisingly, the single-threaded CPU code is outperformed by its competitors, but it's worth noting that it reaches its maximum performance already at small systems, while the other two require larger systems to achieve a measure of efficiency. Table 1 shows the maximum performance we've managed to achieve and the theoretical peak performance that should be achieved. The latter was calculated from the number of floating point operations per interaction and the reported peak flops of the hardware. The GPU requires a large problem (approx. 20,000 particles) to reach its peak performance, and we have unfortunately been unable to run a sufficiently large problem with the chosen application make a final comparison. This is likely to due to a problem in the application itself, not the GPU library.

The reason why the GPU code performs relatively poorly compared to peak at the problem sizes we've been able to run is communication. The application is structured in a way that requires data to be transferred to the device for every Monte Carlo iteration. The latency of this data transfer is on the order of 10 microseconds, comparable to the computation time for small problem sizes.

We've already discussed in Sect. 2.1 the structure that simulation codes must follow in order for the library to be meaningful. Here we'd like comment on the difficulty of the task of rewriting a typical program. Of the three programs we worked with, one was already compatible, requiring only changes in a handful of lines of code. The other two required more substantial rewriting, most of which involved restructuring the way the energy calculations follow changes in system conformation. While the absolute amount of work varies a lot depending on the complexity of the code and the original design, it's a relatively small task compared to a complete rewrite or port in a new language.

Our experiences with writing GPU code for this application compared to Cell code can be summarized roughly as follows: we prefer GPU programming to Cell. We find that while the programming concepts and the algorithm design are very similar for both architectures, the scheduler makes GPU programming much easier and more effective. With the Cell, a lot of programmer time is spent on writing program structures that are unnecessary with a scheduler that distributes the computational workload and hides the communication automatically, as well as code to make vector processing and memory alignment issues work properly. Unfortunately, we did not think to measure the time spent on each code separately, so we have no quantitative measure to support this opinion.

There are two important weaknesses that harm the usefulness of our work in a research environment. The first is that the library code, most notably the GPU component, must be hand-tuned to fully utilize the hardware. Having to hand-tune it is a significant drawback to the usability of a library, because it takes time and is error-prone. Fortunately, much progress has been made recently in the field of auto-tuning for GPU architectures [2][8][3], and we feel that this is a necessary feature for a library like this if it is to see widespread use and longevity.

The second weakness is that our library explicitly requires the use of only point-charges or small spherical charge distributions and the Coulomb potential. If another charge distribution and/or another potential is desired, the user would have to make changes in the particle-particle interaction function in the library and recompile. This is because CUDA doesn't support function pointers on the GPU. With OpenCL's just-in-time compilation, however, there is the possibility of providing user-defined interaction functions and create a library with even more flexibility than the current design.

4   Conclusions

In this paper, we have presented two implementations of a commonly used algorithm in an attempt to provide a useful resource to domain scientists, to give a hint as to what works, and to highlight the obstacles that may hinder non-specialists from using hardware architectures like the Cell and GPU.

Using GPUs or the Cell/BE is generally more difficult than writing a program for a single-core or even a multicore CPU. Domain scientists are better suited at doing their science than spending months learning about and programming for one of these architectures, especially given the rate of change in the industry. Here, a library with a small set of function was general enough to be useful for a range of applications and simple enough for a computer scientist to implement efficiently as a relatively short-term project.

Comparing the experiences of writing very similar GPU and Cell programs yields some small insight into how built-in structures that automate certain aspects of programming (i.e. the GPU scheduler) can reduce the burden on the programmer and make programming more efficient. We believe this is an increasingly important consideration, as both hardware and programming models tend to growing complexity.

We hope that future work can be directed at more general means of supporting the use of heterogeneous multicore architectures in the scientific community. The challenges involved in Cell/BE programming shows most clearly what is needed for this effort. While GPU programming has come a long way since the advent of CUDA and OpenCL, it still displays some of the same problems. In traditional programming workflow, one is able to write a simple code first and optimize later as needed. This makes debugging easier, it makes code easier to maintain, and it focuses programming effort where it's needed. Today, we must often work backward, designing and writing advanced "optimizations" into the first version of the code. While there is value in creating a useful library that circumvents this problem for anyone with a certain type of simulation, a more general solution would make it possible to move back to a more traditional programming workflow.

References

1. Arevalo, A., Matinata, R.M., Pandian, M., Peri, E., Ruby, K., Thomas, F., Almond, C.: Programming for the Cell Broadband Engine. IBM Redbooks (2008)
2. Datta, K., Murphy, M., Volkov, V., Williams, S., Carter, J., Oliker, L., Patterson, D., Shalf, J., Yelick, K.: Stencil Computation Optimization and Auto-tuning on State-of-the-Art Multicore Architectures. In: International Conference for High Performance Computing, Networking, Storage and Analysis (2008)
3. Davidson, A., Owens, J.D.: Toward Techniques for Auto-tuning GPU Algorithms. In: Jónasson, K. (ed.) PARA 2010, Part II. LNCS, vol. 7134, pp. 110–119. Springer, Heidelberg (2012)
4. Farber, R.: Cuda, supercomputing for the masses. Dr Dobbs (2008)
5. Khan, M.O.: Polymer Electrostatics: From DNA to Polyampholytes. PhD thesis, Lund University (2001)
6. Khan, M.O., Kennedy, G., Chan, D.Y.C.: A scalable parallel monte carlo method for free energy simulations of molecular systems. Journal of Computational Chemistry 26(1), 72–77 (2005)
7. Nyland, L., Harris, M., Prins, J.: GPU Gems 3. In: Fast N-Body Simulation with CUDA, ch.31. Pearson Education, Inc. (2007)
8. Williams, S., Carter, J., Oliker, L., Shalf, J., Yelick, K.: Lattice Boltzmann simulation optimization on leading multicore platforms. In: IEEE International Symposium on Parallel and Distributed Processing, Vols. 1-8 (2008)

Algebraic Multigrid Solver
on Clusters of CPUs and GPUs

Aurel Neic, Manfred Liebmann, Gundolf Haase, and Gernot Plank

Institute for Mathematics and Scientific Computing, University of Graz, Austria
aurel.neic@uni-graz.at

Abstract. Solvers for elliptic partial differential equations are needed in a wide area of scientific applications. We will present a highly parallel CPU and GPU implementation of a conjugate gradient solver with an algebraic multigrid pre-conditioner in a package called *Parallel Toolbox*. The solvers operates on fully unstructured discretizations of the PDE. The algorithmic specialities are investigated with respect to many-core architectures and the code is applied to one current application. Benchmark results of computations on clusters of CPUs and GPUs will be presented. They will show that a linear equation system with 25 million unknowns can be solved in about 1 second.

Keywords: Algebraic multigrid, GPU computing, High performance computing.

1 Introduction

As the finite-element approach for solving elliptic partial differential equations is very popular in many scientific applications, fast solvers for the resulting linear equation systems are of great interest. In this paper we want to present a parallel CPU and GPU implementation of a conjugate gradient solver using an algebraic multigrid precondi-tioner (AMG) called *Parallel Toolbox*. The basic ideas behind the parallelization con-cept as well as the differences between the CPU and the GPU implementation will be investigated.

Our code is designed for a fast AMG implementation on clusters of CPUs and GPUs for special problems as the application project described in §2. In contrast, the more general applicable AMG–code BOOMER/HYPRE requires longer setup times and is not suited for GPUs but performs very well on large numbers of CPUs [1,5]. There exist codes for Multigrid on GPUs but all of them require structured grids or at least locally structured grids as in FEAST [10]. Therfore, our implementation is the first multigrid solver on GPUs for fully unstructured discretizations of the underlying problem elliptic partial differential equations.

2 The CARP Project

Let us first introduce the application project where our toolbox has been integrated. The Cardiac Arrhythmias Research Package is a project for the electrophysiological simulation of cardiac tissue which has been developed by Dr. Edward Vigmond and

K. Jónasson (Ed.): PARA 2010, Part II, LNCS 7134, pp. 389–398, 2012.

Dr. Gernot Plank [12]. Because of the great research activity in this area, CARP has a plug-in framework allowing new ionic models to be easily implemented.

By design CARP can run on shared computers as well as on distributed computers. Parallelization is performed within the shared memory model by implementing OpenMP directives and native numerical libraries. For distributed memory parallelization, extensive use of the PETSc parallel library as well as MPI function calls is made. In addition, efforts have been made to implement the *Parallel Toolbox* into CARP.

2.1   Components

The package consists of three main components: a parabolic solver, an ionic current component, and an elliptic solver. Each of these components has a set of APIs for connection to other components. The parabolic solver is responsible for determining the propagation of electrical activity by determining the change in transmembrane voltage from the extracellular electric field and the current state of the transmembrane voltage.

The elliptic solver unit determines extracellular potential from transmembrane voltage at each time step. The ionic model component is computed from a separate library which must be linked in at compile time.

2.2   Mathematical Model

The most complete description of cardiac electricity is given by the bidomain equations (see [12]). The basic bidomain equations relate the intracellular potential ϕ_i to the extracellular potential ϕ_e through the transmembrane current density I_m:

$$\nabla \cdot \bar{\sigma}_i \nabla \phi_i = \beta I_m \tag{1}$$

$$\nabla \cdot \bar{\sigma}_e \nabla \phi_e = -\beta I_m - I_e \tag{2}$$

$$I_m = C_m \frac{\partial V_m}{\partial t} + I_{ion} - I_{trans} \tag{3}$$

where $\bar{\sigma}_i$ and $\bar{\sigma}_e$ are respectively the intracellular and extracellular conductivity tensors, β is the surface to volume ratio of the cardiac cells, I_{trans} denotes the transmembrane current density stimulus as delivered by the intracellular electrode, I_e is an extracellular current density stimulus, C_m is the capacitance per unit area, V_m is the transmembrane voltage which is defined as $\phi_i - \phi_e$, and I_{ion} refers to the current density flowing through the ionic channels. Eqn. 3 is a set of ordinary differential equations which can be solved independently for each node of the spacial discretization.

By adding Eqn. 1 and Eqn. 2 and using the definition of V_m, the equations can be cast in a slightly different form with V_m and ϕ_e as independent variables.

$$\nabla \cdot (\bar{\sigma}_i + \bar{\sigma}_e)\nabla \phi_e = -\nabla \cdot \bar{\sigma}_i \nabla V_m - I_e \tag{4}$$

$$\nabla \cdot \bar{\sigma}_i \nabla V_m = -\nabla \cdot \bar{\sigma}_i \nabla \phi_e + \beta \left(C_m \frac{\partial V_m}{\partial t} + I_{ion} - I_{trans} \right) \tag{5}$$

Eqn. 4 is an elliptic equation and Eqn. 5 is a parabolic equation.

The elliptic equation is discretized by the finite element method using unstructured meshes. The result is a linear equation system of the form

$$K_\ell \bar{u}_\ell = \bar{f}_\ell.$$

This system of equation will be solved by our Parallel Toolbox.

3   The Parallel Toolbox

The Parallel Toolbox is a user friendly C++ toolbox for the parallelization of partial differential equation solvers based on the finite element method.

In scientific computing the finite element approach is used in a variety of areas because of its flexibility towards the geometric complexity of the simulation domain. Different parallelization approaches are possible. One naturally offered by the finite element method evenly distributes the geometric elements on the processing nodes involved in the parallel computation. As a consequence some finite element nodes are on the interface between several subdomains and thus shared between several processes. We call these nodes *shared nodes*.

Data on subdomains can have two different representations, *accumulated* and *distributed*. Accumulated data representation means, that a process is storing the full numerical value of the nodes of its subdomain. Data stored distributed has only full value on nodes uniquely belonging to one process. On a shared node each process owning that node stores only a fraction of the full numerical value.

This definition leads to the conclusion, that accumulated and distributed data representation differ only on the shared nodes of local vectors.

A local accumulated vector $\bar{u}_s$ ($s = 1, \ldots, P$) stores a part of the global vector $\bar{u}$, without changing any numerical values, and both are connected through the mapping operation (or linear map) A_s.

$$\bar{u}_s = A_s \bar{u} \tag{6}$$

$$A_s^{(i,j)} = \begin{cases} 1 & \text{iff global node } \bar{u}_j \text{ is stored locally at } \bar{u}_{s,i} \\ 0 & \text{else} \end{cases} \tag{7}$$

$$dim\, A_s = \#\, \text{local nodes} \times \#\, \text{global nodes}$$

The matrix A_s is not stored as matrix in process s but as vector $l2g$ (local to global node numbering) with $l2g_i := j$. This vector is used to setup the corresponding communication routines. Note that $R := \sum_{s=1}^{P} A_s^T A_s$ results in a diagonal matrix containing the number of subdomains a node belongs to. Therefore, applying the local diagonal matrix $A_s R^{-1} A_s^T$ to a local vector results in the partition of unity.

A local distributed vector holds only a fraction of the data values in its shared nodes. To get the global data values, processes owning a certain node need to add up their values and next neighbour communication is needed. Therefore the global vector $\bar{u}$ can be obtained by applying the transposed mapping operation (A_s^T) on the local vectors $\bar{u}_s$ and adding up the results over all processes.

$$\bar{u} = \sum_{s=1}^{P} A_s^T \bar{u}_s \tag{8}$$

To minimize the communication between the processes, a partitioning strategy is necessary that minimizes the number of boundary nodes between the subdomains. This can be achieved using partitioning tools like METIS, which automatically creates optimal partitions based on the mesh connectivity information. For the parallelization itself, the Parallel Toolbox uses the distributed memory approach via the MPI standard.

3.1 Linear Algebra Operations

The Parallel Toolbox stores the system matrix in distributed data representation. This means that the coefficient matrix K is defined as:

$$\mathsf{K} = \sum_{s=1}^{p} A_s^T \mathsf{K}_s^{FEM} A_s \tag{9}$$

Where A_s is again the global-to-local mapping and K_s^{FEM} is the local finite-element matrix in the subdomain Ω_s.

Depending on the actual algorithm, vectors are represented as distributed or accumulated. However, algorithms are implemented in such a way that the amount of data representation conversion, and thus MPI communication, is minimal[2, §5].

The following list shows the main linear algebra operations and their needed communication:

No communication:
- Matrix-vector multiplication (product of distributed matrix and accumulated vector; result is a distributed vector)
- Vector algebra with vectors of the same data representation
- Converting an accumulated vector into a distributed one

Next neighbour communication:
- Converting a distributed vector into an accumulated one

Global (reduce) communication:
- Computing a vector dot product

3.2 The Communicator Object

It is necessary in parallel linear algebra routines to communicate information that is located on the shared boundary nodes. To make this communication transparent for the user of the toolbox, a communicator object is created.

With the partitioning information of the elements – basically a one-to-one mapping of elements to processors – and the mesh connectivity information it is possible to derive the complete communication setup for parallel algorithms.

In each processing node the constructor of the communicator object only needs the global indices of the local nodes. Then it calculates the local shared nodes by sending its global node numbers to every other processor and intersecting its set of global nodes, with the sets received from the other processors.

After construction, all the relevant operations in parallel linear algebra can be executed using the three communication functions offered by the communicator object:

Accumulate: Converts a distributed vector into a accumulated one, using next neighbour communication to sum up values at nodes shared with neighbouring processes.

Distribute: Converts an accumulated vector into a distributed one without communication.

Collect: Sums globally a scalar value or a vector, e.g. used in a vector dot product.

For further information about the Parallel Toolbox and its parallelization concepts we refer to [6].

3.3 The Algebraic Multigrid Preconditioner

In multigrid theory one has to assume, that there exists a series of regular (finite-element) meshes τ_k, $k = 1, \ldots, \ell$, where the coarser mesh τ_i can be derived from the finer mesh τ_{i+1}. This leads to a series of systems of linear equations

$$K_i \bar{u}_i = \bar{f}_i$$

where K_i is the stiffness matrix of the according finite-element mesh τ_i. The main idea of multigrid methods is, that the error of the solution can be represented as the sum of the eigenfrequencies of the matrix K and lower frequency parts of the error can be reduced faster on a coarser grid.

Algebraic multigrid means, that the hierarchy of meshes τ_i and appropriate operators K_i ($i < \ell$) are created algebraically from the given finest mesh information K_ℓ (and τ_ℓ).

The idea of the coarsening algorithm is to part the set of all nodes I in a set of coarse grid nodes C and a set of fine grid nodes F (see [8]).

$$C \cap F = \varnothing, \quad I = C \cup F \tag{10}$$

The matrix graph is assumed to be symmetric – i.e. originates form a symmetric matrix – and represents the connectivity information of the nodes. An off-diagonal matrix entry can be seen as a connection of one node to another node, for example if $i \neq j$ and $A_{ij} \neq 0$ then node i is connected to node j.

A node connection is called *strong* if

$$|A_{ij}| > \varepsilon |A_{ii}| \tag{11}$$

otherwise it is called *weak*.

The algorithm starts with selecting the first node in the set I and defining it as a coarse grid node. All nodes that are strongly connected to the first node are defined as fine grid nodes. After this initial step the next node that is yet neither fine nor coarse is selected as a coarse grid node and all nodes strongly connected to this node are set as fine grid nodes in the same way as for the first node. The algorithm terminates when all nodes are either fine or coarse grid nodes. The set of coarse grid nodes now defines a coarse grid on which another coarsening algorithm can be applied, thus creating a hierarchy of grids (and linear equation systems). Between two levels in the equation system hierarchy, also proper restriction / interpolation operators and smoothers are needed. Straightforward implementations of the Omega-Jacobi and Gauß-Seidel iterative solvers were used as smoothers. Because both smoothing algorithms need next-neighbour-communication, the AMG performance depends on a fast intercommunication network. In future versions of the Parallel Toolbox, also block-wise multigrid implementations will be available similar to the smoothers used in [3].

Although the algebraic multigrid method can be used as a solver itself, our Parallel Toolbox uses it as a preconditioner for a conjugate gradient solver. For more information about algebraic multigrid methods and their parallelization we refer to [4,2] and [11].

3.4   The GPU Implementation

The main advantage of the GPU consists in the much higher memory bandwidth with respect to the GPU memory and in the better arithmetic performance in comparison to a CPU. The potential speedup of the GPU is at least one order of magnitude but it can only be achieved for massive parallel algorithms taking into account coalesce memory access of the data needed in the single threads. Additionally, unnecessary memory transfer between CPU memory and the limited GPU memory has to be avoided (see [6]).

The GPU algorithms use the same concepts of domain decomposition and data representation as their CPU counterparts. The main difference is, that algorithm parts that would run sequentially on each CPU process – e.g. a matrix-vector product or vector arithmetic – are now implemented in CUDA. By other words, the computationally challenging CPU routines were replaced by GPU routines of higher performance.

In the conjugate gradient solver the equation system, the initial guess and the right-hand-side are transferred to the GPU memory and then all linear algebra operations needed in a conjugate gradient algorithm (matrix-vector product, vector dot product and vector arithmetic) are executed on the GPU. After that, the computed solution is transferred back to CPU memory. If absolute error tolerance is needed, then also an accumulation operation, and therefore MPI communication over the CPU memory, is required.

The algebraic multigrid coarsening is part of the setup and not implemented on GPUs. The interpolation and restriction operations during each solution step are matrix-vector multiplications and can easily be executed on GPUs. For practical reasons, only the Omega-Jacobi smoother is implemented.

Every MPI process uses one GPU (CUDA device) for its computations and the user has basically the same freedoms in his choice of parallelization level and machine topology as on a CPU computation.

 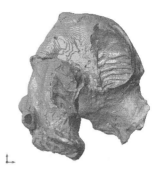

Fig. 1. The wedge_35um (left) and atria (right) geometries

4 Benchmarks

4.1 Geometries

In order to test the performance of the parallel solver we generated a stiffness matrix of an 3D finite-element unstructured mesh and solved the linear equation system $K\bar{u} = \bar{f}$. The meshes were provided by the CARP project (see [7] and [9]).

The mesh for the CPU benchmarks is called wedge_35um. It has approximately 150 million elements and 24.5 million nodes.

On the GPU clusters a geometry named atria is used. It consists of 11 million elements and 2 million nodes.

4.2 Cluster Specifications

The CPU computations were made on the *CINECA SP6* cluster. It features an IBM pSeries 575 architecture with IBM Power6 4.7 GHz processors and 5376 computing cores and a Infiniband x4 DDR network.

The GPU benchmarks were made on the *gpuser* GPU Cluster of the University of Wyoming. It consists of 7 computing nodes connected by Gigabit Ethernet and a total sum of 28 GeForce GTX 295 graphics cards. It should be noted, that the GTX 295 features two GPUs per card, so each computing node contains 4 GTX 295 cards and 8 GPUs.

4.3 Results

The CPU benchmarks were made on the wedge_35um geometry. The linear equation system had 24.5 million unknowns. The error tolerance in the AMG-CG solver was 10^{-6}. Benchmarking started with 4 processes and went until 512 processes, always doubling the last number of processes in each new step. In order to have optimal caching behaviour, hyper-threading was not activated on the Power6 processors, so each MPI process used one processor core.

Table 1. Wall clock solving time of the CPU algorithm

# processes	Solving time [s]	iterations	# AMG levels
4	64.51	19	6
8	40.56	20	6
16	14.51	19	6
32	6.33	19	6
64	2.9	20	6
128	1.77	20	6
256	**1.23**	**21**	**6**
512	1.42	21	6

Table 1 shows the CPU benchmark results. It took 4 processes about 65 seconds to compute the solution. The shortest solution time of 1.23 seconds was achieved with 256 processes. After that, the computation time was stagnating. The reasons for that will be discussed later on.

The parallel efficiency has been computed from these results using 4 processes as baseline, i.e. the efficiency for p processors is calculated as $\frac{4\,t_4}{p\,t_p}$.

An optimal linear scalability occurs, when the amount of processes doubles and the computation time halves, i.e. the efficiency is one. The efficiency in each parallelization step can be seen in figure 2 on the left. Between 4 and 256 processes the scalability is quite linear and thus the algorithm efficient. Because of caching effects on the IBM Power6 processors, also a superlinear region between 16 and 128 was noticed.

The performance drop on the highest parallelization level can be explained by analysing the amount of network communication in comparison to the local problem sizes. The amount of shared nodes of the geometry increased dramatically on high parallelization levels. If the local problems get too small, the performance gain achieved by reducing the local problem size is nihilated by the increased communication time.

The GPU benchmarks were made on a 2 million unknowns system generated from the `atria` geometry. The error tolerance of the AMG-CG solver was again 10^{-6} and between 2 and 16 processes have been used for the computations, with one MPI process for each GPU.

Table 2. Wall clock solving time of the GPU algorithm

# processes	Solving time [s]	iterations	# AMG levels
2	0.4	20	7
4	0.23	20	7
6	0.17	20	7
8	**0.14**	**20**	**7**
12	1.2	20	7
16	1.3	20	7

As seen in table 2, computation times started at 0.4 seconds with 2 GPUs and reached the minimum of 0.14 seconds on 8 GPUs and on one computing node. After that, a big performance drop occurred. Using more than 8 GPUs requires a second computing node and results in a dramatic performance drop (see table 2) caused by the relatively slow network communication on the Gigabit Ethernet. This bottleneck has to be reduced technically by a faster intercommunication network.

The right part of figure 2 shows the computation efficiency. Again the performance bottleneck can be seen when more than one computing node is used.

For a direct comparison between GPU and CPU performance, the `atria` example was also used for a CPU benchmark. The solving time was between 9.2 seconds on 2 cores and 3.6 seconds on 16 cores. On 6 CPU cores the solution was computed in 5.8 seconds whereas on the same amount of GPUs only 0.17 seconds were needed. This resulted in a peak speedup of 34.

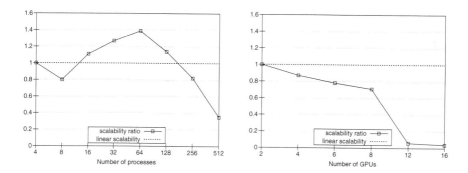

Fig. 2. Efficiency of the CPU algorithm for 24.5 mill. unknowns (left) and efficiency of the GPU algorithm for 2 mill. unknowns (right).

5 Conclusion

We presented the concepts of our domain decomposition and data representation, which are the basis of the parallel linear algebra implementations. Further we investigated the algebraic multigrid method and the differences between the CPU and the GPU implementation. Finally we presented benchmarks on two geometries from the CARP project and both the CPU and GPU benchmarks showed good efficiency on most parallelization levels.

The CPU algorithm proved a very good efficiency up to 256 processes. The peak performance was on 256 processes with a solution time for a 24.5 million unknowns problem of 1.23 seconds. Only on the highest parallelization level of 512 cores the performance dropped. The reason for that was the already tiny local problem size and thus the bad ratio between computation and communication.

The GPU benchmarks showed, that data transfer is vital in GPU computations. Even though the data transfer from the GPU memory to CPU memory and from one MPI-host to another MPI-host are overlapping, performance dropped dramatically when the benchmark was run on more then one computing node (more then 8 MPI-processes were used). In that case the relatively low Gigabit network bandwidth proved to be the performance bottleneck. In any case it should be noted, that a dramatic increase in performance, compared to CPU computations, was shown. The fastest solution time of a 2 million unknowns system was 0.14 seconds and the peak speedup compared to a CPU computation was 34.

GPU clusters with a very high network bandwidth and low latencies(e.g. Infiniband) or with direct GPU to GPU data transfer technology, would help to improve parallel GPU efficiency. Although the supreme computation performance of modern GPUs will probably always render data transfer speed problematic for keeping efficiency rates high.

Using block solvers and block preconditioner, where operations needing more network bandwidth are only used locally on the computing nodes and between the several nodes only operations of lesser network demands are used, could be a solution for the network limitations. This hypothesis is a subject of future research.

References

1. Brezina, M., Cleary, A.J., Falgout, R.B., Henson, V.E., Jones, J.E., Manteuffel, T.A., McCormick, S.F., Ruge, J.W.: Algebraic multigrid based on element interpolation AMGe. SIAM J. Sci. Comput. 22(5), 1570–1592 (2000)
2. Douglas, C., Haase, G., Langer, U.: A Tutorial on Elliptic PDE Solvers and Their Parallelization. Software, Environments, and Tools. SIAM, Philadelphia (2003)
3. Göddeke, D.: Fast and Accurate Finite-Element Multigrid Solvers for PDE Simulations on GPU Clusters. PhD thesis, Technische Universität Dortmund, Fakultät für Mathematik (May 2010), http://hdl.handle.net/2003/27243
4. Haase, G., Kuhn, M., Reitzinger, S.: Parallel AMG on distributed memory computers. SIAM SISC 24(2), 410–427 (2002)
5. Henson, V.E., Yang, U.M.: BoomerAMG: a Parallel Algebraic Multigrid Solver and Preconditioner. Applied Numerical Mathematics 41, 155–177 (2002)
6. Liebmann, M.: Efficient PDE Solvers on Modern Hardware with Applications in Medical and Technical Sciences. PhD in natural sciences, Institute of Mathematics and Scientific Computing – Karl Franzens University Graz (2009)
7. Prassl, A.J., Kickinger, F., Ahammer, H., Grau, V., Schneider, J.E., Hofer, E., Vigmond, E.J., Trayanova, N.A., Plank, G.: Automatically generated, anatomically accurate meshes for cardiac electrophysiology problems. IEEE Trans. Biomed. Eng. 56(5), 1318–1330 (2009)
8. Ruge, J.W., Stüben, K.: Efficient solution of finite difference and finite element equations by algebraic multigrid (amg). In: Paddon, D.J., Holstein, H. (eds.) Multigrid methods for integral and differential equations. The Institute of Mathematics and Its Applications Conference Series, pp. 169–212 (1985)
9. Seemann, G., Höper, C., Sachse, F.B., Dössel, O., Holden, A.V., Zhang, H.: Heterogeneous three-dimensional anatomical and electrophysiological model of human atria. Philos. Transact. A Math. Phys. Eng. Sci. 364(1843), 1465–1481 (2006)
10. Turek, S., Göddeke, D., Becker, C., Buijssen, S.H., Wobker, H.: FEAST – Realisation of hardware-oriented numerics for HPC simulations with finite elements. Concurrency and Computation: Practice and Expecience 22(6), 2247–2265 (2010); Special Issue Proceedings of ISC 2008
11. Vassilevski, P.S.: Multilevel Block Factorization Preconditioners: Matrix-based Analysis and Algorithms for Solving Finite Element Equations, 1st edn. Springer, New York (2008)
12. Vigmond, E.J., Hughes, M., Plank, G., Leon, L.: Computational tools for modeling electrical activity in cardiac tissue. Journal of Electrocardiology 36, 69–74 (2003)

Solution of Identification Problems in Computational Mechanics – Parallel Processing Aspects

Radim Blaheta, Roman Kohut, and Ondřej Jakl

Institute of Geonics, Academy of Sciences of the Czech Republic
Studentská 1768, 708 00 Ostrava, Czech Republic
{blaheta,kohut,jakl}@ugn.cas.cz
http://www.ugn.cas.cz

Abstract. Problems of identification of material parameters (mostly parameters appearing in constitutive relations) have applications in many fields of engineering including investigation of processes in a rock mass. This paper outlines the structure of parameter identification problems, methods for their solution and describes an identification (calibration) problem from geotechnics, which will serve as a realistic benchmark problem for illustration of the behaviour of selected parameter identification methods.

Keywords: inverse problems, identification of material parameters, Nelder-Mead method, genetic algorithms, rock mechanics.

1 Introduction

Generally, the identification problems appear in investigation of physical processes in material environment. The processes are described by the state variables u and driven by the control variables f. The material is characterized by parameters $\kappa \in \mathcal{K} \subset R^p$.

Direct problems focus on computation of $u = u_h(\kappa) = u_h(\kappa, x, t)$, where $(x,\, t)$ gives space and time localization, if f and κ are known. On the opposite, identification problems use the knowledge of f and some partial apriori knowledge on the state variable u for (partial or full) determination of κ.

If the apriori information about the state variable u is given by the vector $d = (d_i) \in R^m$ of measured values, then the search for the unknown material parameters can be formulated as the following minimization problem

$$F(\kappa) = \| \mathcal{M}u_h(\kappa) - d \| \longrightarrow \min_{\kappa \in \mathcal{K}}. \tag{1}$$

Above, $\mathcal{M}$ is an observation operator, which computes from u_h values corresponding to the measured data from d. In the simplest case, it just select the values $u(x_i, t_i)$ corresponding to d_i.

In contrary to direct problems, it is known that some identification problems are not well posed [4], which means that some of the following properties can be violated:

K. Jónasson (Ed.): PARA 2010, Part II, LNCS 7134, pp. 399–409, 2012.

- there exists a solution of the problem,
- the solution is unique,
- the solution is stable under small changes of input data.

Although the properties of the minimization problems can be difficult to analyse, a lot of different iterative techniques can be used for the minimization (1) (mostly without theoretical proof of convergence). The range of applicable methods includes

- gradient methods, e.g. Gauss-Newton, Levenberg-Marquardt, conjugate gradients, see [3], [4], [6], [7],
- gradient-free direct method, e.g. Nelder-Mead simplex method [3],
- stochastic methods e.g. [5], genetic algorithms e.g. [6], [9] and [8].

In this paper, we discuss the use of these methods also from the point of view of parallelization. Some of the approaches are illustrated by numerical experiments, implementation of the other methods is in progress. The problem we are interested in concerns the temperature distribution determined by the solution of the nonstacionary linear heat equation

$$c\rho\frac{\partial\tau}{\partial t} = \lambda\sum_i \frac{\partial^2 u}{\partial x_i^2} + Q(t) \qquad \text{in} \quad \Omega \times (0,T) \qquad (2)$$

which is to be fulfilled by the temperature $u: \ \Omega \times (0,T) \to R$ with the corresponding boundary and initial conditions. Here, c is the specific heat, ρ is the density of material, λ are the coefficients of the heat conductivity and Q is the density of the heat source. The time interval under consideration is denoted $(0,T)$.

Note also that the identification problem is very close to the calibration of a mathematical model. The difference is in stressing either the computed material parameters or the coincidence of values predicted by the model with measured data.

2   A Benchmark Problem

In order to introduce our benchmark problem we present the in-situ Äspö Pillar Stability Experiment (APSE) that has been performed at SKBs Äspö Hard Rock Laboratory in south eastern Sweden with the aid of investigation of granite mass damage due to mechanical and thermal loading. The measured data are now used for validation of mathematical models within the DECOVALEX 2011 international project. APSE used electrical heaters to increase temperatures and induce stresses in a rock pillar between deposition holes (Fig. 1) until its partial failure. To determine accurately the temperature changes, a heat flow model is formulated and monitored temperatures are used for identification of heat flow parameters (heat capacity, heat conduction coefficient, heat convection into the holes). The identification should provide parameters taking into account water bearing fractures and water flow and calibrate the model. More details and another approach to the model calibration can be found in [1].

The modelling was realized by the GEM software [2], which can be characterized as a non-commercial 3-dimensional finite-element package oriented on the solution of linear problems arising in geosciences (elasticity, plasticity, thermoelasticity). GEM serves both research purposes and practical applications and its development is mainly problem-driven, reflecting the needs of the current research and practice.

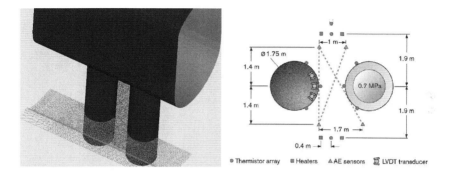

Fig. 1. The APSE model – detail of the FE grid around the pillar created in the GEM software [2] (*left*) and plan view on the pillar, holes, location of heaters and points of temperature measurement (*right*)

The exploited APSE model considers domain of $105 \times 125 \times 118$ m and $99 \times 105 \times 59$ nodes. The grid is refined around the pillar, see Fig. 1. The heaters are producing heat which varies in time. The model assumes original temperature $14.5°$C on the outer boundaries, zero flux onto the tunnel and nonzero flux given the convection onto the holes. The initial condition is given again by the temperature $14.5°$C.

Monitoring of the temperatures during two month heating phase of APSE is essential for calibration of the thermal model. There are 14 temperature monitoring positions and temperatures are measured in 12 time moments. Altogether 168 values of temperature measurement (vector d) are used for parameter identification, which according to (1) can be written as follows

$$F = F(\lambda_1, c_1, \lambda_2, c_2, \lambda_3, c_3, H_1, H_2, H_3)$$
$$= \left(\sum_i [u_h(x_i, t_i) - d_i]^2\right)^{0.5} \longrightarrow \min. \tag{3}$$

The material parameters represent different conductivity λ and heat capacity c for dry and wet side of model (according to Fig. 1). The rock in the right hole had yielded from a depth of approximately 0.5 m down to 3 m which motivates to introduce a third type of material with different λ and c for the damaged part of the pillar. We supposed heat conduction between rock and air in excavated holes determined by different values of the heat conduction coefficient H for individual holes with third coefficient corresponding to surface for the above mentioned damaged part of the pillar. It gives nine material parameters of the cost functional F in (3).

3   Nelder-Mead Optimization

The first optimization algorithm, which we describe in this paper, is the Nelder-Mead algorithm, which maintains a simplex $S^{(k)}$ in the space of parameter vectors. This simplex locally approximates the cost function F and serves for getting information about its behaviour and getting approximation to the optimal point. If $F = F(\kappa)$ and $\kappa \in R^p$ then the k-th step simplex $S^{(k)}$ is determined by $p + 1$ vectors of parameters (vertices) $\kappa^{(k, 1)}, \ldots, \kappa^{(k, p+1)}$. We assume that the cost function values are evaluated and vertices are sorted, so that

$$F(\kappa^{(k, 1)}) \leq F(\kappa^{(k, 2)}) \leq \ldots \leq F(\kappa^{(k, p+1)}).$$

The k-th step then continues by evaluation of the stop criterion and if the approximation is not found to be satisfactory, then the worst vertex $\kappa^{(k, p+1)}$ is replaced by a new one or, in a specific case, the whole simplex is shrunk.

In any case, first, the new vertex is sought in the form

$$\kappa(\mu) = (1 + \mu)\bar{\kappa} - \mu\kappa^{(k, p+1)},$$

where $\bar{\kappa} = \left((\kappa^{(k, 1)} + \ldots + \kappa^{(k, p)})/p\right)$ is the barycentre and μ is equal to $\mu_r = 1$ for reflection, $\mu_e = 2$ for extension, $\mu_{oc} = 1/2$ for outer contraction and $\mu_{ic} = -1/2$ for inner contraction.

The k-th step always begins with evaluation of $F_r = F(\kappa(\mu_r))$. If $F(\kappa^{(k, 1)}) \leq F_r < F(\kappa^{(k, p)})$ then we take $\kappa(\mu_r)$ as the new point, otherwise we gradually test for the expansion, outside and inside contraction and take the selected case. It means that the k-th step typically contains one or two evaluations of the cost functions. In the case of contractions, we can also decide for shrinking the simplex, which is more expensive and costs p evaluations of the cost function. The details can be found in [3].

The optimization is stopped when both decrease of the cost functional F is small (below ε_F) and changes of parameters are small (below ε_p) or if too many evaluations of the cost function are required.

As concerns the parallelization, the Nelder-Mead method is in principle sequential, which means that parallelization can be involved only in evaluation of the cost function itself and eventually in realization of the shrinking of simplex in some steps. In the GEM software mentioned above, the parallel solver, employed for providing the cost values, makes use of the OpenMP standard, i.e. its parallelization is based on the shared variables model. We shall see some parallel performance indicators in Section 5.

In the rest of the section, let us discuss the numerical solution of the benchmark problem from Section 2, i.e. searching for various heat transfer coefficients,

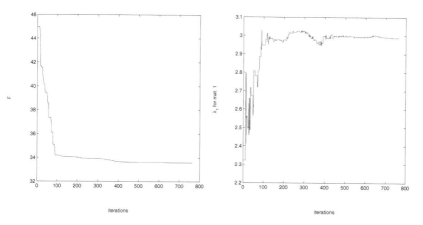

Fig. 2. The convergence of the cost functional F (*left*) and one of the parameters (λ_1) (*right*)

by the Nelder-Mead method. Note that we use the method of unconstrained optimization, but to guarantee the positivity of the parameters, we use exponential transformation, i.e. finding x such that $\kappa = e^x$ is the required parameter. As the parameters have quite different orders, we scale the heat capacity c for having all parameters in order of units.

To find a very accurate approximation of the parameters, we stop iterations with $\varepsilon_F = 0.001$ and $\varepsilon_p = 0.01$. With a physically motivated initial guess, the stop occurred after 764 iterations, for a non physical (more or less random) initial guess surprisingly less iterations were required. The convergence behaviour is illustrated in Fig. 2. But the stopping test could be fulfilled much earlier (say after 100 iterations) if we weaken the requirement on small changes in all parameters. This is also due to the fact that the cost function depends only mildly on some of the parameters, see [11]. The use of weaker stopping parameter ε_p is natural for solving calibration problems.

Note also that computation of $u = u(\kappa, x, t)$ represents here the solution of an evolution parabolic heat transfer problem, which is solved by linear finite element (FE) discretization in space and backward Euler method in time. Linear systems appearing in each time step are solved iteratively by conjugate gradient method preconditioned by one-level additive Schwarz method, which is efficient it this case, see [10]. Of course parallel processing can be also used for assembling the FE matrices.

4   Stochastic and Evolution Methods

To get a larger space for parallelization we shall deal with stochastic and evolution methods for optimization. In this section, we shall consider a constrained search space for parameters, i.e. $\kappa \in \mathcal{K} = \prod_1^p \langle \kappa_{i,\min}, \kappa_{i,\max} \rangle$. The simplest stochastic (Monte Carlo) algorithm is then as follows

MC algorithm with $N = N_{MC}$ individuals

(1)   generate N random vectors $\kappa^{(i)} \in \mathcal{K}$, $i = 1, \ldots, N$,
(2)   evaluate (in parallel) $F(\kappa^{(i)})$, $i = 1, \ldots, N$,
(3)   select $\kappa = \mathrm{argmin}_i F(\kappa^{(i)})$.

Genetic algorithms (GA) enrich the selection by operations of crossing and mutation. It provides the following algorithm

GA with $N = N_{GA}$ individuals

(1)   generate N random vectors $\kappa^{(i)} \in \mathcal{K}$, $i = 1, \ldots, N$
(2)   for given generation, evaluate (in parallel)
      $F_i = F(\kappa^{(i)})$, if F_i is not known yet,
(3)   select γN parameter vectors[1] $\kappa^{(i)}$ with smallest values
      F_i; so called parents. Then create $(1 - \gamma)N$ new
      vectors (children) by crossing randomly selected
      parents,
(4)   create a new generation by taking the selected
      parents and created children with mutating some of
      them,
(5)   evaluate stopping test and GOTO (2) if results are
      still not satisfactory.

In our case, the crossing and mutation acts on parameter vectors and can be described as algebraic (not binary) rule, see e.g. [8], [9]. For example:
   Crossing of vectors x and y is a new vector z, which can be given by

$$z_i = x_i + \alpha_i(y_i - x_i),$$

where for discrete crossing α_i is selected from $\{0, 1\}$ with probability $1/2$, but also α_i can be selected randomly in the range $\langle -\delta, 1 + \delta \rangle$ for e.g. $\delta = 0.25$,
   Mutation of the vector x concerns its components x_i. Each component is mutated with probability, which is usually $1/p$. Mutation uses a range Δ_i, for $x_i \in \langle \kappa_{i,\min}, \kappa_{i,\max} \rangle$ it is typically $\Delta_i = 0.1(\kappa_{i,\max} - \kappa_{i,\min})$. Mutation of x then gives a new vector z, e.g.

$$z_i = x_i \pm \Delta_i 2^{-k\alpha},$$

where α is chosen uniformly in $\langle 0, 1 \rangle$ and k is so called precision constant depending on the problem.
   For more details on GA, we refer to [9] and [8].
   In our context, the GA approach is attractive for a large space for parallelization and still reasonable efficiency, see [6]. The expectations for the case of an ideal parallelization can be derived from (sequential) numerical experiments in Table 1. The results are based on the solution of the APSE optimization problem described in Section 2. Ideal parallelization here means that the evaluation

---

[1] γ is normally chosen in the range 0.5 to 0.1 [9].

Table 1. The APSE model optimization problem – computation results (see text for explanation)

Algorithm	# Function Evaluations	Ideal Parallel Comp. Time	λ_1	c_1	λ_2	c_2	$F - F_{best}$
NM	150	150 TU	2.9996	2.4046	5.9994	4.8041	0.0025
GA-DR	740	37 TU	3.0653	2.4672	5.9338	4.7471	0.0404
GA-EIR	1220	69 TU	3.0158	2.4209	5.9809	4.7868	0.0119
GA-ELR	890	47 TU	3.0828	2.4628	5.9308	4.7629	0.0506
high acc. solution	–		3.0000	2.4030	6.0000	4.8060	0

of the cost function dominates in GA and is parallelized with 100 % efficiency, which is not unrealistic with regards to the mutual independence of the calculations. TU is a time unit — the time needed for one cost function evaluation. Recall that the FE system solution is responsible for 99 % of the cost function computation. In the table we can also see the differences in selected computed material parameters (conductivity λ and heat capacity c) provided by different optimization algorithms. The accuracy of different methods is also compared in terms of difference between the cost function value for given experiment and the best obtained cost function value. Several variants of genetic algorithms are considered, namely

GA-DR (discrete recombination) Let $x = (x_1, \ldots, x_n)$ and $y = (y_1, \ldots, y_n)$ are the parents strings. Then the offspring $z = (z_1, \ldots, z_n)$ is computed by $z_i = x_i$ or y_i, where x_i or y_i are chosen with probability 0.5.

GA-EIR (extended intermediate recombination) $z_i = x_i + \alpha_i(y_i - x_i)$, $i = 1, \ldots, n$; α_i is chosen uniform randomly in the interval $< -0.25, 1.25 >$; z_i must be in the searching interval, $a_i \leq z_i \leq b_i$.

GA-ELR (extended line recombination) $z_i = x_i + \alpha(y_i - x_i)$, $i = 1, \ldots, n$; α is chosen uniform randomly in the interval $< -0.25, 1.25 >$; z_i must be in the searching interval, $a_i \leq z_i \leq b_i$.

In the GEM framework, we implemented the Breeder GA [9] and subsequently parallelized the evaluation of the computationally dominant cost function $F(\kappa^{(i)})$ in step (2), which is calculated for several vectors $\kappa^{(i)}$ simultaneously. This parallelization was technically realized using the MPI standard, i.e. in the message passing model. A straightforward approach is to replicate p GA (as MPI-processes) on p processors simultaneously, each of which cares only for $1/p$ cost function evaluations, the results of which they afterwards exchange (All-To-All) operation) to get the full set. Since each evaluation of the cost function implies a call of the OpenMP-parallelized solver, the resulting application is hybrid in the sense that it combines MPI and OpenMP parallelization.

5   Numerical Experiments

For the numerical experiments, we had the opportunity to access a multiprocessor machine boasting new twelve-core AMD Opteron 6172 processors. The multiprocessor, let us call it Hubert, had the following basic technical parameters:

- 4 x AMD Opteron 6172 processors (48 cores in total),
- 128 GB of RAM,
- 2 TB RAID-0 disk capacity.

The machine ran Linux (SLES 10) operating system and the code was developed and executed using the Intel Cluster Toolkit with Compilers (ICT) version 4 package, which comprises among others an OpenMP-aware Fortran compiler and MPI implementation.[2]

Our first experiments focused on the performance of the mixed MPI and OpenMP parallelization in the optimization algorithm. We derived a small benchmark problem of 13 754 degrees of freedom, for the solution of which the FE solver needs just several seconds. Running the GA optimization with different parallelization parameters on this benchmark and using defaults in the computing environment, we obtained timings summarized in Table 2.

As one can observe (bold characters), the MPI parallelization of the cost function evaluation is beneficial on our computer when the number of processes does not exceed the number of its processors (sockets), i.e. four. The speedup related to the original code is 2 or more, and increases with decreasing number of threads employed in the FE solver. This is quite understandable with regard to the limits in memory and disk bandwidth. Moreover, with larger number of processes the partition of the computation is not even. In total, the best configuration of the parallel code (4 x 8, employs 32 cores) provides more than ten times shorter solution time compared to the sequential GA optimization.

Next, we focused on the APSE optimization problem, described in Section 2 With this larger problem, having 613 305 degrees of freedom, it turned out soon that on our machine (Hubert) the MPI parallelization of the cost function evaluation in GA may not show any advantage over their sequential evaluation, once the shared resources (memory, disk) are not able to feed several instances of the demanding application with data. Thus, we invested considerable efforts to tailor the run-time environment to our application through adjusting Intel MPI parameters, especially those which control pinning, i.e. how processes and their threads in hybrid applications are allocated on computing elements, e.g. to avoid

---

[2] Our experience indicates that to avoid technical obstacles of hybrid parallel programming, one should make use of cooperating MPI and OpenMP implementations of the same vendor. Because of rapid developments in the software/hardware, the latest versions are to be recommended. (For example, the precedent ICT version 3.2 did not handle the core domains on the new AMD processors correctly.)

Table 2. Benefits of the MPI parallelization added to the GA optimization code and demonstrated on a small benchmark problem

OpenMP code (# threads)	seconds	MPI + OpenMP code (#processes x #threads)	seconds
8	1016	1 x 8	1026
–		2 x 8	633
–		4 x 8	**546**
–		6 x 8	811
4	1586	1 x 4	1604
–		2 x 4	878
–		4 x 4	**704**
–		8 x 4	1049
–		12 x 4	1238
2	2442	1 x 2	2459
–		2 x 2	1494
–		4 x 2	**965**
–		8 x 2	1221
1	5710	1 x 1	5758
–		2 x 1	4098
–		4 x 1	**1989**
–		8 x 1	2134

unnecessary process migration. This was more or less a trial and error procedure and we evaluated the omp:scatter pinning[3] as the best choice.

But even with those optimized run-time parameters, it turned out to be more efficient to increase the number of threads in the cost function computation than to increase the number of functions (i.e. processes) evaluated in parallel. Thus, the parallel GA showed its best execution time with only 2 parallel MPI processes for evaluation of the cost function, each giving rise to 8 threads in the FE solution. However, this best time was only by cca 30 % better than the sequential evaluation of the cost function employing 16 threads.

Finally, let us compare this timings with the Nelder-Mead approach (Section 3). For this purpose, we slightly modified the computation and made such a test arrangement that both codes started their iterative process from the same initial values, generated by random.[4] The stopping criterion was also unified: The (lowest) attained cost function value is to be less than a prescribed constant.

In this direct "competition", the NM algorithm, employing 16 threads in the cost function evaluation, provided his output after 17 311 s execution time and

---

[3] Intel MPI terminology: Domain members, on which threads of each process are allocated, are located as far from each other as possible in terms of common resources (cores, caches, sockets, etc.)

[4] Of course, this is not sufficient for a rigorous comparison of those two approaches, but at least provides some preview.

44 NM iterations. The parallel GA in the (2 processes x 8 threads) constellation finished after 20 935 s and the sequential GA using 16 threads for the cost function evaluation needed 23 488 s. Both GA versions carried out 6 iterations and 200 cost function evaluations. So, there is no substantial difference in the performance of those algorithms.

6   Conclusions

The paper describes the philosophy of the solution of the identification problems and numerical realization of the method. Optimization with the Nelder-Mead and genetic algorithms are discussed in more detail and compared on the basis of solution of a selected real-life optimization problem as a benchmark. The efficiency is increased by parallelization of the solution of the embedded direct method and can be also substantially increased by gradual improvement of the discretization accuracy during optimization.

From practical implementation and testing of the genetic methods we can learn that the high theoretical expectations on nearly optimal efficiency of parallelization in the case of parallel evaluation of the cost function might not be met in some parallel computing environments and especially on entry-level shared memory multiprocessors. We expect however, that the distributed memory architectures, e.g. clusters, will be much more appropriate for our sort of hybrid parallel code (e.g. the memory bottleneck issue will be significantly reduced), and want to verify this in near future. That is why we do not regard the results of our case study negative for the genetic approach.

Comparing the NM a GA algorithms, it was also observed that better cost function values could be reached by the NM algorithm (while GA stagnated). On the other hand, GA are more robust from the point of view of interest in a global solution. Therefore, a combination of both algorithms may be a promising solution.

In this paper, we omit the discussion on the gradient algorithms, which can be efficient, involving some parallelism for computing the Jacobian by either finite differences or a semianalytic approach. For the future, similar identification problems will be applied to other geotechnical problems including the development of in-situ rock mass tests and testing samples of geocomposites in the laboratory scale. There are also another aspects, which will be considered, such as selection of parameters, regularization of the cost function, application to nonlinear problems and computer oriented choice of the optimization method.

Acknowledents. We acknowledge the support of the EEA and Norway Grants through the project No. 049-4V. This work was also supported by the research plan AV0Z30860518 of the Academy of Sciences of the Czech Republic.

References

1. Andersson, J.C., Fälth, B., Kristensson, O.: Äspö pillar stability experiment, TM back calculation. In: Advances on Coupled Thermo-Hydro-Mechanical-Chemical Processes in Geosystems and Engineering, pp. 675–680. HoHai University, Nanjing (2006)

2. Blaheta, R., Jakl, O., Kohut, R., Starý, J.: GEM – A Platform for Advanced Mathematical Geosimulations. In: Wyrzykowski, R., Dongarra, J., Karczewski, K., Wasniewski, J. (eds.) PPAM 2009. LNCS, vol. 6067, pp. 266–275. Springer, Heidelberg (2010)

3. Kelley, C.T.: Iterative Methods for Optimization. SIAM, Philadelphia (1999)

4. Özisik, M.N., Orlande, H.R.B.: Inverse Heat Transfer: Fundamentals and Applications. Taylor and Francis, New York (2000)

5. Mahnken, R.: Identification of Material Parameters for Constitutive Equations. In: Stein, E., de Borst, R., Hughes, T.J.R. (eds.) Encyclopaedia of Computational Mechanics. Solids and Structures, vol. 2. John Wiley, Chichester (2004)

6. Rechea, C., Levasseur, S., Finno, R.: Inverse analysis techniques for parameter identification in simulation of excavation support systems. Computers in Geotechnics 35, 331–345 (2008)

7. Rus, G., Gallego, R.: Optimization algorithms for identification inverse problems with the boundary element method. Eng. Analysis with Boundary Elements 26, 315–327 (2002)

8. Haslinger, J., Jedelský, D., Kozubek, T., Tvrdík, J.: Genetic and Random Search Methods in Optimal Shape Design Problems. Journal of Global Optimization 16, 109–131 (2000)

9. Mühlenbein, H., Schlierkamp-Voosen, D.: Predictive models for the breeder genetic algorithm i. continuous parameter optimization. Evolutionary Computation 1, 25–49 (1993)

10. Blaheta, R., Kohut, R., Neytcheva, M., Starý, J.: Schwarz methods for discrete elliptic and parabolic problems with an application to nuclear waste repository modelling. Mathematics and Computers in Simulation 76, 18–27 (2007)

11. Blaheta, R., Kohut, R.: Parameter identification in heat flow with a geo-application. In: Strakos, Z., Rozloznik, M. (eds.) SNA 2010, Nove Hrady, pp. 29–32 (2010)

ScalaTrace: Tracing, Analysis and Modeling of HPC Codes at Scale

Frank Mueller[1], Xing Wu[1], Martin Schulz[2],
Bronis R. de Supinski[2], and Todd Gamblin[2]

[1] Dept. of Computer Science, North Carolina State University,
Raleigh, NC 27695-7534
mueller@cs.ncsu.edu
[2] Lawrence Livermore National Laboratory,
Center for Applied Scientific Computing, Livermore, CA 94551

Abstract. Characterizing the communication behavior of large-scale applications is a difficult and costly task due to code/system complexity and their long execution times. An alternative to running actual codes is to gather their communication traces and then replay them, which facilitates application tuning and future procurements. While past approaches lacked lossless scalable trace collection, we contribute an approach that provides orders of magnitude smaller, if not near constant-size, communication traces regardless of the number of nodes while preserving structural information. We introduce intra- and inter-node compression techniques of MPI events, we develop a scheme to preserve time and causality of communication events, and we present results of our implementation for BlueGene/L. Given this novel capability, we discuss its impact on communication tuning and on trace extrapolation. To the best of our knowledge, such a concise representation of MPI traces in a scalable manner combined with time-preserving deterministic MPI call replay are without any precedence.

Keywords: High-Performance Computing, Message Passing, Tracing.

1 Introduction

Scalability is one of the main challenges of petascale computing. One central problem lies in a lack of scaling of communication. However, understanding the communication patterns of complex large-scale scientific applications is nontrivial. An array of analysis tools have been developed, both by academia and industry, to aid this process. For example, Vampir is a commercial tool set including a trace generator and GUI to visualize a time line of MPI events [2]. While the trace generation supports filtering, trace files, which are stored locally, grow with the number of MPI events in a non-scalable fashion. Another example is the mpiP tool that uses the profiling layer of MPI to gather user-configurable aggregate metrics for statistical analysis [10]. Locally stored profiling files are constrained in size by the number of unique call sites of MPI events, which is

K. Jónasson (Ed.): PARA 2010, Part II, LNCS 7134, pp. 410–418, 2012.

independent of the number of nodes. However, mpiP does not preserve the structure and temporal ordering of events, which limits its use to high-level analysis. Other communication analysis tools have similar constraints: either their storage requirements do not scale or they are lossy with respect to program structure and temporal ordering.

In contrast to prior work, our work develops a scalable trace-driven approach to analyze MPI communication that can represent lossless, full traces in constant size. We demonstrate in our results that our objective has been achieved for a number of benchmarks. We have further developed tools (a) to replay communication, optionally with widely preserved timing information, (b) to detect inefficiencies in the utilization of the communication API, and (c) to extrapolate traces for strong scaling.

2   Lossless Tracing

Communication analysis tools are currently constrained in that either their storage requirements do not scale or they fall short in tracing all events by only providing aggregate statistics.

In contrast to prior work, ScalaTrace provides a scalable trace-driven approach to analyze MPI communication. While past approaches fail to gather full traces for hundreds of nodes in a scalable manner or only gather aggregate information, we have designed a framework that extracts full communication traces orders of magnitude smaller, if not near constant size, regardless of the number of nodes while preserving structural information and temporal event order.

Our trace-gathering framework utilizes the MPI profiling layer (PMPI) to intercept MPI calls during application execution. Profiling wrappers trace which MPI function was called along with call parameters within each node, such as source and destination of communications, yet without recording the actual message content. This intra-node information (task-level) is compressed on-the-fly. We also perform inter-node compression upon application termination to obtain a single trace file that preserves structural information suitable for lossless replay.

Intra-Node Compression: Within each node, we compress MPI call entries, generally repeated due to a code's loop structure, on-the-fly. To this extent, regular section descriptors (RSDs) are exploited to express MPI events nested in a single loop in constant size [3] while power-RSDs (PRSDs) are utilized to specify recursive RSDs nested in multiple loops [5]. MPI events may occur at any level in PRSDs. For example, the tuple RSD1:<100, MPI_Send1, MPI_Recv1> denotes a loop with 100 iterations of alternating send/receive calls with identical parameters (omitted here), and PRSD1:<1000, RSD1, MPI_Barrier1> denotes 1000 invocations of the former loop (RSD1) followed by a barrier. These construct correspond to the code in Figure 1. The algorithmic details of MPI event compressions over PRSDs can be found elsewhere [7,8].

To efficiently compress events, a set of generic and another set of domain-specific optimizations are performed. (1) Calling sequences are identified by generating a signature derived from a stack walk. Thus, call origins to common routines (*e.g.*, MPI_Send at call site 1) can be distinguished. (2) Communication end-points are encoded in a location-independent manner (relative to the rank of the current MPI task). This fosters identical encoding, *e.g.*, for stencil codes, across nodes. (3) Request handles are identified by a relative index into a handle buffer of constant size, which is dynamically updated. This abstracts from runtime-dependent data structures in a portable manner, *e.g.*, for handles returned by asynchronous communication calls such as MPI_Isend. (4) Iterative constructs with an indeterministic number of repetitions of a common event type are aggregated into a single event. For example, an application may wait for the completion of n asynchronous events using MPI_Waitsome in a loop, yet the MPI call may aggregate multiple completions as a result of each call. This will be abstracted as n calls.

```
for (i = 1; i < 1000; i++) {
   for (k = 1; k < 100; k++) {
      MPI_Send(...); /* send call 1 */
      MPI_Recv(...); /* recv call 1 */
   }
   MPI_Barrier(...); /* barrier call 1 */
}
```

Fig. 1. Sample Code for PRSDs

Inter-Node Compression: Local traces are combined into a single global trace upon application completion within the PMPI wrapper for MPI_Finalize. This approach is in contrast to generating local trace files, which results in linearly increasing disk space requirements and does not scale as traces must be moved to permanent (global) file space. The I/O bandwidth, particularly in systems like BlueGene/L (BG/L) with a limited number of I/O nodes, could severely suffer under such a load. To guarantee scalability, we instead employ cross-node compression, step-wise and in a bottom-up fashion over a binary tree network overlay. To this extent, events and structures (RSD / PRSDs) of nodes are merged when events, parameters, structure and iteration counts match (see [7,8] for algorithmic details).

We again employ a set of generic and domain-specific optimizations: (1) Sequences of nodes/task IDs are encoded using PRSDs, which allows a concise representation even for subsets of nodes as traces from different nodes are merged within the reduction tree. We utilize a radix tree whose encoding fosters efficient PRSD representations of sets of task IDs. (2) Events are temporally reordered when they originate from different nodes (and have no causal relation) to result in a more concise representation.

3   Deterministic Replay

One of the objectives of collecting communication traces is to analyze them off-line. One key virtue of our environment is that one can replay communication traces in a generic manner, even without the availability of application code. All that is needed is the trace itself. Our replay engine will be discussed in more detail in the preliminary results.

We have designed and implemented a replay engine that issues communication calls in the same order that they were originally issued by an application. The input to the replay engine consists of the compressed trace. The replay engine itself does not actually decompress this trace. Instead, it interprets the compressed trace on-the-fly to issue communication calls. In effect, the replay engine implements the inverse functions of the compression algorithms in an interpretative manner. When it encounters an RSD or PRSD, it issues calls iteratively observing the structure, frequency and parameters of communication calls. Hence, our structure-preserving compression scheme is key to a scalable replay methodology, which does not require excess amounts of memory. In fact, its memory requirement is loosely bounded by the size of the *compressed* trace, which is often of constant size.

Using the replay engine, we conducted experiments to verify the correctness of our scalable compression approach. We replayed compressed traces to ensure MPI semantics are preserved, to verify that the aggregate number of MPI events per MPI call matches that of the original code and that the temporal ordering of MPI events within a node are observed. The results of communication replays confirmed the correctness of our approach. During replay, all MPI calls are triggered over the same number of nodes with original payload sizes, yet with a *random* message payload (content). Thus, the replay incurs comparable bandwidth requirements on communication interconnects, albeit with potentially different contention characteristics since event times are not preserved (addressed below). Communication replay also provides an abstraction from compute-bound application performance, which is neither captured nor replayed. This makes the replay mechanism extremely portable, even across platforms, which can benefit rapid prototyping and tuning. It also supports assessing communication needs of future platforms for large-scale procurements.

4   Preserving Time

The objective of trace analysis is generally to find inefficiencies in the code, *e.g.*, as indicated by load imbalance between nodes. Such analysis requires knowledge about the timing between events. Hence, conventional trace techniques simply attach a timestamp to all communication events. Such timestamps also facilitate a time-accurate replay. However, for a scalable compression-based tracing approach like ours, recording the precise timestamp is infeasible. Due to asynchronous event occurrence across nodes, the timing between nodes diverges over time so that absolute time differs. (This holds even if nodes were synchronized at a job start, which is generally not the case). Consequently, absolute times-tamping would require recording the timing information for every single node without being able to compress, which leads to a linearly increasing trace size wrt. the total number of nodes.

Our trace compression scheme and the replay engine support two methods of capturing timing information of different tasks in computational sections (between any two communication calls) [9]. First, a low-cost statistical approach to capture delta times has been designed. Second, to resemble computational imbalance, a variation-preserving recording scheme was devised, still within a constant size trace representation, yet with a higher constant factor. Delta times denote the elapsed time between adjacent trace events. In contrast to absolute time, the relative notion of time makes delta times amenable to compression. Delta times are collected for event pairs. For example, RSD1 has two timing regions that will be captured: (a) from send to receive and (b) from receive to send (between consecutive loop iterations). Each delta time is associated with the terminal communication event, *i.e.*, at the beginning (prologue) of receive for (a) and the beginning of send for (b).

We dynamically create size-limited histograms of delta times suitable for our existing trace compression scheme. Based on a user-defined number of bins, delta times are recorded in an online balancing scheme to equalize bin volumes using a weighted subrange partitioning scheme (algorithm 2 in [9]).

5   Trace Extrapolation

Judging the effect of problem and/or task scaling on performance is a hard problem. While some advances have been made using application modeling or deriving similarities and characteristics from microbenchmarks [4,6,1], we follow a complementary direction. We extrapolate larger communication traces from smaller ones, which can then be used to replay these larger traces and empirically detect communication problems or project system requirements for future procurements of HPC systems. If communication is the impeding factor to scalability, our framework can aid in the analysis of codes and performance projections for existing and future systems.

The extrapolation of system overhead from small application runs to larger ones is a challenging problem that has not been solved. Yet, a number of sub-problems become feasible with ScalaTrace. We consider the problem of strong (task) scaling. Given a *set* of communication traces of, say, 8, 16, 32 and 64 nodes, we can extrapolate the corresponding trace for n nodes of the same application. We devised a method to extrapolate larger communication traces from smaller ones for task (strong) scaling. The key insight is that communication parameters combined with PRSD iterators provide sufficient detail for this approach. *E.g.*, if a communication parameter depends on the number of columns of a matrix whose size is input dependent, then problem scaling will increase this parameter at a certain rate. Given a minimum of three data points, a fitting curve can be constructed to extrapolate the growth rate of this parameter. Thus, payloads can be adjusted according to such fitting curves. A larger problem size can also affect the number of timesteps of a convergence algorithm. By capturing the iterators of timesteps within histograms, such dependencies can be discovered and modeled again *via* curve fitting.

Similar to problem scaling, the size of a group communicator may depend on the total number of nodes, and this growth under task scaling can be extrapolated with curve fitting methods. For such node dependencies, we require $d + 1$ data points (traces) for a d-dimensional Cartesian layout and then apply Gaussian elimination to extrapolate parameters, such as communication end-points, for arbitrary sizes n. *E.g.*, after determining the dimensionality, we can infer the coefficients A, B, C, D for a given layout and solve for values V_i: $A \times n_i + B \times x_i \times y_i + C \times x_i + D = V_i$ for $i \in \{1, ..., d + 1\}$ where $n_i = f(z_i)$ to facilitate the calculation for row-major layouts. We have prototyped a trace extrapolation method for strong scaling that automatically transforms a set of traces of smaller size (number of nodes) to one of arbitrary size. The extrapolated traces have been replayed successfully, as reported in the next section.

6   Results

We assessed the effectiveness of ScalaTrace through experiments with benchmarks and an application on BG/L. Our results confirm the scalability of our on-the-fly MPI trace compression by yielding orders of magnitude smaller or even near constant size traces for processor scaling and problem scaling.

We conducted experiments with the NAS Parallel Benchmark (NPB) codes for class C inputs [12] and UMT2k on BG/L. Figures 2-4 depict the trace file sizes on a logarithmic scale without compression (*none*), with local compression (*intra-node*) and with cross-node compression (*inter*). We identified three categories of codes wrt. inter-node compression efficiency: (1) those that result in near constant-size traces (DT, EP, LU, BT and FT), regardless of the number of nodes, (2) those with sub-linear scaling of trace size as the node count increases (MG and CG) and (3) those that do not scale yet (IS and UMT2k).

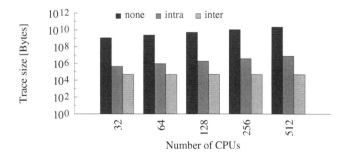

Fig. 2. LU Trace File, Varied # Nodes

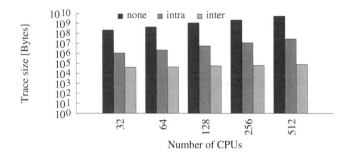

Fig. 3. CG Trace File, Varied # Nodes

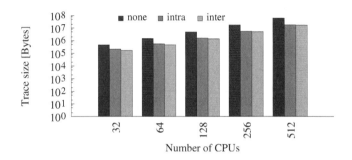

Fig. 4. IS Trace File, Varied # Nodes

The first class, represented by LU (Fig. 2), shows reductions for *intra* but only *inter* delivers constant size traces. The second class, represented by CG (Fig. 3), shows considerable compression at the node level (*intra*) and sub-linear (but not constant) sized traces for inter-node compression. The third class, represented by IS (Fig. 4), shows a similar trend but with a faster than linear growth rate for *inter*.

The next experiment assesses the effectiveness of delta times to resemble application behavior during replay. Figure 5 depicts the wall-clock time for the uninstrumented application, mpiP[10]-instrumented application and three replay options based on uncompressed, intra-node compressed and globally compressed traces. It shows highly accurate replay times irrespective of number of nodes and levels of compression, which is representative for all benchmarks.

We have also verified our extrapolation approach with a subset of the NAS Parallel Benchmark suite [13]. To verify the functional correctness, we replayed the extrapolated traces at the target node size. We instrumented both the replay engine and the original benchmarks with mpiP. We compared the MPI event statistics generated by mpiP. The results demonstrated the communication overheads at the target size are fully captured.

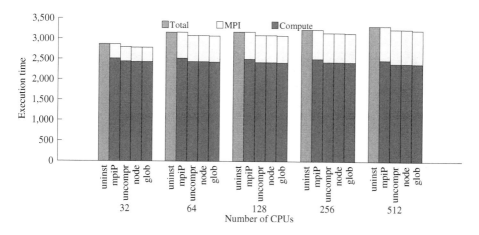

Fig. 5. FT Replay: Aggregate of All Nodes (BG/L)

Besides the functional correctness, experimental results also indicate close resemblance in timing of extrapolated traces when replayed compared to application behavior at scaled size for up to 16k nodes (see Figure 6, dark/maroon bars are extrapolated, scaling is limited by input sizes). We are working on generalizing our approach to a large number of common communication patterns.

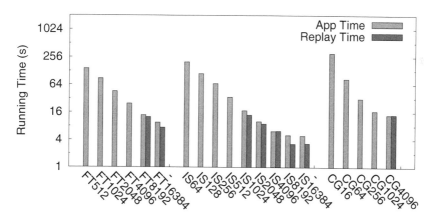

Fig. 6. Replay after Extrapolation on a BG/P

7   Conclusions

This paper gives an updated overview over ScalaTrace. ScalaTrace provides a scalable methodology for event tracing that has been demonstrated for MPI and I/O events. It annotates events with time-preserving information suitable for deterministic MPI call replay. Traces can further be extrapolated in the dimension of number of tasks (nodes) to assess communication scalability and assist procurement decisions for future HPC installations. Further information about ScalaTrace can be found elsewhere [7,9,8,11].

Acknowledgements. This work was supported in part by NSF grants 0410203, 0429653, 0237570 (CAREER), 0937908, and 0958311. Part of this work was performed under the auspices of the U.S. Department of Energy by University of California Lawrence Livermore National Laboratory under contracts DE-AC05-00OR22725, W-7405-Eng-48, and DE-AC52-07NA27344 under LLNL-CONF-427005.

References

1. Bell, R., John, L.: Improved automatic testcase synthesis for performance model validation. In: International Conference on Supercomputing, pp. 111–120 (June 2005)
2. Brunst, H., Hoppe, H.-C., Nagel, W.E., Winkler, M.: Performance Optimization for Large Scale Computing: The Scalable VAMPIR Approach. In: Alexandrov, V.N., Dongarra, J., Juliano, B.A., Renner, R.S., Tan, C.J.K. (eds.) ICCS 2001. LNCS, vol. 2074, pp. 751–760. Springer, Heidelberg (2001)
3. Havlak, P., Kennedy, K.: An implementation of interprocedural bounded regular section analysis. IEEE Transactions on Parallel and Distributed Systems 2(3), 350–360 (1991)
4. Kerbyson, D., Alme, H., Hoisie, A., Petrini, F., Wasserman, H., Gittings, M.: Predictive performance and scalability modeling of a large-scale application. In: Supercomputing (November 2001)
5. Marathe, J., Mueller, F., Mohan, T., de Supinski, B.R., McKee, S.A., Yoo, A.: METRIC: Tracking down inefficiencies in the memory hierarchy via binary rewriting. In: International Symposium on Code Generation and Optimization, pp. 289–300 (March 2003)
6. Marin, G., Mellor-Crummey, J.: Cross architecture performance predictions for scientific applications using parameterized models. In: SIGMETRICS Conference on Measurement and Modeling of Computer Systems (2004)
7. Noeth, M., Mueller, F., Schulz, M., de Supinski, B.R.: Scalable compression and replay of communication traces in massively parallel environments. In: International Parallel and Distributed Processing Symposium (April 2007)
8. Noeth, M., Mueller, F., Schulz, M., de Supinski, B.R.: Scalatrace: Scalable compression and replay of communication traces in high performance computing. Journal of Parallel Distributed Computing 69(8), 710–969 (2009)
9. Ratn, P., Mueller, F., de Supinski, B.R., Schulz, M.: Preserving time in large-scale communication traces. In: International Conference on Supercomputing, pp. 46–55 (June 2008)
10. Vetter, J., McCracken, M.: Statistical scalability analysis of communication operations in distributed applications. In: ACM SIGPLAN Symposium on Principles and Practice of Parallel Programming (2001)
11. Vijayakumar, K., Mueller, F., Ma, X., Roth, P.C.: Scalable multi-level i/o tracing and analysis. In: Petascale Data Storage Workshop (November 2009)
12. Wong, F., Martin, R., Arpaci-Dusseau, R., Culler, D.: Architectural requirements and scalability of the NAS parallel benchmarks. In: Supercomputing (1999)
13. Wu, X., Mueller, F.: Scalaextrap: trace-based communication extrapolation for spmd program. In: PPoPP (2011)

A Lightweight Library
for Building Scalable Tools

Emily R. Jacobson, Michael J. Brim, and Barton P. Miller

Computer Sciences Department, University of Wisconsin
Madison, Wisconsin, USA
{jacobson,mjbrim,bart}@cs.wisc.edu

Abstract. MRNet is a software-based multicast reduction network for building scalable tools. Tools face communication and computation issues when used on large systems; MRNet alleviates these issues by providing multicast communication and data aggregation functionalities. Until now, the MRNet API has been entirely in C++. We present a new, lightweight library that provides a C interface for MRNet back-ends, making MRNet accessible to a wide range of new tools. Further, this library is single threaded to accommodate even more platforms and tools where this is a limitation. This new library provides the same abstractions as the C++ library, using an API that can be derived by applying a standard translation template to the C++ API.

Keywords: scalability, tree-based overlay networks, tools.

1 Introduction

As high performance computers reach processor counts in the hundreds of thousands, or even millions of cores, runtime tools are needed to support the performance profiling and debugging of applications running on these computers. Unfortunately, tools that previously worked at smaller scales do not work effectively on the larger systems. To this end, we have developed a tree-based overlay network infrastructure, called MRNet, for building tools that can scale to the largest of computing platforms. MRNet makes operations such as command and control, and data collection and reduction, efficient at large scale.

Runtime tools are often organized around two main activities: data collection, with data originating from the tool *daemons* or back-ends and traveling to the front-end, and application process control, initiated by a tool's user interface or front-end and directed to the back-end. These are the two areas in which tools pay most costs: computation and communication. Computation is in the form of data collection, aggregation, and analysis, and communication arises from the transfer of data between tool components. Tools have typically been designed with the front-end talking directly to the back-ends, causing the front-end to become a bottleneck for both communication and computation.

MRNet is designed to address many of these issues by providing broadcast and aggregation functionality [7,8]. MRNet uses a tree-based overlay network

K. Jónasson (Ed.): PARA 2010, Part II, LNCS 7134, pp. 419–429, 2012.
© Springer-Verlag Berlin Heidelberg 2012

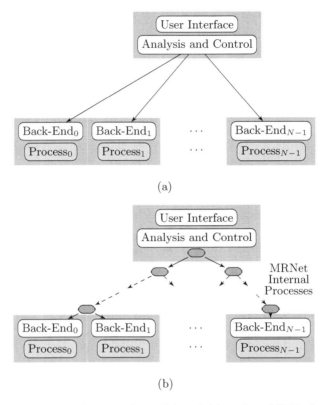

Fig. 1. The components of a typical parallel tool (a) and an MRNet-based tool (b). *Shaded boxes show potential machine boundaries.*

(TBON) of communication processes between the tool front-end and the tool back-ends, as shown in Figure 1. MRNet leverages this structure to distribute computation among internal processes and to support scalable multicast operations. Data can be filtered as it is passed up the tree; such filtering might do data transformation or might simply aggregate packets to be passed to the front-end. Data is transferred between nodes using an efficient, packed binary representation, which provides high-bandwidth communication. Further, the user may designate multiple concurrent data channels, allowing for a variety of types of data processing and aggregation to happen simultaneously. Together, these features all work to mitigate the high costs of communication and computation on large-scale systems. MRNet has been demonstrated on tools running on the largest of existing computing platforms [2,4].

MRNet has two components: `libmrnet`, a library API that is linked into a tool's front-end and back-end components, and `mrnet_commnode`, the program that runs on intermediate nodes that forms the communication processes of the tree-based network interposed between the front-end and back-ends.

The standard MRNet library used in the front- and back-ends [7] provides a C++ interface. While C++ offers many attractive software engineering features, MRNet previously was incompatible with tools written in C. In practice, few parallel applications are written using C++ and not all high-performance computing systems support general-purpose threading. In this paper, we introduce a lightweight back-end library with a pure C interface. The lightweight library is intended for use only within MRNet back-ends and offers a subset of the functionality normally provided by the C++ library to back-ends.

There are two classes of tools for which we anticipate this being useful. The first class is tools written in C; these tools need a C interface with which to interact. The second class are tools that interact with applications written in C, where such a tool might instrument the application with MRNet API calls in order to extract information; the language of the application limits the API that may be used for such purposes. The new library makes MRNet available to many tools that were previously unable to use it. In addition, this lightweight library is single threaded, to accommodate even more platforms and tools where this is a limitation. This lightweight MRNet provides the same abstractions as the C++ library; only the API is different. In most cases, however, there is a direct translation between the C++ API call and the new C version.

We use the term "lightweight" to reflect that new library is less cumbersome to integrate into existing tools as it does not require tool back-ends to use C++. For many tools, a C-based library interface is much easier to develop against, as C binding layers exist for many programming languages. As an example of tools that benefit from MRNet's new lightweight back-end library, consider two MPI application profiling tools, TAU and the CEPBA Tools. The TAU Performance System from the University of Oregon ([5],[9]) uses MRNet to aggregate performance data from parallel processes. TAU inserts a tracing library into the application and instruments the application with trace routines that use MRNet to send collected performance data for processing. A similar approach for MPI profiling is used by the CEPBA-Tools from the Barcelona Supercomputing Center([3]). Although it was previously possible to use standard MRNet in these tracing libraries, it required them to be redeveloped in C++. This introduces a dependency on the `libstdc++` library that many tool and parallel application developers feel is too heavyweight in terms of code size. The new lightweight library makes it much easier to use MRNet in tools for profiling C-based MPI applications with minimal overhead.

The remainder of this paper is organized as follows. In Section 2, we present the basic abstractions of MRNet and describe how these are expressed in the new C-based library. In Section 3, we describe our interface, and in Section 4, we provide an example of a tool that can leverage this new library.

2   Abstractions

The MRNet library, `libmrnet`, allows a tool to use an overlay network of internal processes as a communication and data aggregation substrate between the tool's front-end and back-end processes. MRNet uses a variety of abstractions to support these functions. An MRNet *end-point* represents a tool or application process. In particular, a back-end is a leaf node in the TBON. The front-end can communicate in a multicast fashion with one or more of these endpoints. MRNet uses *communicators* to represent groups of network end-points. Communicators are created and managed by the front-end; currently communication is allowed only between a tool's front-end and its back-ends.

MRNet uses logical channels called *streams* to connect the front-end to the end-points of a particular communicator. Streams can carry data *packets* downstream, from the front-end to back-ends, and upstream, from back-ends toward the front-end. Packets are sets of data elements, where types are specified using a format string similar to that used by C formatted I/O primitive `printf`. For example, a packet whose data is described by the format "%d %f %s" contains an integer, float, and character string.

Data aggregation, the process of transforming input data packets into one or more output packets, is a vital component of MRNet. MRNet uses *filters* to aggregate data packets. When a stream is created, a filter is bound to the stream that defines the aggregation operation to be performed and also the expected packet data format that will be sent on the stream. MRNet supports two types of filters: synchronization filters and transformation filters. Synchronization filters provide a mechanism to deal with asynchronous arrival of packets from child nodes; these filters do no data transformation and operate on packets in a type-independent fashion. MRNet supports three synchronization modes: *Wait For All*, *Time Out*, and *Do Not Wait*. In contrast, transformation filters combine data from multiple packets by performing an aggregation that yields one or more new packets. Several general-use transformation filters are provided, including basic scalar operations like min, max, sum, average, and concatenation operations. Additionally, MRNet allows tool developers to use custom filters. The developer simply writes one or more filter functions, compiles them into a shared library, and loads the filter library in the network. Filters use a standard function signature and can perform arbitrary computations.

The internal processes of the MRNet TBON provide the core functionalities, including the logical channels for control messages and data. Further, these processes perform the data aggregation or reduction operations. When a stream is established, an internal process creates a new *stream manager* and initializes it with a set of end-points to be associated with the stream and the filter(s) to be used on the data packets sent on the stream. Upstream data buffers must be unbatched, demultiplexed, processed, and then rebatched; downstream data is similar, though the data packets may be placed in multiple output buffers because the packet may be destined for multiple back-ends.

Although the new lightweight library provides the same abstractions as the C++ library, there are a few cases in which an abstraction is not applicable in the new library. Communicators are not present in the lightweight library because they are a handle necessary only for the front-end. Standard MRNet allows for both blocking and non-blocking receive operations; because the C-based API is not multi-threaded, only blocking receive is supported. Additionally, there are slight differences in how filters are used. In standard MRNet, filtering is done at every level of the tree, including at the back-end nodes. However, because filtering at the C-based back-end nodes adds an additional level of complexity, we have chosen initially to not filter at the back-ends.

3   Interface

To support the above abstractions, the MRNet API contains `Network`, `NetworkTopology`, `Communicator`, `Stream`, and `Packet` classes. The `Network` class is used to instantiate the TBON and access end-point objects representing tool back-ends. The `NetworkTopology` class provides an interface for discovering topology details of the instantiated `Network`. The `Communicator` class is used to represent a group of end-points when creating a `Stream` for unicast, multicast, or broadcast communication. The `Packet` class encapsulates the data packets that are sent on a `Stream`.

```
C++:                         C:
return_type                  return_type
class:function_name (        class_function_name (
    param1_type param1,          class class_object,
    ...);                        param1_type param1,
                                 ...);
```

Fig. 2. API Translation Template

```
C++:                         C:
int                          int
Stream::send (               Stream_send (
    int tag,                     Stream_t * stream,
    char * fmt_string,           int tag,
    ...);                        char * fmt_string,
                                 ...);
```

Fig. 3. API Translation Example

The lightweight library provides similar functionality for lightweight back-ends, so its public API is comparable to the standard MRNet API. Lightweight API classes are directly translated from the standard API. The translation scheme is shown in Figure 2 and an actual example from the Stream class is provided in Figure 3. In practice, creating a tool that uses the lightweight library will require familiarity with both the C++ and C APIs. However, because they are so similar, this should not be difficult.

Creating the MRNet overlay network is complicated by interactions with various job management systems. In the simplest environments, MRNet launches jobs using facilities like *rsh* or *ssh*. However, in more complex environments it is necessary to submit requests to a job management system. In this case, we are constrained by the operations provided by the job manager. To allow for these models, we currently support two modes of instantiating MRNet-based tools.

In the first mode of process instantiation, MRNet creates the internal and back-end processes, using the specified MRNet topology configuration to determine the hosts on which components should be located.

In the second mode, MRNet relies on a process management system to create some or all of the MRNet processes [1]. This mode accommodates tools that require their back-ends to create, monitor, and control parallel application processes. MRNet creates its internal processes as in the first instantiation mode, but does not instantiate any back-end processes. When the back-ends are started by the process management system, MRNet provides the information necessary to connect the back-ends to the MRNet internal process tree. This information includes the leaf processes' host names and connection port numbers. A tool front-end can extract this information and provide it to the back-ends via the environment, using shared file systems or other information services available on the target system. The new lightweight library supports both methods of instantiation. Additionally, examples of both methods of instantiation are provided with the MRNet source code.

4   Example Tool

MRNet can be used in a wide variety of tools and application. Here, we provide a simple example to demonstrate a few key concepts. At a high level, the tool front-end sends an integer value and a number of iterations to each back-end. During each iteration, or "wave," the back-end sends the integer multiplied by the wave number back up the tree. The values are aggregated using a summation filter and passed to the front-end as a single value, which should be equal to `num_backends` × `val` × `wave_number`.

Figure 4 provides code for a custom filter used in this example. For each packet being aggregated, a value is extracted and added to the current sum (lines 10-13). Then, a new packet containing this summation is created (lines 16-18) and added to the outgoing packets (line 19).

Code for the tool front-end is shown in Figure 5. After several variable definitions in lines 2-7, an instance of the MRNet `Network` is created on line 10, using the topology specified in `topology_file`. The `Network` then loads a filter, queries for the auto-generated broadcast communicator that contains all available end-points, and then establishes a stream that will use this filter. The front-end broadcasts two integers, a value and then number of iterations to complete, on the new stream (line 27). For each iteration, the front-end performs a blocking receive (line 32); it unpacks a single integer and checks the value (lines 33-34). Finally, the front-end sends a message to the rest of the TBON to shut down the network, and then deletes the network itself (lines 38-39).

```
1.    extern "C" {
2.    void Integer_Add(std::vector<PacketPtr> & packets_in,
3.                     std::vector<PacketPtr> & packets_out,
4.                     std::vector<PacketPtr> & packets_out_reverse,
5.                     void ** /* filter state */
6.                     TopologyLocalInfo & /* topology information */)
7.    {
8.      int sum = 0, val;
9.
10.     for (unsigned int i = 0; i < packets_in.size(); i++) {
11.       PacketPtr cur_packet = packets_in[i];
12.       cur_packet->unpack("%d", &val);
13.       sum += val;
14.     }
15.
16.     PacketPtr new_packet (new Packet(packets_in[0]->get_StreamId(),
17.                                     packets_in[0]->getTag(),
18.                                     "%d", sum));
19.     packets_out.push_back(new_packet);
20.   }
21.   } /* extern "C" */
```

Fig. 4. MRNet filter example code

Figures 6 and 7 provides code for the back-ends that reciprocate the actions of the front-end. We provide code both for a standard back-end and a lightweight back-end. It is easy to observe that the code is nearly identical. Each tool back-end first connects to the MRNet network in line 8, using a **Network** constructor that receives its arguments using the program argument vector. While the front-end makes a stream-specific receive call, the back-end uses a stream-anonymous network receive that returns the tag sent by the front-end, the **Packet** with actual data, and a **Stream** object representing stream that the front-end has established (line 12). In both cases, this receive operation is a blocking receive; for the C++ version, this is the default mode, and for C this is the only mode supported. For each iteration, the back-end sends an integer value upstream towards the front-end (line 20).

While this example shows many of the basic concepts of MRNet, including **Network** and **Stream** creation, **Filter** loading, and **Stream** send and receive, these basic ideas allow for many other functionalities. The MRNet API Guide provides a thorough explanation of these core abstractions and their possible uses[6].

```
1.void front_end_main(int argc, char ** argv) {
2.    int send_val=32, recv_val=0;
3.    int tag, retval, filter_id, num_backends, num_iters;
4.    Network * net;
5.    Communicator * comm;
6.    Stream * stream;
7.    PacketPtr pkt;
8.
9.    // Create a new instance of a network
10.   net = Network::CreateNetworkFE(topology_file,
11.                          backend_exe, &dummy_argv);
12.   filter_id =
13.       net->load_FilterFunc(so_file, "Integer_Add");
14.
15.   // A Broadcast communication contains all the backends
16.   comm = net->get_BroadcastCommunicator();
17.
18.   // Create a stream that will use the Integer_Add filter
19.   stream = net->new_Stream(comm, filter_id,
20.                          SFILTER_WAITFORALL);
21.   num_backends = comm->get_EndPoints().size();
22.
23.   // Broadcast a control message to back-ends to send us "num_iters"
24.   // waves of integers
25.   tag = PROT_SUM;
26.   num_iters = 5;
27.   stream->send(tag, "%d %d", send_val, num_iters);
28.
29.   // We expect "num_iters" aggregated respnoses from all back-ends
30.   for (unsigned int i = 0; i < num_iters; i++) {
31.     // Receive and unpack packet containing single int
32.     retval = stream->recv(&tag, pkt);
33.     pkt->unpack("%d", &recv_val);
34.     assert(recv_val == send_val*i*num_backends);
35.   }
36.
37.   // Tell the back-ends to exit; cleanup the network
38.   stream->send(PROT_EXIT, "");
39.   delete stream;
40.   delete net;
41.}
```

Fig. 5. MRNet front-end sample code

```
1.void back_end_main(int argc, char ** argv) {
2.    Stream * stream;
3.    PacketPtr pkt;
4.    int tag = 0, retval = 0, num_iters = 0;
5.
6.    // Create a new instance of a network
7.    Network * net = Network::CreateNetworkBE(argc, argv);
8.
9.    do {
10.     // Anonymous stream receive
11.     net->recv(&tag, pkt, &stream);
12.     switch(tag) {
13.       case PROT_SUM:
14.         // Unpack packet with two integers
15.         pkt->unpack("%d %d", &recv_val, &num_iters);
16.
17.         // Send num_iters waves of integers
18.         for (unsigned int i = 0; i < num_iters; i++) {
19.            stream->send(tag, "%d", recv_val*i);
20.         }
21.         break;
22.       case PROT_EXIT: break;
23.     }
24.   } while (tag != PROT_EXIT)
25.
26.   // Wait for stream to shut down before deleting
27.   while(!stream->is_Closed()) sleep(1);
28.   delete stream;
29.
30.   // Wait for the front-end to shut down the network, then delete
31.   net->waitfor_ShutDown();
32.   delete net;
33.}
```

Fig. 6. MRNet standard back-end sample code

```
1.void back_end_main(int argc, char ** argv) {
2.    Stream_t * stream;
3.    Packet_t * pkt = (Packet_t*)malloc(sizeof(Packet_t));
4.    int tag = 0, retval = 0, num_iters = 0;
5.
6.    // Create a new instance of a network
7.    Network_t * net = Network_CreateNetworkBE(argc, argv);
8.
9.  do {
10.    // Anonymous stream receive
11.    Network_recv(net, &tag, pkt, &stream);
12.    switch(tag) {
13.      case PROT_SUM:
14.        // Unpack packet with two integers
15.        Packet_unpack(pkt, "%d %d", &recv_val, &num_iters);
16.
17.        // Send num_iters waves of integers
18.        for (unsigned int i = 0; i < num_iters; i++) {
19.          Stream_send(stream, tag, "%d", recv_val*i);
20.        }
21.        break;
22.      case PROT_EXIT: break;
23.    }
24.  } while (tag != PROT_EXIT)
25.
26.  // Wait for stream to shut down before deleting
27.  while (!Stream_is_Closed(stream)) sleep(1);
28.  delete_Stream_t(stream);
29.
30.  free(pkt); // Cleanup malloc'd variables
31.
32.  // Wait for the front-end to shut down the network, then delete
33.  Network_waitfor_ShutDown(net);
34.  delete_Network_t(net);
35.}
```

Fig. 7. MRNet lightweight back-end sample code

5   Conclusions

MRNet is a customizable, software-based multicast reduction network for scalable performance and system tools. MRNet reduces the cost of tool activities by leveraging a tree-based overlay network of processes between the tool's front-end and back-ends. MRNet uses this overlay network to distribute communication and computation, reducing analysis time and keeping tool front-end loads manageable. Previously, MRNet had only a C++ interface. We have presented a new lightweight library for MRNet back-ends that is single-threaded and in C. This addition makes MRNet accessible to a wide variety of tools that previously were unable to use MRNet.

Acknowledgements. This work is supported in part by Department of Energy grants DE-SC0004061, 08ER25842, DE-SC0003922, DE-SC0002153, DE-SC0002155, Lawrence Livermore National Lab grant B579934, and a grant from Cray Inc. through the Defense Advanced Research Projects Agency under its Agreement No. HR0011-07-9-0001. Any opinions, findings and conclusions or recommendations expressed in this material are those of the author(s) and do not necessarily reflect the views of the Defense Advanced Research Projects Agency.

The U.S. Government is authorized to reproduce and distribute reprints for Governmental purposes notwithstanding any copyright notation thereon.

References

1. Ahn, D.H., Arnold, D.C., de Supinki, B.R., Lee, G., Miller, B.P., Schulz, M.: Overcoming Scalability Challenges for Tool Daemon Launching. In: 37th International Conference on Parallel Processing (ICPP 2008), Portland, OR, pp. 578–585 (September 2008)
2. Arnold, D.C., Ahn, D.H., de Supinski, B.R., Lee, G., Miller, B.P., Schulz, M.: Stack Trace Analysis for Large Scale Debugging. In: 2007 IEEE International Parallel and Distributed Processing Symposium (IPDPS), Long Beach, CA, pp. 1–10 (March 2007)
3. Llort, G., Gonzalez, J., Servat, H., Gimenez, J., Labarta, J.: On-line detection of large-scale parallel application's structure. In: 2010 IEEE International Parallel and Distributed Processing Symposium (IPDPS), Atlanta, GA, pp. 1–10 (April 2010)
4. Lee, G.L., Ahn, D.H., Arnold, D.C., de Supinski, B.R., Legendre, M., Miller, B.P., Schulz, M., Liblit, B.: Lessons Learned at 208K: Towards Debugging Millions of Cores. In: Supercomputing 2008 (SC 2008), Austin, TX (November 2008)
5. Nataraj, A., Malony, A.D., Morris, A., Arnold, D.C., Miller, B.P.: TAUoverMRNet (ToM): A Framework for Scalable, Parallel Performance Monitoring using TAU and MRNet. In: International Workshop on Scalable Tools for High-End Computing (STHEC 2008), Island of Kos, Greece (June 2008)
6. Paradyn Project. The Multicast/Reduction Network: A User's Guide. University of Wisconsin-Madison, http://www.paradyn.org/mrnet/release_3.0
7. Roth, P.C., Arnold, D.C., Miller, B.P.: MRNet: A Software-Based Multicast/Reduction Network for Scalable Tools. In: Supercomputing 2003 (SC 2003), Phoenix, AZ (November 2003)
8. Roth, P.C., Arnold, D.C., Miller, B.P.: Benchmarking the MRNet Distributed Tool Infrastructure: Lessons Learned. In: 2004 High-Performance Grid Computing Workshop, Held in Conjunction with the 2004 International Parallel and Distributed Processing Symposium (IPDPS), Santa Fe, NM (April 2004)
9. Shende, S., Malony, A.D.: The TAU Parallel Performance System. International Journal of High Performance Computing Applications 20(2), 287–331 (2006)

MATE: Toward Scalable Automated and Dynamic Performance Tuning Environment

Anna Morajko, Andrea Martínez, Eduardo César,
Tomàs Margalef, and Joan Sorribes

Computer Architecture and Operating System Department,
Universitat Autònoma de Barcelona, 08192 Bellaterra, Spain
{Anna.Morajko,Eduardo.Cesar,Tomas.Margalef,Joan.Sorribes}@uab.es
Andrea.Martinez@caos.uab.es

Abstract. The use of parallel/distributed programming increases as it enables high performance computing. There are many tools that help a user in the performance analysis of the application, and that allow to improve the application execution. As there is a high demand of computational power, new systems, such as large scale computer clusters, have become more common and accessible to everyone to solve complex problems. However, these systems generate a new set of problems related to the scalability of current analysis and tuning tools. Our automatic and dynamic tuning environment MATE does not scale well because it has a set of common bottlenecks in its architecture, and hence we have decided to improve the tool for providing dynamic tuning on large scale systems too. For this purpose, we are designing a new tool that introduces a tree-based overlay network infrastructure for scalable metrics collection, and to substitutes the current centralized performance analysis by a distributed one, in order to provide better scalability.

1 Introduction

Nowadays, parallel/distributed applications are used in many science and engineering fields. They may be data intensive and/or may perform complex algorithms. Their main goal is to solve problems as fast as possible. However, development of an efficient parallel application is still a complex task, and parallel applications rarely achieve a good performance immediately. Therefore, a careful performance analysis and optimization is crucial. These tasks are known to be difficult and costly and, in practice, developers must understand both, the application and the analysis/tuning environment behavior.

Moreover, there are many applications that depend on the input data set, or even can vary their behavior during one particular execution according to the data evolution. In such cases, it is not worth to carry out a post-mortem analysis and tuning, since the conclusions based on one execution could be wrong for a new one. It is necessary to carry out a dynamic and automatic tuning of the application during its execution without stopping, recompiling nor rerunning it. In this context, MATE environment was developed.

K. Jónasson (Ed.): PARA 2010, Part II, LNCS 7134, pp. 430–440, 2012.

On the other hand, during last years the hardware evolution has increased significantly, and its cost has sharply decreased. Therefore, new systems as computer clusters became more common and accessible to everyone to solve new classes of scientific problems which have high performance requirements. There are many large scale systems that are composed of thousands of processors. In this case, however, a new problem appears when considering performance analysis and tuning tools. They may present scalability difficulties when executing in large scale systems. These tools may transform themselves in a bottleneck and may cause a parallel application to decrease its performance instead of improve it.

The goal of this work is to describe a set of difficulties that the analysis and tuning tools may present in large scale systems. In general, the main problems are related to the number of processes, the volume of collected data, and the centralized performance analysis. In particular, we focus our work on the MATE environment that automatically and dynamically improves parallel applications performance. It may be executed on many processors; however, its architecture contains certain components that might cause scalability bottlenecks.

In this paper we focus on a scalability problem that appears in the MATE environment. In Section 2, we briefly describe the dynamic and automatic tuning tool MATE. In Section 3, we present general problems of the scalability in large-scale parallel systems. Section 4 shows the proposal architecture of MATE that may allow for reducing its scalability limitations. Section 5 describes the related work in automatic and dynamic tuning. Finally, Section 6 summarizes the conclusions of this work.

2 MATE

MATE (Monitoring, Analysis and Tuning Environment) is a tuning environment for MPI parallel applications [11]. It augments on-line automated performance diagnosis with dynamic code optimization to combine the advantages of both automated analysis and computational steering. MATE does not require program modifications to expose steerable parameters. Instead, it uses dynamic instrumentation to adjust program parameters. With MATE an application is monitored, its performance bottlenecks are detected, their solutions are given, and the application is tuned on the fly to improve its performance. All these steps are performed automatically, dynamically, and continuously during application execution.

MATE uses DynInst [3] to insert instrumentation into running code, collect performance data, and finally tune the application. The fundamental idea is that dynamic analysis and online modifications adapt the application behavior to changing conditions in the execution environment or in the application itself.

MATE consists then of the following components that cooperate to control and improve the execution of the application [14]:

- **Application Controller (AC)** - a daemon-like process that controls the execution and dynamic instrumentation of individual MPI tasks.

- **Dynamic monitoring library (DMLib)** - a library that is dynamically loaded into application tasks to facilitate the performance monitoring and data collection.
- **Analyzer** - a process that carries out the application performance analysis and decides on monitoring and tuning. It automatically detects existing performance problems on the fly and requests appropriate changes to improve the application performance.

MATE uses the functionality required to parse and modify binary executables by means of the DynInst API. It provides a lightweight data collection framework composed of a number of distributed daemon processes and a centralized analyzer process. The centralized performance analyzer is driven by a number of so called tunlets that implement specific performance models and algorithms that evaluate the current behavior and suggest tuning actions of a running application.

MATE has been demonstrated to be an effective and feasible tool to improve performance of real-world applications [13]. An extensive experimental work has been conducted with parallel applications based on master-worker paradigm and automatic tuning of data distribution algorithms like factoring [12]. This algorithm calculates the optimal values of the factoring distribution parameters. These values are later applied to the application using dynamic instrumentation.

MATE is, by design, suitable for any Linux-based cluster environment running MPI applications. In particular, the automatic tuning has been applied to a parallel master-worker implementation of forest fire simulation called XFire developed at UAB [9]. The forest fire propagation is a very complex problem that involves several distinct factors that must be considered: meteorology factors such as temperature, wind, moisture; vegetation features, and terrain topography. The simulation system is used to predict the evolution of the forest fire in different scenarios and help minimize the fire risks and fire damages. Given its complexity, this simulation requires high computing capabilities. The experiments with this highly demanding application and MATE in a cluster environment has proven the benefits from dynamic tuning. However, the tool has been used in this case only on a small size cluster.

3   Scalability Problems

The next step of our research aims to port the existing implementation of MATE to large-scale parallel systems. The objective is to examine and resolve all scalability issues that may appear when running on thousands of processors. The key problems are related to the volume of collected data and the centralized performance analysis [4]. MATE assumes that the performance analysis, based on the global application view, is taking into consideration all the processes and their interactions. Such an approach is feasible for environments with a relatively small number of nodes. However, the centralized analysis becomes a scalability bottleneck if the number of nodes increases. We want to solve this problem by distributing the performance analysis process.

The analyzer component is the main bottleneck of the MATE environment because of the following factors:

- The volume of events to be processed by the analyzer, and the number of connections and AC daemons that have to be managed, increase the tools time response.
- Performance models and thus tuning techniques (tunlets) that we have developed are adequate for a centralized and sequential approach. Although the models are quite simple to make their evaluation easier during execution time, usually the complexity of this evaluation depends on the number of processes. If this dependency is not linear, the scalability will be poor.

Figure 1 presents a master-worker application which is executed with a maximum of 26 workers and 60 iterations under a controlled extra load. Certain variable load is injected in the system to provoke variations in the current conditions and the consequent reactions of MATE to adapt the application to the new conditions. In this case we consider a tunlet to tune the number of workers. As can be seen, the number of workers is adapted as the load pattern changes. As the load in the system increases, the number of workers is changed into a bigger one. Conversely, the number of workers is reduced as the load in the system decreases. However, we can notice that the modifications of the number of workers are delayed several iterations. It is provoked by the amount of workers and events involved in the application execution. Thus, during the whole execution of the application there is a continuous lag between the conditions of the system and the tuning actions.

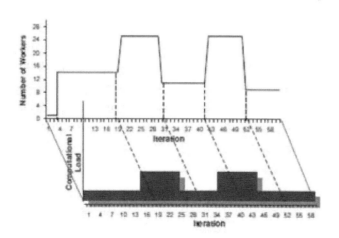

Fig. 1. The scalability problem in MATE

Our proposal is to enhance and extend the use of MATE from two different points of view. First, we want to make MATE scalable with the proposal of overcome the bottleneck presented by MATE when the number of processors involved in the execution of the application increases. Second, we want to automate the creation of tunlets (the inclusion of knowledge in MATE) in order to make the use of this environment easier. We have proposed a solution to this problem based on the definition of a high level specification language and the development of an automatic tunlet generation tool [5].

This paper presents our approach for improving the scalability of MATE by redefining its analysis phase. The novel approach is called the distributed hierarchical collecting-preprocessing approach. As for now, MATE had been following a centralized approach, in which the collection and processing of the information turned into a bottleneck as the amount of processes in the application -and consequently the amount of events- increased.

This approach limited the scalability of MATE. Therefore, we studied different options to provide scalability; however, both distributed and hierarchical approaches presented constraints from the user, the performance model, and the application point of view. Consequently, we selected the best characteristics of each approach in order to provide a viable alternative. The objective of the proposed approach is to overcome the bottleneck of the analyzer, by distributing what can be distributed (the collection of events) and preprocessing what can be processed before the model evaluation. We compare the new approach with the centralized one in order to appreciate the significance of the contribution. We also study the intrusion caused by MATE and the resources requirements.

4   Architecture of Scalable MATE

To overcome these barriers to the scalability, MATE is being developed using overlay networks. An overlay network is a network of computers constructed on top of another network. Nodes in the overlay are connected by virtual or logical links, each of which corresponds to a path in the underlying network. For example, many peer-to-peer networks are overlay networks because they run on top of the Internet. This kind of networks is scalable, flexible, and extensible.

Therefore, to make the MATE environment scalable, we propose to adapt it applying Tree-based Overlay Network inside the MATE infrastructure. TBONs [1] are virtual networks of hierarchically-organized processes that exploit the logarithmic scaling properties of trees to provide efficient data multicast, gather, and data aggregation.

An example implementation of the TBON model is the MRNet framework [17] developed at the University of Wisconsin. MRNet is a software overlay network that provides efficient multicast and reduction communications for parallel and distributed tools and systems. It is useful to build scalable parallel tools as it incorporates a tree hierarchy of processes between the tool's front-end and back-ends to improve group communication performance. These introduced internal processes are also used to distribute many important tool activities that are normally performed by a front-end or tool daemon. As the calculation is distributed and performance data is aggregated, MRNet allows for reducing data analysis time and keeping tool front-end loads manageable.

TBON model allows for developing a new structure of the MATE environment, shown in Figure 2, that will make it scalable.

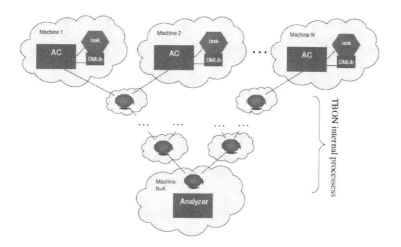

Fig. 2. MATE architecture based on TBON overlay network

MATE sends all events from all processes to the central analyzer. In this case, the data flow will be reduced applying the TBON architecture. A problem related to the number of connections to the global analyzer will be solved by introducing the hierarchy of processes between the ACs and the analyzer. These internal processes will provide event aggregation; each process will be responsible for receiving events generated by a reduced set of daemons, aggregate them, and finally pass them to the superior level of the network. In this way, the aggregated data will arrive to the global analyzer. Each level may aggregate, filter, or sort data, this reduces the data volume and the processing time of the analysis.

All events generated by the application processes are sent through all levels of the TBON (daemon to front-end direction). In the same way, all requests that must be provided to the daemons may be transmitted through this network. For example, a tuning request sent by the global analyzer (front-end) will be sent from a level to an inferior level till the proper AC (daemon) process. Both, AC and analyzer components should be adapted to work with the MRNet library using its API.

However, in case of certain tuning techniques, the TBON usage will not solve all scalability problems. The techniques that require the evaluation of performance metrics calculated over a certain event pattern will remain a bottleneck of the front-end process. As an example, suppose the following situations: calculate the iteration time for each application process as a difference between the start and the end of the iteration or, calculate the delay between a send event from one process and the corresponding receive event on another process. It is impossible to evaluate these metrics directly in the application process as the information required is available only in the global analyzer.

To solve these problems, we propose a new approach based on the distributed evaluation of metrics. The idea is to delegate certain calculations to the internal TBON process (as local analyzers) and thus to unload the global analyzer. Each tunlet must provide arithmetic expressions that characterize the event patterns and distribute these declarations to the TBON nodes. In this way, each TBON node provides a filter detecting required patterns and evaluating given arithmetic expressions. This analysis is distributed and transparent to the global analyzer. Each filter in MRNet may be a dynamically loaded library. Therefore, AC component should be aware of the filter creation (based on the information provided by the tunlet) and its dynamic insertion into TBON nodes.

Moreover, we can apply this solution to other performance models than master-worker. A composition of two (or more) specific patterns gives a solution to a complex parallel application. It implies that the problems existing in one or more of these paradigms may exist in the composed paradigm. Therefore, these problems could be overcome by applying the existing performance models separately for each paradigm. Examples of composed models are: herarchical master-worker and master-worker of pipelines. The first model can be applied in the case of XFire simulator where the data distribution may cause a scalability bottlenecks. The master process may distribute the work to a set of sub-masters and each of them can manage a set of workers. The second model can be applied in a rendering application in order to exploit both functional and data parallelism.

Considering the second one - master-worker of pipelines (where each worker is a pipeline), we can find some parts of the analysis that can be performed independently for each pipeline, and globally for the whole master-worker [8]. Figure 3 presents an example of a master-worker with pipeline composition. It is the typical structure of a master-worker application with five workers, where each worker is a four stages pipeline. The master process sends tasks to each worker and when a task arrives to a worker, this is processed stage by stage of the pipeline. Finally, at the end of the computation, the result is returned to the master process.

In this example we consider a composite performance model that consists of two main levels. The first level is the local analysis, where MATE evaluates the pipeline performance model and tries to improve the execution of each pipeline separately. This part of the optimization uses the metrics provided by each process of the local pipeline and takes into account the resources that each pipeline has assigned. The local analyzer evaluates the performance strategy "Dynamic Pipeline Mapping (DPM)" [15], for the possible reassigment of resources in a pipeline. In this case, each application process is controlled by one AC, all ACs from a pipeline communicates with one TBON node and finally a TBON process with one local analyzer process. Each TBON node provides the collection of events to the local analysis of the corresponding pipeline.

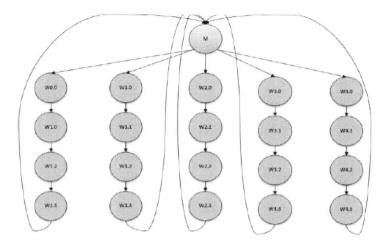

Fig. 3. Composite application example master-worker with pipeline

The second level is a global optimization process, where we can apply a performance model to improve the number of workers required by the application [6]. Then, the local analyzer or TBON processes may transmit all required metrics and results to the superior level. The global analyzer is then responsible for evaluating the given performance model and tune the global number of workers.

Consequently, MATE must be prepared to support this kind of distributed analysis. The architecture of the MATE environment considering the TBON processes and the distributed analysis is presented in Figure 4.

5   Related Work

There is a set of performance analysis tools that manage the scalability problem. However, none of them provides automatic and dynamic tuning of parallel applications. Nevertheless, these tools are already adapted for large scale systems. A well known example is the Scalasca toolset from Jülich Supercomputing Centre [10]. It is a performance analysis tool that has been specifically designed for being used on large-scale systems including Blue Gene and Cray XT. Scalasca integrates both profiling and tracing in a stepwise performance analysis process. It adopts a strategy of refined measurement configurations. It focuses on a feature to identify wait states that occur during the application execution. It was used on more than 65000 processes on the Jugene Blue Gene/P.

The Cray performance analysis tools [7] provide an integrated infrastructure for measurement and analysis of computation, communication, I/O, and memory utilization. It is composed of CrayPat Performance Collector for data capture, and Cray Apprentice2 Performance Analyzer to a post processing data visualization. The Cray performance analysis tools have been used at large scale Cray XT systems with more than 30000 processors.

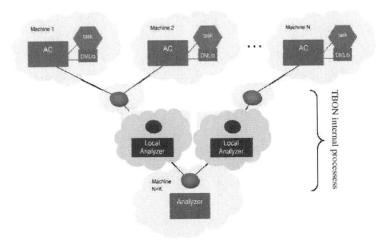

Fig. 4. MATE architecture considering TBON overlay network and distribution analysis

Paradyn was one of the first tools adapted to large scale systems [18]. It is a performance tool for large-scale parallel applications which was developed at University of Wisconsin. It provides monitoring and automatic analysis "on the fly". This tool uses the Dyninst library to instrument the application and then provides automatic performance analysis of the running application. The group that worked on the Paradyn project, developed the MRNet library to improve the scalability of the tool. Paradyn was used in 2006 at large scale systems with more than 1000 nodes at the LLNL.

Periscope [2] is a performance analysis environment which overcomes the scalability barrier performing an automatic distributed online analysis on thousands of processors. Its analysis architecture is based on a hierarchical network of agents. These agents autonomously search performance characteristics on subsets of the parallel application tasks and report the problems back to a master agent.

All these tools manage the scalability problem but do not provide automatic and dynamic tuning of parallel applications. On the other hand, MATE and other tools, such as Autopilot [16] and Active Harmony [19], implement this approach.

Autopilot conducts a performance improvement process using a distributed analysis system of sensors and actuators. Sensors monitor the parallel application obtaining data about its behavior, and actuators make modifications to application variables to improve its performance.

Active Harmony is a framework that allows dynamic adaptation of an application to the network and available resources on which it is running. The main component of its architecture, the Adaptation Controller, is responsible for tuning the application to achieve improved performance. With this objective, it switches algorithms or tunes the parameters that characterize the library used by the application, or directly modifies parameters that characterize the application performance.

However, these tools have features in its architecture that make them not scalable to applications involving hundreds or thousands of processes.

6   Conclusions and Future Work

In this paper, we have presented the automatic and dynamic tuning environment MATE. As in the last years, many large scale systems - composed of thousands of processors - have appeared, a tool scalability problem has arisen. MATE and many other tools are limited in such systems as they are developed using back-end and front-end architecture with many processes controlling the application execution in a centralized manner. Therefore, we decided to indicate the set of bottlenecks of MATE and adapt it to these new large scale systems.

We have demonstrated the possible architecture of MATE considering communication issues. In this case, the utilization of a TBON network is reasonable. We have also presented the further step we plan to follow: a distributed analysis. On the one hand, there are calculations that can be performed locally and then passed to a global analysis. On the other hand, there are certain composed performance paradigms that may be analyzed separately on two levels: local and global (master-worker of pipeline).

Currently we are developing a new version of MATE adapted to large scale systems. The experimentation tests must be performed on thousands of processors. Once the MATE environment is scalable, we have to integrate the distributed analysis components with TBON and perform experimentation using applications with composed paradigms.

Acknowledgments. This research has been supported by the MEC-MICINN Spain under contract TIN2007-64974.

References

1. Arnold, D.C., Pack, G.D., Miller, B.P.: Tree-based Overlay Networks for Scalable Applications. In: 11th International Workshop on High-Level Parallel Programming Models and Supportive Environments (HIPS 2006), Rhodes, Greece (2006)
2. Benedict, S., Petkov, V., Gerndt, M.: PERISCOPE: An Online-based Distributed Performance Analysis Tool. In: Proc. 3rd International Workshop on Parallel Tools for High Performance (2009)
3. Buck, B., Hollingsworth, J.: An API for Runtime Code Patching. International Journal of High Performance Computing Applications 14, 317–329 (2000)
4. Caymes-Scutari, P.: Extending the Usability of a Dynamic Tuning Environment. Ph.D. thesis, Universitat Autònoma de Barcelona (2007)
5. Caymes-Scutari, P., Morajko, A., Margalef, T., Luque, E.: Automatic Generation of Dynamic Tuning Techniques. In: Kermarrec, A.-M., Bougé, L., Priol, T. (eds.) Euro-Par 2007. LNCS, vol. 4641, pp. 13–22. Springer, Heidelberg (2007)

6. César, E., Moreno, A., Sorribes, J., Luque, E.: Modeling Master/Worker Applications for Automatic Performance Tuning. Parallel Computing 32, 568–589 (2006)
7. DeRose, L., Homer, B., Johnson, D., Kaufmann, S., Poxon, H.: Cray Performance Analysis Tools. In: Tools for High Performance Computing, pp. 191–199 (2008)
8. Guevara Quintero, J.: Definition of a Resource Management Strategy for Dynamic Performance Tuning of Complex Applications. Ph.D. thesis, Universitat Autònoma de Barcelona (2010)
9. Jorba, J., Margalef, T., Luque, E., André, J., Viegas, D.: Application of Parallel Computing to the Simulation of Forest Fire Propagation. In: 3rd International Conference in Forest Fire Propagation, vol. 1, pp. 891–900 (1998)
10. Mohr, B., Wylie, B.J.N., Wolf, F.: Performance Measurement and Analysis Tools for Extremely Scalable Systems. Concurrency and Computation: Practice and Experience 22, 2212–2229 (2010)
11. Morajko, A.: Dynamic Tuning of Parallel/Distributed Applications. Ph.D. thesis, Universitat Autònoma de Barcelona (2004)
12. Morajko, A., Caymes, P., Margalef, T., Luque, E.: Automatic Tuning of Data Distribution Using Factoring in Master/Worker Applications. In: Sunderam, V.S., van Albada, G.D., Sloot, P.M.A., Dongarra, J. (eds.) ICCS 2005. LNCS, vol. 3515, pp. 132–139. Springer, Heidelberg (2005)
13. Morajko, A., Caymes-Scutari, P., Margalef, T., Luque, E.: MATE: Monitoring, Analysis and Tuning Environment for parallel/distributed applications. Concurrency and Computation: Practice and Experience 19, 1517–1531 (2007)
14. Morajko, A., Margalef, T., Luque, E.: Design and Implementation of a Dynamic Tuning Environment. Journal of Parallel and Distributed Computing 67(4), 474–490 (2007)
15. Moreno, A., César, E., Guevara, A., Sorribes, J., Margalef, T., Luque, E.: Dynamic Pipeline Mapping (DPM). In: Luque, E., Margalef, T., Benítez, D. (eds.) Euro-Par 2008. LNCS, vol. 5168, pp. 295–304. Springer, Heidelberg (2008), http://dx.doi.org/10.1007/978-3-540-85451-7_32
16. Ribler, R., Vetter, J., Simitci, H., Reed, D.A.: Autopilot: Adaptive Control of Distributed Applications. In: Proc. of IEEE Symposium on HPDC, pp. 172–179 (1998)
17. Roth, P.C., Arnold, D.C., Miller, B.P.: MRNet: A software-based multicast/reduction network for scalable tools. In: Proc. IEEE/ACM Supercomputing 2003, p. 21 (2003)
18. Roth, P.C., Miller, B.P.: On-line automated performance diagnosis on thousands of processes. In: Proc. of the Eleventh ACM SIGPLAN Symposium on Principles and Practice of Parallel Programming, PPoPP 2006, pp. 69–80. ACM, New York (2006), http://doi.acm.org/10.1145/1122971.1122984
19. Tapus, C., Chung, I.H., Hollingsworth, J.: Active Harmony: Towards Automated Performance Tuning. In: Proc. from the Conference on High Performance Networking and Computing, pp. 1–11 (2003)

Improving the Scalability
of Performance Evaluation Tools

Sameer Suresh Shende, Allen D. Malony, and Alan Morris

Performance Research Laboratory
Department of Computer and Information Science
University of Oregon, Eugene, OR, USA
{sameer,malony,amorris}@cs.uoregon.edu

Abstract. Performance evaluation tools play an important role in helping understand application performance, diagnose performance problems
and guide tuning decisions on modern HPC systems. Tools to observe
parallel performance must evolve to keep pace with the ever-increasing
complexity of these systems. In this paper, we describe our experience in
building novel tools and techniques in the TAU Performance System®
to observe application performance effectively and efficiently at scale.
It describes the extensions to TAU to contend with large data volumes
associated with increasing core counts. These changes include new instrumentation choices, efficient handling of disk I/O operations in the
measurement layer, and strategies for visualization of performance data
at scale in TAU's analysis layer, among others. We also describe some
techniques that allow us to fully characterize the performance of applications running on hundreds of thousands of cores.

Keywords: Measurement, instrumentation, analysis, performance tools.

1 Introduction

Tools for parallel performance measurement and analysis are important for evaluating the effectiveness of applications on parallel systems and investigating
opportunities for optimization. Because they executes as part of the parallel program and process performance data that reflects parallel behavior, measurement
and analysis techniques must evolve in their capabilities to address the complexity demands of high-end computing. Scaling in the degree of parallelism is one of
the key driving requirements for next-generation applications. To address scaling
concerns, performance tools can not continue with traditional techniques without considering the impacts of measurement intrusion, increased performance
data size, data analysis complexity, and presentation of performance results.

This paper discusses approaches for improving the scalability of the TAU
Performance System® instrumentation, measurement, and analysis tools. Our
perspective looks generally at the maximizing performance evaluation return
to the tool user. This starts with improving the instrumentation techniques to

K. Jónasson (Ed.): PARA 2010, Part II, LNCS 7134, pp. 441–451, 2012.
© Springer-Verlag Berlin Heidelberg 2012

select key events interest and avoid the problem of blindly generating a lot of low value performance data. Section §2 presents a few of the new techniques in TAU to support more flexible and intelligent instrumentation. Performance measurement presents a scaling challenge for tools as it places hard requirements on overhead and efficiency. Section §3 describes TAU's new measurement capabilities for addressing scale. Scalable performance analysis deals mainly with concerns of reducing large performance data into meaningful forms for the user. Recent additions to TAU performance analysis are discussed in Section §4. The paper concludes with thoughts towards future extreme-scale parallel machines.

2   Instrumentation

2.1   Source Instrumentation

For probe-based performance measurement systems such as TAU, instrumentation is the starting point for thinking about scalability because it is where decisions are made about what to observe. It is also where automation becomes important for tool usability. TAU has traditionally relied on a source-to-source translation tool to instrument the source code. Based on the PDT (Program Database Toolkit) [6] static analysis system, the tau_instrumentor [4] tool can insert instrumentation for routines, outer loops, memory, phases, and I/O operations in the source code. While source code instrumentation provides a robust and a portable mechanism for instrumentation, it does require re-compiling the application to insert the probes. While not directly affected by scaling, source instrumentation can become somewhat cumbersome in optimizing instrumentation for efficient measurement.

In the past, TAU addressed the need to re-instrument the source code by supporting runtime instrumentation (via the tau_run command) using binary editing capabilities of the DyninstAPI [2] package. However, dynamic instrumentation requires runtime support to be efficient at high levels of parallelism since every executable image would need to be modified. Techniques have been developed in ParaDyn to use a multicast reduction network [9] for dynamic instrumentation control, as well as in TAU to use a startup shell script that is deployed on every MPI process and then instruments and spawns an executable image prior to execution [10,7].

2.2   Binary Instrumentation

To address both usability and scalability issues, we implemented binary instrumentation in TAU using re-writing capabilities of DyninstAPI. This allows us to pre-instrument an executable instead of sending an uninstrumented executable to a large number of cores just to instrument it there. By re-writing the executable code using binary editing, probes can be inserted at routine boundaries pre-execution, saving valuable computing resources associated with spawning a DyninstAPI-based instrumentor on each node to instrument the application. The approach improves the startup time and simplifies the usage of TAU as no changes are introduced to the application source code or the build system.

Name	Total ▽	NumSamples	MaxValue	MinValue	MeanValue	Std. Dev.
Bytes Read	217,558	151	4,944	2	1,440.781	995.142
Bytes Written	208,104	119	2,070	8	1,748.773	737.408
Message size for all-reduce	102,400	50	2,048	2,048	2,048	0
Bytes Read <file="socket">	39,016	34	4,944	8	1,147.529	1,194.223
Bytes Read <file="10.1.1.222,port=57864">	36,710	20	2,070	8	1,835.5	605.793
Bytes Written <file="socket">	36,016	26	2,070	8	1,385.231	942.039
Bytes Written <file="10.0.1.222,port=52411">	35,348	19	2,070	8	1,860.421	611.456

Fig. 1. Using a POSIX I/O interposition library, tau_exec shows the volume of I/O in an uninstrumented application

2.3   Instrumentation via Library Wrapping

While this is a solution for binary executables, shared libraries (dynamic shared objects or DSOs) used by the application cannot be re-written at present. If the source code for the DSO is unavailable, we are left with a hole in performance observation. To enable instrumentation of DSOs, we created a new tool, tau_wrap, to automate the generation of wrapper interposition libraries. It takes as input the PDT-parsed representation of the interface of a library (typically provided by a header file) and the name of the runtime library to be instrumented. The tau_wrap tool then automatically generates a wrapper interposition library by creating the source code and the build system for compiling the instrumented library. Each wrapped routine first calls the TAU measurement system and then invokes the DSO routine with the original arguments. The wrapper library may be preloaded in the address space of the application using the tau_exec tool that also supports tracking I/O, memory and communication operations[12]. Preloading of instrumented libraries is now supported on the IBM BG/P and Cray XT5/XE6 architectures.

The ability to enable multiple levels of instrumentation in TAU (as well as runtime measurement options) gives the user powerful control over performance observation. Figure 1 shows how the volume of read and write I/O operations can be tracked by TAU using library-level preloading of the POSIX I/O library in *tau_exec*. If callpath profiling is enabled in TAU measurement, a complete summary of all operations on individual files, sockets, or pipes along a program's callpath can be generated, as shown in Figure 2. Here we have instrumented MPI operations (through the standard MPI wrapper interposition approach). In doing so, we can now see how MPI routines like MPI_Allreduce invoke low-level communications functions, typically unobserved in other performance tools.

The linker is another avenue for re-directing calls for a given routine with a wrapped counterpart. For instance, using the GNU ld --wrap routine_name commandline flag, we can surround a routine with TAU instrumentation. However, this approach requires each wrapped routine to be specified on the link line. While onerous, this could be automated (and TAU does), but one may exceed the system limits for the length of the command line if a lot of routines are desired. Using a combination of wrapped instrumentation libraries with re-linked or re-written binaries provides complete coverage of application and system library routines without access to the source code of the application.

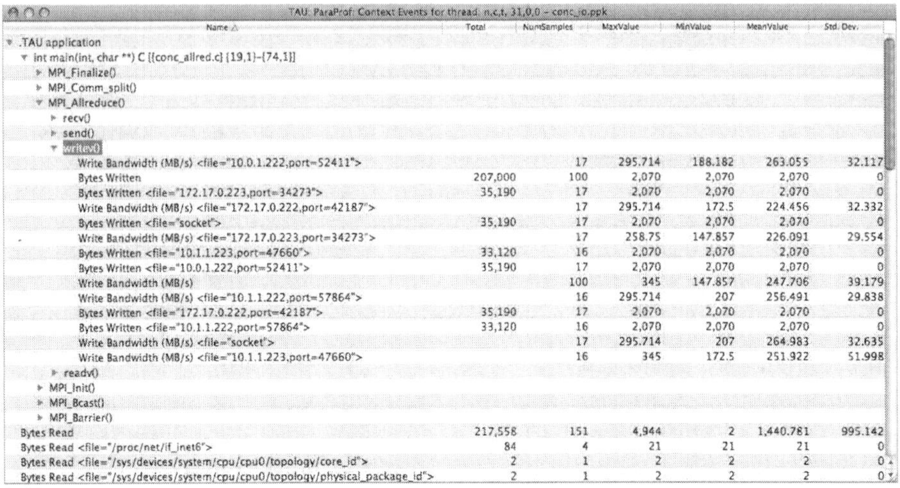

TAU: ParaProf: Context Events for thread: n,c,t, 31,0,0 - conc_io.ppk

Name	Total	NumSamples	MaxValue	MinValue	MeanValue	Std. Dev.
.TAU application						
int main(int, char **) C [{conc_allred.c} {19,1}-{74,1}]						
MPI_Finalize()						
MPI_Comm_split()						
MPI_Allreduce()						
recv()						
send()						
writev()						
Write Bandwidth (MB/s) <file="10.0.1.222,port=52411">		17	295.714	188.182	263.055	32.117
Bytes Written	207,000	100	2,070	2,070	2,070	0
Bytes Written <file="172.17.0.223,port=34273">	35,190	17	2,070	2,070	2,070	0
Write Bandwidth (MB/s) <file="172.17.0.222,port=42187">		17	295.714	172.5	224.456	32.332
Bytes Written <file="socket">	35,190	17	2,070	2,070	2,070	0
Write Bandwidth (MB/s) <file="172.17.0.223,port=34273">		17	258.75	147.857	226.091	29.554
Bytes Written <file="10.1.1.223,port=47660">	33,120	16	2,070	2,070	2,070	0
Bytes Written <file="10.0.1.222,port=52411">	35,190	17	2,070	2,070	2,070	0
Write Bandwidth (MB/s)		100	345	147.857	247.706	39.179
Write Bandwidth (MB/s) <file="10.1.1.222,port=57864">		16	295.714	207	256.491	29.838
Bytes Written <file="172.17.0.222,port=42187">	35,190	17	2,070	2,070	2,070	0
Bytes Written <file="10.1.1.222,port=57864">	33,120	16	2,070	2,070	2,070	0
Write Bandwidth (MB/s) <file="socket">		17	295.714	207	264.983	32.635
Write Bandwidth (MB/s) <file="10.1.1.223,port=47660">		16	345	172.5	251.922	51.998
readv()						
MPI_Init()						
MPI_Bcast()						
MPI_Barrier()						
Bytes Read	217,558	151	4,944	2	1,440.781	995.142
Bytes Read <file="/proc/net/if_inet6">	84	4	21	21	21	0
Bytes Read <file="/sys/devices/system/cpu/cpu0/topology/core_id">	2	1	2	2	2	0
Bytes Read <file="/sys/devices/system/cpu/cpu0/topology/physical_package_id">	2	1	2	2	2	0

Fig. 2. I/O statistics along a calling path reveal the internal workings of MPI library showing the extent of data transfers for each socket and file accessed by the application

2.4 Source and Compiler Instrumentation for Shared Libraries

When the source code is available for instrumentation, direct source instrumentation of static or shared libraries can be done automatically using TAU's compiler scripts (`tau_cxx.sh`, `tau_cc.sh`, and `tau_f90.sh`). The purpose of these scripts is to replace host compilers in the build system without disrupting any of the rest of the build process. TAU also supports compiler-based instrumentation where the compiler emits instrumentation code directly while creating an object file. This is supported for IBM, Intel, PGI, GNU, and Pathscale compilers at present. Source-based instrumentation involves deciphering and injecting routine names as parameters to timer calls in a copy of the original source code. While shared object instrumentation is relatively easy to implement in source-based instrumentation, it poses some unique challenges in identifying routine names for compiler-based instrumentation. Compiler-based instrumentation in statically linked code is easier to implement because the address of routines does not change during execution and is the same across all executing contexts. The address may be mapped to a name using BFD routines at any point during the execution (notably at the end of the execution).

On the other hand, dynamic shared objects, by their very nature, load position-independent object code at addresses that are assigned from an offset using a runtime loader. The same routine may be loaded at a different address in different executing contexts (ranks). Also, as the application executes, different shared objects represented by Python modules may be loaded and unloaded, and the map that correlates addresses to the routine names changes during the execution of the application. This address map (typically stored in the `/proc` file system under Linux) cannot be queried at the end of the execution as the addresses may be re-used and shift as different shared objects are brought in and out of the executing process.

To handle instrumentation of shared objects, we examine the address ranges for the different routines after loading a shared object and determine the mapping of routines names and their addresses, dynamically in each context. This simplifies object code instrumentation in dynamic shared objects and we need only store these address mappings for shared objects that are loaded during execution. During execution, compiler-based instrumentation generates events and calls the measurement library. Events from C and Fortran languages typically map directly to their routine names. C++ events need an additional demangling step. Events from multiple layers co-exist in the executing context and performance data is generated by each context separately.

Providing robust support for selecting events to observe is important for giving optimal visibility of performance. TAU integrates several instrumentation approaches in a cohesive manner allowing a user to slice and examine the performance across multiple application layers at an arbitrary level of detail. By providing access to instrumentation hooks at multiple levels of program transformation, the user can refine the focus of instrumentation to just the relevant part while reducing the overhead by not instrumenting application constructs that may not be pertinent to a given performance experiment, thereby reducing the volume of the performance data generated.

3   Measurement

The scaling of parallel performance measurement must meet critical requirements. Most importantly, it must impact application's performance a little as possible. However the choice of what and how to measure is not that simple. Every performance measurement system will intrude on the execution. It is important then to optimize the balance between the need for performance data and the cost of obtaining it. Our goal in TAU is to provide flexible support for making optimal choices concerning measurement type and degree.

3.1   Parallel Profiling

TAU provides both parallel profiling and tracing in its measurement system. Parallel profiling characterizes the behavior of every application thread in terms of its aggregate performance metrics such as total exclusive time, inclusive time, number of calls, and child calls executed. A rich set of profiling functionality is available in TAU, including callpath profiling, phase profiling, and parameter-based profiling, that offers choices in scope and granularity of performance measurement. Although parallel profiling records minimal temporal information, it is the recommend first measurement choice in TAU because it allows significant performance characterization and runtime performance data is of a fixed size. All profiling measurements take place in a local context of execution and do not involve synchronization or communication. This keeps it lightweight in overhead and intrusion even as the number of threads of execution scales.

However, the largest systems available now have exceeded many of the traditional means of profile data output, collection, and analysis. Tools like TAU have historically written process-local profile files. This method no longer scales to the largest systems since it creates challenges at multiple levels. It can excessively slow down the execution of the application job by creating potentially hundreds of thousands of files. The metadata operations to simply create this number of files have been shown to be a significant bottleneck [3]. After execution, the huge number of files is very difficult to manage and transfer between systems. A more subtle problem is that TAU assigns event identifiers dynamically and locally. This means that the same event can have different IDs in different threads. Event unification has typically been done in TAU in the analysis tools. Unfortunately, this requires verbose and redundant event information to be written with the profiles. Thus, not only do we end up with multiple profile files, they contain excessive information.

The TAU project has been investigating these two issues for the past year. We currently have prototype parallel implementations of *event unification* and *profile merging*. These are built from a MPI-based parallel profile analyzer that runs at the end of the application execution[15]. By using an efficient reduction layer based on a binomial heap, the unification and merging operations are implementation is portable and fast. We have tested it on over 100,000 cores on a Cray XT5 and IBM BG/P. More generally, we are looking to improve the scalability of online profile-based performance measurement. A TAU monitoring system is being implemented that uses scalable infrastructure such as MRNet to provide runtime access to parallel performance data [8,15].

Moving forward, we plan to implement various forms of on-the-fly analysis at the end of application execution, to reduce the burden on the post-mortem analysis tools, and online, to provide data reduction and feedback to the live application. For post-mortem analysis purposes, a new file format will be designed to contain multiple levels of detail and pre-computed derived data (e.g., from the runtime parallel analysis). This will allow the analysis tools the ability to read only the portions of the overall profile that they need for a given analysis or data view. In these way, we are confident that we can address the issues of large scale profile collection and analysis.

3.2   Parallel Tracing

In contrast to profiling, tracing generates a timestamped event log that shows the temporal variation of application performance. TAU traces can be merged and converted to the Vampir's [1] Open Trace Format (OTF), Scalasca's Epilog [5], Paraver [13], or Jumpshot's SLOG2 trace formats. Merging and conversion of trace files is an expensive operation at large core counts. To reduce the time for merging and conversion, and to provide more detailed event information, TAU interfaces with the Scalasca and VampirTrace libraries directly. VampirTrace provides a trace unification phase at the end of execution that requires re-writing binary traces with updated global event identifiers. However, this can be an expensive operation at large scale.

In the near future, TAU will write OTF2 traces natively using the Score-P measurement library from the SILC[14] project. It will feature an efficient trace unification system that only re-writes global event identifier tables instead of re-writing the binary event traces. If the trace visualizer supports the OTF2 format, it will also eliminate the need to convert these large trace files from one format to another. This will improve the scalability of the tracing system.

3.3   Measuring MPI Collective Operations

As applications are re-engineered to run on ever increasing machine sizes, tracking performance of the collective operations on the basis of individual MPI communicators becomes more important. We have recently introduced tracking of MPI communicators in TAU's profiling substrate using its mapping capabilities in parameter-based profiling. TAU partitions the performance data on the basis of its communicator in a collective operation. Each communicator is identified by the list of MPI ranks that belong to it. When multiple communicators use the same set of ranks, the TAU output distinguishes each communicator based on its address. Figure 3 shows the breakdown of the average time spent in the `MPI_Allreduce` routine based on each set of communicators across all 32 ranks in an MPI application. To contend with large core counts, TAU only displays the first eight ranks in a communicator, although this depth may be altered by the user while configuring TAU's measurement library. This is shown for the `MPI_Bcast` call where all ranks participate in the broadcast operation on the `MPI_COMM_WORLD` communicator.

4   ParaProf Scalable Analysis

Scalable performance measurement only produces the performance data. It still needs to be analyzed. Analysis scalability concerns the exploration of potentially large parallel performance datasets. The TAU ParaProf parallel performance analyzer is specifically built for analysis of large scale data from the largest leadership class machines. It can easily analyze full size datasets on common desktop workstations. TAU provides a compressed, normalized, packed data format (ParaProf Packed format, .ppk) as a container for profile data from any supported measurement tool. This makes reading of parallel profiles significantly more efficient in ParaProf.

Analysis in ParaProf takes place in-memory for fast access and to support global aggregation and analysis views. Basic bar charts support large dataset with standard scrollbars allowing the detail for each node/thread to be seen in its own context. Additionally, we present aggregate statistics such as the mean and standard deviation. Aggregate views such as the histogram display allow a simplified view of the entire dataset in a single chart.

TAU: ParaProf: Mean Statistics - conc_allred.32p.ppk

Name	Exclusive TIME	Inclusive TIME	Inclusive TIME / Call
MPI_Allreduce()	17,195.69	17,196.47	343.93
MPI_Allreduce() [<comm> = <ranks: 0, 8, 16, 24> <addr=0x1ab800a0>]	533.94	533.94	341.72
MPI_Allreduce() [<comm> = <ranks: 0, 8, 16, 24> <addr=0x17de0ef0>]	538.41	538.41	344.58
MPI_Allreduce() [<comm> = <ranks: 0, 8, 16, 24> <addr=0xc58d0c0>]	533.66	533.66	341.54
MPI_Allreduce() [<comm> = <ranks: 0, 8, 16, 24> <addr=0xdb64050>]	530.62	530.62	339.60
MPI_Allreduce() [<comm> = <ranks: 1, 9, 17, 25> <addr=0x1a31a070>]	532.47	532.47	340.78
MPI_Allreduce() [<comm> = <ranks: 1, 9, 17, 25> <addr=0x11ee10d0>]	531.00	531.00	339.84
MPI_Allreduce() [<comm> = <ranks: 1, 9, 17, 25> <addr=0xa29ef0>]	534.94	534.94	342.36
MPI_Allreduce() [<comm> = <ranks: 1, 9, 17, 25> <addr=0xdac0fb0>]	538.09	538.09	344.38
MPI_Allreduce() [<comm> = <ranks: 2, 10, 18, 26> <addr=0x10b36fd0>]	541.66	541.66	346.66
MPI_Allreduce() [<comm> = <ranks: 2, 10, 18, 26> <addr=0x16a0def0>]	537.91	537.91	344.26
MPI_Allreduce() [<comm> = <ranks: 2, 10, 18, 26> <addr=0x12438000>]	535.47	535.47	342.70
MPI_Allreduce() [<comm> = <ranks: 2, 10, 18, 26> <addr=0xa6bf070>]	534.12	534.12	341.84
MPI_Allreduce() [<comm> = <ranks: 3, 11, 19, 27> <addr=0x6cd70d0>]	549.62	549.62	351.76
MPI_Allreduce() [<comm> = <ranks: 3, 11, 19, 27> <addr=0x9f5ffd0>]	555.09	555.09	355.26
MPI_Allreduce() [<comm> = <ranks: 3, 11, 19, 27> <addr=0x238c070>]	550.78	550.78	352.50
MPI_Allreduce() [<comm> = <ranks: 3, 11, 19, 27> <addr=0x11146ef0>]	560.97	560.97	359.02
MPI_Allreduce() [<comm> = <ranks: 4, 12, 20, 28> <addr=0x1a83a000>]	516.84	516.84	330.78
MPI_Allreduce() [<comm> = <ranks: 4, 12, 20, 28> <addr=0x1284f070>]	523.38	523.38	334.96
MPI_Allreduce() [<comm> = <ranks: 4, 12, 20, 28> <addr=0xc939fd0>]	523.88	523.88	335.28
MPI_Allreduce() [<comm> = <ranks: 4, 12, 20, 28> <addr=0xe214ef0>]	528.34	528.34	338.14
MPI_Allreduce() [<comm> = <ranks: 5, 13, 21, 29> <addr=0x1b2c8ef0>]	519.31	519.31	332.36
MPI_Allreduce() [<comm> = <ranks: 5, 13, 21, 29> <addr=0x1194f000>]	515.12	515.12	329.68
MPI_Allreduce() [<comm> = <ranks: 5, 13, 21, 29> <addr=0x4003070>]	512.88	512.88	328.24
MPI_Allreduce() [<comm> = <ranks: 5, 13, 21, 29> <addr=0xdf9fd0>]	520.78	520.78	333.30
MPI_Allreduce() [<comm> = <ranks: 6, 14, 22, 30> <addr=0x191b6ef0>]	550.69	550.69	352.44
MPI_Allreduce() [<comm> = <ranks: 6, 14, 22, 30> <addr=0x1406b000>]	548.19	548.19	350.84
MPI_Allreduce() [<comm> = <ranks: 6, 14, 22, 30> <addr=0x4772fd0>]	554.53	554.53	354.90
MPI_Allreduce() [<comm> = <ranks: 6, 14, 22, 30> <addr=0x4003070>]	542.16	542.16	346.98
MPI_Allreduce() [<comm> = <ranks: 7, 15, 23, 31> <addr=0x2f08ef0>]	551.62	551.62	353.04
MPI_Allreduce() [<comm> = <ranks: 7, 15, 23, 31> <addr=0x15f95f50>]	546.59	546.59	349.82
MPI_Allreduce() [<comm> = <ranks: 7, 15, 23, 31> <addr=0x4772fd0>]	554.00	554.00	354.56
MPI_Allreduce() [<comm> = <ranks: 7, 15, 23, 31> <addr=0x6055000>]	549.41	549.41	351.62
MPI_Barrier()	1,610.78	1,610.78	805.39
MPI_Bcast()			
MPI_Bcast() [<comm> = <ranks: 0, 1, 2, 3, 4, 5, 6, 7 ...> <addr=0x683f40>]	2,299.78	2,299.78	2,299.78

Fig. 3. ParaProf's shows the peformance of a collective operation partitioned by the communicator

ParaProf uses OpenGL-based 3D visualization support to enhance the interpretation of large-scale performance data. Here, millions of data elements can be visualized at once and be manipulated in real time. We provide triangle mesh displays, 3d bar plots, and scatterplots, all with width, height, depth, and color to provide 4 axes of differentiable data values. For instance, Figure 4 shows a ParaProf 3D view of the entire parallel profile for the XBEC application on 128K core of an IBM BG/P. Figure 5(left) is an example of ParaProf's new 3D communication matrix view showing the volume of point-to-point interprocessor communication between sender and receiver tasks. Although this is for a smaller execution, parallel programs larger than 2k processors will necessarily require such a 3D communications perspective.

Internally, the performance data representation in ParaProf is kept as minimally as possible. Rather than store NxM tables of performance data for each region and node, we keep sparse lists to allow for differing regions on each node. Our goal is to apply this to all visualization options where complete information is being rendered. However, it is also possible to conduct various forms of data dimensionality analysis and reduction. We have implemented several scalable analysis operations, including averaging, histogramming, and clustering.

To validate the scalability of TAU's paraprof profile browser, we synthesized a large one million core profile dataset by replicating a 32k core count dataset

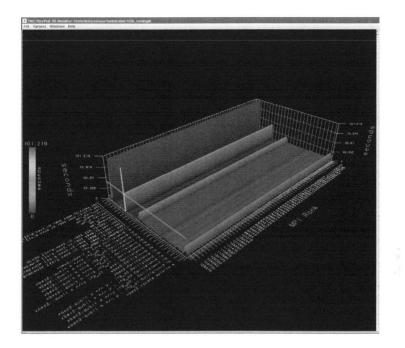

Fig. 4. ParaProf 3D browser shows the profile of a code running on 128k cores

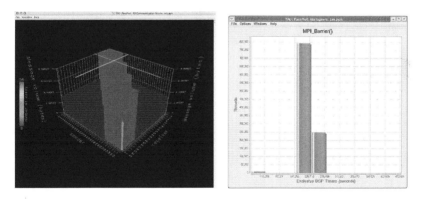

Fig. 5. Left: ParaProf's 3D communication matrix shows the volume of communication between a pair of communicating tasks. Right: ParaProf's histogram display showing the performance of MPI_Barrier in a synthesized 1 million core count profile dataset.

repeatedly. While it is cumbersome to scroll through a million lines representing individual MPI ranks, TAU's histogram displays are useful in highlighting the performance variation of a routine across multiple cores. Figure 5(right) shows a histogram display of the distribution of threads based on their MPI_Barrier execution time. The number of bins partitions the range of the chose performance metric for an event, and this can be selected by the user. Our goal here was to

```
                        TAU: ParaProf: n,c,t 0,0,0 - mat1k.ppk
Metric: TIME
Value: Exclusive percent

68.044% ▇▇▇▇▇▇▇▇
        24.417% ▇▇▇▇        __pgi_cu_launch multiply_matrices (pgi_kernel_7,gx=32,gy=32,gz=1,bx=16,by=16,bz=1) [{mm2.f90}{15}]
            3.206% ▇                __pgi_cu_init multiply_matrices [{mm2.f90}{9}]
            1.572% ▇                __pgi_cu_download2 multiply_matrices var=a [{mm2.f90}{20}]
            1.572% ▯                __pgi_cu_upload2 multiply_matrices var=b [{mm2.f90}{9}]
            0.782% ▮                __pgi_cu_upload2 multiply_matrices var=c [{mm2.f90}{9}]
            0.142%                  mymatrixmultiply [{mmdriv.f90} {1,0}]
            0.122%                  __pgi_cu_launch multiply_matrices (pgi_kernel_2,gx=32,gy=32,gz=1,bx=16,by=16,bz=1) [{mm2.f90}{11}]
            0.12%                   __pgi_cu_free multiply_matrices [{mm2.f90}]
            0.017%                  __pgi_cu_alloc multiply_matrices [{mm2.f90}{9}]
            0.005%                  multiply_matrices [{mm2.f90} {5,0}]
            0.002%                  pgi accelerator region
            2.1E-4%                 __pgi_cu_module multiply_matrices [{mm2.f90}{9}]
            1.2E-4%                 __pgi_cu_module_function multiply_matrices [{mm2.f90}{11}]
            1.2E-4%                 __pgi_cu_paramset multiply_matrices [{mm2.f90}]
                                    __pgi_cu_module_function multiply_matrices [{mm2.f90}{15}]
```

```
                        TAU: ParaProf: Thread Statistics: n,c,t, 0,0,0 - mat1k.ppk
Name                                                                         Exclusive TIME ▽   Inclusive TIME   Calls   Child Calls
__pgi_cu_launch multiply_matrices (pgi_kernel_7,gx=32,gy=32,gz=1,bx=16,by=16,bz=1) [{mm2.f90}{15}]   10.901          10.901           5       0
__pgi_cu_init multiply_matrices [{mm2.f90}{9}]                                3.912          3.912            5       0
__pgi_cu_download2 multiply_matrices var=a   Show Source Code                 0.514          0.514            5       0
__pgi_cu_upload2 multiply_matrices var=b [{   Show Function Bar Chart         0.252          0.252            5       0
__pgi_cu_upload2 multiply_matrices var=c [{   Show Function Histogram         0.252          0.252            5       0
mymatrixmultiply [{mmdriv.f90} {1,0}]         Assign Function Color           0.125          16.021           1       1
__pgi_cu_launch multiply_matrices (pgi_kernel_2,gx=32,gy=32,gz=1,bx=16,bz=1) [{mm2.f90}{11}]   0.023    0.023            5       0
__pgi_cu_free multiply_matrices [{mm2.f90}]   Reset to Default Color          0.02           0.02            15       0
__pgi_cu_alloc multiply_matrices [{mm2.f90}{9}]                               0.019          0.019           15       0
multiply_matrices [{mm2.f90} {5,0}]                                           0.003          15.895           5       0
pgi accelerator region                                                        0.001          15.893           5       85
__pgi_cu_module multiply_matrices [{mm2.f90}{9}]                              0              0                5       0
__pgi_cu_module_function multiply_matrices [{mm2.f90}{11}]                     0              0                5       0
__pgi_cu_paramset multiply_matrices [{mm2.f90}]                               0              0               10       0
```

Fig. 6. ParaProf's shows the profile of a kernel executing on a GPGPU using PGI's runtime library instrumentation

ensure that the profile browsers were capable of handling large data volumes and able to handle displays of millions of cores. We do not have access to machines with a million cores at present, but such large scale machines are being built and will be available in the near future.

5   Conclusions and Future Work

Scaling will continue to be a dominant concern in high-end computing, especially as attention turns towards exascale platforms for science and engineering applications. High levels of concurrent execution on upwards of one million cores are being forecast by the community. Parallel performance tools must continue to be enhanced, re-engineered, and optimized to meet these scaling challenges in instrumentation, measurement, and analysis.

Scaling is not the only concern. Future HPC systems will likely rely on heterogeneous architectures comprised of accelerator components (GPGPU). This will require development of performance measurement and analysis infrastructure to understand parallel efficiency of the application at all levels of execution. We are working closely with compiler vendors (such as PGI and CAPS Entreprise HMPP) to target instrumentation of accelerator kernels at the runtime system level. Using weak bindings of key runtime library events, TAU can intercept and track the time spent in key events as they execute on the host. For instance, Figure 6 shows the time spent in launching individual kernels on the GPGPU as well as the time spent in transferring data from the host memory to the memory of the GPGPU. Variable names as well as source locations are shown in the profile display.

However, in general, the heterogeneous environment will dictate what is possible for performance observation. The challenge for heterogeneous performance tools will be to capture performance data at all levels of execution and integrate that information into consistent, coherent representation of performance for analysis purposes. Heterogeneity introduces issues such as asynchronous, overlapped concurrency between the CPU and accelerator devices, and potentially limited performance measurement visibility, making solutions to this challenge difficult.

References

1. Brunst, H., Kranzlmüller, D., Nagel, W.E.: Tools for Scalable Parallel Program Analysis - Vampir NG and DeWiz. In: Distributed and Parallel Systems, Cluster and Grid Computing, vol. 777 (2004)
2. Buck, B., Hollingsworth, J.: An API for Runtime Code Patching. Journal of High Performance Computing Applications 14(4), 317–329 (2000)
3. Frings, W., Wolf, F., Petkov, V.: Scalable Massively Parallel I/O to Task-Local Files. In: Proc. SC 2009 Conference (2009)
4. Geimer, M., Shende, S.S., Malony, A.D., Wolf, F.: A Generic and Configurable Source-Code Instrumentation Component. In: Allen, G., Nabrzyski, J., Seidel, E., van Albada, G.D., Dongarra, J., Sloot, P.M.A. (eds.) ICCS 2009, Part II. LNCS, vol. 5545, pp. 696–705. Springer, Heidelberg (2009)
5. Geimer, M., Wolf, F., Wylie, B., Brian, J.N., Ábrahám, E., Becker, D., Mohr, B.: The SCALASCA Performance Toolset Architecture. In: Proc. of the International Workshop on Scalable Tools for High-End Computing (STHEC), pp. 51–65 (2008)
6. Lindlan, K.A., Cuny, J., Malony, A.D., Shende, S., Mohr, B., Rivenburgh, R., Rasmussen, C.: A Tool Framework for Static and Dynamic Analysis of Object-Oriented Software with Templates. In: Proc. of SC 2000 Conference (2000)
7. Mucci, P.: Dynaprof (2010), http://www.cs.utk.edu/~mucci/dynaprof
8. Nataraj, A., Malony, A., Morris, A., Arnold, D., Miller, B.: In Search of Sweet-Spots in Parallel Performance Monitoring. In: Proc. IEEE International Conference on Cluster Computing (2008)
9. Roth, P., Arnold, D., Miller, B.: Proc. High-Performance Grid Computing Workshop, IPDPS (2004)
10. Shende, S., Malony, A., Ansell-Bell, R.: Instrumentation and Measurement Strategies for Flexible and Portable Empirical Performance Evaluation. In: Proc. Tools and Techniques for Performance Evaluation Workshop, PDPTA. CSREA, pp. 1150–1156 (2001)
11. Shende, S., Malony, A.D.: The TAU Parallel Performance System. The International Journal of High Performance Computing Applications 20(2), 287–311 (2006)
12. Shende, S., Malony, A.D., Morris, A.: Simplifying Memory, I/O, and Communication Performance Assessment using TAU. In: Proc. DoD UGC 2010 Conference. IEEE Computer Society (2010)
13. Barcelona Supercomputing Center, "Paraver" (2010), http://www.bsc.es/paraver
14. VI-HPS, "SILC" (2010), http://www.vi-hps.org/projects/silc
15. Lee, C.W., Malony, A.D., Morris, A.: TAUmon: Scalable Online Performance Data Analysis in TAU. In: Guarracino, M.R., Vivien, F., Träff, J.L., Cannataro, M., Danelutto, M., Hast, A., Perla, F., Knüpfer, A., Di Martino, B., Alexander, M. (eds.) Euro-Par-Workshop 2010. LNCS, vol. 6586, pp. 493–499. Springer, Heidelberg (2011)

Automatic Performance Analysis of OpenMP Codes on a Scalable Shared Memory System Using Periscope*

Shajulin Benedict and Michael Gerndt

Technische Universität München
Fakultät für Informatik I10, Boltzmannstr. 3, 85748 Garching, Germany

Abstract. OpenMP is a successful interface for programming parallel applications on shared memory systems. It is widely applied on small scale shared memory systems such as multicore processors, but also in hybrid programming on large supercomputers. This paper presents performance properties for OpenMP and their automatic detection by Periscope. We evaluate Periscope's OpenMP analysis strategy in the context of the Altix 4700 supercomputer at Leibniz Computing Center (LRZ) in Garching. On this unique machine OpenMP scales up to 500 cores, one partition of in total 19 partitions. We present results for the NAS parallel benchmarks and for a large hybrid scientific application.

Keywords: Memory accesses analysis, OpenMP, Performance analysis, Supercomputers.

1 Introduction

OpenMP, a directive-based programming interface for multi-threaded parallelism, is a widely accepted de facto standard for programming scientific applications since 1997. The basic goal behind OpenMP is to express parallelism in an easy way. Although, OpenMP succeeded in simplifying writing portable parallel applications, it requires careful tuning, e.g., with respect to load balancing and distribution of threads and data. It is crucial to have tools that reveal performance problems so that the tuning can be carried out by the programmer.

Periscope [4] is a performance analysis tool that searches performance properties, e.g., stall cycles due to cache misses, in a distributed fashion based on agents. Each of the analysis agents, i.e., the nodes of the agent hierarchy, searches autonomously for inefficiencies in a subset of the application processes or threads.

In this paper, we define properties formalizing the OpenMP performance problems. The concept of performance properties was first introduced by the European-American working group APART (www.fz-juelich.de/apart) on automatic performance analysis tools. A new search strategy named *OpenMPAnalysis*

---

* This work is partially funded by BMBF under the ISAR project, grant 01IH08005A and the SILC project, grant 01IH08006E.

K. Jónasson (Ed.): PARA 2010, Part II, LNCS 7134, pp. 452–462, 2012.

which is responsible for evaluating the OpenMP regions for performance bottlenecks is presented. We evaluated the new OpenMP strategy on the Altix 4700
supercomputer which supports OpenMP runs with up to 512 cores in a single partition. The Altix is the only machine where pure OpenMP codes can
be scaled to that size. For hybrid codes combining MPI and OpenMP, such as
the Gyrokinetic Electromagnetic Numerical Experiment (GENE) code, OpenMP
usage spreads over thousands of cores. The standard support of Periscope for
scaling to that size, i.e., runtime evaluation of raw data and online combination of performance properties from different processes, apply to the OpenMP
analysis in the same way as to pure MPI codes.

The rest of the paper is organized as follows. Section 2 presents related work
and Section 3 explains Periscope and its OpenMP search strategy. In Section 4,
we discuss the definition for various OpenMP performance properties. Section 5
discusses experimental results. Finally, Sections 6 presents a few conclusions.

2   Related Works

Performance analysis and tuning of threaded programs is one of the big challenges of the multi-core era. There exists a few performance analysis tools that
help the user to identify whether or not their application is running efficiently
on the computing resources available.

Performance analysis tools, such as, ompP, Intel Thread Profiler, and TAU,
analyze OpenMP based on a profiling approach, whereas, Scalasca and Vampir
undergo analysis using traces.

'ompP' [6] is a text-based profiling tool for OpenMP applications. It relies on
OPARI for source-to-source instrumentation. It is a measurement-based profiler
and does not use program counter sampling. An advantage of this tool is its
simplicity. ompP performs an overhead analysis based on synchronization, load
imbalance, thread management, and limited parallelism.

Intel Thread Profiler [5] supports applications threaded with OpenMP, Windows API, or POSIX threads (Pthreads). Thread Profiler is used to identify
bottlenecks, synchronization delays, stalled threads, excessive blocking time and
so forth. It is a plug-in to the VTune Performance Analyzer. It provides results
in the form of graphical displays.

TAU [9] supports trace-based and profiling-based performance analysis. It
performs an off-line analysis and provides graphical and text-based displays. It
uses Vampir to display traces of OpenMP executions.

Scalasca [3] performs an offline automatic analysis of OpenMP codes based
on profiling data. OpenMP events are also inserted into a trace and can be
visualized with Vampir.

Vampir [1] is exclusively using traces of OpenMP programs and presents
analysis results via a rich set of different chart representations, such as, state
diagrams, statistics, and timelines of events.

Compared with other performance analysis tools, Periscope enables an automatic performance analysis that is based on a formal specification of high-level performance properties. Instead of presenting raw performance data it delivers the main bottlenecks including their severity. The search is performed online by a set of autonomous agents distributed over the parallel system. In this way, performance data could be analyzed locally and only severe performance problems will be communicated to the central master agent which interacts with the user.

3   Periscope and the OpenMP Search Strategy

Periscope already had different search strategies for single node and MPI performance bottlenecks when we developed a new search strategy named *OpenMP-Analysis*, exclusively for detecting OpenMP-based performance problems. In this section, we discuss Periscope's overall agent-based architecture and the newly developed OpenMP search strategy.

3.1   Overall Architecture

The overall architecture of Periscope consists of four major entities, namely, User-Interface, Frontend, Analysis Agent Network, and MRI monitors. All entities have their own obligations to finally identify the performance problems in parallel applications. A more detailed description of the architecture can be found in [8].

The *User-Interface* displays the results of the runtime analysis by directly mapping the detected properties to the source code. The *Frontend* starts the application and the analysis agents based on the specifications provided by the user, namely, number of processes and threads, search strategy, and so on. The *analysis agent network* consists of three different agent types, namely, master agent, communication agent and analysis agent. The *master agent* forwards commands from the frontend to the analysis agents and receives the found performance properties from the individual analysis agents and forwards them to the frontend. The *communication agents* combine similar properties found in their sibling agents and forward only the combined properties. The *analysis agents* are responsible for performing the automated search for performance properties based on the search strategy selected by the user. The *MRI monitors* linked to the application provide an application control interface. They communicate with the analysis agents, control the application's execution, and measure performance data.

The search of the analysis agents is based on the phase concept. Scientific applications are iterative and each iteration executes the same code, which is called the phase region in Periscope. The agents analyze the performance for an execution of the phase region, and, if necessary refine the search for performance problems in subsequent executions.

3.2 OpenMP Search Strategy

The new OpenMPAnalysis strategy to search for performance properties related to i) extensive startup and shutdown overhead for fine-grained parallel regions, ii) load imbalance, iii) sequentialization, and iv) synchronization was implemented.

The OpenMPAnalysis search strategy executes the following steps :

1. *Create initial candidates set.* At this stage, the strategy first creates OpenMP candidate properties for every OpenMP region in the application. This is based on the static program information created by Periscope's source code instrumentation tool.
2. *Configure performance measurements.* Based on these candidate properties it requests the measurements of the required performance data to prove whether the properties exist in the code. Each property provides information about which data are required.
3. *Execution of experiment.* The agents release the application which then executes the phase region. The measurements are performed automatically via the MRI monitor.
4. *Evaluate candidate properties.* The application stops at the end of the phase region and the agents retrieve the performance data. All the candidate properties are then evaluated whether their condition is fulfilled and the severity of the performance problems are computed. Finally, the found property set is checked whether further refinement is appropriate.

4 OpenMP Performance Properties

The OpenMP performance properties are specified as C++ classes. The properties provide methods to determine the required information, to calculate the condition as well as to compute the severity. The notations used here for defining OpenMP properties are as follows:

- *Severity* : Significance of the performance problem
- *reg* : Region name
- *k* : thread number
- *n* : number of threads
- T_0 : execution time for the master thread
- $T_{1...(n-1)}$: execution time for the team members - other than the master thread
- *phaseCycles* : time spent in executing the entire phase

In the following, we present the individual properties that are currently included in the OpenMPAnalysis strategy.

4.1   Parallel Region Startup and Shutdown Overhead Property

For each execution of a parallel region in OpenMP, the master thread may fork multiple threads and destroy those threads at the end. The master thread continues execution after the team of threads was started. In addition, initialization of thread private data and aggregation of reduction variables has to be executed. Writing too many parallel regions in an application causes overhead due to the startup and shutdown process. This performance property is designed to expose parallel region startup and shutdown overhead in parallel regions.

To calculate the severity of the *parallel region startup and shutdown overhead property*, we measure the following:

- the parallel region execution time for the master thread T_0
- the execution time for the parallel region body for thread k, T_k

The severity is calculated using the formula given below:

$$Severity(reg) = \frac{T_0 - \sum_{k=1...n}(T_k/(n-1))}{phaseCycles} * 100 \tag{1}$$

4.2   Load Imbalance in OpenMP Regions

Load imbalance emerges in OpenMP regions from an uneven distribution of work to the threads. It manifests at global synchronization points, e.g., at the implicit barrier of parallel regions, worksharing regions, and explicit barriers. Load imbalance is a major performance problem leading to the under-utilization of resources.

Load Imbalance in Parallel Region Property. The *load imbalance in parallel region property* is reported when threads have an imbalanced execution in a parallel region. In order to calculate the severity, we measure implicit barrier wait time W and calculate the difference of the observed unbalanced time UT and the optimized balanced time BT.

$$Severity(reg) = \frac{UT - BT}{phaseCycles} * 100 \tag{2}$$

UT and BT are represented in equations 3a and 3b.

$$UT = \max\{W_0..W_n\} \tag{3a}$$

$$BT = \overline{Work} + \min\{W_0..W_n\} \tag{3b}$$

where, $\overline{Work}$ is the average computational work of all the threads executed during the maximum barrier wait time.

$$\overline{Work} = \sum_{0 \leq k \leq n} (\max\{W_0..W_n\} - W_k) \tag{4}$$

Equation 2 is common for most of the load imbalance OpenMP properties.

Load Imbalance in Parallel Loop Property. The parallel loop region distributes the iterations to different threads. While the scheduling strategies determine the distribution of iterations, the application developer can tune the application by selecting the best strategy. Often, choosing a better scheduling strategy with correct chunk size is a question mark because even most experienced developers are new to this programming sphere. A sub-optimal strategy might thus lead to load imbalance. The measurements are done similar to the *load imbalance in parallel region property*. Measurement of the load imbalance based on the barrier time is only possible if the parallel loop is not annotated with the *nowait* clause.

Load Imbalance in Parallel Sections Property. In OpenMP, the sections construct allows the programmer to execute independent code parts in parallel. A load imbalance manifests at the implicit barrier region, similar to the parallel region, and determines the under-utilization of resources.

The *load imbalance in parallel sections property* is further refined into two sub properties as below:

- *load imbalance due to not enough sections property*, is reported when the number of OpenMP threads is greater than the number of parallel sections. In this case, a few threads do not participate in the computation.
- *load imbalance due to uneven sections property*, identifies the load imbalance which is due to the fact that threads execute different numbers of sections.

However, the calculation for the severity is quite similar to Equation 2 except that additional static information (number of sections in the construct) and the number of sections assigned to the threads at runtime are checked.

Load Imbalance in Explicit Barrier Property. Application developers often use explicit barriers to synchronize threads, so that they avoid race conditions. Early threads reaching the barrier have to wait until all threads reach it before proceeding further. The severity is calculated after measuring the explicit barrier time for each thread using the Equation 2. To note, the wait time in this property is the execution time of the explicit barrier.

4.3   Sequential Computation in Parallel Regions

In general, if parallel codes spend too much time in sequential regions, this will severely limit scalability according to the famous Amdahl's law. Sequential regions within parallel regions are coded in form of master, single, and ordered regions.

Sequential Computation in Master Region Property. If a master region is computationally expensive it limits scalability. The severity of *sequential in* *master region property* is the percentage of the execution time of the phase region spent in the master region.

$$Severity(reg) = \frac{T_0}{phaseCycles} * 100 \qquad (5)$$

Sequential Computation in Single Region Property. The underlying principle of a single region is similar to the master region: code wrapped by the OMP MASTER directive is only intended for the master thread, code wrapped by the OMP SINGLE directive is intended for one and only one thread but not necessarily to be the master thread. Thus, the code is executed sequentially. The severity is calculated in the same way as for the master region.

Sequential Computation in Ordered Loop Property. An ORDERED region in the body of a parallel loop specifies that this code is executed in the order of the sequential execution of the loop. Thus, the code is executed sequentially. This performance property is modeled in a way to measure the performance loss due to this ordered execution constraint. The severity is computed based on the SUM of the time spent in the ordered region O_k in all the threads.

$$Severity(reg) = \frac{\sum_{k=0..n} O_k}{phaseCycles} * 100 \qquad (6)$$

4.4   OpenMP Synchronization Properties

In addition to the above mentioned OpenMP properties, we have defined properties that are specific to synchronization in OpenMP, namely, critical sections and atomic regions.

Critical Section Overhead Property. For critical sections two aspects are important. The first is the contention for the lock guarding its execution. The second is the wait time of other threads while a thread is executing within a critical region. The severity of the *Critical Section Overhead property* is calculated by taking the maximum value of critical section overhead CSO among the threads. The CSO is the difference between the critical section region's execution time C and the execution time of the critical section's body CB.

$$Severity(reg) = \frac{\max\{CSO_0..CSO_n\}}{phaseCycles} * 100 \qquad (7)$$
$$where, CSO_k = C_k - CB_k$$

Frequent Atomic Property. To eliminate the possibility of race conditions, the ATOMIC directive specifies that a memory location will be updated atomically. Those who are familiar with POSIX threads are aware of the overhead for operating system calls to use semaphores for this purpose. Similarly, too many atomic operations have a negative effect on the performance. Thus, the severity of this property is the percentage of time spent in an atomic region with respect to the phase region.

5 Experimental Results

The Periscope performance analysis environment with the OpenMP strategy has been tested with several benchmarks and a real-world application on our Altix 4700 at Leibniz Rechenzentrum (LRZ) in Garching. The Altix supercomputer consists of 19 NUMA-link4 interconnected shared memory partitions with over 9600 Itanium 2 cores with an aggregated peak performance of over 60 TFlops. OpenMP can be used across one partition with 512 cores.

5.1 OpenMP Analysis of NAS Parallel Benchmarks (NPB)

In this paper we present results for LU, LU-HP, Block Tridiagonal (BT), and Scalar Pentadiagonal (SP) from NPB version 3.2. with Class C. The LU benchmark is a simulated CFD application. SP and BT benchmarks solve partial differential equations using different finite difference methods.

The tests were carried out in batch mode reserving 128 CPUs in a single node. For test runs with 2, 4, 8, 16, 32, 64. and 128 threads, Periscope identified OpenMP performance properties as shown in Table 1.

Table 1 shows the region name, file name, region first line number, property name, and the severity of the performance properties obtained for different runs. A few points could be observed as follows: 1) The highest severity values were found for load imbalance in BT, 2) startup and shutdown overhead was found only for large thread numbers and even for 128 threads it is only 0.01 in the parallel region in line 26 in initialize.f in BT, and 3) the severity of most properties increase with the number of threads. But in a few cases, the property's severity was reduced for larger thread numbers, e.g., in BT for load imbalance with 128 threads.

5.2 GENE Analysis

The GENE code [2] is an iterative solver for a non-linear gyrokinetic equations in a 5-dimensional phase space to identify the turbulence in magnetized fusion plasmas. It was developed in the Max Planck Institute for Plasma Physics in Garching. GENE consists of 47 source files with 16,258 lines. The code is written in Fortran 95 with MPI-based and OpenMP-based parallelization.

To check for OpenMP performance properties we ran GENE in a hybrid mode with 2 to 32 threads per process. All the MPI processes showed the same OpenMP performance behavior and the agents were able to combine the found performance properties while propagating the properties in the agent tree for output to the master agent.

Table 1. Identified OpenMP properties in NPB benchmarks. A '*' indicates that the property was not reported. 'L.I.x' identifies load imbalance in parallel region and loop as well as barrier region. 'S.S.O' identifies parallel region startup and shutdown overhead.

NAS	File Name	Region Name	LineNo.	Property Name	2	4	8	16	32	64	128
	initialize.f	Par.Reg.	36	S.S.O	*	*	*	*	*	*	0.01
BT	x_solve.f	Par.Do.Reg.	55	L.I.L	1.37	*	5.14	11.92	14.74	15.81	8.5
	y_solve.f	Par.Do.Reg.	52	L.I.L	1.34	2.63	3.32	15.21	19.16	20.3	11.6
	z_solve.f	Par.Do.Reg.	52	L.I.L	1.34	2.63	3.32	15.21	19.16	20.37	11.63
	ssor.f	Bar.Reg.	211	L.I.B	*	*	*	1.56	2.48	3.6	4.38
LU	ssor.f	Par.Reg.	120	L.I.P	*	*	*	1.9	1.1	2.19	1.13
	setbv.f	Par.Reg.	27	S.S.O	*	*	*	*	*	0.05	0.10
	jacld.f	Par.Reg.	35	S.S.O	*	*	*	*	*	*	8.38
	jacld.f	Par.Reg.	35	L.I.P	1.45	1.78	3.06	3.95	5.8	8.5	7.54
	jacu.f	Par.Reg.	35	S.S.O	*	*	*	*	*	*	8.51
LU-HP	jacu.f	Par.Reg.	35	L.I.P	1.22	1.15	2.16	2.76	5.3	7.4	7.71
	blts.f	Par.Do.Reg.	54	S.S.O	*	*	*	*	*	5.63	8.64
	blts.f	Par.Do.Reg.	54	L.I.L	2.08	1.01	1.53	1.52	2.06	3.46	5.3
	buts.f	Par.Do.Reg.	54	L.I.L	2.47	1.24	2.05	2.44	3.69	4.56	5.61
	tzetar.f	Par.Do.Reg.	26	L.I.L	*	*	1.06	1.4	1.4	1.79	1.87
SP	x_solve.f	Par.Do.Reg.	31	L.I.L	*	1.76	3.6	3.64	2.85	1.73	1.04
	y_solve.f	Par.Do.Reg.	31	L.I.L	*	1.09	1.9	2.17	1.68	2.05	1.66
	z_solve.f	Par.Do.Reg.	35	L.I.L	*	2.33	3.4	3.1	2.9	4.0	2.66

Table 2. OpenMP properties found in GENE

File Name	Line Number	Property	2	4	8	16	32
vel_space.F90	244	L.I.L	0.08	0.098	0.15	*	*
boundary.F90	140	L.I.L	0.002	0.038	0.0007	0.0019	0.00053
boundary.F90	140	S.S.O	0.001	*	0.001	0.0019	0.002
CalFullrhs_kxky.F90	100	L.I.L	0.001	0.0006	0.002	0.0066	0.0089
CalFullrhs_kxky.F90	84	L.I.L	0.0003	0.0005	0.0002	0.006	0.0009

From the results for GENE in Table 2 we can see that there are no significant OpenMP performance bottlenecks for those runs. However, it can be seen that the severity of the *parallel region startup and shutdown overhead property* increases steadily when the number of threads is increased. It was observed that there were slight load imbalance problems in various code regions, such as, velspace.F90, boundary.F90, and CalFullrhs_kxky.F90.

6   Conclusions

The increasing need to develop applications for multi-core architectures based on programming models such as OpenMP, leads to an increasing interest in tools that can pinpoint and quantify performance problems on the fly. Such tools require formalized OpenMP performance properties to identify those problems. In this work, we extended Periscope with a catalogue of OpenMP performance properties and a new search strategy named OpenMPAnalysis. Periscope's analysis agents are now able to search for OpenMP performance properties in pure OpenMP codes as well as hybrid codes.

To evaluate our OpenMP support in Periscope, we experimented with four NPB representatives, namely, LU, LU-HP, BT, SP, and GENE - a real world scientific application. Our study revealed OpenMP performance problems and the severities pinpoint those regions that would benefit from tuning transformations. In addition, the OpenMP performance analysis was demonstrated for the GENE code. In addition to these applications we validated our OpenMP properties and analysis with the kernels of the APART Test Suite (ATS) (www.fz-juelich.de/apart).

We ran the experiments on a scalable shared memory system, the Altix 4700, exploiting the scalability support built into Periscope. Please note, that Periscope's OpenMP analysis is not limited to large scale machines but can also be used to tune OpenMP code on single multicore processors.

References

1. Knüpfer, A., Brunst, H., Doleschal, J., Jurenz, M., Lieber, M., Mickler, H., Müller, M.S., Nagel, W.E.: The Vampir Performance Analysis Tool-Set. In: Proc. of the 2nd Int. Work. on Parallel Tools for HPC, HLRS, Stuttgart, pp. 139–155. Springer, Heidelberg (2008)
2. Chen, Y., Parker, S.E.: A δf particle method for gyrokinetic simulations with kinetic electrons and electromagnetic perturbations. Comput. Phys. 189(2), 463–475 (2003), http://dx.doi.org/10.1016/S0021-9991(03)00228-6
3. Geimer, M., Wolf, F., Wylie, B.J.N., Abraham, E., Becker, D., Mohr, B.: The Scalasca performance toolset architecture. In: Concurrency & Computation:Practice and Experience, pp. 702–719. Wiley Interscience (2010), http://dx.doi.org/10.1002/cpe.v22:6
4. Gerndt, M., Kereku, E.: Search Strategies for Automatic Performance Analysis Tools. In: Kermarrec, A.-M., Bougé, L., Priol, T. (eds.) Euro-Par 2007. LNCS, vol. 4641, pp. 129–138. Springer, Heidelberg (2007)
5. Intel Thread Profiler, preprint (2009), http://www.intel.com/software/products/threading/tp/
6. Fürlinger, K., Gerndt, M.: ompP: A Profiling Tool for OpenMP. In: Mueller, M.S., Chapman, B.M., de Supinski, B.R., Malony, A.D., Voss, M. (eds.) IWOMP 2005 and IWOMP 2006. LNCS, vol. 4315, pp. 15–23. Springer, Heidelberg (2008)
7. DeRose, L., Mohr, B., Seelam, S.: Profiling and Tracing OpenMP Applications with POMP Based Monitoring Libraries. In: Danelutto, M., Vanneschi, M., Laforenza, D. (eds.) Euro-Par 2004. LNCS, vol. 3149, pp. 39–46. Springer, Heidelberg (2004)

8. Benedict, S., Brehm, M., Gerndt, M., Guillen, C., Hesse, W., Petkov, V.: Automatic Performance Analysis of Large Scale Simulations. In: Lin, H.-X., Alexander, M., Forsell, M., Knüpfer, A., Prodan, R., Sousa, L., Streit, A. (eds.) Euro-Par 2009. LNCS, vol. 6043, pp. 199–207. Springer, Heidelberg (2010)
9. Shende, S.S., Malony, A.D.: The TAU parallel performance system. International Journal of High Performance Computing Applications, 287–311 (2006)
10. Wolf, F., Mohr, B.: Automatic performance analysis of hybrid MPI/OpenMP applications. In: Proceedings of the 11th Euromicro Conference on Parallel, Distributed and Network-Based Processing (PDP 2003), pp. 13–22 (2003)

Further Improving the Scalability
of the Scalasca Toolset

Markus Geimer[1], Pavel Saviankou[1], Alexandre Strube[1,2],
Zoltán Szebenyi[1,3], Felix Wolf[1,3,4], and Brian J.N. Wylie[1]

[1] Jülich Supercomputing Centre, 52425 Jülich, Germany
{m.geimer,p.saviankou,a.strube,z.szebenyi,b.wylie}@fz-juelich.de
[2] Universitat Autònoma de Barcelona, 08193 Barcelona, Spain
[3] RWTH Aachen University, 52056 Aachen, Germany
[4] German Research School for Simulation Sciences, 52062 Aachen, Germany
f.wolf@grs-sim.de

Abstract. Scalasca is an open-source toolset that can be used to analyze the performance behavior of parallel applications and to identify
opportunities for optimization. Target applications include simulation
codes from science and engineering based on the parallel programming
interfaces MPI and/or OpenMP. Scalasca, which has been specifically
designed for use on large-scale machines such as IBM Blue Gene and
Cray XT, integrates runtime summaries suitable to obtain a performance
overview with in-depth studies of concurrent behavior via event tracing.
Although Scalasca was already successfully used with codes running with
294,912 cores on a 72-rack Blue Gene/P system, the current software design shows scalability limitations that adversely affect user experience
and that will present a serious obstacle on the way to mastering larger
scales in the future. In this paper, we outline how to address the two
most important ones, namely the unification of local identifiers at measurement finalization as well as collating and displaying analysis reports.

Keywords: Scalasca, scalability.

1 Introduction

Driven by growing application requirements and accelerated by current trends
in microprocessor design, the number of processor cores on modern supercomputers increases from generation to generation. With today's leadership systems featuring more than a hundred thousand cores, writing efficient codes that
exploit all the available parallelism becomes increasingly difficult and requires
adequate tool support for performance analysis. Unfortunately, increased concurrency levels impose higher scalability demands not only on applications but
also on the software tools needed for their development. When applied to larger
numbers of processors, familiar tools often cease to work in a satisfactory manner
(e.g., due to serialized operations, escalating memory requirements, limited I/O
bandwidth, or failed renderings).

K. Jónasson (Ed.): PARA 2010, Part II, LNCS 7134, pp. 463–473, 2012.

Scalasca [2] is an open-source toolset that can be used to analyze the performance behavior of parallel applications and to identify opportunities for optimization. Target applications include simulation codes from science and engineering written in C, C++ and Fortran and based on the parallel programming interfaces MPI and/or OpenMP. Scalasca has been specifically designed for use on large-scale systems including IBM Blue Gene and Cray XT, but is also well suited for a wide range of small- and medium-scale HPC platforms. Scalasca combines runtime summaries suitable to obtain a performance overview with in-depth studies of concurrent behavior via event tracing. The traces are analyzed to identify wait states that occur, for example, as a result of unevenly distributed workloads. Especially when trying to scale communication-intensive applications to large processor counts, such wait states can present serious challenges to achieving good performance. Thanks to a novel parallel trace-analysis scheme, the search can be performed even for very large numbers of cores. Internally, runtime summarization and tracing are tightly integrated, allowing the user to switch between the two modes via environment variables and even to apply them simultaneously.

Although Scalasca's scalable design already facilitated performance analyses of application runs on a Blue Gene/P system with 294,912 (288k) cores [6], the architecture underlying version 1.3.0 (released March 2010) still shows scalability limitations, which primarily occur (i) during the unification of local identifiers at measurement finalization and (ii) while collating and displaying analysis reports. Both limitations adversely influence user experience in the form of either increased time needed for performance-data acquisition or prolonged response times during the interactive exploration of analysis results. Since they will present serious obstacles on the way to mastering higher scales in the future, we outline in this paper how they can be effectively addressed to ensure scalability even when hundreds of thousands of cores are employed. In the remainder of the paper, we explain each of the two challenges in more detail along with the progress achieved so far, followed by related work and an outlook on what still needs to be done to match our objective of substantially increasing Scalasca's scalability.

2   Unification of Local Identifiers

In Scalasca, event data measured as part of the traces refer to objects such as source-code regions, call paths, or communicators. Motivated by the desire to minimize storage requirements and avoid redundancy in traces, events reference these objects using numerical identifiers, while the objects themselves are defined separately. However, to establish a global view of the program behavior during analysis, these identifiers must be consistent across all processes. Unfortunately, generating global identifiers purely locally as a hash function of some definition key would pose the danger of global conflicts, which are very hard to resolve. For this reason, each process may use a different local identifier to denote the same object. However, ultimately a global set of unique object definitions must be created and local identifiers mapped onto global identifiers in a consistent manner.

This procedure, which is called *unification* and which requires exchanging and comparing definitions among different processes, is performed during application finalization to avoid perturbation during measurement. Since runtime summarization and tracing share the same set of definitions, unification is also needed in summary mode. Definitions always refer to processes with potentially multiple threads, which is why both the previous and the new unification algorithms equally apply to pure MPI as well as hybrid OpenMP/MPI applications.

Although Scalasca's current unification algorithm already takes advantage of message communication to facilitate the efficient exchange of object definitions and the generation of local-to-global identifier mapping tables, it is still predominantly sequential, posing a serious scalability limitation. To overcome this situation, this sequential step was parallelized using a hierarchical algorithm so that it can be efficiently performed for codes running on hundreds of thousands of cores. It consists of the following three steps:

- generation of a unified set of global definitions,
- generation of local-to-global identifier mappings for each process, and
- writing the global set of definitions as well as the identifier mappings to disk.

Note that the last step is only required in tracing mode, since the identifier mapping can already be applied at run time in summarization mode. In the following paragraphs, all three steps will be described in more detail.

2.1   Definition Unification

Unifying object definitions is a data-reduction procedure that combines the local definitions from each process by first removing duplicates and then assigning a unique global identifier to each of the remaining definitions. In the new hierarchical scheme, this is done in several iterations as can be seen in Figure 1. During the first iteration, processes with odd rank numbers send their definitions to their neighbor processes with even rank numbers, which unify them with their own definitions. During subsequent iterations, these partially unified definitions are exchanged in a similar way, doubling the rank offset between sender and receiver in each step. In the end – after $\lceil \log_2 P \rceil$ iterations – the unified set of global definitions is available at rank zero.

During this step, it is essential to use efficient data structures and algorithms for identifying whether a particular definition object has already been defined or not. Depending on the types of the definition objects and their expected number, we use hash tables (e.g., for character strings), trees (e.g., for call paths), and vectors (e.g., for Cartesian topology definitions). Moreover, definition objects which are known to be unique across all processes, such as so-called "locations" referring to a thread of execution or Cartesian topology coordinates, are simply aggregated to avoid unnecessary search costs.

Note that an important prerequisite for using the hierarchical algorithm is that the semantics of the individual attributes of a definition object remain stable, which unfortunately was not the case for Cartesian topology definitions

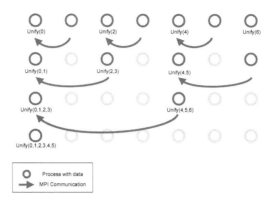

Fig. 1. Hierarchical definition unification algorithm

in the original scheme. Here the topology identifier assigned by the measurement system was either the identifier of the associated communicator for MPI Cartesians or a special constant for user-defined and hardware topologies. These would have been mapped to identifiers in the range $[0, n - 1]$ during the first iteration of the unification, communicator or the special constant and creating a need for a different unification algorithm during subsequent iterations. Since the communicator identifier is also available via a second attribute, this issue could be resolved by having the measurement system directly assign local topology identifiers in the range $[0, n - 1]$. As a positive side effect, this change now also allows us to handle more than one user-defined or hardware topology.

2.2   Generation of Local-to-Global Identifier Mappings

With the global set of definitions at hand, a translation table, in the following called *mapping*, can be created that maps the local identifiers used by each process onto the identifiers of the global definitions. This is done by broadcasting the unified global definitions from rank zero to all other processes, which subsequently compare their local definitions with the global ones to determine the identifier mapping.

Here it is crucial to exclude the above-mentioned definition objects from the broadcast that are unique to each process, since their data volume may dominate the overall size of the global definitions even at relatively small scales (512–1024 processes) and lead to prohibitively large data sets to be transferred at larger scales. However, unique identifiers for location objects and an associated mapping are still required. Due to the deterministic nature of the hierarchical unification scheme presented in Section 2.1, the global location identifiers can be locally reconstructed by determining the identifier offset via an exclusive prefix sum over the number of locations per process (emulated using MPI_Scan and a local subtraction for compatibility with non-MPI 2.x compliant MPI implementations).

2.3 Writing Definition Data and Mappings to Disk

To avoid expensive rewriting, traces still contain the local identifiers when they are written to disk. Therefore, the per-process identifier mappings need to be stored on disk as well if tracing mode is configured, from where they are later retrieved to correctly resolve the still existing local-identifier references in the traces before they are processed by Scalasca's parallel trace analyzer. Currently, all mappings are sequentially written to a single mapping file, with the file offset for each rank stored in the global definitions file.

Ideally, all processes would write their information in parallel to disjoint sections of the single mapping file, however, the amount of data per process is rather small so that lock conflicts on the file-system block level would significantly degrade I/O performance in this scenario. Therefore, the mapping information is gathered in chunks of 4 MB on a small set of processes, with the root process within each group defined as the first rank providing data for a particular chunk being gathered. Since there is no such "multi-root gather" operation provided by the MPI standard, it has been implemented using point-to-point messages and a hierarchical gather algorithm, very similar to the one used for the definition unification.

Using parallel I/O from this small set of gather processes was evaluated on the IBM Blue Gene/P system Jugene at Jülich Supercomputing Centre using a GPFS parallel file system. However, we found that the setup costs grew significantly when more than one I/O node was involved, which could not be amortized while writing the relatively small amount of data. Our current solution therefore lets rank zero write the mappings incrementally after receiving the 4 MB chunks already collected during the multi-root gather operations. In addition, rank zero gathers the map file offsets for each rank and writes them to the global definitions file. The offsets are again calculated using an exclusive prefix sum over the amount of mapping data generated by each process.

2.4 Experimental Results

Figure 2(a) shows a comparison of the original serial version of the unification algorithm and the new parallel algorithm for the definition data produced by a fully compiler-instrumented binary of the SMG2000 benchmark code [1] at various scales up to 288k processes, measured on the IBM Blue Gene/P system Jugene at the Jülich Supercomputing Centre with no run-time filtering applied. Although still exhibiting a similar scaling behavior, it can be seen that the new algorithm shows a remarkable improvement over the previous sequential version, reducing the time to unify the definitions at large scales by several orders of magnitude.

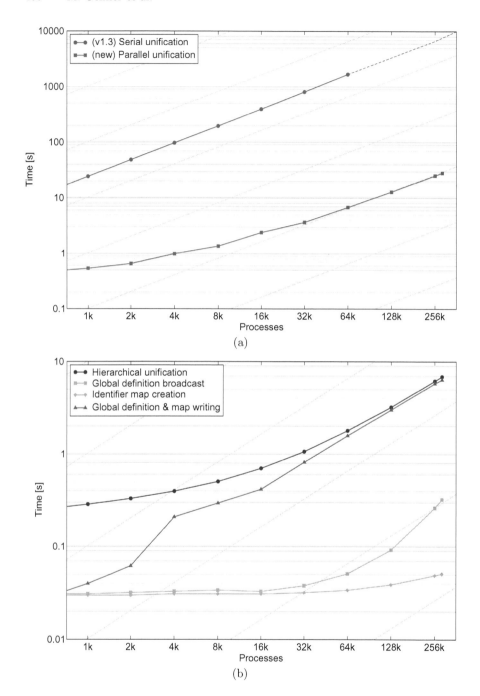

Fig. 2. Execution times of the new parallel and the previous serial unification schemes (a) as well as a breakdown of the individual steps of the parallel scheme (b) for the definition data of the SMG2000 benchmark code. (Note the different y-axis scales.)

Figure 2(b) shows a detailed breakdown of the unification runtime into the different steps involved, allowing the scalability of each step to be individually judged. As could be expected, the time to write the global definition data and mapping information is growing linearly at larger scales. This also applies to the hierarchical unification, where the scale-dependent data (i.e., the definition objects which are unique for each process as well as MPI communicator definitions) start to dominate the overall amount of data that is collected. Although almost all of this kind of data is not included in the broadcast and hence not considered during the identifier map creation, this does not apply to the definitions of communicators.

Currently, Scalasca encodes each communicator as a bitstring, with the i-th bit indicating whether the global rank i is part of the communicator or not. Therefore, the size of a communicator definition is linearly dependent on the number of processes, causing also communicator definitions to become a dominant part of the overall data volume sent during the broadcast and processed during the identifier map creation at some point. This suggests that a revised representation of communicators is required to further improve the scalability of these two steps (as well as the hierarchical unification). The design of a more space-efficient distributed scheme to record the constituency of a communicator is already in progress.

3   Collating and Displaying Analysis Reports

At the end of runtime summary measurements and after automated parallel trace analysis, Scalasca produces intermediate analysis reports from the unified definitions and the metric values for each call path collated from each process. The definitions of the measured and analyzed metrics, program call tree, and system configuration constitute metadata describing the experiment, where only the latter vary with the number of processes and threads. On the other hand, the amount of metric value data increases linearly. Intermediate analysis reports are sequentially post-processed to derive additional metrics and create a structured metric hierarchy prior to examination with textual or graphical tools.

Scalasca 1.3 saves analysis reports as single files using XML syntax, where 'exclusive' metric values are stored for each call path (explained in Section 3.3). This means that exclusive values first have to be calculated and then the numeric values have to be converted to formatted text when writing. Since values which are not present in reports default to zero when reports are read, a 'sparse' representation can be exploited which avoids writing vectors consisting entirely of zeroes. Furthermore, the final report is typically compressed when writing, which is advantageous in faster writing (and reading) time as well as much smaller archival size. To calculate inclusive metric values, all of the exclusive metric values must be read and aggregated.

Scaling of Scalasca analysis report collation time and associated analysis report explorer GUI loading time and memory requirements for Sweep3D trace analyses on Jugene [6] are shown in Figure 3.

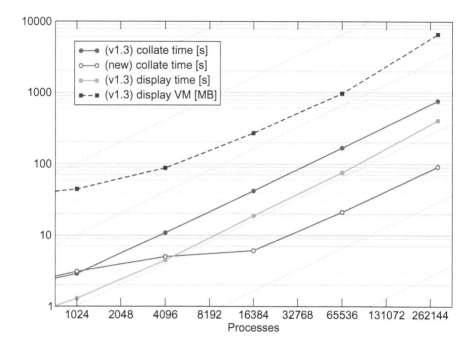

Fig. 3. Scalasca Sweep3D trace analysis report collation time, with associated loading time and memory requirements for the analysis report explorer GUI

3.1   Storing Metric Values in Binary Format

To reduce the collation time, we designed a new file format for storing experiment data, keeping the XML format essentially unchanged for storing metadata, while introducing a new binary format and file organization for storing metric values. In this way, third party tools such as TAU [5] that implement their own reader and/or writer for Scalasca analysis reports can keep their implementations for the more complicated metadata part of the report and only re-implement readers for the metric index and value files.

Figure 3 shows that when using the new format at smaller scales the collation time is dominated by writing the metadata, whereas beyond 16 thousand processes the writing time scales linearly with the amount of metric values to be written (regardless of the format). As in Section 2.3, writing of metric values in parallel from a configurable number of compute nodes was investigated, and again, best performance was obtained using a single writer. Collation time for the 288k-process analysis report is reduced from 13 minutes to 90 seconds. The sizes of both old and new analysis report formats grow linearly with the number of processes (as well as with the number of metrics and measured call paths). While the uncompressed intermediate reports are comparable in size between both format versions, the currently uncompressed set of binary files is typically four times larger than the (compressed) XML report, in this case 1530 MB vs.

404 MB for the 288k-process trace analysis report. When the file-system bandwidth is low compared to the processor speed, however, the performance gained from compression when writing intermediate or post-processed reports may justify the effort required to run the compression algorithm. For this reason, we are contemplating a configurable solution with optional compression.

3.2   File Organization

Instead of writing the entire report to a single XML file as before, we split the file into an XML part for metadata and a binary part for metric values. In anticipation of the dynamic-loading capabilities described later, each metric is stored in a separate file. This file is accompanied by an index file that specifies the data layout, allowing a sparse data representation if needed. A trace analysis report such as the ones considered here now consists of 197 files in total.

3.3   Dynamic Loading

Figure 3 shows that the time to load an entire analysis report and the amount of memory required increase slightly worse than linearly with the number of processes. For the 288k-process Sweep3D trace analysis report, loading takes 7 minutes on the 4.2 GHz Power6 Jugene BG/P front-end system and requires 6.5 GB of memory. (Naturally, a 64-bit version of the GUI and other report tools is necessary for such large amounts of analysis data.) Although paging to disk is fortunately avoided here, each GUI interaction still requires several minutes, seriously impairing interactive analysis report exploration. Command-line utilities to post-process analysis reports are found to require up to 15 GB memory and more than 30 minutes execution time. Since the binary analysis reports are much larger than the compressed XML reports, utilities using the new format are actually 10% slower when reading the entire report.

The solution to these problems is to avoid reading all the data into memory at once. With the new format, it is now possible to read any part of the data on-demand. In a typical usage scenario, the GUI would initially show only the root of the call tree (i.e., main) plus associated metric values. As soon as the user expands the root node, only the data of its direct children are loaded. The children of a node can easily be stored in consecutive order, in which case this will likely require just a single file system access. Of course, the indices for the metrics are small enough to be kept in memory, allowing for efficient data lookup.

Another major source of prolonged response times is that metric values for individual call-tree nodes (i.e., call paths) are stored with exclusive semantics. Exclusive means that the value stored for a given call-tree node refers to that node only – its children excluded – as opposed to an inclusive value, which is the sum of the exclusive values in the subtree anchored at the node. Typically we want to display both exclusive and inclusive values for a given node and if we store one, we can calculate the other. Calculating inclusive values from exclusive ones is very expensive, requiring a complete traversal of the corresponding subtree and adding up all values after loading them into memory, whereas converting from inclusive to exclusive is cheap. One just subtracts the values of the children from the value of the current node. In spite of losing some of the sparsity that can be exploited to save disk space, inclusive semantics are preferable with dynamic-loading. This would reduce the memory usage of the GUI to just the data being currently displayed on screen (plus optionally some additional caching). Of course, incremental reading can be easily extended to incremental rewriting, limiting the memory footprint of most post-processing tools to basically a single row of data at a time. At the time of writing, a quantitative comparison was not yet available.

3.4   Index Structures for Sparse Data

While a few of the metrics such as time and the number of times a call path has been visited (and optional hardware counter metrics) are dense, having non-zero values for each call path, many of the metrics are very sparse. For example, MPI point-to-point communication time has non-zero values only for the corresponding MPI calls (and their parent call paths when stored with inclusive semantics). For these metrics, we use a sparse representation, that is, we refrain from writing the rows of data that would only contain zeros in this metric. To keep track of which rows are written and allow for efficient lookup of the data straight from the binary file, the index file contains a list of the identifiers of the call paths whose data rows were actually stored. Keeping this index in memory, we can find the appropriate file offset for any call path in logarithmic time. Utilities which process analysis reports and change call path identifiers, such as remap and cut, therefore need to update the primary XML metadata, any secondary indices that refer to modified call-path identifiers, and potentially also the files containing the associated metric data.

4   Related Work

Obviously, the problem of unifying process-local definitions is not specific to the Scalasca toolset. For example, the VampirTrace measurement library [3] performs the unification as a serial post-processing step, which in addition also rewrites the generated trace files using the global identifiers. A second example is the TAU performance system [5], which serially unifies call-path information on-the-fly while loading profile data into the graphical profile browser ParaProf. Recently, a hierarchical unification algorithm for call paths has been implemented to support on-line visualization of so-called snapshot profiles. Treating unification as a global reduction problem puts it into a larger category of global reduction operations used in parallel performance analysis, many of them applied online [4].

Each performance tool typically has its own native format for profile data. VampirTrace produces OTF profiles consisting of separate binary files for each MPI process rank and several additional metadata files, which are subsequently integrated into a plain text profile. TAU also produces individual textual profiles for each MPI rank by default, and can then combine these into a single-file packed binary format, or alternatively, it can concatenate the individual profiles from each process rank at the end of measurement into a single textual file for each performance metric. TAU's profile viewer can read all of these formats and many more, including Scalasca integrated XML profiles and individual gprof textual profiles.

5   Conclusion and Outlook

In this paper, we showed how to address the most serious impediments to achieving further scalability in Scalasca. The previously serial unification of local identifiers created to reference mostly global objects was parallelized using a hierarchical reduction scheme, accelerating the procedure by almost a factor of 250 on 64k cores (and an estimated factor of about 350 on 288k cores) on an IBM Blue Gene/P. Subsequent tests on other large-scale HPC systems, such as Cray XT, showed similar benefits. Therefore, the parallel unification was integrated into the Scalasca measurement system and is already part of the latest 1.3.2 release (November 2010). Further improvements can be achieved by revising the handling and representation of MPI communicators as well as optimizing the identifier mappings, which still contain a significant amount of redundancy.

A speedup of more than 7 was observed for collating and writing analysis reports on 288k cores after replacing the XML file format used to store metric values with a binary alternative. Further optimizations of the file layout and the underlying data model as well as dynamic loading capabilities are likely to improve the interactive response times of the report explorer GUI.

References

1. Accelerated Strategic Computing Initiative: The ASC SMG2000 benchmark code (2001), http://www.llnl.gov/asc/purple/benchmarks/limited/smg/
2. Geimer, M., Wolf, F., Wylie, B.J.N., Ábrahám, E., Becker, D., Mohr, B.: The Scalasca performance toolset architecture. Concurrency and Computation: Practice and Experience 22(6), 702–719 (2010)
3. Knüpfer, A., Brunst, H., Doleschal, J., Jurenz, M., Lieber, M., Mickler, H., Müller, M.S., Nagel, W.E.: The Vampir performance analysis tool set. In: Resch, M., Keller, R., Himmler, V., Krammer, B., Schulz, A. (eds.) Tools for High Performance Computing, pp. 139–155. Springer, Heidelberg (2008)
4. Roth, P.C., Arnold, D.C., Miller, B.P.: MRNet: A software-based multicast/reduction network for scalable tools. In: Proc. Supercomputing Conference, SC 2003, Phoenix, AZ, USA (November 2003)
5. Shende, S.S., Malony, A.D.: The TAU parallel performance system. International Journal of High Performance Computing Applications 20(2), 287–331 (2006)
6. Wylie, B.J.N., Böhme, D., Mohr, B., Szebenyi, Z., Wolf, F.: Performance analysis of Sweep3D on Blue Gene/P with the Scalasca toolset. In: Proc. International Parallel & Distributed Processing Symposium, Workshop on Large-Scale Parallel Processing IPDPS–LSPP, Atlanta, GA, USA. IEEE Computer Society (April 2010)

Author Index

Printed by Publishers' Graphics LLC